The IMA Volumes in Mathematics and its Applications

Volume 97

Series Editors
Avner Friedman   Robert Gulliver

Springer
New York
Berlin
Heidelberg
Barcelona
Budapest
Hong Kong
London
Milan
Paris
Santa Clara
Singapore
Tokyo

The IMA Volumes in Mathematics and its Applications

Current Volumes:

1 **Homogenization and Effective Moduli of Materials and Media** J. Ericksen, D. Kinderlehrer, R. Kohn, and J.-L. Lions (eds.)

2 **Oscillation Theory, Computation, and Methods of Compensated Compactness** C. Dafermos, J. Ericksen, D. Kinderlehrer, and M. Slemrod (eds.)

3 **Metastability and Incompletely Posed Problems** S. Antman, J. Ericksen, D. Kinderlehrer, and I. Muller (eds.)

4 **Dynamical Problems in Continuum Physics** J. Bona, C. Dafermos, J. Ericksen, and D. Kinderlehrer (eds.)

5 **Theory and Applications of Liquid Crystals** J. Ericksen and D. Kinderlehrer (eds.)

6 **Amorphous Polymers and Non-Newtonian Fluids** C. Dafermos, J. Ericksen, and D. Kinderlehrer (eds.)

7 **Random Media** G. Papanicolaou (ed.)

8 **Percolation Theory and Ergodic Theory of Infinite Particle Systems** H. Kesten (ed.)

9 **Hydrodynamic Behavior and Interacting Particle Systems** G. Papanicolaou (ed.)

10 **Stochastic Differential Systems, Stochastic Control Theory, and Applications** W. Fleming and P.-L. Lions (eds.)

11 **Numerical Simulation in Oil Recovery** M.F. Wheeler (ed.)

12 **Computational Fluid Dynamics and Reacting Gas Flows** B. Engquist, M. Luskin, and A. Majda (eds.)

13 **Numerical Algorithms for Parallel Computer Architectures** M.H. Schultz (ed.)

14 **Mathematical Aspects of Scientific Software** J.R. Rice (ed.)

15 **Mathematical Frontiers in Computational Chemical Physics** D. Truhlar (ed.)

16 **Mathematics in Industrial Problems** A. Friedman

17 **Applications of Combinatorics and Graph Theory to the Biological and Social Sciences** F. Roberts (ed.)

18 ***q*-Series and Partitions** D. Stanton (ed.)

19 **Invariant Theory and Tableaux** D. Stanton (ed.)

20 **Coding Theory and Design Theory Part I: Coding Theory** D. Ray-Chaudhuri (ed.)

21 **Coding Theory and Design Theory Part II: Design Theory** D. Ray-Chaudhuri (ed.)

22 **Signal Processing Part I: Signal Processing Theory** L. Auslander, F.A. Grünbaum, J.W. Helton, T. Kailath, P. Khargonekar, and S. Mitter (eds.)

John Goutsias Ronald P.S. Mahler
Hung T. Nguyen
Editors

Random Sets

Theory and Applications

With 39 Illustrations

Springer

John Goutsias
Department of Electrical and
Computer Engineering
The Johns Hopkins University
Baltimore, MD 21218, USA

Ronald P.S. Mahler
Lockheed Martin Corporation
Tactical Defense Systems
Eagan, MN 55121, USA

Hung T. Nguyen
Department of Mathematical Sciences
New Mexico State University
Las Cruces, NM 88003, USA

Series Editors:
Avner Friedman
Robert Gulliver
Institute for Mathematics and its
Applications
University of Minnesota
Minneapolis, MN 55455, USA

Mathematics Subject Classifications (1991): 03B52, 04A72, 60D05, 60G35, 68T35, 68U10, 94A15

Library of Congress Cataloging-in-Publication Data
Random sets : theory and applications / John Goutsias, Ronald P.S. Mahler, Hung T. Nguyen, editors.
p. cm. — (The IMA volumes in mathematics and its applications ; 97)
ISBN 0-387-98345-7 (hardcover : alk. paper)
1. Random sets. I. Goutsias, John. II. Mahler, Ronald P.S. III. Nguyen, Hung T., 1944- . IV. Series: IMA volumes in mathematics and its applications ; v. 97.
QA273.5.R36 1997
519.2—dc21 97-34138

Printed on acid-free paper.

Production managed by Karina Mikhli; manufacturing supervised by Thomas King.
Camera-ready copy prepared by the IMA.
Printed and bound by Braun-Brumfield, Inc., Ann Arbor, MI.
Printed in the United States of America.

9 8 7 6 5 4 3 2 1

ISBN 0-387-98345-7 Springer-Verlag New York Berlin Heidelberg SPIN 10644864

FOREWORD

This IMA Volume in Mathematics and its Applications

RANDOM SETS: THEORY AND APPLICATIONS

is based on the proceedings of a very successful 1996 three-day Summer Program on "Application and Theory of Random Sets." We would like to thank the scientific organizers: John Goutsias (Johns Hopkins University), Ronald P.S. Mahler (Lockheed Martin), and Hung T. Nguyen (New Mexico State University) for their excellent work as organizers of the meeting and for editing the proceedings. We also take this opportunity to thank the Army Research Office (ARO), the Office of Naval Research (ONR), and the Eagan, Minnesota Engineering Center of Lockheed Martin Tactical Defense Systems, whose financial support made the summer program possible.

Avner Friedman

Robert Gulliver

PREFACE

"Later generations will regard set theory as a disease from which one has recovered."

– Henri Poincaré

Random set theory was independently conceived by D.G. Kendall and G. Matheron in connection with stochastic geometry. It was however G. Choquet with his work on capacities and later G. Matheron with his influential book on *Random Sets and Integral Geometry* (John Wiley, 1975), who laid down the theoretical foundations of what is now known as the theory of *random closed sets.* This theory is based on studying probability measures on the space of *closed subsets* of a locally compact, Hausdorff, and separable base space, endowed with a special topology, known as the *hit-or-miss topology.* Random closed sets are just random elements on these spaces of closed subsets. The mathematical foundation of random closed sets is essentially based on *Choquet's capacity theorem*, which characterizes distribution of these set-valued random elements as nonadditive set functions or "nonadditive measures." In theoretical statistics and stochastic geometry such nonadditive measures are known as *infinitely monotone*, *alternating capacities of infinite order*, or *Choquet capacities*, whereas in expert systems theory they are more commonly known as *belief measures*, *plausibility measures*, *possibility measures*, etc. The study of random sets is, consequently, inseparable from the study of nonadditive measures.

Random set theory, to the extent that is familiar to the broader technical community at all, is often regarded as an obscure and rather exotic branch of pure mathematics. In recent years, however, various aspects of the theory have emerged as promising new theoretical paradigms for several areas of academic, industrial, and defense-related R&D. These areas include stochastic geometry, stereology, and image processing and analysis; expert systems theory; an emerging military technology known as "information fusion;" and theoretical statistics.

Random set theory provides a solid theoretical foundation for certain image processing and analysis problems. As a simple example, Fig. 1 illustrates an image of an object (a cube), corrupted by various noise processes, such as clutter and occlusions. Images, as well as noise processes, can be modeled as random sets. Nonlinear algorithms, known collectively as *morphological operators*, may be used here in order to provide a means of recovering the object from noise and clutter. Random set theory, in conjunction with *mathematical morphology*, provides a rigorous statistical foundation for nonlinear image processing and analysis problems that is analogous to that of conventional linear statistical signal processing. For example, it allows one to demonstrate that there exist optimal algorithms that recover images from certain types of noise processes.

In *expert systems theory*, random sets provide a means of modeling and

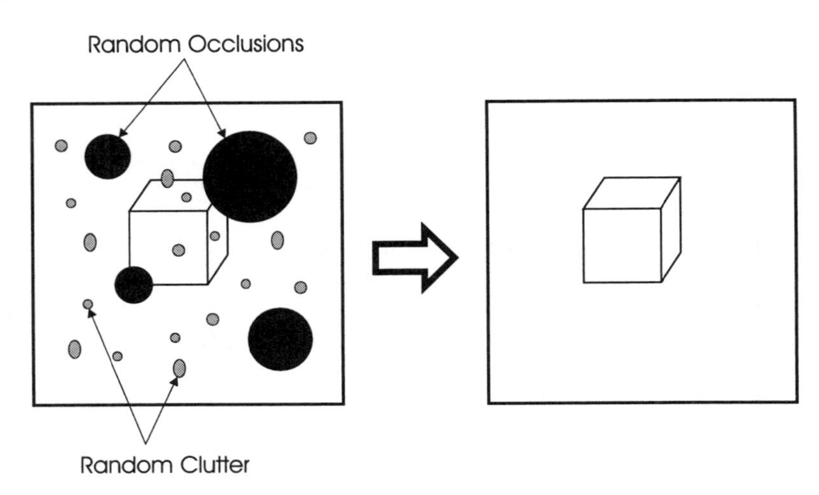

FIG. 1. *Random sets and image processing.*

manipulating evidence that is imprecise (e.g., poorly characterized sensor signatures), vague or fuzzy (e.g., natural language statements), or contingent (e.g., rules). In Fig. 2, for example, we see an illustration of a natural-language statement such as "*Gustav is NEAR the tower.*" Each of the four (closed) ellipses represents a plausible interpretation of the concept "*NEAR the tower,*" and the numbers p_1, p_2, p_3, p_4 represent the respective beliefs that these interpretations of the concept are valid. A discrete random variable that takes the four ellipses as its values, and which has respective probabilities p_1, p_2, p_3, p_4 of attaining those values, is a random set representative of the concept.

Random sets provide also a convenient mathematical foundation for a statistical theory that supports *multisensor, multitarget information fusion.* In Fig. 3, for example, an unknown number of unknown targets are being interrogated by several sensors whose respective observations can be of very diverse type, ranging from statistical measurements generated by radars to English-language statements supplied by human observers. If the sensor suite is interpreted as a single sensor, if the target set is interpreted as a single target, and if the observations are interpreted as a single *finite-set* observation, then it turns out that problems of this kind can be attacked using direct generalizations of standard statistical techniques by means of the theory of random sets.

Finally, random set theory is playing an increasingly important role in *theoretical statistics.* For example, suppose that a continuous but random voltage is being measured using a digital voltmeter and that, on the basis of the measured data, we wish to derive bounds on the expected value of the original random variable, see Fig. 4. The observed quantity is a random subset (specifically, a random interval) and the bounds can be expressed in terms of certain nonlinear integrals, called *Choquet integrals*, computed

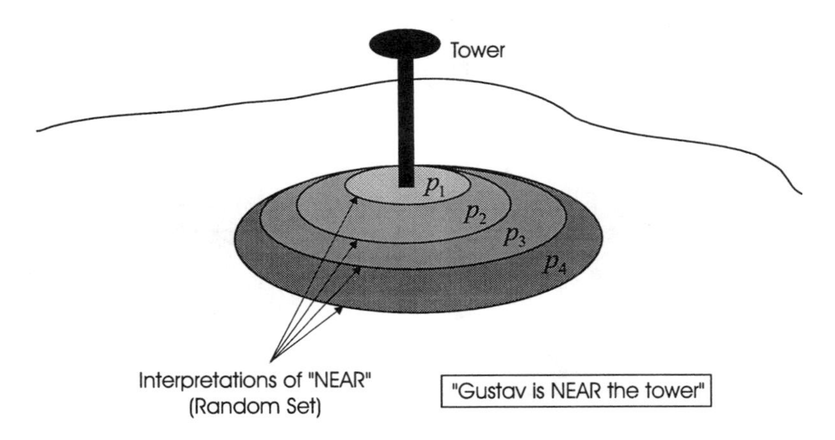

FIG. 2. *Random sets and expert systems.*

with respect to nonadditive measures associated with that random subset.

On August 22–24, 1996, an international group of researchers convened under the auspices of the *Institute for Mathematics and Its Applications* (IMA), in Minneapolis, Minnesota, for a scientific workshop on the "Applications and Theory of Random Sets." To the best of our knowledge this was the first scientific gathering in the United States, devoted primarily to the subject of random sets and allied concepts. The immediate purpose of the workshop was to bring together researchers and other parties from academia, industry, and the U.S. Government who were interested in the potential application of random set theory to practical problems of both industrial and government interest. The long-term purpose of the workshop was expected to be the enhancement of imaging, information fusion, and expert system technologies and the more efficient dissemination of these technologies to industry, the U.S. Government, and academia.

To accomplish these two purposes we tried to bring together, and encourage creative interdisciplinary cross-fertilization between, three communities of random-set researchers which seem to have been largely unaware of each other: theoretical statisticians, those involved in imaging applications, and those involved in information fusion and expert system applications. Rather than "rounding up the usual suspects"–a common, if incestuous, practice in organizing scientific workshops–we attempted to mix experienced researchers and practitioners having complementary interests but who, up until that time, did not have the opportunity for scientific interchange.

The result was, at least for a scientific workshop, an unusually diverse group of researchers: theoretical statisticians; academics involved in applied research; personnel from government organizations and laboratories, such as the National Institutes of Health, Naval Research and Development, U.S. Army Research Office, and USAF Wright Labs, as well as industrial R&D engineers from large and small companies, such as Applied Biomath-

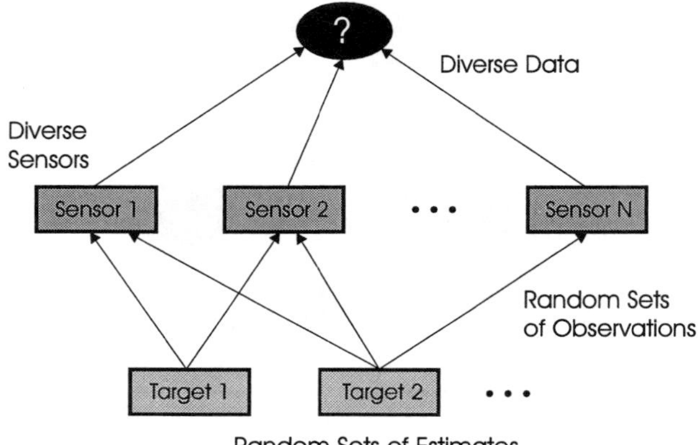

FIG. 3. *Random sets and information fusion.*

ematics, Data Fusion Corporation, Lockheed Martin, Neptune and Company, Oasis Research Center, Raytheon, Texas Instruments, and Xerox. The papers in this volume reflect this diversity. A few papers are tutorial in nature, some are detailed mathematical treatises, some are summary overviews of an entire subject, and still others are investigations rooted in practical engineering intuition.

The workshop was structured into three sessions, devoted respectively to the following topic areas, each organized and chaired by one of the editors:

- *Image Modeling and Analysis* (J. Goutsias).
- *Information/Data Fusion and Expert Systems* (R.P.S. Mahler).
- *Theoretical Statistics and Expert Systems* (H.T. Nguyen).

Each session was preceded by a plenary presentation given by a researcher of world standing:

- Ilya Molchanov, University of Glasgow, Scotland.
- Jean-Yves Jaffray, University of Paris VI, France.
- Ulrich Hőhle, Bergische Universität, Germany.

The following institutions kindly extended their support to this workshop:

- *U.S. Office of Naval Research*, Mathematical, Computer, and Information Sciences Division.
- *U.S. Army Research Office*, Electronics Division.
- *Lockheed Martin*, Eagan, Minnesota Engineering Center.

The editors wish to express their appreciation for the generosity of these sponsors. They also extend their special gratitude to the following individuals for their help in ensuring success of the workshop: Avner Friedman,

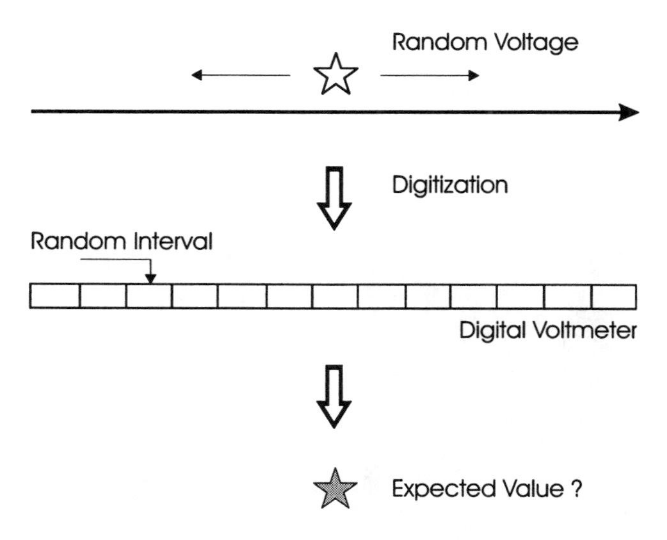

FIG. 4. *Random sets and theoretical statistics.*

IMA, Director; Julia Abrahams, Office of Naval Research; William Sander, Army Research Office; Wesley Snyder, North Carolina State University; Marjorie Hahn, Tufts University; Larry Wasserman, Carnegie-Mellon University; Charles Mills, Lockheed Martin, Director of Engineering; Amy Cavanaugh, IMA, Workshop Coordinator; and John Schepers, IMA, Workshop Financial Coordinator.

In committing these proceedings to the attention of the larger scientific and engineering community, the editors hope that the workshop will have thereby contributed to one of the primary goals of IMA: facilitating creative interchange between statisticians, scientists, and academic and industrial engineers in technical domains of potential practical significance.

John Goutsias
Ronald P.S. Mahler
Hung T. Nguyen

Workshop on
Applications and Theory of Random Sets
Institute for Mathematics and its Applications (IMA)
University of Minnesota
Minneapolis, Minnesota
August 22–24, 1996

Participants
(From left-to-right)

Lower Row: Scott Ferson, Wesley Snyder, Yidong Chen, Bert Fristedt, John Handley, Sinan Batman, Edward Dougherty, Nikolaos Sidiropoulos, Dan Schonfeld, I.R. Goodman, Wolfgang Kober, Stan Music, Ronald Mahler, Jean-Yves Jaffray, Elbert Walker, Carol Walker, Hung Nguyen, John Goutsias

Upper Row: Robert Launer, Paul Black, Tonghui Wang, Shozo Mori, Robert Taylor, Ulrich Hőehle, Ilya Molchanov, Michael Stein, Krishnamoorthy Sivakumar, Fred Daum, Teddy Seidenfeld

CONTENTS

PART I
Image Modeling and Analysis

MORPHOLOGICAL ANALYSIS OF RANDOM SETS AN INTRODUCTION

JOHN GOUTSIAS*

Abstract. This paper provides a brief introduction to the problem of processing random shapes by means of mathematical morphology. Compatibility issues with mathematical morphology suggest that shapes should be modeled as random closed sets. This approach however is limited by theoretical and practical difficulties. Morphological sampling is used to transform a random closed set into a much simpler discrete random set. It is argued that morphological sampling of a random closed set is a sensible thing to do in practical situations. The paper concludes by reviewing three useful random set models.

Key words. Capacity Functional, Discretization, Mathematical Morphology, Random Sets, Shape Processing and Analysis.

AMS(MOS) subject classifications. 60D05, 60K35, 68U10

1. Introduction.

Development of stochastic techniques for image processing and analysis is an important area of investigation. Consider, for example, the problem of analyzing microscopic images of cells, like the ones depicted in the first row of Fig. 1. Image analysis consists of obtaining measurements characteristic to the images under consideration. When we are only interested in geometric measurements (e.g., object location, orientation, area, perimeter length, etc.), and in order to simplify our problem, we may decide to reduce gray-scale images into binary images by means of thresholding, thus obtaining shapes, like the ones depicted in the second row of Fig. 1. Since shape information is frequently random, as is clear from Fig. 1, binary microscopic images of cells may be conceived as realizations of a two-dimensional *random set* model. In this case, measurements are considered to be estimates of random variables, and statistical analysis of such random variables may lead to successful shape analysis.

There are other reasons why stochastic techniques are important for shape processing and analysis. In many instances, shape information is not directly observed. For example, it is quite common that a three-dimensional object (e.g., a metal or a mineral) is partially observed through an imaging system that is only capable of producing two-dimensional pictures of cross sections. The problem here is to infer geometric properties of the three-dimensional object under consideration by means of measurements obtained from the two-dimensional cross sections (this is the main theme in *stereology* [1], an important branch of *stochastic geometry* [2]). Another example is the problem of restoring shape information corrupted by sensor

* Department of Electrical and Computer Engineering, Image Analysis and Communications Laboratory, The Johns Hopkins University, Baltimore, MD 21218 USA. This work was supported by the Office of Naval Research, Mathematical, Computer, and Information Sciences Division, under ONR Grant N00060-96-1376.

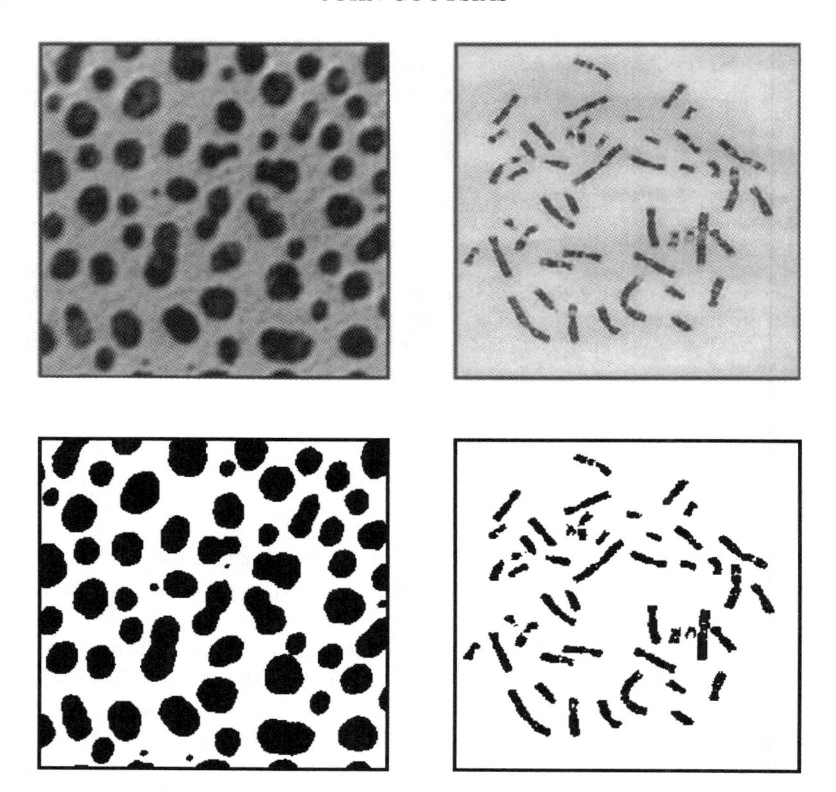

FIG. 1. *Gray-scale (first row) microscopic images of cells and their binary counterparts (second row) obtained after thresholding. Geometric features of interest (e.g., location, orientation, area, perimeter length, etc.) are usually preserved after thresholding.*

noise and clutter. This task is very important in military target detection problems, where targets are frequently imaged through hostile environments by means of imperfect imaging sensors. Both problems consist of recovering shape information from imperfectly or partially observed data and are clearly *ill-posed inverse problems* that need proper *regularization*.

A popular approach to regularizing inverse problems is by means of *stochastic regularization* techniques. A random model is assumed for the images under consideration and statistical techniques are then employed for recovering lost information from available measurements. This approach frequently leads to robust and highly effective algorithms for shape recovery. To be more precise, let us consider the problem of restoring shape information from degraded data. Shapes are usually combined by means of set union or intersection (or set difference, since $A \setminus B = A \cap B^c$). It is therefore reasonable to model shapes as sets (and more precisely as random sets) and assume that data $\mathbf{Y}$ are described by means of a *degradation equation* of the form:

$$\mathbf{Y} = (\mathbf{X} \setminus \mathbf{N}_1) \cup \mathbf{N}_2 \,, \tag{1.1}$$

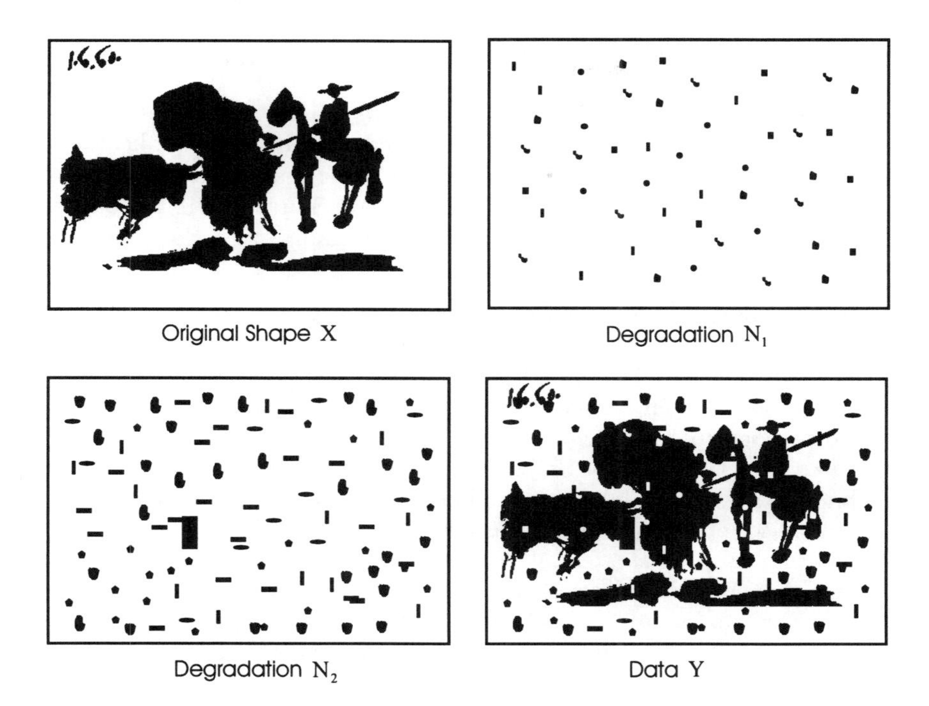

FIG. 2. *A binary Picasso image* $\mathbf{X}$*, the degradation noises* $\mathbf{N}_1$*,* $\mathbf{N}_2$*, and data* $\mathbf{Y}$*, obtained by means of (1.1).*

where $\mathbf{X}$ is a random set that models the shape under consideration and $\mathbf{N}_1$, $\mathbf{N}_2$ are two random sets that model degradation. In particular, $\mathbf{N}_1$ may model incomplete data collection, whereas, $\mathbf{N}_2$ may model degradation due to sensor noise and clutter. Figure 2 depicts the effect of degradations $\mathbf{N}_1$, $\mathbf{N}_2$ on a binary Picasso image[1] by means of (1.1). The problem of shape restoration consists of designing a set operator Ψ such that

$$\hat{\mathbf{X}} = \Psi(\mathbf{Y}) = \Psi((\mathbf{X} \setminus \mathbf{N}_1) \cup \mathbf{N}_2) \, , \tag{1.2}$$

is "optimally" close to $\mathbf{X}$, in some sense. Refer to [3], [4] for more information on this subject and for specific "optimal" techniques for shape restoration by means of random set modeling.

Since shapes are combined by means of unions and intersections, it is natural to consider *morphological image operators* Ψ in (1.2) [5]. This leads to a popular technique for shape processing and analysis, known as *mathematical morphology*, that is briefly described in Section 2. Our main purpose here is to provide an introduction to the problem of processing random shapes by means of mathematical morphology. Compatibility issues with mathematical morphology suggest that shapes should be modeled as

[1] Pablo Picasso, *Pass with the Cape*, 1960.

random closed sets [6]. However, this approach is limited by theoretical and practical difficulties, as explained in Section 3. In Section 4, *morphological sampling* is employed so as to transform a random closed set into a much simpler *discrete random set*. It is argued that, in many applications, morphological sampling of a random closed set is a sensible thing to do. Discrete random sets are introduced in Section 5. Three useful random set models are then presented in Section 6, and concluding remarks are finally presented in Section 7.

2. Mathematical morphology. A popular technique for shape processing and analysis is mathematical morphology. This technique was originally introduced by Matheron [6] and Serra [7] as a tool for investigating geometric structure in binary images. Although extensions of mathematical morphology to grayscale and other images (e.g., multispectral images) exist (e.g., see the book by Heijmans [5]), we limit our exposition here to the binary case. In the following, a binary image X will be first considered to be a subset of the two-dimensional Euclidean space $\mathbb{R}^2$.

Morphological shape operators are defined by means of a *structuring element* $A \subset \mathbb{R}^2$ (shape mask) which interacts with a binary image X so as to enhance or extract useful information. The type of interaction is determined by testing whether the translated structuring element $A_v = \{a+v \mid a \in A\}$ *hits* or *misses* X; i.e., testing whether $X \cap A_v \neq \emptyset$ (A_v hits X) or $X \cap A_v = \emptyset$ (A_v misses X). This is the main idea behind the most fundamental morphological operator, known as the *hit-or-miss transform*, given by

$$X \otimes (A, C) = \{v \in \mathbb{R}^2 \mid A_v \subseteq X, X \cap C_v = \emptyset\}, \tag{2.1}$$

where A, C are two structuring elements such that $A \cap C = \emptyset$. Although the hit-or-miss transform satisfies a number of useful properties, perhaps the most striking one is the fact that any *translation invariant* shape operator Ψ (i.e., an operator for which $\Psi(X_v) = [\Psi(X)]_v$, for every $v \in \mathbb{R}^2$) can be written as a union of hit-or-miss transforms (e.g., see [5]).

When $C = \emptyset$ in (2.1), the hit-or-miss transform reduces to a morphological operator known as *erosion*. The erosion of a binary image X by a structuring element A is given by

$$X \ominus A = \{v \in \mathbb{R}^2 \mid A_v \subseteq X\}.$$

It is clear that erosion comprises of all points v of $\mathbb{R}^2$ for which the structuring element A_v, located at v, fits inside X. The dual of erosion, with respect to set complement, is known as *dilation*. The dilation of a binary image X by a structuring element A is given by

$$X \oplus A = (X^c \ominus \check{A})^c = \{v \in \mathbb{R}^2 \mid X \cap \check{A}_v \neq \emptyset\}, \tag{2.2}$$

where $\check{A} = \{-v \mid v \in A\}$ is the *reflection* of A about the origin. Therefore, dilation comprises of all points v of $\mathbb{R}^2$ for which the translated structuring

Original Shape

Erosion

Dilation

FIG. 3. *Erosion and dilation of the Picasso image X depicted in Fig. 2 by means of a 5×5 SQUARE structuring element A. Notice that erosion comprises of all pixels v of X for which the translated structuring element A_v fits inside X, whereas dilation comprises of all pixels v in $\mathbb{R}^2$ for which the translated structuring element A_v hits X.*

element $\check{A}_v$ hits X. From (2.1) and (2.2) it is clear that $X \otimes (\emptyset, \check{A}) = (X \oplus A)^c$, and dilation is therefore the set complement of the hit-or-miss transform of X by $(\emptyset, \check{A})$. It can be shown that erosion is *increasing* (i.e., $X \subseteq Y \Rightarrow X \ominus A \subseteq Y \ominus A$) and *distributes* over intersection (i.e., $(\bigcap_{i \in I} X_i) \ominus A = \bigcap_{i \in I}(X_i \ominus A)$), whereas, dilation is increasing and distributes over union (i.e., $(\bigcup_{i \in I} X_i) \oplus A = \bigcup_{i \in I}(X_i \oplus A)$). Furthermore, if A contains the origin, then erosion is *anti-extensive* (i.e., $X \ominus A \subseteq X$) whereas dilation is *extensive* (i.e., $X \subseteq X \oplus A$). The effects of erosion and dilation on the binary Picasso image are illustrated in Fig. 3.

Suitable composition of erosions and dilations generates more complicated morphological operators. One such composition produces two useful morphological operators known as *opening* and *closing*. The opening of a binary image X by a structuring element A is given by

$$X \bigcirc A = (X \ominus A) \oplus A \,,$$

whereas the closing is given by

$$X \bullet A = (X \oplus A) \ominus A \,.$$

It is not difficult to show that $X \bigcirc A = \bigcup_{v:\, A_v \subseteq X} A_v$ and, therefore, the

Original Shape

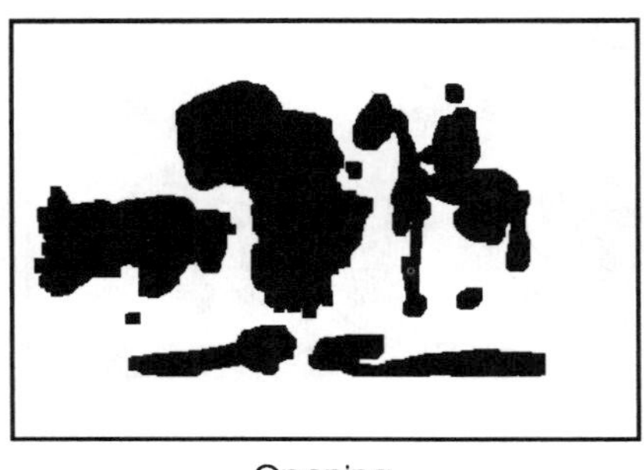

Opening

Closing

FIG. 4. *Opening and closing of the Picasso image X depicted in Fig. 2, by means of a 5×5 SQUARE structuring element A. Notice that opening comprises of the union of all translated structuring elements A_v that fit inside X, whereas closing is the set complement of the union of all translated structuring elements A_v that fit inside X^c.*

opening $X \circ A$ is the union of all translated structuring elements A_v that fit inside shape X. On the other hand, the closing is the dual of opening in the sense that $X \bullet A = (X^c \circ \check{A})^c$. It can be shown that opening is increasing, anti-extensive, and *idempotent* (i.e., $(X \circ A) \circ A = X \circ A$), whereas closing is increasing, extensive, and idempotent. Figure 4 depicts the effects of opening and closing on the binary Picasso image. Notice that the opening $X \circ A$ behaves as a *shape filter*, in the sense that it eliminates all components of X that cannot contain a translated copy of A. In fact, opening and closing are special cases of *morphological filters* [8]. By definition, a morphological filter is any image operator that is increasing and idempotent. Clearly, opening and closing are morphological filters whereas erosion and dilation are not, since they are not idempotent.

3. Random sets. A *random set* (RS) on $\mathbb{R}^2$ is a random element that takes values in a collection $\mathcal{S}$ of subsets of $\mathbb{R}^2$. If $(\Omega, \Sigma(\Omega), \mu)$ is a probability space [9], then a RS $\mathbf{X}$ is a measurable mapping from Ω into $\mathcal{S}$, that is

$$\{\omega \in \Omega \mid \mathbf{X}(\omega) \in \mathcal{A}\} \in \Sigma(\Omega), \quad \forall \mathcal{A} \in \Sigma(\mathcal{S}) \ ,$$

where $\Sigma(\mathcal{S})$ is an appropriate σ-field in $\mathcal{S}$. The RS $\mathbf{X}$ defines a *probability distribution* P_X on $\Sigma(\mathcal{S})$ by

$$P_X[\mathcal{A}] = \mu[\{\omega \in \Omega \mid \mathbf{X}(\omega) \in \mathcal{A}\}], \quad \forall \mathcal{A} \in \Sigma(\mathcal{S}).$$

A common choice for $\mathcal{S}$ is the *power set* $\mathcal{P} = \mathcal{P}(\mathbb{R}^2)$ of $\mathbb{R}^2$ (i.e., the collection of all subsets of $\mathbb{R}^2$) with $\Sigma(\mathcal{S}) = \Sigma(\mathcal{P})$ being the σ-field in $\mathcal{P}$ generated by sets of the form $\{X \in \mathcal{P} \mid v_i \notin X, i = 1,2,...,m;\ w_j \in X, j = 1,2,...,n\}$, where v_i, $w_j \in \mathbb{R}^2$, and $m, n \geq 0$ are integers. It is worthwhile noticing here that $\Sigma(\mathcal{P})$ is also generated by the simple family $\{\{X \in \mathcal{P} \mid X \cap \{v\} = \emptyset\}, v \in \mathbb{R}^2\}$. Consider the *finite-dimensional distribution functions* of RS $\mathbf{X}$, given by

$$p_{v_1,v_2,...,v_n}(x_1, x_2, ..., x_n) = P_X[\, \mathrm{I}_{\mathbf{X}}(v_i) = x_i,\ i = 1,2,...,n\,],$$

where

$$\mathrm{I}_{\mathbf{X}}(v) = \begin{cases} 1, & \text{if } v \in \mathbf{X} \\ 0, & \text{otherwise} \end{cases},$$

is the *indicator function* of $\mathbf{X}$, and $v_i \in \mathbb{R}^2$, $x_i \in \{0,1\}$. As a direct consequence of *Kolmogorov's theorem* [9], the probability distribution of a RS $\mathbf{X}$: $\Omega \to \mathcal{P}$ is uniquely determined from a collection of finite-dimensional distribution functions $\{p_{v_1,v_2,...,v_n}(x_1, x_2, ..., x_n);\ v_i \in \mathbb{R}^2, x_i \in \{0,1\}, n \geq 1\}$ that satisfy Kolmogorov's conditions of *symmetry* and *consistency* [9] (see also [10]). Therefore, a random set $\mathbf{X}$: $\Omega \to \mathcal{P}$ is uniquely specified by means of its finite-dimensional distribution functions.

A question that immediately arises here is whether the previous choices for $\mathcal{S}$ and $\Sigma(\mathcal{S})$ lead to a definition for a RS that is compatible with mathematical morphology. To be more precise, let us concentrate on the problem of transforming a RS by means of a morphological operator Ψ (that, in this paper, is limited to an erosion, dilation, opening, or closing). If $\mathbf{X}$ is a random set, it is expected that $\Psi(\mathbf{X})$ will also be a random set. This translates to the requirement that morphological operators need to be measurable with respect to $\Sigma(\mathcal{S})$. For example, if $\mathbf{X}$ is a RS, we expect that the dilation $\mathbf{X} \oplus K$ of $\mathbf{X}$ by a *compact* (i.e., topologically *closed* and *bounded*) structuring element K is a RS as well. However, it is not difficult to verify that $\{X \in \mathcal{P} \mid v \notin X \oplus K\} = \{X \in \mathcal{P} \mid X \cap (\check{K} \oplus \{v\}) = \emptyset\}$, which is clearly not an element of $\Sigma(\mathcal{P})$, since K is not necessarily finite. Hence, it is not in general possible to determine the probability $P_{X \oplus K}[\mathrm{I}_{X \oplus K}(v) = 0]$ that the dilated RS $\mathbf{X} \oplus K$ does not contain point v from the probability distribution of RS $\mathbf{X}$. In other words, the previous probabilistic description of RS $\mathbf{X}$ is not sufficiently rich to determine the probability distribution of a morphologically transformed RS $\mathbf{X} \oplus K$. Therefore, the previously discussed choices for $\mathcal{S}$ and $\Sigma(\mathcal{S})$ are not compatible with mathematical morphology. If we assume that shapes include their boundary (which is the most common case in practice), then we can set $\mathcal{S} = \mathcal{F}$, where $\mathcal{F}$ is the collection of all closed subsets of $\mathbb{R}^2$, and consider a σ-field $\Sigma(\mathcal{F})$ containing sets of the form $\{X \in \mathcal{F} \mid X \cap K = \emptyset\}$, for $K \in \mathcal{K}$, where $\mathcal{K}$ is the collection of all compact subsets of $\mathbb{R}^2$. It can be shown that the smallest such σ-field is the one generated by the family $\{\{X \in \mathcal{F} \mid X \cap K = \emptyset\}, K \in \mathcal{K}\}$ as well as by the family $\{\{X \in \mathcal{F} \mid X \cap G \neq \emptyset\}, G \in \mathcal{G}\}$, where $\mathcal{G}$ denotes the collection of all (topologically) *open* subsets of $\mathbb{R}^2$. This leads to modeling random shapes by means of *random closed sets* (RACS). A RACS $\mathbf{X}$ is a measurable mapping from Ω into $\mathcal{F}$, that is [6], [11]

$$\{\omega \in \Omega \mid \mathbf{X}(\omega) \in \mathcal{A}\} \in \Sigma(\Omega), \quad \forall \mathcal{A} \in \Sigma(\mathcal{F}) .$$

The RACS $\mathbf{X}$ defines a probability distribution P_X on $\Sigma(\mathcal{F})$ by

$$P_X[\mathcal{A}] = \mu[\{\omega \in \Omega \mid \mathbf{X}(\omega) \in \mathcal{A}\}] , \quad \forall \mathcal{A} \in \Sigma(\mathcal{F}) .$$

An alternative to specifying a RACS by means of a probability distribution, that is defined over classes of sets in $\Sigma(\mathcal{F})$, is to specify the RACS by means of its *capacity functional*, defined over compact subsets of $\mathbb{R}^2$. The capacity functional T_X of a RACS $\mathbf{X}$ is defined by

$$T_X(K) = P_X[\mathbf{X} \cap K \neq \emptyset] , \quad \forall K \in \mathcal{K} .$$

This functional satisfies the following five properties:

PROPERTY 3.1. Since no closed set hits the empty set, $T_X(\emptyset) = 0$.

PROPERTY 3.2. Being a probability, T_X satisfies $0 \leq T_X(K) \leq 1$, for every $K \in \mathcal{K}$.

PROPERTY 3.3. The capacity functional is increasing on $\mathcal{K}$; i.e.,

$$K_1, K_2 \in \mathcal{K} \text{ and } K_1 \subseteq K_2 \Rightarrow T_X(K_1) \leq T_X(K_2) .$$

PROPERTY 3.4. The capacity functional is *upper semi-continuous* (u.s.c.) on $\mathcal{K}$, which is equivalent to

$$K_n \downarrow K \text{ in } \mathcal{K} \Rightarrow T_X(K_n) \downarrow T_X(K) ,$$

where $A_n \downarrow A$ means that $\{A_n\}$ is a decreasing sequence such that $\inf A_n = A$.

PROPERTY 3.5. If, for $K, K_1, K_2, ... \in \mathcal{K}$,

$$Q_X^{(0)}(K) = Q_X(K) = P_X[\mathbf{X} \cap K = \emptyset] = 1 - T_X(K) , \tag{3.1}$$

and

$$Q_X^{(n)}(K; K_1, K_2, ..., K_n) = Q_X^{(n-1)}(K; K_1, K_2, ..., K_{n-1}) - Q_X^{(n-1)}(K \cup K_n; K_1, K_2, ..., K_{n-1}) ,$$

for $n = 1, 2, ...,$ then

$$0 \leq Q_X^{(n)}(K; K_1, K_2, ..., K_n) \\ = P_X[\mathbf{X} \cap K = \emptyset;\ \mathbf{X} \cap K_i \neq \emptyset,\ i = 1, 2, ..., n] \leq 1 , \tag{3.2}$$

for every $n \geq 1$.

A functional T_X that satisfies properties 3.3–3.5 above is known as an *alternating capacity of infinite order* or a *Choquet capacity* [12]. Therefore, the capacity functional of a RACS is a Choquet capacity that in addition satisfies properties 3.1 and 3.2. As a direct consequence of the *Choquet–Kendall–Matheron theorem* [6], [12], [13], the probability distribution of a RACS $\mathbf{X}$ is uniquely determined from a Choquet capacity $T_X(K), K \in \mathcal{K}$, that satisfies properties 3.1 and 3.2. It can be shown that

$$T_X(K_1 \cup K_2) \ \leq \ T_X(K_1) + T_X(K_2) \ , \ \ \forall K_1, K_2 : K_1 \cap K_2 = \emptyset \ .$$

The capacity functional is therefore only subadditive and hence not a measure. However, knowledge of $T_X(K)$, for every $K \in \mathcal{K}$, allows us determine the probability distribution of $\mathbf{X}$. Functional $Q_X(K)$ in (3.1) is known as the *generating functional* of RACS $\mathbf{X}$, whereas, functional $Q_X^{(n)}(K; K_1, K_2, ..., K_n)$ is the probability that the RACS $\mathbf{X}$ misses K and hits K_i, $i = 1, 2, ..., n$ (see (3.2)).

Let us now consider the problem of morphologically transforming a RACS. As we mentioned before, if Ψ: $\mathcal{F} \rightarrow \mathcal{F}$ is a measurable operator with respect to $\Sigma(\mathcal{F})$, then $\Psi(\mathbf{X})$ will be a RACS, provided that $\mathbf{X}$ is a RACS. In simple words, the probability distribution of $\Psi(\mathbf{X})$ can be in principle determined from the probability distribution of $\mathbf{X}$ and knowledge of operator Ψ. It can be shown that erosion, dilation, opening, and closing of a closed set, by means of a compact structuring element, are all measurable with respect to $\Sigma(\mathcal{F})$. Therefore, erosion, dilation, opening, and closing of a RACS, by means of a compact structuring element, is also a RACS. Understanding the effects that morphological transformations have on random sets requires statistical analysis. We would therefore need to relate statistics of $\Psi(\mathbf{X})$ with statistics of $\mathbf{X}$. This can be done by relating the capacity functional $T_{\Psi(X)}$ of $\Psi(\mathbf{X})$ with the capacity functional T_X of $\mathbf{X}$. In general, a simple closed-form relationship is feasible only in the case of dilation, in which case [6], [14]

$$T_{X \oplus A}(K) \ = \ T_X(K \oplus \check{A}) \ , \ \ \forall A, K \in \mathcal{K} \ . \tag{3.3}$$

However, it can be shown that [15]

$$T_{X \ominus A}(K) \ = \ 1 - \sum_{K' \subseteq K} (-1)^{|K'|} R_X(K' \oplus A) \ , \ \forall A, K \in \mathcal{K}_o \ , \tag{3.4}$$

with

$$R_X(K) = P_X[\mathbf{X} \supseteq K] = \sum_{K' \subseteq K} (-1)^{|K'|} [\, 1 - T_X(K') \,] \ , \ \ \forall K \in \mathcal{K}_o, \tag{3.5}$$

where $\mathcal{K}_o \subset \mathcal{K}$ is the collection of all *finite* subsets of $\mathbb{R}^2$. Therefore, a closed-form relationship between $T_{X \ominus A}(K)$ and $T_X(K)$ can be obtained, by means of (3.4), (3.5), when both A and K are finite. Furthermore, and as a direct consequence of (3.3)–(3.5), we have that [15]

$$T_{X \bigcirc A}(K) = 1 - \sum_{K' \subseteq K \oplus \check{A}} (-1)^{|K'|} R_X(K' \oplus A) \ , \ \forall A, K \in \mathcal{K}_o \ , \tag{3.6}$$

whereas

$$T_{X \bullet A}(K) = 1 - \sum_{K' \subseteq K} (-1)^{|K'|} R_{X \oplus A}(K' \oplus A) \ , \ \forall A, K \in \mathcal{K}_o \ , \tag{3.7}$$

with

$$R_{X \oplus A}(K) = \sum_{K' \subseteq K} (-1)^{|K'|} [\, 1 - T_X(K' \oplus \check{A}) \,] \ , \ \forall K \in \mathcal{K}_o \, . \tag{3.8}$$

It is worthwhile noticing that (3.4)–(3.8) are related to the well known *Möbius transform* of combinatorics (e.g., see [16]). If W is a finite set, $\mathcal{P}(W)$ its power set, and $U(K)$ is a real-valued functional on $\mathcal{P}(W)$, then the Möbius transform of U is a functional $V(K)$ on $\mathcal{P}(W)$, given by

$$V(K) = \sum_{K' \subseteq K} U(K') \ , \ \forall K \in \mathcal{P}(W) \, . \tag{3.9}$$

Referring to (3.4), (3.5), it is clear that $1 - T_{X \ominus A}(K)$ is the Möbius transform of functional $(-1)^{|K|} R_X(K \oplus A)$, whereas $R_X(K)$ is the Möbius transform of functional $(-1)^{|K|}[1 - T_X(K)]$. Similar remarks hold for (3.6)–(3.8). Notice that $U(K)$ can be recovered from $V(K)$ by means of the *inverse Möbius transform*, given by

$$U(K) = \sum_{K' \subseteq K} (-1)^{|K \setminus K'|} V(K') \ , \ \forall K \in \mathcal{P}(W) \, . \tag{3.10}$$

Direct implementation of (3.4)–(3.8) is hampered by substantial storage and computational requirements. However, the storage scheme and the *fast Möbius transform* introduced in [17] can be effectively employed here so as to ease such requirements. We should also point-out here that the capacity functional of a RACS is the same as the *plausibility functional* used in the theory of evidence in expert systems [18], and that $R_X(K)$ in (3.5) is known as the *commonality functional* [17]. Finally, there is a close relationship between random set theory and expert systems, as is nicely explained by Nguyen and Wang in [19] and Nguyen and Nguyen in [20].

4. Discretization of RACSs. From our previous discussion, it is clear that the capacity functionals $T_{X \ominus A}(K)$, $T_{X \bigcirc A}(K)$, and $T_{X \bullet A}(K)$ can be evaluated from the capacity functional $T_X(K)$ only when A and

K are finite. It is therefore desirable to: (a) consider finite structuring elements A, and (b) make sure that RACSs $\mathbf{X} \ominus A$, $\mathbf{X} \circ A$, and $\mathbf{X} \bullet A$ are uniquely specified by means of their capacity functionals only over $\mathcal{K}_o$. Requirement (b) is not true in general, even if A is finite. However, we may consider discretizing $\mathbf{X}$, by sampling it over a sampling grid S, in order to obtain a *discrete random set* (DRS) $\mathbf{X}_d = \sigma(\mathbf{X})$, where σ is a sampling operator. It will soon become apparent that a DRS is uniquely specified by means of its capacity functional only over finite subsets of $\mathbb{R}^2$. It is therefore required that erosion, dilation, opening, or closing of a RACS $\mathbf{X}$ by a compact structuring element A be discretized. Moreover, it is desirable that the resulting discretization produces an erosion, dilation, opening, or closing of a DRS $\mathbf{X}_d = \sigma(\mathbf{X})$ by a finite structuring element $A_d = \sigma(A)$. In this case, the discretized morphological transformations $\mathbf{X}_d \ominus A_d$, $\mathbf{X}_d \oplus A_d$, $\mathbf{X}_d \circ A_d$, and $\mathbf{X}_d \bullet A_d$ will be DRSs, whose capacity functional can be evaluated from the capacity functional of $\mathbf{X}_d$, by means of (3.3)–(3.8). Notice however that this procedure should be done in such a way that the resulting discretization is a good approximation (in some sense) of the original continuous problem. We study these issues next.

Let S be a *sampling grid* in $\mathbb{R}^2$, such that

$$S = \{k_1 e_1 + k_2 e_2 \mid k_1, k_2 \in \mathbb{Z}\} ,$$

where $e_1 = (1,0)$, $e_2 = (0,1)$ are the two linearly independent unit vectors in $\mathbb{R}^2$ along the two coordinate directions and $\mathbb{Z}$ is the set of all integers. Consider a bounded open set C, given by

$$C = \{x_1 e_1 + x_2 e_2 \mid -1 < x_1, x_2 < 1\} ,$$

known as the *sampling element*. Let $\mathcal{P}(S)$ be the power set of S. Then, an operator σ: $\mathcal{F} \rightarrow \mathcal{P}(S)$, known as the *sampling operator*, is defined by

$$\sigma(X) = \{s \in S \mid C_s \cap X \neq \emptyset\} = (X \oplus C) \cap S , \quad X \in \mathcal{F} , \tag{4.1}$$

whereas, an operator ρ: $\mathcal{P}(S) \rightarrow \mathcal{F}$, known as the *reconstruction operator*, is defined by

$$\rho(V) = \{v \in \mathbb{R}^2 \mid C_v \cap S \subseteq V\} , \quad V \in \mathcal{P}(S) . \tag{4.2}$$

See [5], [21], [22] for more details. The combined operator $\pi = \rho\sigma$ is known as the *approximation operator*. When operator σ is applied on a closed set $X \in \mathcal{F}$ it produces a discrete set $\sigma(X)$ on S. On the other hand, application of operator ρ on a discrete set $\sigma(X)$ produces a closed set $\pi(X) = \rho\sigma(X)$ that approximates X. The effects that operators σ, ρ, and π have on a closed set X are illustrated in Fig. 5.

Whether or not a closed set X is well approximated by $\pi(X)$ depends on how fine X is sampled by the sampling operator σ. To mathematically

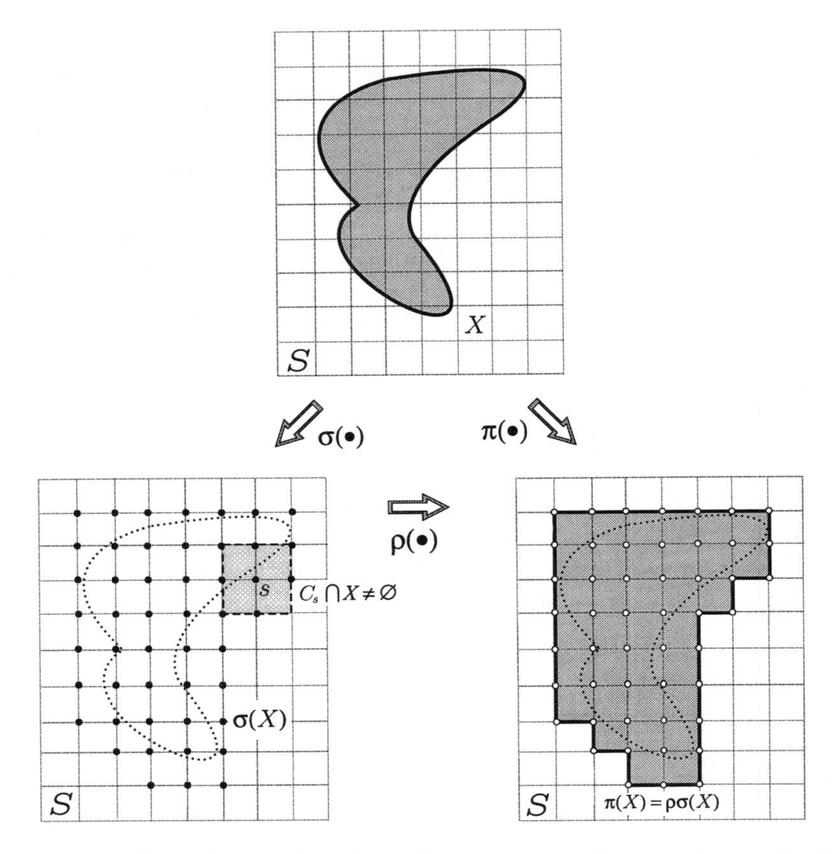

FIG. 5. *The effects of morphological sampling σ, reconstruction ρ, and approximation π on a closed subset X of $\mathbb{R}^2$.*

quantify this, consider sequences $\{S_n\}_{n\geq 1}$ and $\{C_n\}_{n\geq 1}$ of sampling grids and sampling elements, such that

$$S_{n+1} = \frac{1}{2}\, S_n,\ n \geq 1,\ S_1 = S \quad \text{and} \quad C_{n+1} = \frac{1}{2} C_n,\ n \geq 1,\ C_1 = C\,,$$

where $cX = \{cx \mid x \in X\}$. We then define sampling and reconstruction operators σ_n and ρ_n, by replacing S and C in (4.1), (4.2), by S_n and C_n. This determines a sequence of increasingly fine discretizations of a closed set X, denoted by $\mathcal{D} = \{S_n, \sigma_n, \rho_n\}_{n\geq 1}$, that is known as the *covering discretization* [5], [22]. It can be shown that, for $X \in \mathcal{F}$,

$$X \subseteq \cdots \subseteq \pi_{n+1}(X) \subseteq \pi_n(X) \subseteq \pi_{n-1}(X) \subseteq \cdots \subseteq \pi_1(X)\,,$$

and

$$\bigcap_{n\geq 1} \pi_n(X) \;=\; X\,,$$

which means that the approximation $\pi_n(X)$ of X *monotonically* converges to X from above (this is also denoted by $\pi_n(X) \downarrow X$), which implies that $\pi_n(X) \xrightarrow{\mathcal{F}} X$, where $\xrightarrow{\mathcal{F}}$ denotes convergence in the *hit-or-miss topology* (see [5], [6] for more information about the hit-or-miss topology).

For every $n = 1, 2, \ldots$, define a sequence $\mathbf{X}_{d,n}$ by (recall (4.1))

$$\mathbf{X}_{d,n} = \sigma_n(\mathbf{X}) = (\mathbf{X} \oplus C_n) \cap S_n ,$$

where $\mathbf{X}$ is a RACS, and a sequence $\mathbf{X}_n$ by (recall (4.2))

$$\mathbf{X}_n = \rho_n(\mathbf{X}_{d,n}) = \rho_n\sigma_n(\mathbf{X}) = \pi_n(\mathbf{X}) = \{v \in \mathbb{R}^2 \mid (C_n)_v \cap S_n \subseteq \mathbf{X}_{d,n}\} .$$

$\mathbf{X}_{d,n}$ *almost surely* (a.s.) contains a countable number of points and is therefore a DRS. On the other hand, it is known that $\mathbf{X}_n$ is an a.s. closed set, whereas it has been shown in [10] that π_n is a measurable mapping; therefore, $\mathbf{X}_n$ is a RACS. In fact, it is not difficult to show that $\mathbf{X}_n \downarrow \mathbf{X}$, a.s., which implies that $\mathbf{X}_n \xrightarrow{\mathcal{F}} \mathbf{X}$, a.s., as well. Furthermore, if A is a *$\mathcal{D}$–regular compact structuring element*, for which

$$A = \pi_N(A) , \quad \text{for some } 1 \leq N < \infty , \tag{4.3}$$

then it can be shown that [10], [23]

$$\rho_n(\sigma_n(\mathbf{X}) \ominus \sigma_n(A)) \xrightarrow{\mathcal{F}} \mathbf{X} \ominus A , \quad a.s. ,$$

$$\rho_n(\sigma_n(\mathbf{X}) \oplus \sigma_n(A)) \xrightarrow{\mathcal{F}} \mathbf{X} \oplus A , \quad a.s. ,$$

$$\rho_n(\sigma_n(\mathbf{X}) \circ \sigma_n(A)) \xrightarrow{\mathcal{F}} \mathbf{X} \circ A , \quad a.s. ,$$

$$\rho_n(\sigma_n(\mathbf{X}) \bullet \sigma_n(A)) \xrightarrow{\mathcal{F}} \mathbf{X} \bullet A , \quad a.s. .$$

This means that the covering discretization guarantees that erosion, dilation, opening, or closing of a RACS $\mathbf{X}$ by a $\mathcal{D}$–regular compact structuring element A (i.e., a structuring element that satisfies (4.3)) can be well approximated by an erosion, dilation, opening, or closing, respectively, of a DRS $\sigma_n(\mathbf{X})$ by a finite structuring element $\sigma_n(B)$, for some large n, as is desirable. The requirement that A is a $\mathcal{D}$-regular compact structuring element is not a serious limitation since a wide collection of structuring elements may satisfy this property [23].

The previous results focus on the a.s. convergence of discrete morphological operators to their continuous counterparts. However, results concerning convergence of the associated capacity functionals also exist. It has been shown in [10] that the capacity functional of the approximating RACS $\pi_n(\mathbf{X})$ monotonically converges from above to the capacity functional of RACS $\mathbf{X}$; i.e.,

$$T_{\pi_n(X)}(K) \downarrow T_X(K) , \quad \forall K \in \mathcal{K} .$$

Furthermore, it has been shown that the capacity functional of RACS $\rho_n(\sigma_n(\mathbf{X}) \ominus \sigma_n(A))$ converges to the capacity functional of RACS $\mathbf{X} \ominus A$, in the limit, as $n \to \infty$, provided that A is a $\mathcal{D}$–regular compact structuring element, with a similar convergence result being true for the case of dilation, opening, and closing. Finally, it has been shown that

$$\lim_{n \to \infty} T_{X_{d,n}}(\sigma_n(K)) = T_X(K), \quad \forall K \in \mathcal{K},$$

and

$$\lim_{n \to \infty} T_{X_{d,n} \ominus \sigma_n(A)}(\sigma_n(K)) = T_{X \ominus A}(K), \quad \forall K \in \mathcal{K},$$

provided that A is a $\mathcal{D}$–regular compact structuring element, with a similar convergence result being true for dilation, opening, and closing. Therefore, and for sufficiently large n, the continuous morphological transformations $\mathbf{X} \ominus A$, $\mathbf{X} \oplus A$, $\mathbf{X} \bigcirc A$, and $\mathbf{X} \bullet A$, can be well approximated by the discrete morphological transformations $\mathbf{X}_{d,n} \ominus \sigma_n(A)$, $\mathbf{X}_{d,n} \oplus \sigma_n(A)$, $\mathbf{X}_{d,n} \bigcirc \sigma_n(A)$, and $\mathbf{X}_{d,n} \bullet \sigma_n(A)$, respectively, provided that A is a $\mathcal{D}$–regular compact structuring element. This shows that, in most practical situations, it will be sufficient enough to limit our interest to a DRS $\mathbf{X}_d = \sigma(\mathbf{X}) = (\mathbf{X} \oplus C) \cap S$ instead of RACS $\mathbf{X}$, for a sufficiently fine sampling grid S, with the benefit (among some other benefits) of relating the capacity functional of a morphologically transformed DRS $\Psi(\mathbf{X}_d)$ to the capacity functional of $\mathbf{X}_d$, by means of (3.3)–(3.8). It can be shown that

$$T_{X_d}(K) = P_X[\mathbf{X} \cap ((K \cap S) \oplus C) \neq \emptyset], \quad \forall K \in \mathcal{K},$$

which shows that the capacity functional of the DRS $\mathbf{X}_d$ need to be known only over finite subsets of S. Finally, it has been shown in [10] that

$$T_{X_d}(B) = \sup\{T_X(K);\ K \in \mathcal{K},\ K \subset B \oplus C\}, \quad \forall B \in \mathcal{I},$$

where $\mathcal{I}$ is the collection of all *bounded* subsets of S, which relates the capacity functional of the DRS $\mathbf{X}_d = \sigma(\mathbf{X})$ with the capacity functional of RACS $\mathbf{X}$.

5. Discrete random sets. Following our previous discussion, given a probability space $(\Omega, \Sigma(\Omega), \mu)$, a DRS $\mathbf{X}$ on $\mathbb{Z}^2$ is a measurable mapping from Ω into $\mathcal{Z}$, the power set of $\mathbb{Z}^2$, that is

$$\{\omega \in \Omega \mid \mathbf{X}(\omega) \in \mathcal{A}\} \in \Sigma(\Omega), \quad \forall \mathcal{A} \in \Sigma(\mathcal{Z}),$$

where $\Sigma(\mathcal{Z})$ is the σ-field in $\mathcal{Z}$ generated by the simple family $\{\{X \in \mathcal{Z} \mid X \cap B = \emptyset\}, B \in \mathcal{B}\}$, where $\mathcal{B}$ is the collection of all *bounded* subsets of $\mathbb{Z}^2$. A DRS $\mathbf{X}$ defines a probability distribution P_X on $\Sigma(\mathcal{Z})$ by

$$P_X[\mathcal{A}] = \mu[\{\omega \in \Omega \mid \mathbf{X}(\omega) \in \mathcal{A}\}], \quad \forall \mathcal{A} \in \Sigma(\mathcal{Z}).$$

The *discrete capacity functional* of a DRS $\mathbf{X}$ is defined by

$$T_X(B) = P_X[\mathbf{X} \cap B \neq \emptyset] , \quad \forall B \in \mathcal{B} .$$

This functional satisfies the following four properties:

PROPERTY 5.1. Since no set hits the empty set, $T_X(\emptyset) = 0$.

PROPERTY 5.2. Being a probability, T_X satisfies $0 \leq T_X(B) \leq 1$, for every $B \in \mathcal{B}$.

PROPERTY 5.3. The discrete capacity functional is increasing on $\mathcal{B}$; i.e.,

$$B_1, B_2 \in \mathcal{B} \text{ and } B_1 \subseteq B_2 \;\Rightarrow\; T_X(B_1) \leq T_X(B_2) .$$

PROPERTY 5.4. If, for $B, B_1, B_2, ... \in \mathcal{B}$,

$$Q_X^{(0)}(B) = Q_X(B) = P_X[\mathbf{X} \cap B = \emptyset] = 1 - T_X(B) , \tag{5.1}$$

and

$$Q_X^{(n)}(B; B_1, B_2, ..., B_n) = Q_X^{(n-1)}(B; B_1, B_2, ..., B_{n-1}) - Q_X^{(n-1)}(B \cup B_n; B_1, B_2, ..., B_{n-1}) ,$$

for $n = 1, 2, ...$, then

$$0 \leq Q_X^{(n)}(B; B_1, B_2, ..., B_n) = P_X[\mathbf{X} \cap B = \emptyset;\; \mathbf{X} \cap B_i \neq \emptyset,\; i = 1, 2, ..., n] \leq 1 , \tag{5.2}$$

for every $n \geq 1$.

As a special case of the Choquet-Kendall-Matheron theorem, the probability distribution of a DRS is uniquely determined by a discrete capacity functional $T_X(B)$, $B \in \mathcal{B}$, that satisfies properties 5.1–5.4 above. Functional $Q_X(B)$, $B \in \mathcal{B}$, in (5.1) is known as the *discrete generating functional* of $\mathbf{X}$, whereas, functional $Q_X^{(n)}(B; B_1, B_2, ..., B_n)$, B, B_1, B_2, ..., $B_n \in \mathcal{B}$, is the probability that the DRS $\mathbf{X}$ misses B and hits B_i, $i = 1, 2, ..., n$ (see (5.2)).

In practice, images are observed through a finite-size window W, $|W| < \infty$, where $|A|$ denotes the *cardinality* (or *area*) of set A. Therefore, it seems reasonable to consider DRSs whose realizations are limited within W. Let $\mathcal{B}_W$ be the collection of all (bounded) subsets of $\mathbb{Z}^2$ that are included in W. A DRS $\mathbf{X}$ is called an *a.s. W-bounded DRS* if $P_X[\mathbf{X} \in \mathcal{B}_W] = 1$. It is not difficult to see that an a.s. W-bounded DRS is uniquely specified by means of a discrete capacity functional $T_X(B)$, $B \in \mathcal{B}_W$. Furthermore, if $M_X(X)$, $X \in \mathcal{B}_W$, is the *probability mass function* of $\mathbf{X}$, i.e., if

$$M_X(X) = P_X[\mathbf{X} = X] , \quad X \in \mathcal{B}_W ,$$

and $L_X(B) = P_X[\mathbf{X} \subseteq B]$, $B \in \mathcal{B}_W$, is the so-called *belief functional* of $\mathbf{X}$ (e.g., see [17], [19]), then L_X is the Möbius transform of the probability mass function (recall (3.9)); i.e.,

$$L_X(B) = \sum_{X \subseteq B} M_X(X) , \quad \forall B \in \mathcal{B}_W , \tag{5.3}$$

which leads to

$$M_X(X) = \sum_{B \subseteq X} (-1)^{|X \setminus B|} L_X(B) , \quad \forall X \in \mathcal{B}_W , \tag{5.4}$$

by Möbius inversion (recall (3.10)). It is also not difficult to see that

$$L_X(B) = 1 - T_X(B^c) , \quad \forall B \in \mathcal{B}_W , \tag{5.5}$$

where $B^c = W \setminus B$. The previous equations show that the probability mass function of a DRS $\mathbf{X}$ can be computed from the discrete capacity functional by first calculating the belief functional $L_X(B)$, by means of (5.5), and by then obtaining $M_X(X)$ as the Möbius inverse of $L_X(B)$, by means of (5.4). On the other hand, the discrete capacity functional of a DRS $\mathbf{X}$ can be computed from the probability mass function by first calculating the belief functional L_X as the Möbius transform of M_X, by means of (5.3), and by then setting $T_X(B) = 1 - L_X(B^c)$.

These simple observations may have some important consequences for the statistical analysis of random sets. For example, if a model (i.e., a probability mass function) is to be optimally fit to given data, then mathematical morphology may be effectively used to accomplish this task. For example, given a realization X of a DRS $\mathbf{X}$, we may consider the empirical estimate

$$\hat{T}_{X,W'}(B) = \frac{|(X \oplus \check{B}) \cap (W' \ominus B)|}{|W' \ominus B|} , \quad B \in \mathcal{B} ,$$

obtained by means of erosion and dilation, for the discrete capacity functional of $\mathbf{X}$ (see also eq. (2.3) in [24]). Under suitable stationarity and ergodicity assumptions, this estimator converges a.s. to $T_X(B)$, as $W' \uparrow \mathbb{Z}^2$, for each $B \in \mathcal{B}$. Assuming that W' is large enough, so as to obtain a good approximation $\hat{T}_{X,W'}(B)$ of $T_X(B)$, for every $B \in \mathcal{B}_W$, $W \subset W'$, we can approximately calculate the probability mass function of the a.s. bounded DRS $\mathbf{X} \cap W$, by replacing $T_X(B)$ by $\hat{T}_{X,W'}(B)$ in (5.5), and by then calculating the Möbius transform of L_X in (5.4).

6. Models for random sets. A number of interesting random set models have been proposed in the literature and have found useful application in a number of image analysis problems. In this section, we briefly review some of these models, namely the *Boolean model*, *quermass weighted random closed sets*, and *morphologically constrained discrete random sets*. Additional models may be found in [2].

6.1. The Boolean model. Consider a RACS Ξ, defined by

$$\Xi = \bigcup_{k=1}^{\infty} (\Xi_k)_{v_k} , \tag{6.1}$$

where $\{v_1, v_2, ...\}$ is a collection of random points in $\mathbb{R}^2$ and $\{\Xi_1, \Xi_2, ...\}$ is a collection of non-empty a.s. bounded RACSs. Ξ is known as the *germ-grain model* [2], [25]. The points $\{v_1, v_2, ...\}$ are the *germs*, whereas the RACSs $\{\Xi_1, \Xi_2, ...\}$ are the *primary grains* of Ξ. The simplest model for the germs is a stationary Poisson point process in $\mathbb{R}^2$ with intensity λ. If the grains are taken to be *independent and identically distributed* (i.i.d.) a.s. bounded RACSs with capacity functional $T_{\Xi_o}(K)$, $K \in \mathcal{K}$, and independent of the germ process, then the germ-grain RACS is known as the *Boolean model.* The degradation images depicted in Fig. 2 are realizations of Boolean models. Notice that the Boolean model models random scattering of a "typical" random shape. It can be shown that the capacity functional of the Boolean model is given by

$$T_\Xi(K) = 1 - \exp\{-\lambda \mathrm{E}[|\Xi_1 \oplus \check{K}|]\} , \quad \forall K \in \mathcal{K} ,$$

provided that $\mathrm{E}[|\Xi_1 \oplus \check{K}|] < \infty$, for every $K \in \mathcal{K}$.

The Boolean model is of fundamental interest in random set theory, but most often it constitutes only a rough approximation of real data (however, the Boolean model may be a good statistical model for the binary images depicted in the second row of Fig. 1). Nevertheless, it has been successfully applied in a number of practical situations (e.g., see [26]–[30]). More information about the Boolean model may be found in [2], [31]. Discrete versions of the Boolean model have appeared in [32], [33].

6.2. Quermass weighted random closed sets. A new family of random set models have been recently proposed by Baddeley, Kendall, and van Lieshout in [34] that are based on weighting a Boolean model. Consider an observation window $W \subset \mathbb{R}^2$, $|W| < \infty$, and a Poisson point process in W of finite intensity λ (to be referred to as the *benchmark Poisson process*). A *quermass interaction point process* is a point process in W that has density $f(\mathbf{x})$ (for the configuration $\mathbf{x} = \{x_1, x_2, ..., x_n\}$), with respect to the benchmark Poisson point process, given by

$$f(\mathbf{x}) = \frac{1}{Z(\lambda, \gamma)} \gamma^{-W_r^2(\Xi)} , \tag{6.2}$$

where Ξ is the Boolean model (6.1). In (6.2), γ is a real-valued positive parameter, W_r^2 are the *quermass integrals* (or *Minkowski functionals*) in two dimensions [1], and $Z(\lambda, \gamma)$ is a normalizing constant (so that $f(\mathbf{x})$ is a probability density). Notice that $W_0^2(X)$ is the *area* of set X (in which case (6.2) leads to the area-interaction point processes in [35]), $W_1^2(X)$ is proportional to the length of the *perimeter* of X, whereas, W_2^2, is proportional to the *Euler functional* that equals the number of components in X reduced by the number of holes.

The quermass interaction point process may be viewed as the point process generated by the germs of a Boolean model Ξ under weighting $\gamma^{-W_r^2(\Xi)}$. When $\gamma > 1$, clustering or attraction between the points is observed, whereas, $\gamma < 1$ produces ordered patterns or repulsion. Finally, when $\gamma = 1$, the process is simply Poisson. The quermass interaction point process is not Poisson for $\gamma \neq 1$ and can be used in (6.1) to define a *quermass weighted random closed set* that generalizes the Boolean model.

6.3. Morphologically constrained DRSs. Consider an a.s. W-bounded DRS $\mathbf{X}$ whose probability mass function $M_{\mathbf{X}}(X)$ satisfies the *positivity condition*

$$M_{\mathbf{X}}(X) > 0 , \quad \forall X \in \mathcal{B}_W .$$

By setting

$$U(X) = -T \ln \frac{M_{\mathbf{X}}(X)}{M_{\mathbf{X}}(\emptyset)} , \quad \forall X \in \mathcal{B}_W ,$$

it can be shown that

$$M_{\mathbf{X}}(X) = \frac{1}{Z(T)} \exp\{-\frac{1}{T} U(X)\} , \quad \forall X \in \mathcal{B} , \tag{6.3}$$

where

$$Z(T) = \sum_{X \in \mathcal{B}_W} \exp\{-\frac{1}{T} U(X)\} , \tag{6.4}$$

is a normalizing constant known as the *partition function.* In (6.3) and (6.4), U is a real-valued functional on $\mathcal{B}_W$, known as the *energy function*, such that $U(\emptyset) = 0$, whereas T is a real-valued nonnegative constant, known as the *temperature*. An a.s. W-bounded DRS $\mathbf{X}$ with probability mass function $M_{\mathbf{X}}(X)$, given by (6.3), (6.4), is a *Gibbs random set* (or a binary *Gibbs random field*) [36]. Gibbs random sets have been extensively used in a variety of image processing and analysis tasks, including texture synthesis and analysis [37] and image restoration [38].

It can be shown that, in the limit, as $T \to \infty$, a Gibbs random set $\mathbf{X}$ assigns equal probability to all possible realizations, in which case

$$\lim_{T \to \infty} M_{\mathbf{X}}(X) = \frac{1}{|\mathcal{B}_W|} , \quad \forall X \in \mathcal{B}_W .$$

On the other hand, if

$$\mathcal{U} = \{X^* \in \mathcal{B}_W \mid U(X^*) \leq U(X), \ \forall X \in \mathcal{B}_W\} ,$$

then

$$\lim_{T \to 0^+} M_X(X) = \begin{cases} 1/|\mathcal{U}|, & \text{for every } X \in \mathcal{B}_W \\ 0, & \text{otherwise} \end{cases} .$$

Therefore, and in the limit, as the temperature tends to zero, a Gibbs random set $\mathbf{X}$ is uniformly distributed over all global minima of its energy function, known as the *ground states.*

The ground states are the most important realizations of a Gibbs random set. A typical ground state may be well structured and is frequently representative of real data obtained in a particular application. Since we are mainly concerned with statistical models which generate well structured realizations that satisfy certain geometric properties, it has been suggested in [39] that mathematical morphology can be effectively used for designing an appropriate energy function in (6.3). This simple idea leads to a new family of random sets on $\mathbb{Z}^2$, known as *morphologically constrained DRSs.* The probability mass function of a morphologically constrained DRS is given by

$$M_X(X) = \frac{1}{Z(\theta, T)} \exp\{-\frac{1}{T} \theta \cdot |\Psi(X)|\} , \quad \forall X \in \mathcal{B}_W , \tag{6.5}$$

where

$$Z(\theta, T) = \sum_{X \in \mathcal{B}_W} \exp\{-\frac{1}{T} \theta \cdot |\Psi(X)|\} . \tag{6.6}$$

In (6.5) and (6.6), θ is a (possibly vector) parameter and Ψ is a (possibly vector) morphological operator. In this case, $U(X) = \theta \cdot |\Psi(X)|$.

A simple example of a morphologically constrained DRS is to consider a vector operator Ψ of the form

$$\Psi(X) = (X, X \ominus A_1, X \ominus A_2, ..., X \ominus A_K) , \quad \forall X \in \mathcal{B}_W ,$$

where $\{A_1, A_2, ..., A_K\}$ is a collection of structuring elements in $\mathcal{B}_W$, and real-valued parameters θ, given by

$$\theta = (\alpha, \beta_1, \beta_2, ..., \beta_K) ,$$

in which case

$$U(X) = \alpha|X| + \sum_{k=1}^{K} \beta_k |X \ominus A_k| , \quad \forall X \in \mathcal{B}_W . \tag{6.7}$$

For simplicity, if we consider $K = 1$, $\alpha = -1$, and $\beta_1 = 1$ in (6.7), it is clear that, at low enough temperatures, the most favorable realizations X will be the ones of maximum area $|X|$, under the constraint of minimizing the area $|X \ominus A_1|$ of erosion $X \ominus A_1$; i.e., the ground states in this case

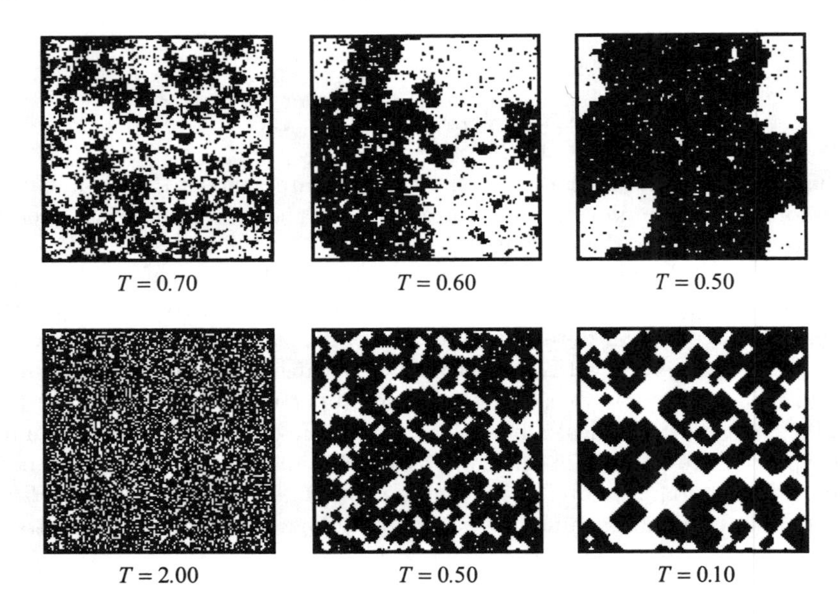

FIG. 6. *Realizations of morphologically constrained DRSs with energy functions given by (6.8)–first row and (6.9)–second row, at three different temperatures. Notice that, at high temperatures, realizations are rather random, whereas, as the temperature decreases, realizations become more structured.*

will have the largest possible area and the smallest number of components that may contain structuring element A_1.

An important special case of (6.7) is when $K = 2$, $A_1 = \{(0,0),(1,0)\}$, and $A_2 = \{(0,0),(0,1)\}$, in which case

$$U(X) \;=\; \alpha|X| + \beta_1|X \ominus A_1| + \beta_2|X \ominus A_2|\,, \quad \forall X \in \mathcal{B}_W\,. \tag{6.8}$$

This choice results in the *Ising model*, a well known random set model of statistical mechanics (e.g., see [40]). Realizations of this model, at three different temperatures and for the case when $\alpha = 2.2$, $\beta_1 = -1.2$, and $\beta_2 = -1.0$, are depicted in the first row of Fig. 6. At high temperatures, realizations are random. However, as the temperature decreases to zero, realizations become well structured.

It has been suggested in [39] that a useful energy function is the one given by

$$\begin{aligned} U(X) \;=\; \alpha|X| \;&+\; \sum_{i=0}^{I} \beta_i |X \circ iB \setminus X \circ (i+1)B| \\ &+\; \sum_{j=1}^{J} \gamma_j |X \bullet jB \setminus X \bullet (j-1)B|\,, \quad \forall X \in \mathcal{B}_W\,, \end{aligned} \tag{6.9}$$

with α, $\beta_i, i = 0,1,\ldots,I$, and $\gamma_j, j = 1,2,\ldots,J$ being real-valued parameters, where $0B = \{(0,0)\}$ and $nB = (n-1)B \oplus B$, for $n = 1,2,\ldots$. At low enough

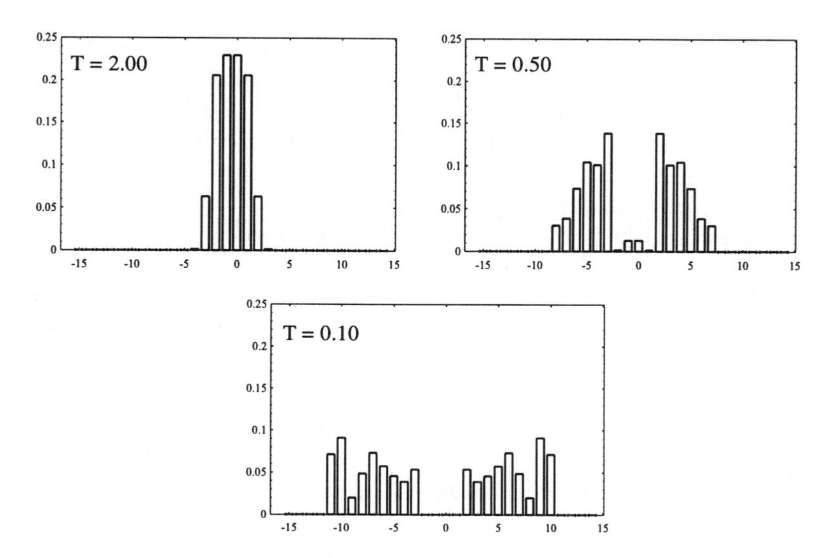

FIG. 7. *The size densities associated with the realizations depicted in the second row of Fig. 6.*

temperatures, the resulting morphologically constrained DRS $\mathbf{X}$ may favor realizations whose size density [6]

$$s_X(n) = \frac{1}{|W|} \begin{cases} \mathrm{E}[|\mathbf{X} \circ nB \setminus \mathbf{X} \circ (n+1)B|], & \text{for } n \geq 0 \\ \mathrm{E}[|\mathbf{X} \bullet |n|B \setminus \mathbf{X} \bullet (|n|-1)B|], & \text{for } n \leq -1 \end{cases},$$

is directly influenced by the particular values of α, β, and γ, see [39]. Notice that, in this case,

$$\begin{aligned} \Psi(X) = {} & (X, X \setminus X \circ B, X \circ B \setminus X \circ 2B, ..., X \circ IB \setminus X \circ (I+1)B, \\ & X \bullet B \setminus X, X \bullet 2B \setminus X \bullet B, ..., X \bullet JB \setminus X \bullet (J-1)B) , \end{aligned}$$

whereas

$$\theta = (\alpha, \beta_0, \beta_1, ..., \beta_I, \gamma_1, \gamma_2, ..., \gamma_J) .$$

The second row of Fig. 6 depicts realizations of this model (with $I = 1$, $J = 2$, $\alpha = 0$, $\beta_0 = \beta_1 = \gamma_1 = \gamma_2 = 1$) at three different temperatures and for the case when B is the RHOMBUS structuring element $\{(0,0), (1,0), (-1,0), (0,1), (0,-1)\}$, whereas Fig. 7 depicts the associated size densities. Notice that at high temperatures, realizations are rather random with the size density values (estimated by means of the Monte Carlo estimation approach discussed in [3], [4]) clustered around size 0. However, as the temperature decreases to zero, the size density tends to zero at small sizes, as expected (for this example, it can be shown that $\lim_{T \to 0^+} s_X(n) = 0$, for $-2 \leq n \leq 1$).

Preliminary work indicates that morphologically constrained DRSs are powerful models capable of solving a number of important image processing and analysis problems [39]. By comparing (6.5), (6.6) with (6.2), it is not difficult to see that morphologically constrained DRSs may be related to the quermass weighted random sets, discussed in Section 6.2. It may therefore be possible to extend these models to the continuous case as well. This extension was suggested to the author by Adrian Baddeley, of the University of Western Australia, Perth, Australia, and Marie-Colette Lieshout, of the University of Warwick, Coventry, United Kingdom, during the 1996 International Symposium on Advances in Theory and Applications of Random Sets, Fontainebleau, France, and is currently under investigation.

7. Concluding remarks. Morphological analysis of random images modeled as random sets is an exciting area of research with great potential in solving some difficult problems of image processing and analysis. However, a number of difficult issues need to be resolved and a number of questions be answered before a useful practical tool is developed. Among other things, the author believes that is very important to develop new and more powerful random set models (both continuous and discrete) that take into account geometric structure in images. Another important problem is the development of a statistical theory for random sets, complete with appropriate estimators and statistical tests. Although such a theory was virtually non-existent only a few years ago, recent developments demonstrate that it is possible to develop such a theory with potentially great consequences in the areas of mathematical statistics, stochastic geometry, stereology, as well as image processing and analysis.

REFERENCES

[1] I. Saxl, *Stereology of Objects with Internal Structure*, Elsevier, Amsterdam, The Netherlands, 1989.

[2] D. Stoyan, W.S. Kendall, and J. Mecke, *Stochastic Geometry and its Applications*, Second Edition, John Wiley, Chichester, England, 1995.

[3] K. Sivakumar and J. Goutsias, *On estimating granulometric discrete size distributions of random sets*, This Volume, pp. 47–71.

[4] K. Sivakumar and J. Goutsias, *Discrete morphological size distributions and densities: Estimation techniques and applications*, Journal of Electronic Imaging, Special Issue on Random Models in Imaging, 6 (1997), pp. 31–53.

[5] H.J.A.M. Heijmans, *Morphological Image Operators*, Academic Press, Boston, Massachusetts, 1994.

[6] G. Matheron, *Random Sets and Integral Geometry*, John Wiley, New York City, New York, 1975.

[7] J. Serra, *Image Analysis and Mathematical Morphology*, Academic Press, London, England, 1982.

[8] J. Serra and L. Vincent, *An overview of morphological filtering*, Circuits, Systems and Signal Processing, 11 (1992), pp. 47–108.

[9] P. Billingsley, *Probability and Measure*, John Wiley, New York City, New York, 1986.

[10] K. Sivakumar and J. Goutsias, *Binary random fields, random closed sets, and morphological sampling*, IEEE Transactions on Image Processing, 5 (1996), pp. 899–912.
[11] G. Matheron, *Théorie des ensembles aléatoires*, Ecole Nationale Superieure des Mines des Paris, Fontainebleau, France, 1969.
[12] G. Choquet, *Theory of capacities*, Annales de L'Institut Fourier, 5 (1953–1954), pp. 131–295.
[13] D.G. Kendall, *Foundations of a theory of random sets*, Stochastic Geometry (E.F. Harding and D.G. Kendall, eds.), John Wiley, London, United Kingdom, 1974, pp. 322–376.
[14] J. Goutsias, *Modeling random shapes: An introduction to random closed set theory*, Mathematical Morphology: Theory and Hardware (R.M. Haralick, ed.), Oxford University Press, New York City, New York, 1997.
[15] J. Goutsias, *Morphological analysis of discrete random shapes*, Journal of Mathematical Imaging and Vision, 2 (1992), pp. 193–215.
[16] M. Aigner, *Combinatorial Theory*, Springer–Verlag, New York City, New York, 1979.
[17] H.M. Thoma, *Belief function computation*, Conditional Logic in Expert Systems (I.R. Goodman, M.M. Gupta, H.T. Nguyen, and G.S. Rogers, eds.), Elsevier, North Holland, 1991.
[18] G. Shafer, *A Mathematical Theory of Evidence*, Princeton University Press, Princeton, New Jersey, 1976.
[19] H.T. Nguyen and T. Wang, *Belief functions and random sets*, This Volume, pp. 243–255.
[20] H.T. Nguyen and N.T. Nguyen, *Random sets in decision making*, This Volume, pp. 297–320.
[21] H.J.A.M. Heijmans and A. Toet, *Morphological sampling*, Computer Vision, Graphics, and Image Processing, 54 (1991), pp. 384–400.
[22] H.J.A.M. Heijmans, *Discretization of morphological image operators*, Journal of Visual Communication and Image Representation, 3 (1992), pp. 182–193.
[23] K. Sivakumar and J. Goutsias, *On the discretization of morphological operators*, Journal of Visual Communication and Image Representation, 8 (1997), pp. 39–49.
[24] I. Molchanov, *Statistical problems for random sets*, This Volume, pp. 27–45.
[25] K.-H. Hanisch, *On classes of random sets and point process models*, Elektronische Informationsverarbeitung und Kybernetik, 16 (1980), pp. 498–502.
[26] P.J. Diggle, *Binary mosaics and the spatial pattern of heather*, Biometrics, 37 (1981), pp. 531–539.
[27] J. Masounave, A.L. Rollin, and R. Denis, *Prediction of permeability of non–woven geotextiles from morphometry analysis*, Journal of Microscopy, 121 (1981), pp. 99–110.
[28] U. Bindrich and D. Stoyan, *Stereology for pores in wheat bread: Statistical analyses for the Boolean model by serial sections*, Journal of Microscopy, 162 (1991), pp. 231–239.
[29] J.-L. Quenec'h, M. Coster, J.-L. Chermant, and D. Jeulin, *Study of the liquid-phase sintering process by probabilistic models: Application to the coarsening of WC-Co cermets*, Journal of Microscopy, 168 (1992), pp. 3–14.
[30] N. Cressie and F.L. Hulting, *A spatial statistical analysis of tumor growth*, Journal of the American Statistical Association, 87 (1992), pp. 272–283.
[31] I.S. Molchanov, *Statistics of the Boolean Model for Practitioners and Mathematicians*, John Wiley, Chichester, England, 1997.
[32] N.D. Sidiropoulos, J.S. Baras, and C.A. Berenstein, *Algebraic analysis of the generating functional for discrete random sets and statistical inference for intensity in the discrete Boolean random–set model*, Journal of Mathematical Imaging and Vision, 4 (1994), pp. 273–290.
[33] J.C. Handley and E.R. Dougherty, *Maximum–likelihood estimation for discrete*

Boolean models using linear samples, Journal of Microscopy, 182 (1996), pp. 67–78.

[34] A.J. Baddeley, W.S. Kendall, and M.N.M. van Lieshout, *Quermass-interaction processes*, Technical Report 293, Department of Statistics, University of Warwick, England, 1996.

[35] A.J. Baddeley and M.N.M. van Lieshout, *Area-interaction point processes*, Annals of the Institute of Statistical Mathematics, 47 (1995), pp. 601–619.

[36] J. Besag, *Spatial interaction and the statistical analysis of lattice systems (with discussion)*, Journal of the Royal Statistical Society, Series B, 36 (1974), pp. 192–236.

[37] G.R. Cross and A.K. Jain, *Markov random field texture models*, IEEE Transactions on Pattern Analysis and Machine Intelligence, 5 (1983), pp. 25–39.

[38] S. Geman and D. Geman, *Stochastic relaxation, Gibbs distributions, and the Bayesian restoration of images*, IEEE Transactions on Pattern Analysis and Machine Intelligence, 6 (1984), pp. 721–741.

[39] K. Sivakumar and J. Goutsias, *Morphologically constrained discrete random sets*, Advances in Theory and Applications of Random Sets (D. Jeulin, ed.), World Scientific, Singapore, 1997, pp. 49–66.

[40] J.J. Binney, N.J. Dowrick, A.J. Fisher, and M.E.J. Newman, *The Theory of Critical Phenomena: An Introduction to the Renormalization Group*, Oxford University Press, Oxford, England, 1992.

STATISTICAL PROBLEMS FOR RANDOM SETS

ILYA MOLCHANOV*

Abstract. This paper surveys different statistical issues that involve random closed sets. The main topics include the estimation of capacity functionals, averaging and expectations of random sets, models of compact sets, and statistics of the Boolean model.

Key words. Aumann Expectation, Averaging, Boolean Model, Capacity Functional, Distance Function, Empirical Distribution, Point Process, Random Polygon, Set-Valued Expectation, Spatial Process.

AMS(MOS) subject classifications. 60D05, 60G55, 62M30

1. Introduction.

A random closed set X is a random element with values in the family $\mathcal{F}$ of closed subsets of a basic carrier space E. This carrier space is usually assumed to be the d-dimensional Euclidean space $\mathbb{R}^d$ or a discrete grid $\mathbb{Z}^d$ in $\mathbb{R}^d$. We refer to [35,55] for measurability details and remind that the event $\{X \cap K \neq \emptyset\}$ is measurable for each compact set K. The functional

$$T_X(K) = \mathbf{P}\{X \cap K \neq \emptyset\}\,, \quad K \in \mathcal{K}\,,$$

defined on the family $\mathcal{K}$ of compact sets in E, is said to be the *capacity functional* of X. It is important to note that $T_X(\cdot)$ is not a measure, since it is not additive, but just subadditive. However, $T_X(\cdot)$ determines the probability distribution of X, i.e. the corresponding probability measure on $\mathcal{F}$. Note that it is very complicated to deal with such a measure directly, since its values are defined on classes of sets. In view of this, the capacity functional is more convenient, since it is defined on sets rather than classes of sets, although we have to give up the additivity property.

The general problem of parametric inference for random closed sets can be outlined as the problem of estimating a parameter θ of the distribution that belongs to a parametric family $T(K;\theta) = \mathbf{P}_\theta\{X \cap K \neq \emptyset\}$. If this parameter θ takes values from an infinite-dimensional space or is a set itself, we actually deal with the non-parametric setup. For example, let $X = \xi M$, where ξ is a random variable with a known expectation and M is a closed subset of $\mathbb{R}^d$. Then M is the parameter, which however can be estimated rather easily.

Since $T_X(\cdot)$ is not a measure, it is difficult to define the likelihood and find a reasonable analogue of the probability density function. Some derivatives for capacities [21,28] are not directly applicable for statistical purposes related to random sets. An approach based on the point processes suggested in [32] only works for several special models.

---

* Department of Statistics, University of Glasgow, Glasgow G12 8QW, Scotland, U.K.

The likelihood approach can be developed for discrete random sets. If E is finite, then X can take no more than a finite number of values. As a result, $\mathbf{P}\{X = K\}$ can be used instead of a density of X with subsequent application of the classical maximum likelihood principle [54]. If X is "continuous" (X is a general subset of $\mathbb{R}^d$), then it can be discretized to apply this method. Unfortunately, it is quite difficult to compute $\mathbf{P}\{X = K\}$ in terms of the capacity functional of X, since this involves the computationally intensive Möbius inversion formula from combinatorial theory [1]. At large, the discrete approach ignores geometry, since it is difficult to come up with efficient geometric concepts for subsets of the grid. Another reason is that the geometry is ignored, when only the probabilities $\mathbf{P}\{X = K\}$ are considered.

We confine ourselves to the continuous setup, where the following basic approaches to statistics are known as yet.

1. Handling with samples of general sets and their *empirical distributions.* This is related to the theory of empirical measures and is similar to statistics of general point processes without any reference to Poisson or other special assumptions [29].
2. Statistics of i.i.d. (independent identically distributed) samples of *compact* sets. This involves models of compact random sets and estimation of their parameters, averaging of sets, characterization of shapes of random sets, and the corresponding probability models.
3. Statistics of *stationary* random sets is conceptually related to spatial statistics, statistics of Poisson-related point processes and statistics of random fields.

The structure of the paper follows the above classification. Note that different approaches presume different tools to be applied. The first approach goes back to the theory of empirical measures [53], the second one refers to geometric probability, extreme values, statistical theory of shape and multivariate statistics. The third approach is essentially based on mathematical tools borrowed from integral and convex geometry.

2. Empirical distributions and capacity functionals.

2.1. Samples of i.i.d. random sets. Let $X_1, \ldots, X_n$ be i.i.d. random closed sets with the same distribution as the random closed set X. These sets can be used to estimate the capacity functional $T_X(K)$ by

$$\hat{T}_{X,n}(K) = \frac{1}{n} \sum_{i=1}^{n} \mathbf{1}_{X_i \cap K \neq \emptyset},$$

where

$$\mathbf{1}_A = \begin{cases} 1, & \text{if } A \text{ is true,} \\ 0, & \text{otherwise.} \end{cases}$$

By the strong law of large numbers, the *empirical capacity functional* $\hat{T}_{X,n}(K)$ converges to $T_X(K)$ almost surely (a.s.) as $n \to \infty$ for each $K \in \mathcal{K}$. If

$$\sup_{K \in \mathcal{M},\, K \subseteq K_0} |\hat{T}_{X,n}(K) - T(K)| \to 0 \quad \text{a.s. as} \quad n \to \infty, \tag{2.1}$$

for some $\mathcal{M} \subseteq \mathcal{K}$ and each $K_0 \in \mathcal{K}$, then X is said to satisfy the *Glivenko-Cantelli theorem* over the class $\mathcal{M}$. The classical Glivenko-Cantelli theorem (for random variables) can be obtained for the case when $X = (-\infty, \xi]$, for a random variable ξ, and $\mathcal{M}$ is a class of singletons. However, there are peculiar examples of random sets which do not satisfy (2.1) even for a simple class $\mathcal{M}$.

EXAMPLE 2.1. Let $X = \xi + M \subset \mathbb{R}$, where ξ is a Gaussian random variable, and M is a nowhere dense subset of $[0,1]$ having positive Lebesgue measure. The complete metric space $[0,1]$ cannot be covered by a countable union of nowhere dense sets, so that, for each $n \geq 1$, there exists a point $x_n \in [0,1] \setminus \cup_{i=1}^{n} X_i$. Thus, $\hat{T}_{X,n}(\{x_n\}) = 0$, while $T_X(\{x_n\}) \geq \varepsilon > 0$, $n \geq 1$, and (2.1) does not hold for $\mathcal{M} = \{\{x\}\colon x \in [0,1]\}$. See [38] for other examples. □

In order to prove (2.1) we have to impose some conditions on realizations of random sets. The random set X is called *regular closed* if X almost surely coincides with the closure of its interior.

THEOREM 2.1. ([36]) Let X be a regular closed random set, and let

$$T_X(K) = \mathbf{P}\{(\operatorname{Int} X) \cap K \neq \emptyset\} \tag{2.2}$$

for all $K \in \mathcal{M}$. Then (2.1) holds for each $K_0 \in \mathcal{K}$. If $\mathcal{M} = \mathcal{K}$, then (2.2) is also a necessary condition for (2.1) provided X is a.s. continuous, i.e. $\mathbf{P}\{X \in \partial X\} = 0$ for all $x \in \mathbb{R}^d$.

2.2. Stationary random sets. If X is a *stationary random set* (which means that X and $X+x$ have identical distributions for each $x \in \mathbb{R}^d$), then different translations of X can be used instead of the sample $X_1, \ldots, X_n$. For this, X must be *ergodic* [12,25], which means that spatial averages can be used to estimate the corresponding probabilities. These spatial averages are computed within an observation window W which expands unboundedly in all directions. The latter is denoted by $W \uparrow \mathbb{R}^d$. It is convenient to view W as an element of a family W_s, $s > 0$, of windows, satisfying several natural assumptions [12]. The natural choice, $W_s = sW_1$, $s > 0$, satisfies these conditions for a convex compact set W_1 containing the origin.

If X is ergodic, the spatial average of $\mathbf{1}_{(K+x)\cap X \neq \emptyset}$ for $x \in W$ estimates the corresponding hitting probability $\mathbf{P}\{(K+x) \cap X \neq \emptyset\}$ that does not depend on x and equals the capacity functional $T_X(K)$. The corresponding estimator is given by

$$\hat{T}_{X,W}(K) = \frac{\mu((X \oplus \check{K}) \cap (W \ominus K))}{\mu(W \ominus K)}, \quad K \in \mathcal{K}, \tag{2.3}$$

where μ is the Lebesgue measure in $\mathbb{R}^d$ and $X \oplus \check{K} = \{x - y\colon x \in X, y \in K\}$. If X is only observed inside the window W, then it is not possible to check whether X and $K + x$ have a non-empty intersection if $K + x$ stretches beyond W. Note that (2.3) uses the so-called "minus-sampling" correction for these edge effects, that is (2.3) refers to information contained "sufficiently deep" in the window W. The set $W \ominus K = \{x : x + K \subset W\}$ is called the *erosion* of W by K, see [52,24]. In the simplest case K is a singleton $\{x\}$, and $\hat{T}_{X,W}(\{x\})$ equals the quotient of $\mu(X \cap W)$ and $\mu(W)$.

The conditions of the Glivenko-Cantelli theorem are especially simple if X is a stationary random closed set. Then (2.1) holds for $\mathcal{M} = \mathcal{K}$ if X is regular closed (and only if, provided X is a.s. continuous) [37]. The technique of Mase [34] can be applied to investigate asymptotic properties of estimator (2.3).

2.3. General minimum contrast estimators. It is possible to formulate a general *minimum contrast approach* for the empirical capacity functionals [39]. Let the capacity functional $T(K;\theta)$ depend on a parameter $\theta \in \Theta$. If the capacity functional of X is $T(K;\theta_0)$, then the minimum contrast estimator of θ_0 is given by

$$\hat{\theta}_n = \arg\inf_{\theta \in \Theta} \sup_{K \in \mathcal{M},\, K \subseteq K_0} |T(K;\theta) - \hat{T}_{X,n}(K)| \tag{2.4}$$

for a certain compact K_0 and an appropriate class $\mathcal{M} \subseteq \mathcal{K}$. Clearly, a similar formula is applicable for the empirical capacity functional $\hat{T}_{X,W}$ if X is stationary.

THEOREM 2.2. Let $T(K;\theta)$ depend on the parameter θ taking values in the compact metric space Θ. Assume that for all disjoint $\theta_1, \theta_2 \in \Theta$, $T(K;\theta_1) \neq T(K;\theta_2)$ for at least one $K \in \mathcal{M}$, and $T(K;\theta_n) \to T(K;\theta)$ for each $K \in \mathcal{M}$ as soon as $\theta_n \to \theta$ in Θ. If the random set X with the capacity functional $T(K;\theta_0)$ satisfies the Glivenko-Cantelli theorem (2.1) over the class $\mathcal{M}$, then the minimum contrast estimator $\hat{\theta}_n$ from (2.4) is strongly consistent, i.e. $\hat{\theta}_n \to \theta_0$ in Θ.

PROOF. Without loss of generality assume that $\hat{\theta}_n \to \theta' \neq \theta_0$ a.s. as $n \to \infty$. Then, by (2.1) and (2.4),

$$\begin{aligned}
&\sup_{K \in \mathcal{M}} |T(K;\hat{\theta}_n) - T(K;\theta_0)| \\
&\qquad \leq \sup_{K \in \mathcal{M}} |T(K;\hat{\theta}_n) - \hat{T}_{X,n}(K)| + \sup_{K \in \mathcal{M}} |\hat{T}_{X,n}(K) - T(K;\theta_0)| \\
&\qquad \leq 2 \sup_{K \in \mathcal{M}} |\hat{T}_{X,n}(K) - T(K;\theta_0)| \to 0 \quad \text{a.s. as} \quad n \to \infty .
\end{aligned}$$

On the other hand, $T(K';\theta') \neq T(K';\theta_0)$ for some $K' \in \mathcal{M}$, so that

$$\begin{aligned}\sup_{K\in\mathcal{M}} |T(K;\hat\theta_n) - T(K;\theta_0)| &\geq |T(K';\hat\theta_n) - T(K';\theta_0)| \\ &\to |T(K';\theta') - T(K';\theta_0)| > 0.\end{aligned}$$

The obtained contradiction proves the theorem. □

3. Averaging and expectations. In this section we consider *compact* random sets, which means that their realizations belong to $\mathcal{K}$ almost surely. For simplicity, all sets are assumed to be almost surely non-empty. A simple example of a random compact set

$$X = \begin{cases} [0,1] & \text{with probability } 1/2 \\ \{0,1\} & \text{with probability } 1/2 \end{cases}$$

shows that a "reasonable" expectation of X is not easy to define. Perhaps, in some cases it is even impossible to come up with a reasonable concept of average or expectation.

The main difficulty in defining expectations is explained by the fact that the space $\mathcal{F}$ is not linear. The usual trick is to "linearize" $\mathcal{F}$, i.e. to embed $\mathcal{F}$ into a linear space. Then one defines expectation in this linear space and, if possible, "maps" it back to $\mathcal{F}$.

3.1. Aumann expectation. The Aumann expectation was first defined in [4] in the context of integration of set-valued functions, while its random set meaning was discovered in [3]. A random point ξ is said to be a *selection* of X if $\xi \in X$ almost surely. It is well known from set-valued analysis [62] that there exists at least one selection of X. If the norm of X

$$\|X\| = \sup\{\|x\| : x \in X\}$$

has a finite expectation (X is said to be *integrable*), then $\mathbf{E}\xi$ exists, so that X possesses at least one integrable selection. The *Aumann expectation* of X is defined as the set of expectations of all integrable selections, i.e.

$$\mathbf{E}_{\mathrm{A}}X = \{\mathbf{E}\xi : \xi \text{ is a selection of } X\}.$$

The family of all selections can be very large even for simple (and deterministic) sets. For instance, if $X = \{0,1\}$ almost surely, then the selections of X can be obtained as $\xi(\omega) = \mathbf{1}_{\Omega''}(\omega)$ for all measurable partitions $\Omega = \Omega' \cup \Omega''$. Hence $\mathbf{E}\xi = \mathbf{P}(\Omega'')$, so that the set of expectations of $\xi(\omega)$ depends on the atomic structure of the underlying probability space. This fact explains why two random sets with the same distribution (but defined on different probability spaces) may have different Aumann expectations.

THEOREM 3.1. ([4,49]) If the basic probability space $(\Omega, \Sigma, \mathbf{P})$ contains no atoms and $\mathbf{E}\|X\| < \infty$, then $\mathbf{E}_{\mathrm{A}}X$ is convex and, moreover, $\mathbf{E}_{\mathrm{A}}X = \mathbf{E}_{\mathrm{A}}\mathrm{conv}(X)$.

It is possible to modify the definition of the Aumann expectation to get sometimes non-convex results. This can be done by using the *canonical* probability space $\Omega = \mathcal{F}$, so that the distribution of X is a measure on $\mathcal{F}$. The corresponding expectation is denoted by $\mathbf{E}_{\mathrm{A-red}}X$. For instance, if X takes values $K_1, \dots, K_n$ with probabilities $p_1, \dots, p_n$ respectively, then

$$\mathbf{E}_{\mathrm{A-red}}X = p_1 K_1 \oplus \cdots \oplus p_n K_n .$$

If the distribution of X is non-atomic, then $\mathbf{E}_{\mathrm{A-red}}X = \mathbf{E}X$, while in general $\mathbf{E}_{\mathrm{A-red}}X \subseteq \mathbf{E}_{\mathrm{A}}X$, see [60].

Remember that a random convex compact set X corresponds uniquely to its *support function*

$$h(X,u) = \sup\{\langle x, u\rangle : \; x \in X\}, \quad u \in \mathbb{R}^d ,$$

where $\langle x, u\rangle$ is the scalar product in $\mathbb{R}^d$. If $\mathbf{E}\|X\| < \infty$, then $h(X,u)$ has finite expectation for all u. The support function is characterized by its sublinear property which means that $h(X,\cdot)$ is homogeneous ($h(X,cu) = ch(X,u)$ for all positive c) and subadditive ($h(X,u+v) \le h(X,u)+h(X,v)$ for all $u, v \in \mathbb{R}^d$). This property is kept after taking expectation, so that $\mathbf{E}h(X,u)$, $u \in \mathbb{R}^d$, is again the support function of a convex set, which is exactly the Aumann expectation of X, if the probability space Ω is non-atomic.

THEOREM 3.2. ([4,49,59]) If $\mathbf{E}\|X\| < \infty$, then $\mathbf{E}h(X,u) = h(\mathbf{E}_{\mathrm{A}}X, u)$ for all $u \in \mathbb{R}^d$.

The convexity of $\mathbf{E}_{\mathrm{A}}X$ is the major drawback of the Aumann expectation, which severely restricts its application, e.g. in image analysis. Note that the Aumann expectation stems from the Hörmander embedding theorem, which states that the family of compact closed subsets of a Banach space E can be embedded, by an isometry j, into the space of all bounded functionals on the unit ball in the adjoint space E^*. This isometry $j(K)$ is given by the support function $h(K,\cdot)$.

3.2. Fréchet expectation. This expectation is a specialization of the general concept available for metric spaces. Remember that $\mathcal{K}$ can be metrized by the Hausdorff metric

$$\rho_{\mathrm{H}}(K_1, K_2) = \inf\{\varepsilon > 0 : \; K_1 \subseteq K_2^{\varepsilon},\, K_2 \subseteq K_1^{\varepsilon}\},$$

where K^{ε} is the ε-neighborhood of $K \in \mathcal{K}$. In application to the metric space $(\mathcal{K}, \rho_{\mathrm{H}})$, find the set $K_0 \in \mathcal{K}$ which minimizes $\mathbf{E}\rho_{\mathrm{H}}(X,K)^2$ for $K \in \mathcal{K}$. K_0 is said to be the *Fréchet expectation* of X, and $\mathbf{E}\rho_{\mathrm{H}}(X,K_0)^2$ is called the variance of X, see [19]. Unfortunately, in most practical cases it is not possible to solve the basic minimization problem, since the parameter space $\mathcal{K}$ is too rich. Also the Fréchet expectation can be non-unique.

3.3. Doss expectation. Remember that $\rho_{\mathrm{H}}(\{x\}, K) = \sup\{\rho(x,y) : y \in K\}$ is the Hausdorff distance between $\{x\}$ and $K \in \mathcal{K}$. The *Doss expectation* of the random set X is defined by

$$\mathbf{E}_{\mathrm{D}}X = \{y : \rho(x,y) \leq \mathbf{E}\rho_{\mathrm{H}}(\{x\}, X) \text{ for all } x \in \mathbb{R}^d\},$$

see [27]. This expectation can also be defined for random elements in general metric spaces [16]. Clearly, $\mathbf{E}_{\mathrm{D}}X$ is the intersection of all balls centered at $x \in \mathbb{R}^d$ with radius $\mathbf{E}\rho_{\mathrm{H}}(x, F)$ for $x \in \mathbb{R}^d$, so that $\mathbf{E}_{\mathrm{D}}X$ is convex. Note also that the expectation of a singleton $\{\xi\}$ in $\mathbb{R}^d$ is $\{\mathbf{E}\xi\}$ given by the "usual" expectation of ξ.

3.4. Vorob'ev expectation. If $\mathbf{1}_X(x)$ is the *indicator function* of X, i.e.,

$$\mathbf{1}_X(x) = \begin{cases} 1, & \text{if } x \in X \\ 0, & \text{otherwise} \end{cases},$$

then $\mathbf{E}\mathbf{1}_X(x) = p_X(x) = \mathbf{P}\{x \in X\}$ is called the *coverage function*. Assume that $\mathbf{E}\mu(X) < \infty$. A set-theoretic mean $\mathbf{E}_{\mathrm{V}}X$ is defined in [61] by $L_p = \{x \in \mathbb{R}^d : p_X(x) \geq p\}$ for p which is determined from the inequality

$$\mu(L_q) \leq \mathbf{E}\mu(X) \leq \mu(L_p), \quad \text{for all} \quad q > p.$$

The set $L_{1/2} = \{x \in \mathbb{R}^d : p_X(x) \geq 1/2\}$ has properties of a *median*, see [58].

This approach considers indicator functions as elements of $L^2(\mathbb{R}^d)$. Hence singletons as well as sets of almost surely vanishing Lebesgue measure are considered as uninteresting, since the corresponding indicator function $\mathbf{1}_X(x)$ vanishes almost surely for all x as soon as the corresponding distribution is atomless.

It is possible to prove that for all Borel sets B with $\mu(B) = \mathbf{E}\mu(X)$, $\mathbf{E}\mu(X \triangle \mathbf{E}_{\mathrm{V}}X) \leq \mathbf{E}\mu(X \triangle B)$, where '$\triangle$' denotes the symmetric difference, see [58]. Furthermore, $\mathbf{E}\mu(X \triangle L_{1/2}) \leq \mathbf{E}\mu(X \triangle B)$ for each bounded B. This latter property is similar to the classical property of the median which minimizes the first absolute central moment.

3.5. Radius-vector mean. Let X be *shrinkable* with respect to the origin 0, i.e. let $[0,1)X \subset \operatorname{Int} X$. (A shrinkable set is also star-shaped.) Let r_X be the *radius-vector function* of X defined by

$$r_X(u) = \sup\{t : tu \in X, t \geq 0\}, \quad u \in \mathbb{S}^{d-1},$$

where $\mathbb{S}^{d-1}$ is the unit sphere in $\mathbb{R}^d$. The expected values $\mathbf{E}r_X(u)$, $u \in \mathbb{S}^{d-1}$, define a function which is the radius-vector function of a deterministic shrinkable set. This set is called the *radius-vector mean* of X [58]. Radius-vector functions are very popular in the engineering literature, where it is usual to apply Fourier methods for shape description [6,48]. The major shortcomings are the necessity to work with star-shaped sets and the non-linearity with respect to translations of the sets, since the radius-vector function depends non-linearly on the location of the origin within the set. On the other hand, in many cases, it is very difficult to identify the "natural" location of a star-shaped set.

3.6. Distance average. Let $\rho(x, X)$ be the *distance function* of X, i.e. $\rho(x, X)$ equals the Euclidean distance from x to the nearest point of X. A suitable level set of the expected distance function $\bar{d}(x) = \mathbf{E}\rho(x, X)$ serves as the mean of X. To find this suitable level, we threshold $\bar{d}(x)$ to get a family of sets $X(\varepsilon) = \{x\colon\ \bar{d}(x) \geq \varepsilon\}$, $\varepsilon > 0$. Then the *distance average* $\bar{X}$ is the set $X(\varepsilon)$, where ε is chosen to minimize

$$\|\bar{d}(\cdot) - \rho(\cdot, X(\varepsilon))\|_\infty = \sup_{x \in \mathbb{R}^d} |\bar{d}(x) - \rho(x, X(\varepsilon))|\,.$$

See [5] for details and further generalizations. This approach allows us to deal with sets of zero Lebesgue measure since, even in this case, the distance functions are non-trivial (in contrast to indicator functions used to define the Vorob'ev expectation).

3.7. Axiomatic properties of expectations. It is possible to formulate several basic properties or axioms that should hold for a "reasonable" expectation $\mathbf{E}X$ of X. The first group of the axioms is related to inclusion relationships.

A1 If X is deterministic, then $\mathbf{E}X = X$.
A2 If $K \subseteq X$ a.s., where K is deterministic, then $K \subseteq \mathbf{E}X$.
A3 If $X \subseteq W$ a.s. for a deterministic set W (perhaps, from some special family), then $\mathbf{E}X \subseteq W$.
A4 If $X \subseteq Y$ a.s., then $\mathbf{E}X \subseteq \mathbf{E}Y$.

The second group consists of the properties related to invariance with respect to some transformation.

B1 If X is distribution-invariant with respect to a certain group G (which means that gX and X have the same distribution for each $g \in G$), then the expectation of X must be invariant with respect to G.
B2 Translation-equivariance: $\mathbf{E}(X + x) = \mathbf{E}X + x$.
B3 Homogeneity: $\mathbf{E}(cX) = c\mathbf{E}(X)$.

The third group of properties relates expectations of sets and "usual" expectations of random variables and vectors.

C1 If $X = \{\xi\}$ is a random singleton, then $\mathbf{E}X = \{\mathbf{E}\xi\}$.
C2 If $X = B_\eta(\xi)$ is a ball of random radius η and center ξ, then $\mathbf{E}X = B_{\mathbf{E}\eta}(\mathbf{E}\xi)$.
C3 If $X = \operatorname{conv}(\xi_1, \ldots, \xi_n)$ is the convex hull of a finite number of random points, then $\mathbf{E}X = \operatorname{conv}(\mathbf{E}\xi_1, \ldots, \mathbf{E}\xi_n)$.

A reasonable expectation must as well appear as a solution of some minimization problem and as a limit in an appropriate strong law of large numbers. Note that some of these natural properties are non-compatible and have far-reaching consequences. For example, **A4** implies that $\mathbf{E}X \ni \mathbf{E}\xi$ for each selection $\xi \in X$, so that $\mathbf{E}X$ should contain the Aumann expectation of X. However, detail discussion of these properties is beyond the scope of this paper.

4. Models of compact sets. Statistical studies of compact random sets are difficult due to lack of models (or distributions of random sets) which allow evaluations and provide sets of sufficiently variable shape. Clearly, simple random sets, like random singletons and balls, are ruled out, if we would like to have models with really *random shapes*. We start with models of planar random polygons.

4.1. Typical polygons in tessellations. A homogeneous and isotropic Poisson line process in $\mathbb{R}^2$ corresponds to the Poisson point process in the space $[0, 2\pi) \times [0, \infty)$ with intensity measure $\lambda\mu$, where the coordinates provide a natural parameterization of the lines through their normals and the distances to the origin. These lines divide the plane into convex polygons. These are shifted in such a way that their centers of gravity lay in the origin and are interpreted as realizations of the "typical" random polygon X called the *Poisson polygon* [35,55,58].

Consider the Dirichlet (or Voronoi) mosaic generated by a stationary Poisson point process of intensity λ. For each point x_i we construct the open set consisting of all points of the plane whose distance to x_i is less than the distances to other points. If shifted by x_i, the closures of these sets give realizations of the so-called *Poisson-Dirichlet polygon*, see [58].

The distributions of both polygons depend on only one parameter, which characterizes the polygons' mean size. For example, the Poisson-Dirichlet polygon X has the same distribution as $\lambda^{-1/2}X_1$, where X_1 is the Poisson-Dirichlet polygon obtained by a point process with unit intensity. Several important numerical parameters of the Poisson polygon and the Poisson-Dirichlet polygon are known either theoretically or obtained by numerical integration or simulations [58]. It is however not clear how to compute the capacity functionals for both random polygons.

The corresponding statistical estimation is not difficult, since the only parameter λ (the intensity of the line or point process) is related to the mean area (or mean perimeter) of the polygons.

One can use random polygons to obtain further models of random sets. For instance, a *rounded polygon* is defined by $Y = X \oplus B_\xi(0)$, where ξ is a positive random variable, and $B_\xi(0)$ is the disk of radius ξ centered at the origin.

4.2. Finite convex hulls. Another model of a random isotropic polygon is the convex hull of N independent points uniformly distributed within the disk $B_r(0)$ of radius r centered at the origin, see [58]. Clearly,

$$\mathbf{P}\{X \subset K\} = (\mu(K)/\mu(B_r(0)))^N ,$$

where K is a convex subset of $B_r(0)$. However, further exact distributional characteristics are not so easy to find; mostly only asymptotic properties for large N can be investigated [50].

4.3. Convex-stable sets. Convex-stable random sets appear as weak limits (in the sense of the convergence in distribution) of scaled convex hulls

$$a_n^{-1}\mathrm{conv}(Z_1 \cup \cdots \cup Z_n) \tag{4.1}$$

of i.i.d. random compact sets $Z_1, Z_2, \ldots$ having a regularly varying distribution in a certain sense. We will deal only with the simplest model where the Z_i's are random singletons, $Z_i = \{\xi_i\}$. This case has been studied in [2,9,10], while statistical applications have been considered in [46]. In order to obtain non-degenerate limit distributions, the probability density f of the ξ_i's must be *regular varying* in $\mathbb{R}^d$, i.e. for some vector $e \neq 0$,

$$f(tu_t)/f(te) \to \phi(u) \neq 0, \infty \tag{4.2}$$

as soon as $u_t \to u \neq 0$ as $t \to \infty$. In the isotropic case one can use the density

$$f(u) = \frac{c}{c_1 + \|u\|^{\alpha+d}}, \quad u \in \mathbb{R}^d,$$

for some $\alpha > 0$. The constant c is a scaling parameter, while the normalizing parameter c_1 is chosen in such a way that f is a probability density function. Then (4.2) holds, $a_n \sim n^{1/\alpha}$ as $n \to \infty$, and, for $\|e\|^{\alpha+d} = c$, (4.2) yields $\phi(u) = c\|u\|^{-\alpha-d}$. Thus, for $a_n = n^{1/\alpha}$, the random set $a_n^{-1}\mathrm{conv}(\xi_1, \ldots, \xi_n)$ converges in distribution as $n \to \infty$ to the isotropic convex-stable random set X such that

$$\mathbf{P}\{X \subset K\} = \exp\{-\int_{\mathbb{R}^d \setminus K} \phi(u)\, du\}.$$

The model has two parameters: the *size parameter* c and the *shape parameter* α. The choice of a *positive* α implies that X is almost surely a compact convex polygon, see [13]. The limiting random set X can also be interpreted as the convex hull of a *scale invariant Poisson point process* [12].

Statistical inference for convex-stable sets (and for other models of compact random sets) is based on the method of moments applied to some functionals, see [46]. For instance, if $\mathsf{A}(X)$ (resp. $\mathsf{U}(X)$) is the area (resp. perimeter) of $X \subset \mathbb{R}^2$, then the equation

$$\mathbf{E}\mathsf{U}(X)/\sqrt{\mathbf{E}\mathsf{A}(X)} = 2\pi\Gamma(1-\alpha^{-1})\left(\pi\alpha\Gamma(2-2\alpha^{-1})/(\alpha-1)\right)^{-1/2}$$

yields an estimate of α.

More complicated convex-stable random sets appear if $Z_1, Z_2, \ldots$ in (4.1) are random balls or other compact random sets [40]. Convex-stable random polygons have been applied in [33] to model more complicated star-shaped random sets that appear as planar sections of human lungs.

Other possible models for compact random sets include radius-vector perturbed sets [33,56], weak limits of intersections of half-planes [40], and set-valued growth processes [58]. It is typical also to consider models of random fractal sets, see [18,58]. Usually they are defined by some iterative random procedure or through level sets or graphs of random functions.

5. Sets or figures. Typically, the starting point is a sample of i.i.d realizations of a random compact set. If positions of the sets are known, then we speak about statistics of *sets*, in difference to statistics of *figures* when locations/orientations of sets are not specified. This means that the positions of the sets are irrelevant for the problem and the aim is to find the average shape of the sets in the sample. Such a situation appears in studies of particles (dust powder, sand grains, abrasives etc.).

5.1. Shape ratios. At the first approximation, one can characterize shape of a compact set X by numerical parameters, called *shape ratios*, see [58]. For instance, the *area-perimeter ratio* (or compacity) is given by $4\pi\mathsf{A}(X)/\mathsf{U}(X)^2$, *circularity shape ratio* is the quotient of the diameter of the circle with area $\mathsf{A}(X)$ and the diameter of the minimum circumscribed circle of X. All these shape ratios are motion- and scale-invariant, so that their values do not depend on translations/rotations and scale transformations of X.

In the engineering literature [47] it is usual to perform statistical analysis of a sample of sets $X_1, \ldots, X_n$ by computing several shape ratios for each set from the sample. This yields a multivariate sample, which can be analyzed using methods from multivariate statistics. The "only" problem is that distributions of shape ratios are not known theoretically for most models of random sets, and even their first moments are known only in very few special cases.

5.2. Averaging of figures. If an observer deals with a sample of figures rather than sets, then the definitions of mean value of a random compact set are not directly applicable. For instance, the images of particles are isotropic sets, whence the corresponding set-expectations are balls or discs.

The approach below can be found in [57]. It is inspired by the studies of shapes and landmark configurations, see [7,17,31,65]. Landmarks are characteristic points of planar objects, such as the tips of the nose and the chin, if human profiles are studied. However, for the study of particles such landmarks are not natural. Perhaps they could be points of extremal curvature or other interesting points on the boundary, but for a useful application of the landmark method the number of landmarks per object has to be constant, and this may lead to difficulties or unnatural restrictions.

To come to the space of figures we introduce an equivalence relationship on the space $\mathcal{K}$ of compact sets. Namely, two compact sets are equivalent if they can be superimposed by a rigid motion (scale transformations are excluded). Since the space $\mathcal{K}$ is already non-linear, taking factor space $\mathcal{K}/_\sim$ worsens the situation even more than in the theory of shape and landmarks. For the practically important problem of determining an empirical mean of a sample of figures the following general idea [20,57] seems to be natural:

> Give the figures particular locations and orientations such that they are in a certain sense "close together;" then consider the new sample as a sample of sets and, finally, determine a set-theoretic mean.

As we have seen, we usually linearize $\mathcal{K}$, i.e. replace sets by some functions. Then motions of sets correspond to transformations of functions considered to be elements of a general Hilbert space [57]. For example, if convex compact sets are described by support functions, then the group of proper motions of sets corresponds to some group acting on the space of support functions. It was proven in [57] that the "optimal" translations of convex sets are given by their Steiner points

$$s(K) = \frac{1}{b_d} \int\limits_{\mathbb{S}^{d-1}} h(K,u)u\,du\,,$$

so that the Steiner points $s(K_1),\dots,s(K_n)$ of all figures must be superimposed. These translations bring the support functions "close together" in the L^2-metric. However, finding their optimal rotations is a difficult computational problem.

6. Statistics of stationary random sets.

6.1. Definition of the Boolean model. A simple example of a stationary random set is a Poisson point process in $\mathbb{R}^d$. This is a random set of points Π_λ such that each bounded domain K contains a Poisson number of points with parameter $\lambda\mu(K)$, and the numbers of points in disjoint domains are independent.

This simple concept gives rise to the most important model of stationary random sets called the *Boolean model.* Consider a sequence $\Xi_1, \Xi_2, \dots$ of i.i.d. compact random sets and shift each of the Ξ_i's using the enumerated points $\{x_1, x_2, \dots\}$ of the Poisson point process Π_λ in $\mathbb{R}^d$. The Boolean model Ξ is defined to be the union of all these shifted sets, so that

$$\Xi = \bigcup_{i:x_i \in \Pi_\lambda} (\Xi_i + x_i)\,, \tag{6.1}$$

see [23,35,44,55]. This definition formalizes the heuristic notion of a clump and can be used to produce sets of complicated shape from "simple" components. The points x_i are called "germs" (so that λ is the intensity of the germs or the intensity of the Boolean model), and the random set Ξ_0 is called the (typical) "grain." Usually the typical grain is chosen to be the deterministic set (perhaps, randomly rotated), random ball (ellipse) or random polygon. We assume that Ξ_0 is almost surely convex.

The key object of statistical inference theory for the Boolean model can be described as follows.

> Given an observation of a Boolean model estimate its parameters: the intensity of the germ process and the distribution of the grain.

As in statistics of point processes, all properties of estimators are formulated for an observation window W growing without bounds (although, in practice, the window is fixed). The major difficulty is caused by the fact that the particular grains are not observable because of overlaps, especially if the intensity is high. The grains are occluded so that the corresponding germ points are not observable at all and it is not easy to infer how many grains produce a clump and also if there are some grains totally covered by other.

By the Choquet theorem, the distribution of Ξ_0 is determined by the corresponding capacity functional $T_{\Xi_0}(K)$. The capacity functional of the Boolean model Ξ can be evaluated as

$$T_{\Xi}(K) = \mathbf{P}\{\Xi \cap K \neq \emptyset\} = 1 - \exp\{-\lambda \mathbf{E}\mu(\Xi_0 \oplus \check{K})\}, \tag{6.2}$$

see [23,35,55]. Here $\check{K} = \{-x\colon x \in K\}$. By the Fubini theorem, we get from (6.2)

$$T_{\Xi}(K) = 1 - \exp\{-\lambda \int_{\mathbb{R}^d} T_{\Xi_0}(K+x)dx\}. \tag{6.3}$$

The functional in the left-hand side of (6.3) is determined by the whole set Ξ and can be estimated from observations of Ξ. However, it is unlikely that the integral equation (6.3) can be solved directly in order to find λ and the capacity functional T_{Ξ_0} which are of interest.

6.2. Estimation methods for the Boolean model. Statistics of the Boolean model began with estimation of λ and mean values of the Minkowski functionals of the grain (mean area, volume, perimeter, etc.). In general, these values do not suffice to retrieve the distribution of the grain Ξ_0, although sometimes it is possible to find an appropriate distribution if some parametric family is given. Clearly, without the distribution of the grain it is impossible to simulate the underlying Boolean model, and, therefore, to use tests based on simulations. For example, if the grain is a ball, then, in general, it is impossible to determine its radius distribution by the corresponding moments up to the dth order. However, if the distribution belongs to some parametric family, say log-normal, then it is determined by these moments. Even in the parametric setup most studies end with proof of strong consistency. Results concerning asymptotic normality are still rather exceptional [26,45], and there are no theoretical studies of efficiency.

Below we give a short review of some known estimation methods. Mostly, we will present only relations between *observable* characteristics and parameters of the Boolean model, bearing in mind that replacing these observable characteristics by their empirical counterparts provides estimators for the corresponding parameters.

In the simplest case, the accessible information about Ξ is the volume fraction p covered by Ξ. Because of stationarity and (6.2)

$$p = T(\{0\}) = 1 - \exp\{-\lambda \mathbf{E}\mu(\Xi_0)\}\,, \tag{6.4}$$

so that $\lambda\mathbf{E}\mu(\Xi_0)$ can be estimated. Estimators of p can be defined as, e.g., $\hat{p} = \mu(\Xi \cap W)/\mu(W)$ or $\hat{p} = \mathrm{card}(\Xi \cap W \cap Z)/\mathrm{card}(W \cap Z)$, where W is a window and Z is a lattice in $\mathbb{R}^d$, see [34]. These estimators of p are strongly consistent and asymptotically normal (if the second moments of both $\mu(\Xi_0)$ and $\|\Xi_0\|$ are finite).

Two-point covering probabilities determine the *covariance* of Ξ

$$C(v) = \mathbf{P}\{\{0, v\} \subset \Xi\} = 2p - T(\{0, v\})\,.$$

Then the function

$$q(v) = 1 + \frac{C(v) - p^2}{(1-p)^2} = \exp\{\lambda \mathbf{E}\mu(\Xi_0 \cap (\Xi_0 - v))\} \tag{6.5}$$

yields the set-covariance function of the grain $\gamma_{\Xi_0}(v) = \mathbf{E}\mu(\Xi_0 \cap (\Xi_0 - v))$, see [58]. This fact has been used in [41] to estimate the shape of a deterministic convex grain. It should be noted that the covariance is quite flexible in applications, since it only depends on the capacity functional on two-point compacts. For example, almost all covariance-based estimators are easy to reformulate for censored observations, see [39]. Applications of the covariance function to statistical estimation of the Boolean model parameters were discussed also in [23,52,55]. Sometimes the covariance approach leads to unstable integral equations of the first kind. Recently, it has been shown [8] that such equations can be solved in an efficient way using the stochastic regularization ideas.

Historically, the first statistical method for the Boolean model was the *minimum contrast method* for contact distribution functions or the covariance, see [52], where a number of references to the works of the Fontainebleau school in the seventies are given, and also [15]. Its essence is to determine the values of T_Ξ for some sub-families of compact sets (balls, segments or two-point sets). Then the right-hand side of (6.2) can be expressed by means of known integral geometric formulae. In particular, when $K = B_r(0)$ is a ball of radius r, then the Steiner formula [35,55,51] gives the expansion of $\mathbf{E}\mu(\Xi_0 \oplus B_r(0))$ as a polynomial of dth order whose coefficients are expressed through the Minkowski functionals of Ξ_0. In the planar case we get

$$T_\Xi(B_r(0)) = 1 - \exp\left\{-\lambda \mathbf{E}\left[\mathsf{A}(\Xi_0) + \mathsf{U}(\Xi_0)r + \pi r^2\right]\right\},$$

so that

$$-\frac{1}{r}\log\left(\frac{1 - T_\Xi(B_r(0))}{1-p}\right) = \lambda \mathbf{E}\mathsf{U}(\Xi_0) + \lambda\pi r.$$

The next step is to replace the left-hand side of (6.2) by its empirical counterpart and approximate it by a polynomial, see [11,23]. The corresponding estimators have been thoroughly investigated in [26]. Although it is possible to prove their strong consistency and asymptotic normality, it is usually a formidable task to calculate their variances.

6.3. Intensities and spatial averages. Now consider another estimation method which could be named the *method of intensities*. First, note that (6.4) is an equality relating some spatial average to parameters of the Boolean model. The volume fraction is the simplest spatial average. It is possible to consider other spatial averages, for example, mean surface area per unit volume, the specific Euler-Poincaré characteristics or, more generally, densities of extended Minkowski measures, see [64]. According to the method of intensities, estimators are chosen as solutions of equations relating these spatial averages to estimated parameters.

A particular implementation of the method of intensities depends on the way the Minkowski functionals are extended onto the extended convex ring (the family of locally finite unions of convex compact sets). The *additive extension* [51] was used in [30,63]. For this, a functional ϕ defined on convex sets is extended onto the family of finite unions of convex sets by additivity, so that

$$\phi(K_1 \cup K_2) = \phi(K_1) + \phi(K_2) - \phi(K_1 \cap K_2)$$

for each two convex sets K_1 and K_2, etc. For example, if $\phi(K)$ is 1 for non-empty convex K and vanishes if $K = \emptyset$, then the additive extension of ϕ is the Euler-Poincaré characteristics χ. Now the spatial average (or *intensity*)

$$D_\phi = \lim_{W \uparrow \mathbb{R}^d} \frac{\phi(\Xi \cap W)}{\mu(W)} \tag{6.6}$$

is related to λ and the expected values of some functionals on Ξ_0. Since D_ϕ is observable, this allows us to estimate these values by the method of moments. For example, in the planar case for the isotropic typical grain

Ξ_0, we get the system of equations

$$L_A = \lambda(1-p)\mathbf{E}\mathsf{U}(\Xi_0), \tag{6.7}$$

$$\chi_A = (1-p)\left(\lambda - \frac{\lambda^2}{4\pi}(\mathbf{E}\mathsf{U}(\Xi_0))^2\right), \tag{6.8}$$

where L_A is the specific boundary length of Ξ and χ_A is the specific Euler-Poincarè characteristics (which appears if ϕ in (6.6) is the boundary length and $\phi(K) = \mathbf{1}_{K\neq\emptyset}$ respectively). Equations (6.7) and (6.8) together with (6.4) can be used to estimate λ and the expected area and perimeter of Ξ_0. So far the second-order asymptotic properties of the corresponding estimators are unknown.

Another technique, the so-called *positive extension* [35,51], has been applied to the Boolean model in [11,52,55]. It goes back to [22,14]. According to this approach, in the planar case (6.4), (6.7), and

$$N_A^+ = (1-p)\lambda \tag{6.9}$$

are used to express the intensity λ, the mean perimeter of the grain $\mathbf{E}\mathsf{U}(\Xi_0)$ and the mean area $\mathbf{E}\mathsf{A}(\Xi_0)$ through the following *observable* values: the area fraction p, the specific boundary length L_A and the intensity N_A^+ of the point process of (say lower) positive tangent points. For the latter, we define the lower tangent point for each of the grains (if there are several points one point can be chosen according to some fixed rule) and thin out those points that are covered by other grains. The rest form a point process N^+ of exposed tangent points. Note that N_A^+ can be estimated by

$$\hat{N}_{A,W}^+ = \frac{\text{number of exposed tangent points in } W}{\mu(W)}.$$

This method yields biased but strong consistent and asymptotically normal estimators [45]. For instance, the intensity estimator

$$\hat{\lambda} = \frac{\hat{N}_{A,W}^+}{1-\hat{p}_W}$$

is asymptotically normal, so that $\mu(W)^{1/2}(\hat{\lambda}_W - \lambda)$ converges weakly to a centered normal distribution with variance $\lambda/(1-p)$. This variance does not depend on the dimension of the space nor on the shape of the typical grain. Higher-order characteristics of the point process of tangent points are ingredients to construct estimators of the grain distribution, see [42]. For example, both the covariance and the points of N^+ contain information which suffices to estimate all parameters of the Boolean model, even including the distribution (capacity functional) of the typical grain.

Further estimators for the Aumann expectation of the set $\Xi_0 \oplus \check{\Xi}_0$ (which equals to $2\Xi_0$ if Ξ_0 is centrally symmetric) are suggested in [43].

7. Concluding remarks. Currently, statistics of random sets raises more problems than gives final answers. By now, basic estimators are known for some models together with first asymptotic results. It should be noted that further asymptotic results are quite difficult to obtain because of complicated spatial dependencies. Mathematical results in statistics of the Boolean model are accompanied by a number of simulation studies and development of the relevant software.

The most important open problems in statistics of random sets are related to the development of new models of compact sets, new approaches to averaging sets and figures, likelihood approach for continuous random sets and lower bounds for the estimators' variances, statistics of general germ-grain models (those without the Poisson assumption) and reliable statistical tests.

REFERENCES

[1] M. AIGNER, *Combinatorial Theory*, Springer, New York, 1979.

[2] D. ALDOUS, B. FRISTEDT, PH.S. GRIFFIN, AND W.E. PRUITT, *The number of extreme points in the convex hull of a random sample*, J. Appl. Probab., 28 (1991), pp. 287–304.

[3] Z. ARTSTEIN AND R.A. VITALE, *A strong law of large numbers for random compact sets*, Ann. Probab., 3 (1975), pp. 879–882.

[4] R.J. AUMANN, *Integrals of set-valued functions*, J. Math. Anal. Appl., 12 (1965), pp. 1–12.

[5] A.J. BADDELEY AND I.S. MOLCHANOV, *Averaging of random sets based on their distance functions*, J. Math. Imaging and Vision., (1997). To Appear.

[6] J.K. BEDDOW AND T.P. MELLOY, *Testing and Characterization of Powder and Fine Particles*, Heyden & Sons, London, England, 1980.

[7] F.L. BOOKSTEIN, *Size and shape spaces for landmark data in two dimensions (with discussion)*, Statist. Sci., 1 (1986), pp. 181–242.

[8] D. BORTNIK, *Stochastische Regularisierung und ihre Anwendung auf Stochastisch-Geometrische Schätzprobleme*, Ph.D. Thesis, Westfälische Wilhelms-Universität Münster, Münster, 1996.

[9] H. BROZIUS, *Convergence in mean of some characteristics of the convex hull*, Adv. in Appl. Probab., 21 (1989), pp. 526–542.

[10] H. BROZIUS AND L. DE HAAN, *On limiting laws for the convex hull of a sample*, J. Appl. Probab., 24 (1987), pp. 852–862.

[11] N.A.C. CRESSIE, *Statistics for Spatial Data*, Wiley, New York, 1991.

[12] D.J. DALEY AND D. VERE-JONES, *An Introduction to the Theory of Point Processes*, Springer, New York, 1988.

[13] R.A. DAVIS, E. MULROW, AND S.I. RESNICK, *The convex hull of a random sample in* $\mathbb{R}^2$, Comm. Statist. Stochastic Models, 3 (1987), pp. 1–27.

[14] R.T. DEHOFF, *The quantitative estimation of mean surface curvature*, Transactions of the American Institute of Mining, Metallurgical and Petroleum Engineering, 239 (1967), p. 617.

[15] P.J. DIGGLE, *Binary mosaics and the spatial pattern of heather*, Biometrics, 37 (1981), pp. 531–539.

[16] S. DOSS, *Sur la moyenne d'un élément aléatoire dans un espace distancié*, Bull. Sci. Math., 73 (1949), pp. 48–72.

[17] I.L. DRYDEN AND K.V. MARDIA, *Multivariate shape analysis*, Sankhya A, 55 (1993), pp. 460–480.

[18] K.J. FALCONER, *Random fractals*, Math. Proc. Cambridge Philos. Soc., 100 (1986),

pp. 559–582.
[19] M. FRÉCHET, *Les éléments aléatoires de nature quelconque dans un espace distancié*, Ann. Inst. H. Poincaré, Sect. B, Prob. et Stat., 10 (1948), pp. 235–310.
[20] L.A. GALWAY, *Statistical Analysis of Star-Shaped Sets*, Ph.D. Thesis, Carnegie-Mellon University, Pittsburgh, Pennsylvania, 1987.
[21] S. GRAF, *A Radon-Nikodym theorem for capacities*, J. Reine Angew. Math., 320 (1980), pp. 192–214.
[22] A. HAAS, G. MATHERON AND J. SERRA, *Morphologie mathematique et granulometries en place*, Ann. Mines, 11–12 (1967), pp. 736–753, 767–782.
[23] P. HALL *Introduction to the Theory of Coverage Processes*, Wiley, New York, 1988.
[24] H.J.A.M. HEIJMANS, *Morphological Image Operators*, Academic Press, Boston, 1994.
[25] L. HEINRICH, *On existence and mixing properties of germ–grain models*, Statistics, 23 (1992), pp. 271–286.
[26] L. HEINRICH, *Asymptotic properties of minimum contrast estimators for parameters of Boolean models*, Metrika, 31 (1993), pp. 349–360.
[27] W. HERER, *Esperance mathematique au sens de Doss d'une variable aleatoire a valeurs dans un espace metrique*, C. R. Acad. Sci., Paris, Ser. I, 302 (1986), pp. 131–134.
[28] P.J. HUBER, *Robust Statistics*, Wiley, New York, 1981.
[29] A.F. KARR, *Point Processes and Their Statistical Inference*, Marcel Dekker, New York, 2nd edn., 1991.
[30] A.M. KELLERER, *Counting figures in planar random configurations*, J. Appl. Probab., 22 (1985), pp. 68–81.
[31] H. LE AND D.G. KENDALL, *The Riemannian structure of Euclidean shape space: A novel environment for statistics*, Ann. Statist., 21 (1993), pp. 1225–1271.
[32] M.N.L. VAN LIESHOUT, *On likelihood for Markov random sets and Boolean models*, Advances in Theory and Applications of Random Sets (D. Jeulin, ed.), World Scientific Publishing Co., Singapore, 1997, pp. 121–136.
[33] A. MANCHAM AND I.S. MOLCHANOV, *Stochastic models of randomly perturbed images and related estimation problems*, Image Fusion and Shape Variability Techniques (K.V. Mardia and C.A. Gill, eds.), Leeds, England: Leeds University Press, 1996, pp. 44–49.
[34] S. MASE, *Asymptotic properties of stereological estimators of volume fraction for stationary random sets*, J. Appl. Probab., 19 (1982), pp. 111–126.
[35] G. MATHERON, *Random Sets and Integral Geometry*, Wiley, New York, 1975.
[36] I.S. MOLCHANOV, *Uniform laws of large numbers for empirical associated functionals of random closed sets*, Theory Probab. Appl., 32 (1987), pp. 556–559.
[37] I.S. MOLCHANOV, *On convergence of empirical accompanying functionals of stationary random sets*, Theory Probab. Math. Statist., 38 (1989), pp. 107–109.
[38] I.S. MOLCHANOV, *A characterization of the universal classes in the Glivenko–Cantelli theorem for random closed sets*, Theory Probab. Math. Statist., 41 (1990), pp. 85–89.
[39] I.S. MOLCHANOV, *Handling with spatial censored observations in statistics of Boolean models of random sets*, Biometrical J., 34 (1992), pp. 617–631.
[40] I.S. MOLCHANOV, *Limit Theorems for Unions of Random Closed Sets*, vol. 1561 of Lect. Notes Math., Springer, Berlin, 1993.
[41] I.S. MOLCHANOV, *On statistical analysis of Boolean models with non-random grains*, Scand. J. Statist., 21 (1994), pp. 73–82.
[42] I.S. MOLCHANOV, *Statistics of the Boolean model: From the estimation of means to the estimation of distributions*, Adv. in Appl. Probab., 27 (1995), pp. 63–86.
[43] I.S. MOLCHANOV, *Set-valued estimators for mean bodies related to Boolean models*, Statistics, 28 (1996), pp. 43–56.
[44] I.S. MOLCHANOV, *Statistics of the Boolean Model for Practitioners and Mathematicians*, Wiley, Chichester, 1997.
[45] I.S. MOLCHANOV AND D. STOYAN, *Asymptotic properties of estimators for param-*

eters of the Boolean model, Adv. in Appl. Probab., 26 (1994), pp. 301–323.

[46] I.S. MOLCHANOV AND D. STOYAN, *Statistical models of random polyhedra*, Comm. Statist. Stochastic Models, 12 (1996), pp. 199–214.

[47] E. PIRARD, *Shape processing and analysis using the calypter*, J. Microscopy, 175 (1994), pp. 214–221.

[48] T. RÉTI AND I. CZINEGE, *Shape characterization of particles via generalised Fourier analysis*, J. Microscopy, 156 (1989), pp. 15–32.

[49] H. RICHTER, *Verallgemeinerung eines in der Statistik benötigten Satzes der Maßtheorie*, Math. Ann., 150 (1963), pp. 85–90, 440–441.

[50] R. SCHNEIDER, *Random approximations of convex sets*, J. Microscopy, 151 (1988), pp. 211–227.

[51] R. SCHNEIDER, *Convex Bodies. The Brunn–Minkowski Theory*, Cambridge University Press, Cambridge, England, 1993.

[52] J. SERRA, *Image Analysis and Mathematical Morphology*, Academic Press, London, England, 1982.

[53] G.R. SHORACK AND J.A. WELLNER, *Empirical Processes with Applications to Statistics*, Wiley, New York, 1986.

[54] N.D. SIDIROPOULOS, J.S. BARAS, AND C.A. BERENSTEIN, *Algebraic analysis of the generating functional for discrete random sets and statistical inference for intensity in the discrete Boolean random-set model*, Journal of Mathematical Imaging and Vision, 4 (1994), pp. 273–290.

[55] D. STOYAN, W.S. KENDALL, AND J. MECKE, *Stochastic Geometry and Its Applications*, Wiley, Chichester, 2nd edn., 1995.

[56] D. STOYAN AND G. LIPPMANN, *Models of stochastic geometry – A survey*, Z. Oper. Res., 38 (1993), pp. 235–260.

[57] D. STOYAN AND I.S. MOLCHANOV, *Set-valued means of random particles*, Journal of Mathematical Imaging and Vision, 27 (1997), pp. 111–121.

[58] D. STOYAN AND H. STOYAN, *Fractals, Random Shapes and Point Fields*, Wiley, Chichester, 1994.

[59] R.A. VITALE, *An alternate formulation of mean value for random geometric figures*, J. Microscopy, 151 (1988), pp. 197–204.

[60] R.A. VITALE, *The Brunn-Minkowski inequality for random sets*, J. Multivariate Anal., 33 (1990), pp. 286–293.

[61] O.YU VOROB'EV, *Srednemernoje Modelirovanie (Mean-Measure Modelling)*, Nauka, Moscow, 1984. In Russian.

[62] D. WAGNER, *Survey of measurable selection theorem*, SIAM J. Control Optim., 15 (1977), pp. 859–903.

[63] W. WEIL, *Expectation formulas and isoperimetric properties for non-isotropic Boolean models*, J. Microscopy, 151 (1988), pp. 235–245.

[64] W. WEIL AND J.A. WIEACKER, *Densities for stationary random sets and point processes*, Adv. in Appl. Probab., 16 (1984), pp. 324–346.

[65] H. ZIEZOLD, *Mean figures and mean shapes applied to biological figures and shape distributions in the plane*, Biometrical J., 36 (1994), pp. 491–510.

ON ESTIMATING GRANULOMETRIC DISCRETE SIZE DISTRIBUTIONS OF RANDOM SETS*

KRISHNAMOORTHY SIVAKUMAR† AND JOHN GOUTSIAS‡

Abstract. Morphological granulometries, and the associated size distributions and densities, are important shape/size summaries for random sets. They have been successfully employed in a number of image processing and analysis tasks, including shape analysis, multiscale shape representation, texture classification, and noise filtering. For most random set models however it is not possible to analytically compute the size distribution. In this contribution, we investigate the problem of estimating the granulometric (discrete) size distribution and size density of a discrete random set. We propose a Monte Carlo estimator and compare its properties with that of an empirical estimator. Theoretical and experimental results demonstrate superiority of the Monte Carlo estimation approach. The Monte Carlo estimator is then used to demonstrate existence of phase transitions in a popular discrete random set model known as a binary Markov random field, as well as a tool for designing "optimal" filters for binary image restoration.

Key words. Granulometries, Image Restoration, Markov Random Fields, Mathematical Morphology, Monte Carlo Estimation, Random Sets, Size Distribution, Phase Transition.

AMS(MOS) subject classifications. 60D05, 60K35, 68U10, 82B20, 82B26, 82B27

1. Introduction. Granulometric size distributions provide a wealth of information about image structure and are frequently used as "shape/size" summaries [1]. They have been employed in a number of image processing and analysis tasks such as shape and texture analysis [2], multiscale shape representation [3], morphological shape filtering and restoration [4]–[7], analysis, segmentation, and classification of texture [8]–[12], and as a goodness-of-fit tool for testing whether an observed image has been generated by a given random set model [13].

Application of this concept in real life problems requires that computation of the size distribution is feasible. As noted by Matheron however it is in general very difficult to analytically compute the granulometric size distribution [1]. In [4] and [9] the size distribution is empirically estimated from a single image realization, whereas in [5] and [8] is analytically obtained for a simple (and restrictive) random set model. In order to effectively employ size distributions, it is imperative to estimate these quantities for more general random set models.

* This work was supported by the Office of Naval Research, Mathematical, Computer, and Information Sciences Division, under ONR Grant N00060-96-1376.

† Texas Center for Applied Technology and Department of Electrical Engineering, Texas A&M University, College Station, TX 77843-3407.

‡ Department of Electrical and Computer Engineering, Image Analysis and Communications Laboratory, The Johns Hopkins University, Baltimore, MD 21218.

A natural approach to estimating the size distribution is to employ an *empirical estimator*. In this case, no particular image model is assumed except that the image is a realization of a stationary random set, viewed through a finite observation window. The empirical estimator is naturally *unbiased* and, under suitable covariance assumptions, it is also *consistent*, in the limit as the observation window grows to infinity. In practice, however, the size of the observation window is fixed. Hence, we do not have much control over the *mean-squared-error* (MSE) of the empirical estimator. As a general rule, one can get good estimates of the size distribution only at "small sizes." In order to overcome this problem, we need to assume a random set model for the images under consideration.

In this contribution, we review a number of Monte Carlo approaches, first proposed in [14] and [15], to estimating the size distribution and the associated size density of a random set model. Three unbiased and consistent estimators are considered. The first estimator is applicable when *independent and identically distributed* (i.i.d.) random set realizations are available [14]. However, this is not always possible. For example, one of the most popular image modeling assumptions is to consider images as being realizations of a particular *discrete random set* (DRS) model known as binary *Markov random field* (MRF) (e.g., see [16] and [17]). In general, obtaining i.i.d. samples is not possible in this case. Dependent MRF realizations may however be obtained by means of a *Markov Chain Monte Carlo* (MCMC) approach (see [18]–[20]) and an alternative Monte Carlo estimator for the size distribution can be obtained in this case [15]. Implementation of such an estimator however may require substantial computations. To ameliorate this problem, a third Monte Carlo estimator has been proposed that enjoys better computational performance [15]. We would like to mention here the related work of Sand and Dougherty [21] who have derived asymptotic expressions for the moments of the size distribution (see also [22] and [23]). Their work, however, is based on a rather restrictive image model. Bettoli and Dougherty [24] as well as Dougherty and Sand [23] have derived explicit expressions for size distribution moments. They have however assumed a deterministic image corrupted by random pixel noise and have only considered linear granulometries.

2. Mathematical morphology.

Mathematical Morphology has become a popular approach to many image processing and analysis problems, due primarily to the fact that it considers geometric structure in binary images (shapes) [1], [2]. Binary images are represented as subsets of a two-dimensional space, usually the two-dimensional Euclidean plane $\mathbb{R}^2$ or the two-dimensional discrete plane $\mathbb{Z}^2$ or some subset of these. Geometric information is extracted from a binary image by "probing" it with a small elementary shape (e.g., a line segment, circle, etc.), known as the *structuring element*.

Two basic morphological image operators are *erosion* and *dilation*. The erosion $X \ominus B$ of a binary image X by a structuring element B is given by

$$X \ominus B = \bigcap_{v \in \check{B}} X_v = \{v \mid B_v \subseteq X\}, \tag{2.1}$$

whereas the dilation $X \oplus B$ is given by

$$X \oplus B = \bigcup_{v \in B} X_v = \{v \mid \check{B}_v \cap X \neq \emptyset\}, \tag{2.2}$$

where $\check{B} = \{-v \mid v \in B\}$ is the *reflection* of B about the origin and B_v is the *translate* of B by v. It is easy to verify that erosion is *increasing* (i.e., $X \subseteq Y \Rightarrow X \ominus B \subseteq Y \ominus B$) and *distributes* over intersection (i.e., $(\cap_{i \in I} X_i) \ominus B = \cap_{i \in I}(X_i \ominus B)$). On the other hand, dilation is increasing and distributes over union (i.e., $(\cup_{i \in I} X_i) \oplus B = \cup_{i \in I}(X_i \oplus B)$). Moreover, erosion and dilation are duals with respect to set complementation (i.e. $X \oplus B = (X^c \ominus \check{B})^c$) and satisfy the so called *adjunctional* relationship $X \oplus B \subseteq Y \Leftrightarrow X \subseteq Y \ominus B$. Finally, if B contains the origin, then dilation is *extensive* (i.e., $X \subseteq X \oplus B$) whereas erosion is *anti-extensive* (i.e., $X \ominus B \subseteq X$).

Opening and *Closing* are two secondary operators obtained by suitably composing the basic morphological operators. In particular, the opening $X \circ B$ of a binary image X by a structuring element B is given by

$$X \circ B = (X \ominus B) \oplus B = \bigcup_{v:\, B_v \subseteq X} B_v, \tag{2.3}$$

whereas the closing $X \bullet B$ is given by

$$X \bullet B = (X \oplus B) \ominus B = \bigcap_{v:\, \check{B}_v \cap X = \emptyset} \check{B}_v^c. \tag{2.4}$$

It is easy to verify that opening is increasing, anti-extensive, and *idempotent* (i.e., $(X \circ B) \circ B = X \circ B$), whereas closing is increasing, extensive, and idempotent. Furthermore, opening and closing are duals with respect to set complementation (i.e., $X \circ B = (X^c \bullet \check{B})^c$). In general, any increasing, anti-extensive, and idempotent operator is called an opening, whereas any increasing, extensive, and idempotent operator is called a closing.

3. Markov random fields. Let $W = \{w \in \mathbb{Z}^2 \mid w = (m,n), 0 \leq m \leq M-1, 0 \leq n \leq N-1\}$, $M, N < +\infty$, be a two-dimensional finite collection of $M \times N$ *sites* in $\mathbb{Z}^2$. With each site w, we associate a subset $\mathcal{N}_w \subset W$ such that

$$w \notin \mathcal{N}_w \quad \text{and} \quad w \in \mathcal{N}_v \Leftrightarrow v \in \mathcal{N}_w . \tag{3.1}$$

$\mathcal{N}_w$ is known as the *neighborhood* of site w, whereas a pair (v, w) of sites in W that satisfy (3.1) are called *neighbors*. In most applications, $\mathcal{N}_w =$

$\mathcal{N} \oplus \{w\}$, where $\mathcal{N}$ is a structuring element centered at $(0,0)$ such that $(0,0) \notin \mathcal{N}$. This is the only choice to be considered in this paper. Let $\mathbf{x}(w)$ be a binary random variable assigned at site $w \in W$; i.e. $\mathbf{x}(w)$ takes values in $\{0,1\}$. The collection $\mathbf{x} = \{\mathbf{x}(w), w \in W\}$ defines a two-dimensional *binary random field* (BRF) on W that assumes realizations $x = \{x(w), w \in W\}$ in the Cartesian product $\mathcal{S} = \{0,1\}^{MN}$. If: (a) the probability mass function $\Pr[\mathbf{x} = x]$ of $\mathbf{x}$ is strictly positive, for every $x \in \mathcal{S}$, and (b) for every $w \in W$, the conditional probability of $\mathbf{x}(w)$, given the values of $\mathbf{x}$ at all sites in $W \setminus \{w\}$, depends only on the values of $\mathbf{x}$ at sites in $\mathcal{N}_w$, i.e., if

$$\begin{aligned} &\Pr[\mathbf{x}(w) = x(w) \mid \mathbf{x}(v) = x(v), v \in W \setminus \{w\}] \\ &\qquad = \Pr[\mathbf{x}(w) = x(w) \mid \mathbf{x}(v) = x(v), v \in \mathcal{N}_w], \end{aligned} \tag{3.2}$$

for every $w \in W$, then $\mathbf{x}$ is called a (binary) MRF on W with neighborhood $\{\mathcal{N}_w, w \in W\}$. In (3.2), $A \setminus B = A \cap B^c$ is the *set difference* between sets A and B. The conditional probabilities on the right-hand side of (3.2) are known as the *local characteristics* of $\mathbf{x}$.

In the following, we denote by $\mathbf{X} = \{v \in W \mid \mathbf{x}(v) = 1\}$ the DRS [25], [26] corresponding to the binary random field $\mathbf{x}$. Observe that $\mathbf{x}$ is given by the indicator function of $\mathbf{X}$; i.e.

$$\mathbf{x}(v) = I_{\mathbf{X}}(v) = \begin{cases} 1, & v \in \mathbf{X} \\ 0, & v \notin \mathbf{X} \end{cases}.$$

Henceforth, we use upper-case symbols to denote a set and the corresponding lower-case symbols to denote the associated binary function. Notice that these are two equivalent representations.

A site w close to the boundary of W needs special attention since it may be that $\mathcal{N}_w \not\subset W$. In this case, either $\mathcal{N}_w$ is replaced by $\mathcal{N}_w \cap W$ or neighborhood $\tilde{\mathcal{N}}_w$ replaces $\mathcal{N}_w$, where $\tilde{\mathcal{N}}_w$ is formed by mapping each site $(m,n) \in \mathcal{N}_w$ to site $((m)_M, (n)_N)$, where $(m)_M$ means that m is evaluated modulo M. The first choice is known as a *free boundary condition* whereas the second choice is known as a *toroidal boundary condition.* Although both boundary conditions are important, the free boundary condition is more natural in practice. However, the toroidal boundary condition frequently simplifies analysis and computations.

It can be easily shown that, if the probability mass function $\Pr[\mathbf{x} = x]$ of $\mathbf{x}$ satisfies the *positivity condition* (a) above, then

$$\Pr[\mathbf{x} = x] = \pi(x) = \frac{1}{Z} \exp\{-\frac{1}{T} U(x)\}, \quad \text{for every } x \in \mathcal{S}, \tag{3.3}$$

where

$$Z = \sum_{x \in \mathcal{S}} \exp\{-\frac{1}{T} U(x)\}, \tag{3.4}$$

is a normalizing constant known as the *partition function.* In (3.3) and (3.4), U is a real-valued functional on $\mathcal{S}$, known as the *energy function*, whereas T is a real-valued positive constant, known as the *temperature.* A probability mass function of the exponential form (3.3) is known as the *Gibbs distribution*, whereas any random field $\mathbf{x}$ whose probability mass function is of the form (3.3) and (3.4) is known as a *Gibbs random field* (GRF). A MRF is a special case of a GRF when, in addition to the positivity condition, the *Markovian condition* (3.2) is satisfied.

EXAMPLE 3.1. A simple MRF of some interest in image processing is the *Ising model.* This is a BRF for which

$$U(x) = \alpha \sum_{w \in W} (2x(w) - 1) + \sum_{w \in W} (2x(w) - 1)[\beta_1(2x(w_1) - 1) + \beta_2(2x(w_2) - 1)], \; x \in \mathcal{S}, \tag{3.5}$$

with appropriate modifications at the boundary of W, where $w_1 = (m, n-1)$ and $w_2 = (m-1, n)$, provided that $w = (m, n)$, and α, β_1, and β_2 are real-valued parameters. In this case, $\mathcal{N}_w = \{(m-1, n), (m+1, n), (m, n-1), (m, n+1)\}$. The importance of this model stems from the fact that a number of quantities (e.g., the partition function) can be analytically calculated. This is not however true for more general MRF's. □

Direct simulation of a MRF is not possible in general, primarily due to the lack of analytical tools for calculating its partition function. A popular indirect technique for simulating a MRF is based on MCMC. According to this technique, an *ergodic* Markov chain $\{\mathbf{x}_k, k = 1, 2, ...\}$ is generated, with state-space $\mathcal{S}$, whose equilibrium distribution is the Gibbs distribution π under consideration; i.e., such that

$$\lim_{k \to +\infty} \Pr[\mathbf{x}_k = x \mid \mathbf{x}_1 = x_o] = \pi(x), \text{ for every } x, x_o \in \mathcal{S}.$$

In this case, and for large enough k, any state of the Markov chain $\{\mathbf{x}_k, k = 1, 2, ...\}$ will approximately be a sample drawn from π. One of the most commonly used MCMC techniques is based on the so-called *Metropolis's algorithm* with random site updating. In this case, the transition probability $p_k(i, j) = \Pr[\mathbf{x}_{k+1} = x_j \mid \mathbf{x}_k = x_i]$ associated with Markov chain $\{\mathbf{x}_k, k = 1, 2, ...\}$ is given by

$$p_k(i, j) = \frac{1}{MN} \begin{cases} 1, & \text{when } \pi(x_i) \leq \pi(x_j) \\ \pi(x_j)/\pi(x_i), & \text{when } \pi(x_i) > \pi(x_j), \text{for } i \neq j, \end{cases}$$

provided that x_i and x_j differ at only one site, this site being chosen randomly among all sites in W (otherwise $p_k(i, j) = 0$), and

$$p_k(i, i) = 1 - \sum_{j=1, j \neq i}^{|\mathcal{S}|} p_k(i, j),$$

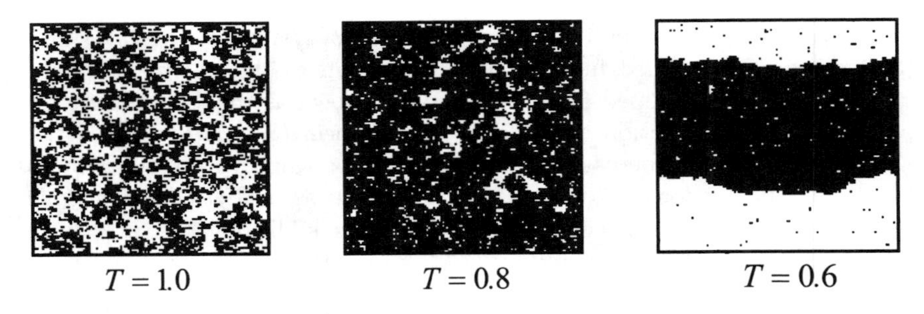

FIG. 1. *Realizations of an Ising model at three different temperatures. The critical temperature is* $T_c = 0.82$.

where $|A|$ denotes the *cardinality* of a set A (i.e., the total number of its elements). It is a well known fact that, in the limit, as $T \to +\infty$, a GRF $\mathbf{x}$ becomes an i.i.d. random field; i.e.,

$$\lim_{T \to +\infty} \pi(x) = \frac{1}{|\mathcal{S}|}, \quad \text{for every } x \in \mathcal{S}. \tag{3.6}$$

On the other hand, if

$$\mathcal{U} = \{x^* \in \mathcal{S} \mid U(x^*) \leq U(x), \text{ for every } x \in \mathcal{S}\},$$

then

$$\lim_{T \to 0+} \pi(x) = \lim_{T \to 0+} \frac{1}{Z} \exp\{-\frac{1}{T} U(x)\} = \begin{cases} 1/|\mathcal{U}|, & \text{for } x \in \mathcal{U} \\ 0, & \text{otherwise} . \end{cases}$$

Therefore, the probability mass function of a GRF becomes uniform, over all global minima of its energy function (known as the *ground states*), in the limit, as the temperature decreases to zero.

As a direct consequence of (3.6), only short range pixel interactions are possible at high enough temperatures and a typical MRF realization will be dominated by fine structures at such temperatures. In this case, the MRF under consideration is said to be at a *fine phase*. On the other hand, at low enough temperatures, long range pixel interactions are possible (depending on the set of ground states of $\mathbf{x}$) and a typical MRF realization will be dominated by large structures at such temperatures. In this case, the MRF under consideration is said to be at a *coarse phase*. In general, as the temperature decreases to zero, the transition from fine to coarse phase is not smooth. In most cases, there exists a temperature T_c, known as the *critical temperature*, at which abrupt change from fine to coarse phase occurs. Around the critical temperature, a typical MRF realization is characterized by coexistence of structures of a wide range of sizes. This phenomenon, known as *phase transition*, is typical in many natural systems (e.g., see [27]). This is illustrated in Fig. 1 and for the Ising model discussed in Example 3.1, defined over a 128×128 rectangular grid, with $\alpha = 0$,

$\beta_1 = -0.25$, and $\beta_2 = -0.5$ (see also (3.5)). This MRF experiences phase transition at a critical temperature $T_c = 0.82$, computed as solution to (3.7) below. Figure 1 depicts typical realizations of the Ising model under consideration, obtained by means of a Metropolis's MCMC algorithm with random site updating, at three different temperatures $T = 1.0, 0.8, 0.6$.

Mathematically, phase transition is due to the fact that, around T_c, the partition function in (3.4) is not an analytic function of the temperature, as the data window W grows to infinity. In most cases, the *specific heat* c_T, given by

$$c_T = \frac{2T}{|W|}\frac{\partial \ln Z}{\partial T} + \frac{T^2}{|W|}\frac{\partial^2 \ln Z}{\partial T^2} = \frac{1}{|W|\,T^2}\,\mathrm{Var}[U(\mathbf{X})],$$

where Var[ ] denotes variance, diverges as $T \to T_c$, when $W \to +\infty$ (in the sense of Van Hove – see [28]), usually due to the divergence of the second partial derivative of the logarithm of the partition function with respect to T; i.e.,

$$\lim_{T \to T_c} \lim_{W \to +\infty} c_T = +\infty.$$

It can be shown that the Ising model, discussed in Example 3.1, experiences phase transition at a critical temperature T_c that satisfies

$$\sinh\left(-\frac{2\beta_1}{T_c}\right) \sinh\left(-\frac{2\beta_2}{T_c}\right) = 1, \tag{3.7}$$

provided that $\alpha = 0$ in (3.5) and a toroidal boundary condition is assumed. Unfortunately, no analytical results are available for more general cases, and the problem of phase transition is one of the most intriguing unsolved problems of statistical mechanics.

4. Granulometries, size distributions, and size densities. We now discuss the notions of (discrete) granulometry and anti-granulometry as well as the notions of (discrete) size distribution and size density, associated with a DRS $\mathbf{X}$ (see also [1] and [29]). In the following, we denote by $\mathcal{P}$ all subsets of the two-dimensional discrete space $\mathbb{Z}^2$.

DEFINITION 4.1. A *granulometry* on $\mathcal{P}$ is a parameterized family $\{\gamma_s\}_{s=0,1,\ldots}$ of mappings, from $\mathcal{P}$ into itself, such that:

1. γ_0 is the identity mapping on $\mathcal{P}$; i.e., $\gamma_0(X) = X$, for every $X \in \mathcal{P}$
2. γ_s is increasing; i.e., $X_1 \subseteq X_2 \Rightarrow \gamma_s(X_1) \subseteq \gamma_s(X_2)$, $s = 1, 2, \ldots$, $X_1, X_2 \in \mathcal{P}$
3. γ_s is anti-extensive; i.e., $\gamma_s(X) \subseteq X$, $s = 1, 2, \ldots$, for every $X \in \mathcal{P}$
4. $\gamma_s \gamma_r = \gamma_r \gamma_s = \gamma_{\max(s,r)}$, $s, r = 1, 2, \ldots$

It can be shown that $\{\gamma_s\}_{s=0,1,\ldots}$ is a granulometry on $\mathcal{P}$ if and only if γ_0 is the identity mapping on $\mathcal{P}$ and $\{\gamma_s\}_{s=1,2,\ldots}$ is a family of openings such that $r \geq s \Rightarrow \gamma_r \subseteq \gamma_s$. Given a granulometry $\{\gamma_s\}_{s=0,1,\ldots}$ on $\mathcal{P}$, the dual parameterized family $\{\phi_s\}_{s=0,1,\ldots}$ of mappings on $\mathcal{P}$, with $\phi_s(X) = (\gamma_s(X^c))^c$, $s = 0, 1, \ldots$, $X \in \mathcal{P}$, is known as the *anti-granulometry* associated with $\{\gamma_s\}_{s=0,1,\ldots}$. Notice that $\{\phi_s\}_{s=0,1,\ldots}$ is an anti-granulometry on $\mathcal{P}$ if and only if ϕ_0 is the identity operator and $\{\phi_s\}_{s=1,2,\ldots}$ is a family of closings such that $r \geq s \Rightarrow \phi_r \supseteq \phi_s$.

EXAMPLE 4.1. In practice, the most useful granulometries and anti-granulometries are the ones given by $\gamma_0(X) = \phi_0(X) = X$, $\gamma_s(X) = X \circ sB$, $s = 1, 2, \ldots$, and $\phi_s(X) = X \bullet sB$, $s = 1, 2, \ldots$, respectively, where B is a symmetric finite structuring element and sB is recursively defined by $0B = \{(0,0)\}$, and $sB = (s-1)B \oplus B$, for $s = 1, 2, \ldots$. These are known as the *morphological granulometry* and *morphological anti-granulometry*, respectively. □

EXAMPLE 4.2. We may consider the morphological operators $\epsilon_0(X) = \delta_0(X) = X$, $\epsilon_s(X) = X \ominus sB$, $s = 1, 2, \ldots$, and $\delta_s(X) = X \oplus sB$, $s = 1, 2, \ldots$, where B is a symmetric finite structuring element that contains the origin. It is clear that mappings $\{\epsilon_s\}_{s=0,1,\ldots}$ satisfy Conditions 1–3 in Definition 4.1 but violate Condition 4, whereas mapping δ_s is the dual of ϵ_s. Therefore, $\{\epsilon_s\}_{s=0,1,\ldots}$ is not a granulometry in the strict sense of Definition 4.1. We refer to $\{\epsilon_s\}_{s=0,1,\ldots}$ as an (erosion based) *pseudo-granulometry*, with $\{\delta_s\}_{s=0,1,\ldots}$ being the associated *pseudo-anti-granulometry*. □

Let $\mathbf{X}$ be a DRS on $\mathbb{Z}^2$. Furthermore, let $\Lambda_X(w)$ be a function from $\mathbb{Z}^2$ into $\bar{\mathbb{Z}} = \mathbb{Z} \cup \{-\infty, +\infty\}$, given by

$$\Lambda_X(w) = \begin{cases} \sup\{s \geq 0 \mid w \in \gamma_s(\mathbf{X})\}, & \text{for } w \in \mathbf{X} \\ -\inf\{s \geq 1 \mid w \in \phi_s(\mathbf{X})\}, & \text{for } w \in \mathbf{X}^c \end{cases}.$$

This function associates with each point $w \in \mathbf{X}$ the "size" s of the smallest set $\gamma_s(\mathbf{X})$ that contains point w, whereas it associates with each point $w \in \mathbf{X}^c$ the negative "size" $-s$ of the smallest set $\phi_s(\mathbf{X})$ that contains point w [1], [30]. It is not difficult to show that

$$\gamma_s(\mathbf{X}) = \{w \in \mathbb{Z}^2 \mid \Lambda_X(w) \geq s\}, \text{ for every } s \geq 0,$$

whereas

$$\phi_s(\mathbf{X}) = \{w \in \mathbb{Z}^2 \mid \Lambda_X(w) \geq -s\}, \text{ for every } s \geq 0.$$

For a given $w \in \mathbb{Z}^2$, $\Lambda_X(w)$ is a random variable and the function

$$S_X(w,s) = \Pr[\Lambda_X(w) \geq s] = \begin{cases} \Pr[w \in \gamma_s(\mathbf{X})], & \text{for } s \geq 0 \\ \Pr[w \in \phi_{|s|}(\mathbf{X})], & \text{for } s \leq -1 \end{cases}, \tag{4.1}$$

for every $w \in \mathbb{Z}^2$, is called the (discrete) *size distribution* of the DRS $\mathbf{X}$ (with respect to the granulometry $\{\gamma_s\}_{s=0,1,\ldots}$), whereas the function

$$s_X(w,s) = S_X(w,s) - S_X(w,s+1), \; w \in \mathbb{Z}^2, \; s \in \bar{\mathbb{Z}},$$

is called the (discrete) *size density* of $\mathbf{X}$. Since $1 - S_X(w,s)$ is a probability distribution function, we have that, for every $w \in \mathbb{Z}^2$, $S_X(w,s_1) \leq S_X(w,s_2)$, for every $s_1 \geq s_2$ (i.e., $S_X(w,s)$ is a decreasing function of s), $\lim_{s\to-\infty} S_X(w,s) = 1$, and $\lim_{s\to+\infty} S_X(w,s) = 0$, whereas $s_X(w,s) \geq 0$, for every $s \in \bar{\mathbb{Z}}$, $\sum_{s=-\infty}^{+\infty} s_X(w,s) = 1$, and $S_X(w,s) = \sum_{r=s}^{+\infty} s_X(w,r)$, for every $s \in \bar{\mathbb{Z}}$.

In general, $S_X(w,s)$ and $s_X(w,s)$ depend on $w \in \mathbb{Z}^2$, unless $\mathbf{X}$ is a stationary DRS. However, we may consider a "spatial average" of these quantities over a bounded set $W \subset \mathbb{Z}^2$, where W is a data observation window. In this case, we define the spatial averages $S_X(s)$ and $s_X(s)$, by

$$S_X(s) = \frac{1}{|W|} \sum_{w \in W} S_X(w,s) \quad \text{and} \quad s_X(s) = \frac{1}{|W|} \sum_{w \in W} s_X(w,s), \tag{4.2}$$

assuming that $|W| \neq 0$. It is not difficult to show that

$$S_X(s) = \frac{1}{|W|} \begin{cases} E[|\gamma_s(\mathbf{X}) \cap W|], & \text{for } s \geq 0 \\ E[|\phi_{|s|}(\mathbf{X}) \cap W|], & \text{for } s \leq -1 \end{cases}, \tag{4.3}$$

where $E[\ ]$ denotes expectation. Indeed, if $s \geq 0$, we have that

$$\begin{aligned} |W|\, S_X(s) &= \sum_{w \in W} \Pr[w \in \gamma_s(\mathbf{X})] \\ &= \sum_{w \in W} E[\mathrm{I}_{\gamma_s(\mathbf{X})}(w)] \\ &= \sum_{w \in W} E[|\gamma_s(\mathbf{X}) \cap \{w\}|] \\ &= E[\sum_{w \in W} |\gamma_s(\mathbf{X}) \cap \{w\}|\,] \\ &= E[|\bigcup_{w \in W} [\gamma_s(\mathbf{X}) \cap \{w\}]\ |] \\ &= E[|\gamma_s(\mathbf{X}) \bigcap [\bigcup_{w \in W} \{w\}\ ]|] \\ &= E[|\gamma_s(\mathbf{X}) \cap W|], \end{aligned}$$

by virtue of (4.1) and (4.2). A similar proof holds for $s \leq -1$. Furthermore, $s_X(s) = S_X(s) - S_X(s+1)$, for every $s \in \bar{\mathbb{Z}}$, which leads to

$$s_X(s) = \frac{1}{|W|} \begin{cases} E[|(\gamma_s(\mathbf{X}) \setminus \gamma_{s+1}(\mathbf{X})) \cap W|], & \text{for } s \geq 0 \\ E[|(\phi_{|s|}(\mathbf{X}) \setminus \phi_{|s|-1}(\mathbf{X})) \cap W|], & \text{for } s \leq -1 \end{cases}, \tag{4.4}$$

due to the fact that $\gamma_{s+1} \subseteq \gamma_s$ and $\phi_s \subseteq \phi_{s+1}$. Notice that $S_X(s)$ and $s_X(s)$ enjoy similar properties as $S_X(w,s)$ and $s_X(w,s)$, respectively.

5. An empirical size distribution estimator. We now study a useful estimator for $S_X(s)$ (and, therefore, for $s_X(s)$) based on an empirical estimation principle (see also [14]). Let $\mathbf{X}$ be a DRS on $\mathbb{Z}^2$ and let ψ_s be a morphological image operator from $\mathcal{P}$ into itself. For example, if $\psi_s = \gamma_s$, $s = 0, 1, ...$, then $\{\psi_s\}_{s=0,1,...}$ is a granulometry on $\mathcal{P}$, whereas if $\psi_s = \phi_s$, $s = 0, 1, ...$, then $\{\psi_s\}_{s=0,1,...}$ is an anti-granulometry on $\mathcal{P}$. We assume that $\mathbf{X}$ is observed through an increasing sequence $W_1 \subset W_2 \subset \cdots \subset W_r \subset \cdots \subset \mathbb{Z}^2$ of bounded windows, in which case, realizations of $\mathbf{X} \cap W_r$, $r = 1, 2, ...$, will be our observations. Let $W_r'(s) \subseteq W_r$ be the largest non-empty sub-window (assuming that such a sub-window exists) such that

$$\psi_s(X) \cap W_r'(s) = \psi_s(X \cap W_r) \cap W_r'(s), \quad \text{for every } X \in \mathcal{P}. \tag{5.1}$$

Consider the *morphological statistic* $\Delta_X(s, W_r; \psi)$, given by

$$\Delta_X(s, W_r; \psi) = \frac{1}{|W_r'(s)|} |\psi_s(\mathbf{X} \cap W_r) \cap W_r'(s)|. \tag{5.2}$$

From (5.1) and (5.2), observe that

$$\Delta_X(s, W_r; \psi) = \frac{1}{|W_r'(s)|} |\psi_s(\mathbf{X}) \cap W_r'(s)|. \tag{5.3}$$

We now discuss the *almost sure* (a.s.) convergence of $\Delta_X(s, W_r; \psi)$, in the limit as $r \to +\infty$. This problem has been considered by Moore and Archambault [14], and for the case of random closed sets on $\mathbb{R}^2$. Our presentation focuses here on the case of DRS's on $\mathbb{Z}^2$. We first need to define a useful notion of stationarity (see also [14]).

DEFINITION 5.1. A DRS $\mathbf{X}$ is *first-order stationary with respect to (w.r.t.) a morphological operator* ψ: $\mathcal{P} \to \mathcal{P}$, if $E[[\psi(\mathbf{x})](w)]$ is independent of $w \in \mathbb{Z}^2$.

The proof of the following proposition is an immediate consequence of (5.3).

PROPOSITION 5.1. If $\mathbf{X}$ is a first-order stationary DRS w.r.t. a morphological operator ψ_s: $\mathcal{P} \to \mathcal{P}$, then

$$E[\Delta_X(s, W_r; \psi)] = E[[\psi_s(\mathbf{x})](\mathbf{0})] = e_s, \quad \text{for every } r \geq 1,$$

and

$$\begin{aligned}
&\mathrm{Var}[\Delta_X(s, W_r; \psi)] \\
&\quad = \frac{1}{|W_r'(s)|^2} \sum_{v,w \in W_r'(s)} \sum E[[\psi_s(\mathbf{x})](v)[\psi_s(\mathbf{x})](w)] - e_s^2,
\end{aligned}$$

where $\mathbf{0} = \{(0, 0)\}$.

The following proposition is similar to Propositions 2 and 3 in [14].

PROPOSITION 5.2. If $\mathbf{X}$ is a first-order stationary DRS w.r.t. a morphological operator ψ_s: $\mathcal{P} \to \mathcal{P}$, and if $c_s(v,w)$: $\mathbb{Z}^2 \times \mathbb{Z}^2 \to \mathbb{R}$ is a function such that

$$E[[\psi_s(\mathbf{x})](v)[\psi_s(\mathbf{x})](w)] \;\leq\; c_s(v,w) + e_s^2, \quad \text{for every } v,w \in \mathbb{Z}^2, \tag{5.4}$$

and

$$\lim_{r \to +\infty} \frac{1}{|W_r'(s)|^2} \sum\sum_{v,w \in W_r'(s)} c_s(v,w) \;=\; 0, \tag{5.5}$$

then $\Delta_X(s, W_r; \psi)$ converges in probability to e_s, as $r \to +\infty$. Furthermore, if $c_s(v,w)$ is such that

$$\frac{1}{|W_r'(s)|^2} \sum\sum_{v,w \in W_r'(s)} c_s(v,w) \;\leq\; \frac{\alpha}{|W_r'(s)|^\beta}, \tag{5.6}$$

where α and β are positive finite constants, and if $|W_r'(s)| = Kr^2 + O(r)$, where K is a positive finite constant and $O(r)$ is such that $|O(r)/r| \to a < +\infty$, as $r \to +\infty$, then $\lim_{r \to +\infty} \Delta_X(s, W_r; \psi) = e_s$, a.s.

From Propositions 5.1 and 5.2 it is clear that $\Delta_X(s, W_r; \psi)$ is an unbiased (for every $r \geq 1$) and consistent (as $r \to +\infty$) estimator of e_s, provided that $\mathbf{X}$ is a first-order stationary DRS w.r.t. ψ_s, and conditions (5.4), (5.5) are satisfied. If $\mathbf{X}$ is a stationary DRS and if ψ_s is translation invariant (which is the case for the granulometries and anti-granulometries considered here), then $\mathbf{X}$ will be first-order stationary w.r.t. ψ_s, and the *empirical estimator*

$$\hat{S}_X(s, W_r) \;=\; \begin{cases} \Delta_X(s, W_r; \gamma), & \text{for } s \geq 0 \\ \Delta_X(|s|, W_r; \phi), & \text{for } s \leq -1 \end{cases}, \tag{5.7}$$

will be an unbiased (for every $r \geq 1$) and consistent (as $r \to +\infty$) estimator of the size distribution $S_X(s)$ in (4.3), provided that conditions (5.4) and (5.5) are satisfied (for the proper choice of ψ_s). Additionally, the empirical estimator

$$\hat{s}_X(s, W_r) \;=\; \hat{S}_X(s, W_r) - \hat{S}_X(s+1, W_r), \tag{5.8}$$

will be an unbiased (for every $r \geq 1$) and consistent (as $r \to +\infty$) estimator of the size density $s_X(s)$ in (4.4), provided that conditions (5.4) and (5.5) are satisfied (for the proper choice of ψ_s), at both sizes s and $s+1$. Unfortunately, these conditions are difficult to verify in practice (however, see [14] for a few examples for which this can be done).

The previous results indicate that estimators (5.7) and (5.8) are reliable only when data are observed through large enough windows (i.e., for large enough r). Since, in practice, the observation window is fixed, estimators (5.7) and (5.8) are not expected to be reliable for large values of $|s|$.

EXAMPLE 5.1. In the case of morphological granulometries (see Example 4.1), it is easy to verify that $W'_r(s) = W_r \ominus 2sB$ satisfies (5.1). This leads to the empirical estimator

$$\hat{S}_X^{\circ}(s, W_r) = \frac{1}{|W_r \ominus 2|s|B|} \begin{cases} |[(\mathbf{X} \cap W_r) \circ sB] \cap (W_r \ominus 2sB)|, & \text{for } s \geq 0 \\ |[(\mathbf{X} \cap W_r) \bullet |s|B] \cap (W_r \ominus 2|s|B)|, & \text{for } s \leq -1 \end{cases} \tag{5.9}$$

for the (discrete) size distribution that is based on a morphological granulometry, provided that $|W_r \ominus 2|s|B| \neq 0$. The resulting estimator for the size density is clearly the *pattern spectrum* proposed by Maragos in [3]. □

EXAMPLE 5.2. In the case of the (erosion based) pseudo-granulometries of Example 4.2, it is easy to verify that $W'_r(s) = W_r \ominus sB$ satisfies (5.1). This leads to the empirical estimator

$$\hat{S}_X^{\ominus}(s, W_r) = \frac{1}{|W_r \ominus |s|B|} \begin{cases} |[(\mathbf{X} \cap W_r) \ominus sB] \cap (W_r \ominus sB)|, & \text{for } s \geq 0 \\ |[(\mathbf{X} \cap W_r) \oplus |s|B] \cap (W_r \ominus |s|B)|, & \text{for } s \leq -1 \end{cases} \tag{5.10}$$

for the size distribution that is based on an (erosion based) pseudo-granulometry, provided that $|W_r \ominus |s|B| \neq 0$. This distribution contains useful shape information, as is clear from the work of [31] (see also [32]). □

6. Monte Carlo size distribution estimators. We now study three estimators for the size distribution $S_X(s)$ (and size density $s_X(s)$) which are based on a Monte Carlo estimation principle. These estimators are shown to be unbiased and consistent, as the number of Monte Carlo iterations approach infinity. Therefore, their statistical behavior *is not* controlled by the data size but by the computer program used for their implementation. Furthermore, no difficult to verify assumptions, similar to (5.4)–(5.6), are needed here.

Suppose that $\mathbf{X}$ is a DRS on $\mathbb{Z}^2$ such that it is possible to draw i.i.d. samples from its probability mass function inside a window W; i.e., we can draw samples $X^{(1)} \cap W, X^{(2)} \cap W, \ldots$, where $\{X^{(k)}, k = 1, 2, \ldots\}$ is a collection of i.i.d. realizations of $\mathbf{X}$. A *Monte Carlo estimator* of the size distribution $S_X(s)$ in (4.3) is now given by (see also [14])

$$\hat{S}_{X,1}(s, K_s) = \frac{1}{K_s} \frac{1}{|W|} \begin{cases} \sum_{k=1}^{K_s} |\gamma_s(\mathbf{X}^{(k)}) \cap W|, & \text{for } s \geq 0 \\ \sum_{k=1}^{K_s} |\phi_{|s|}(\mathbf{X}^{(k)}) \cap W|, & \text{for } s \leq -1 \end{cases} . \tag{6.1}$$

It can be easily verified that

$$E[\hat{S}_{X,1}(s, K_s)] = S_X(s), \tag{6.2}$$

and that the relative MSE satisfies

$$\frac{E[(\hat{S}_{X,1}(s, K_s) - S_X(s))^2]}{S_X^2(s)} \leq \frac{1}{K_s} \frac{1 - S_X(s)}{S_X(s)}, \tag{6.3}$$

provided that $S_X(s) \neq 0$. Thus, $\hat{S}_{X,1}(s, K_s)$ is an unbiased (for every $K_s \geq 1$) and consistent (as $K_s \to +\infty$) estimator of the size distribution $S_X(s)$. Additionally, the Monte Carlo estimator

$$\hat{s}_{X,1}(s, K_s, K_{s+1}) = \hat{S}_{X,1}(s, K_s) - \hat{S}_{X,1}(s+1, K_{s+1}),$$

will be an unbiased (for every $K_s, K_{s+1} \geq 1$) and consistent (as $K_s, K_{s+1} \to +\infty$) estimator of the size density $s_X(s)$ in (4.4).

The relative MSE (6.3) is directly controlled by the underlying number K_s of Monte Carlo iterations, regardless of the size of the observation window W. In fact, the relative MSE decreases to zero at a rate inversely proportional to the number of Monte Carlo iterations. As is clear from (6.3), it suffices to set

$$K_s = \frac{1}{\epsilon} \frac{1 - S_X(s)}{S_X(s)}, \quad \text{for every } s \in \bar{\mathbb{Z}}, \tag{6.4}$$

so as to uniformly obtain (over all s) a relative MSE of no more than ϵ. The value of K_s in (6.4) directly depends on the particular value of $S_X(s)$. As expected, and in order to obtain the same relative MSE for all s, small values of $S_X(s)$ require more Monte Carlo iterations than larger ones. In view of the fact that $S_X(s)$ is not known a-priori and the fact that $K_s \to +\infty$, as $S_X(s) \to 0^+$, we may in practice decide to estimate only size distribution values that are above a pre-defined threshold $a > 0$, with relative MSE of no more than ϵ. In this case, we may set $K_s = (1-a)/\epsilon a$, for every $s \in \bar{\mathbb{Z}}$. Notice finally that the numerical implementation of (6.1) may require replacing window W by a smaller window W', so that (see also (5.1)) $\gamma_s(X) \cap W' = \gamma_s(X \cap W) \cap W'$, for every $X \in \mathcal{P}$, when $s \geq 0$, with a similar replacement when $s \leq -1$. This is clearly required in order to avoid "boundary effects" when computing $\gamma_s(X^{(k)})$ and $\phi_s(X^{(k)})$ on W.

When $\mathbf{X}$ is a MRF, independent samples of $\mathbf{X}$ can be generated by means of a "*Many-Short-Runs*" MCMC approach (e.g., see [18], [33]). K_s identical and statistically independent ergodic Markov chains are generated that approximately converge to probability π after k_o steps. The first k_o samples are discarded and the $(k_o+1)^{st}$ sample is retained. If $\mathbf{X}^{(k)}$ is the $(k_o+1)^{st}$ state of the kth chain, then $\{\mathbf{X}^{(1)}, \mathbf{X}^{(2)}, ..., \mathbf{X}^{(K_s)}\}$ will be a collection of independent DRS's, whose distribution is approximately that of $\mathbf{X}$. Unfortunately, the resulting estimator requires $(k_o+1)K_s$ Monte Carlo iterations and is asymptotically unbiased and consistent only in the limit as $k_o, K_s \to +\infty$ (e.g., see [34], [35]). Therefore, implementation of a "Many-Short-Runs" MCMC approach may require substantial number of Monte Carlo iterations before a "reasonable" estimate is achieved.

Alternatively, a "*Single-Long-Run*" MCMC approach can be employed. In this case, a single ergodic Markov chain $\mathbf{X}_1, \mathbf{X}_2, ..., \mathbf{X}_{k_o+(K_s-1)l+1}$ is generated of $k_o + (K_s-1)l+1$ samples. The first k_o samples are discarded, in order to guarantee that the chain has approximately reached equilibrium, whereas the remaining samples are subsampled at every l^{th}– step, in order to reduce correlation between consecutive samples. In this case, $\mathbf{X}^{(k)} = \mathbf{X}_{k_o+(k-1)l+1}$, for $k = 1, 2, ..., K_s$, will be a collection of dependent DRS's, whose distribution is approximately that of $\mathbf{X}$ (e.g., see [18]–[20]). This technique, together with (6.1), produces an estimator $\hat{S}_{X,2}(s, K_s)$ of $S_X(s)$ whose bias can be shown to satisfy (compare with (6.2))

$$| E[\hat{S}_{X,2}(s, K_s)] - S_X(s) | \leq \frac{1}{K_s} \frac{\delta_{k_o}}{1 - \rho_l} S_X(s), \tag{6.5}$$

where $\rho_l = \lambda_{\max}^l$ and $\delta_{k_o} = \lambda_{\max}^{k_o} / \pi_{\min}$, with $\lambda_{\max} < 1$ being the second largest (in absolute value) eigenvalue of the transition probability matrix of the underlying Markov chain, and $\pi_{\min} = \min\{\pi(\mathbf{x}), \mathbf{x} \in \mathcal{S}\} > 0$ (e.g., see [34], [35]). Furthermore, the relative MSE satisfies (compare with (6.3))

$$\frac{E[(\hat{S}_{X,2}(s, K_s) - S_X(s))^2]}{S_X^2(s)} \lessapprox \frac{1}{K_s} \left[\frac{1 - S_X(s)}{S_X(s)} + \frac{2\rho_l}{1 - \rho_l} \frac{1}{\pi_{\min}} \right], \tag{6.6}$$

for large values of K_s, provided that $S_X(s) \neq 0$. It is now clear that $\hat{S}_{X,2}(s, K_s)$ is an asymptotically unbiased and consistent estimator of $S_X(s)$, as $K_s \to +\infty$, regardless of the values of k_o, l, $\lambda_{\max}$, and $\pi_{\min}$. The same is also true (as $K_s, K_{s+1} \to +\infty$) for the Monte Carlo estimator

$$\hat{s}_{X,2}(s, K_s, K_{s+1}) = \hat{S}_{X,2}(s, K_s) - \hat{S}_{X,2}(s + 1, K_{s+1}),$$

of the size density $s_X(s)$. However, it is not possible to calculate the upper bounds on the bias and the relative MSE in (6.5) and (6.6), since $\lambda_{\max}$ and $\pi_{\min}$ cannot be computed in general.

EXAMPLE 6.1. In analogy to Examples 5.1 and 5.2, a Monte Carlo estimator of an opening based size distribution is given by (recall (6.1))

$$\hat{S}^{\circ}_{X,i}(s, K_s) = \frac{1}{K_s} \frac{1}{|W \ominus 2|s|B|} \begin{cases} \sum_{k=1}^{K_s} |[(\mathbf{X}^{(k)} \cap W) \circ sB] \cap (W \ominus 2sB)|, & s \geq 0 \\ \sum_{k=1}^{K_s} |[(\mathbf{X}^{(k)} \cap W) \bullet |s|B] \cap (W \ominus 2|s|B)|, & s \leq -1 \end{cases} \tag{6.7}$$

provided that $|W \ominus 2|s|B| \neq 0$, where $i = 1, 2$, depending on whether or not $\mathbf{X}^{(k)}, k = 1, 2, ..., K_s$, are i.i.d. DRS's (compare with (5.9)). On the other hand, a Monte Carlo estimator of an erosion based size distribution is given by

$$\hat{S}^{\ominus}_{X,i}(s, K_s) = \frac{1}{K_s} \frac{1}{|W \ominus |s|B|} \begin{cases} \sum_{k=1}^{K_s} |[(\mathbf{X}^{(k)} \cap W) \ominus sB] \cap (W \ominus sB)|, & s \geq 0 \\ \sum_{k=1}^{K_s} |[(\mathbf{X}^{(k)} \cap W) \oplus |s|B] \cap (W \ominus |s|B)|, & s \leq -1 \end{cases} \tag{6.8}$$

provided that $|W \ominus |s|B| \neq 0$, where $i = 1, 2$ (compare with (5.10)). □

Estimators (6.7) and (6.8) are the most useful in practice. They may however turn out to be computationally intensive. This is due to the fact that (6.7) and (6.8) require computation of openings and closings (in the case of (6.7)), or erosions and dilations (in the case of (6.8)), over the entire sub-windows $W \ominus 2|s|B$ and $W \ominus |s|B$, respectively, which need to be re-evaluated at each Monte Carlo iteration. In order to ameliorate this problem, we may assume that $\mathbf{X}$ is a first-order stationary DRS w.r.t. γ_s and ϕ_s, in which case $S_X(w, s) = S_X(\mathbf{0}, s)$ and the size distribution is therefore independent of w (recall (4.1) and Definition 5.1). It can be easily shown that (compare with (4.3))

$$S_X(s) = \begin{cases} E[[\gamma_s(\mathbf{x})](\mathbf{0})], & \text{for } s \geq 0 \\ E[[\phi_{|s|}(\mathbf{x})](\mathbf{0})], & \text{for } s \leq -1 \end{cases}.$$

This leads to the following Monte Carlo estimator of an opening based size distribution (recall (2.1)–(2.4)):

$$\hat{\Sigma}^{\circ}_{X,i}(s, K_s) = \frac{1}{K_s} \begin{cases} \sum_{k=1}^{K_s} \bigvee_{v \in sB} (\bigwedge_{w \in sB \oplus \{v\}} \mathbf{x}^{(k)}(w)), & \text{for } s \geq 0 \\ \sum_{k=1}^{K_s} \bigwedge_{v \in |s|B} (\bigvee_{w \in |s|B \oplus \{v\}} \mathbf{x}^{(k)}(w)), & \text{for } s \leq -1 \end{cases}, \tag{6.9}$$

where $i = 1, 2$, depending on whether or not $\mathbf{X}^{(k)}, k = 1, 2, ..., K_s$, are i.i.d. DRS's (compare with (6.7)). In (6.9), $\vee$ and $\wedge$ denote maximum and minimum, respectively. On the other hand, a Monte Carlo estimator of an erosion based size distribution is given by (recall (2.1) and (2.2))

$$\hat{\Sigma}^{\ominus}_{X,i}(s, K_s) = \frac{1}{K_s} \begin{cases} \sum_{k=1}^{K_s} \bigwedge_{w \in sB} \mathbf{x}^{(k)}(w), & \text{for } s \geq 0 \\ \sum_{k=1}^{K_s} \bigvee_{w \in |s|B} \mathbf{x}^{(k)}(w), & \text{for } s \leq -1 \end{cases}, \tag{6.10}$$

where $i = 1, 2$ (compare with (6.8)). If we now consider estimator (6.10), for $s \geq 0$, then, at each Monte Carlo iteration, we only need to calculate the local minimum of $\mathbf{x}^{(k)}$ over sB. Therefore, estimator (6.10), for $s \geq 0$, is expected to be computationally more efficient than estimator (6.8), for $s \geq 0$. The same remark is also true for estimator (6.10), with $s \leq -1$, as compared to estimator (6.8), with $s \leq -1$, and for estimator (6.9) as compared to estimator (6.7).

It is easy to verify that estimator (6.9), with $i = 1$, is an unbiased (for every $K_s \geq 1$) and consistent (as $K_s \to +\infty$) estimator of the size distribution $S_X(s)$, with a relative MSE bounded from above by the same upper bound as in (6.3). Furthermore, estimator (6.9), with $i = 2$, is an asymptotically unbiased and consistent estimator of $S_X(s)$, as $K_s \to +\infty$, with the bias and the relative MSE bounded from above by the same upper bounds as in (6.5) and (6.6), respectively. Similar upper bounds can be obtained for estimators (6.8) and (6.10). Finally, the corresponding estimators for the size density are given by

$$\hat{\sigma}^{\circ}_{X,i}(s, K_s, K_{s+1}) = \hat{\Sigma}^{\circ}_{X,i}(s, K_s) - \hat{\Sigma}^{\circ}_{X,i}(s+1, K_{s+1}), \tag{6.11}$$

and

$$\hat{\sigma}^{\ominus}_{X,i}(s, K_s, K_{s+1}) = \hat{\Sigma}^{\ominus}_{X,i}(s, K_s) - \hat{\Sigma}^{\ominus}_{X,i}(s+1, K_{s+1}). \tag{6.12}$$

To conclude this section, we should point-out that a MRF $\mathbf{X}$ on W is first-order stationary w.r.t. γ_s and ϕ_s, if the associated energy function $U(\mathbf{x})$ as well as γ_s and ϕ_s are invariant to *circular shifts*, in the sense that

$$U(\mathbf{x}_w) = U(\mathbf{x}), \quad \gamma_s(\mathbf{x}_w) = [\gamma_s(\mathbf{x})]_w, \quad \text{and} \quad \phi_s(\mathbf{x}_w) = [\phi_s(\mathbf{x})]_w,$$

for every $w = (i,j) \in W$, with $x_w(m,n) = x((m-i)_M, (n-j)_N)$. The Ising model discussed in Example 3.1 is first-order stationary w.r.t. γ_s and ϕ_s, provided that γ_s and ϕ_s are translation invariant on $\mathbb{Z}^2$ and a toroidal boundary condition is assumed for W.

7. Applications.

7.1. Phase transition. We now compare the performance of the empirical and Monte Carlo estimators for estimating the size density of a particular MRF model, known as the Ising model. Our results also demonstrate the phenomenon of phase transition exhibited by this model. We consider the Ising model discussed in Example 3.1, defined over a 128×128 rectangular grid, with $\alpha = 0$, $\beta_1 = -0.25$, and $\beta_2 = -0.5$ (see also (3.5)). This MRF experiences phase transition at a critical temperature $T_c = 0.82$, computed as solution to (3.7). Figure 1 depicts typical realizations of the Ising model under consideration, obtained by means of a Metropolis's MCMC algorithm with random site updating, at three different temperatures $T = 1.0, 0.8, 0.6$. The Monte Carlo estimates of the erosion and opening based size densities (w.r.t. to a RHOMBUS structuring element $B = \{(0,0), (-1,0), (1,0), (0,-1), (0,1)\}$) are depicted in the first row of Figs. 2 and 3, respectively. To obtain these estimates, we have assumed a toroidal boundary condition and we have employed estimators (6.10), (6.12) (for Fig. 2) and (6.9), (6.11) (for Fig. 3). On the other hand, the second row of Figs. 2 and 3 depict the corresponding empirical estimates, obtained by means of (5.10) and (5.9), respectively. Since the Ising model under consideration is *self-dual* (i.e., $\pi(X) = \pi(X^c)$), the size density values will be symmetric, i.e., $s_X(s) = s_X(-s-1)$, for $s = 0, 1, \ldots$ (recall the

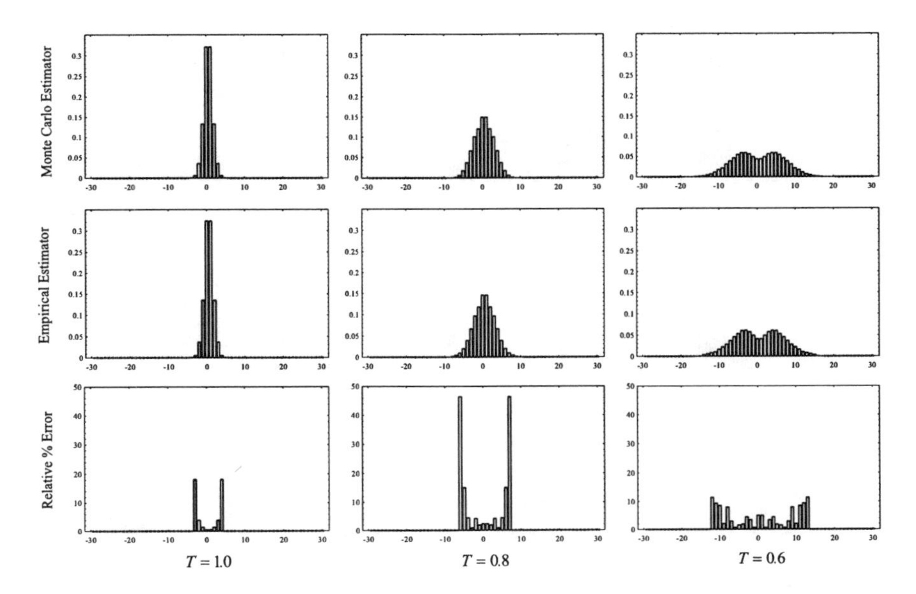

FIG. 2. *Monte Carlo (first row) and empirical (second row) estimates of the erosion based size density of the Ising model of Fig. 1 at three different temperatures. The third row depicts the relative (%) error between the empirical and Monte Carlo estimates.*

duality property of erosion-dilation and opening-closing). Therefore, we only need to estimate the size density values at non-negative sizes. From Fig. 3, it is clear how the size density characterizes the granular structure of a MRF at different temperatures. Above the critical temperature (e.g., at $T = 1.0$) most non-negligible values of the size density are at small sizes, which verifies the presence of a fine phase. Below the critical temperature (e.g., at $T = 0.6$) the size density spreads out to larger sizes, whereas it is negligible at small sizes, which verifies the presence of a coarse phase. Close to the critical temperature (e.g., at $T = 0.8$) the size density takes non-negligible values at a wide range of sizes, which verifies the presence of phase transition (i.e., the coexistence of both fine and coarse phases). As is clear from the third rows of Figs. 2 and 3, the empirical estimators (5.9) and (5.10) work reasonably well at high temperatures. However, these estimators are unreliable at low temperatures, especially for large sizes. This is partially due to the fact that this estimator uses a single realization of the image, which is observed only through a finite window. On the other hand, conditions (5.4)–(5.6) may not be satisfied here, in which case the empirical estimators may not be consistent. It is worthwhile to note that the performance of the empirical estimator for the erosion-based size density is relatively better as compared with that for the opening-based size density. This is due to the fact that the sub-window $W'_r(s) \subseteq W_r$ satisfying (5.1) is larger for the case of erosion and dilation than for the case of opening and closing (see also Examples 5.1 and 5.2).

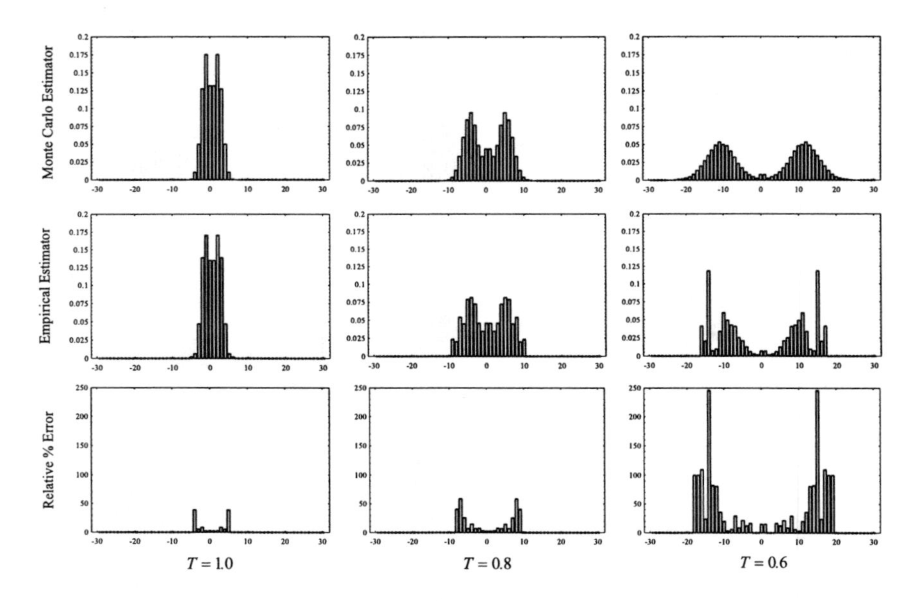

FIG. 3. *Monte Carlo (first row) and empirical (second row) estimates of the opening based size density of the Ising model of Fig. 1 at three different temperatures. The third row depicts the relative (%) error between the empirical and Monte Carlo estimates.*

7.2. Shape restoration. In many image processing applications, binary image data Y are frequently corrupted by noise and clutter. It is therefore of great interest to design a set operator Ψ that, when applied on Y, *optimally* removes noise and clutter. Solutions to this problem, known as *shape restoration*, have been recently obtained by means of mathematical morphology [4]–[6], [36]. Image data Y are considered to be realizations of a DRS $\mathbf{Y}$, which is mathematically described by means of a *degradation equation* of the form:

$$\mathbf{Y} = (\mathbf{X} \setminus \mathbf{N}_1) \cup \mathbf{N}_2, \tag{7.1}$$

where $\mathbf{X}$ is a DRS that models the shape under consideration and $\mathbf{N}_1$, $\mathbf{N}_2$ are two DRS's that model degradation [4]. In particular, $\mathbf{N}_1$ models degradation due to sensor noise, incomplete data collection, or occlusion, whereas $\mathbf{N}_2$ models degradation due to the presence of clutter or sensor noise. The problem of optimal shape restoration consists of designing a set operator Ψ such that

$$\hat{X} = \Psi(Y) = \Psi((X \setminus N_1) \cup N_2), \tag{7.2}$$

is "optimally close" to X, in the sense of minimizing a *distance metric* $D(X, \hat{X})$.

It has been shown in [6] that if:[1] (a) $\mathbf{N}_1 = \emptyset$, a.s., (b) $\mathbf{X}$ and $\mathbf{N}_2$ are

[1] In fact, the assumptions made in [6] are less restrictive than the ones stated here.

generated by a random set model of the form

$$\bigcup_{m=1}^{M} \bigcup_{l=1}^{L_m} (mB)_{v_{ml}}, \quad v_{ml} \in \mathbb{Z}^2, \tag{7.3}$$

where $L_1, L_2, ..., L_M$ are multinomial distributed random numbers with $\sum_{m=1}^{M} L_m$ being Poisson distributed and $(mB)_{v_{ml}}$ are non-interfering[2] randomly translated structuring elements, (c) $\mathbf{X}$ and $\mathbf{N}_2$ are a.s. non-interfering, and (d) we limit our interest to operators Ψ of the form

$$\hat{X} = \Psi_\circ(Y) = \bigcup_{s \in S_+} [Y \circ sB \setminus Y \circ (s+1)B], \tag{7.4}$$

for some index set S_+, then (7.4) is the solution to the optimal binary image restoration problem, in the sense of minimizing the *expected symmetric difference metric*

$$D(X, \hat{X}) = E[|(\mathbf{X} \setminus \hat{\mathbf{X}}) \cup (\hat{\mathbf{X}} \setminus \mathbf{X})|], \tag{7.5}$$

provided that

$$S_+ = \{s \geq 0 \mid s_{N_2}(s) < s_X(s)\}. \tag{7.6}$$

It is worthwhile noticing here that, if $S_+ = \{n, n+1, ...\}$, for some $n \geq 0$, then $\Psi_\circ(Y) = Y \circ nB$, whereas if $S_+ = \{0, 1, ..., n-1\}$, for some $n \geq 1$, then $\Psi_\circ(Y) = Y \setminus Y \circ nB$. By duality (since $A \setminus B = A \cap B^c$), and due to the particular form of (7.5), it can be shown that if: (a) $\mathbf{N}_2 = \emptyset$, a.s., (b) $\mathbf{X}^c$ and $\mathbf{N}_1$ are generated by the random set model (7.3), (c) $\mathbf{X}^c$ and $\mathbf{N}_1$ are a.s. non-interfering, and (d) we limit our interest to operators Ψ of the form

$$\hat{X} = \Psi_\bullet(Y) = \bigcap_{s \in S_-} [Y \bullet |s|B \setminus Y \bullet (|s|-1)B]^c, \tag{7.7}$$

for some index set S_-, then (7.7) is the solution to the optimal binary image restoration problem, in the sense of minimizing the expected symmetric difference metric (7.5), provided that

$$S_- = \{s \leq -1 \mid s_{N_1^c}(s) < s_X(s)\}, \tag{7.8}$$

and B is a symmetric structuring element. Notice that $\Psi_\bullet(Y) = [\Psi_\circ(Y^c)]^c$. Furthermore, if $S_- = \{..., -n-2, -n-1\}$, for some $n \geq 0$, then $\Psi_\bullet(Y) = Y \bullet nB$, whereas if $S_- = \{-n, -n+1, ..., -1\}$, for some $n \geq 1$, then $\Psi_\bullet(Y) = [Y \bullet nB \setminus Y]^c$.

[2] Two sets A and B are called non-interfering if for every connected component C of $A \cup B$, $C \subseteq A$ or $C \subseteq B$.

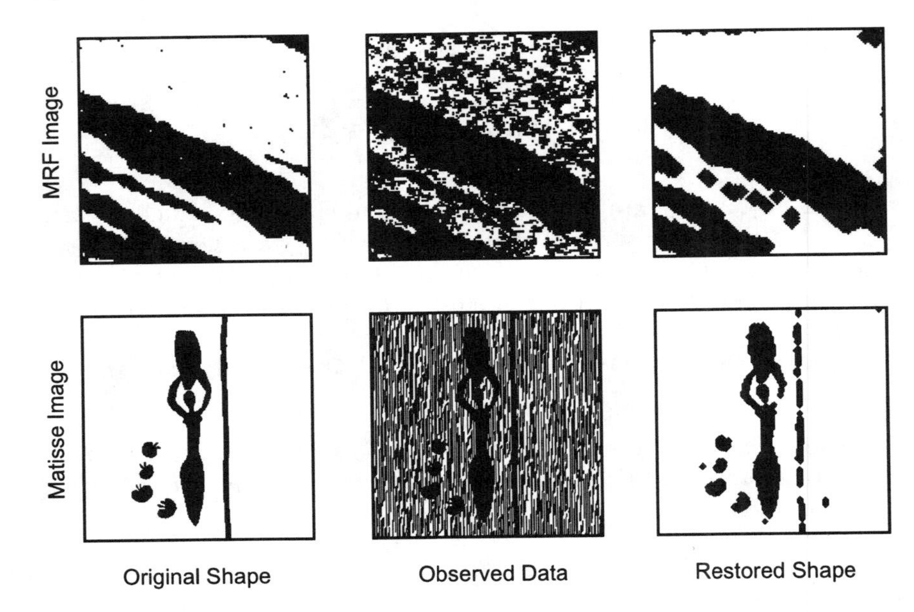

FIG. 4. *Shape restoration for a MRF image (first row) and a Matisse image (second row) corrupted by union noise. Although the observed data experience 31% and 45% error, for the MRF and the Matisse image, respectively, shape restoration by means of (7.4), (7.5) produces images with only 5.34% and 2.28% error, respectively.*

We may relax the modeling and non-interfering assumptions for $\mathbf{X}$, $\mathbf{X}^c$, $\mathbf{N}_1$, and $\mathbf{N}_2$, and we may consider in (7.2) an operator $\Psi = \Psi_\bullet \Psi_\circ$. In this case, $\Psi(Y)$ will be a *suboptimal*, but nevertheless useful, solution to the shape restoration problem under consideration. If we assume that $\mathbf{X}$ is a DRS with size density $s_X(s)$, and if N_1, N_2 are two realizations of a DRS $\mathbf{N}$ with size density $s_N(s)$, then (recall (7.4), (7.6), (7.7), and (7.8))

$$\hat{X} = \Psi_\bullet(\tilde{X}) = \bigcap_{s \in S_-} [\tilde{X} \bullet |s|B \setminus \tilde{X} \bullet (|s|-1)B]^c, \tag{7.9}$$

where

$$\tilde{X} = \Psi_\circ(Y) = \bigcup_{s \in S_+} [Y \circ sB \setminus Y \circ (s+1)B]. \tag{7.10}$$

In (7.9) and (7.10),

$$\begin{aligned} S_- &= \{s \leq -1 \mid s_{N^c}(s) < s_X(s)\} \\ &= \{s \leq -1 \mid s_N(|s|-1) < s_X(s)\}, \end{aligned} \tag{7.11}$$

whereas

$$S_+ = \{s \geq 0 \mid s_N(s) < s_{X \setminus N}(s)\}. \tag{7.12}$$

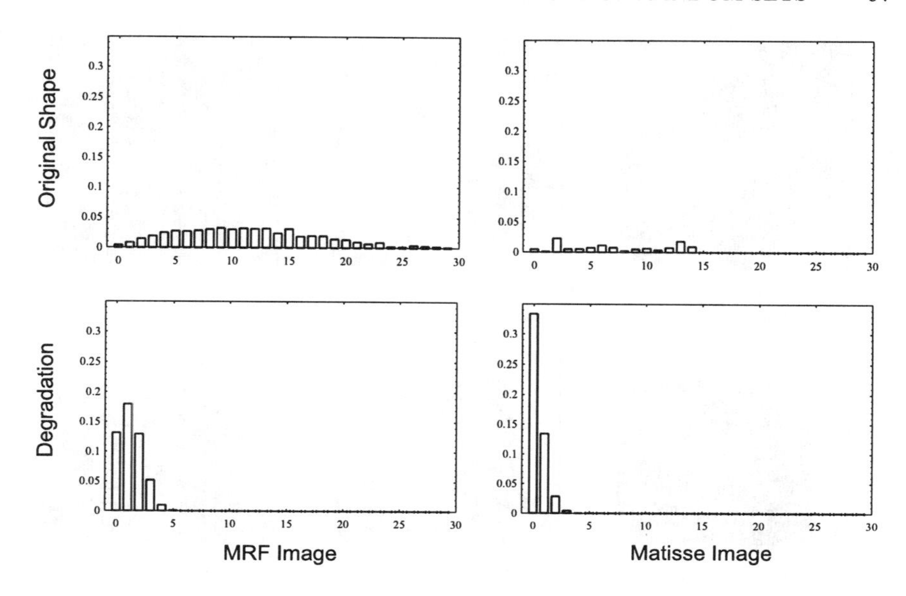

FIG. 5. *Estimated size densities of the original shape and degradation for the MRF (first column) and the Matisse (second column) images depicted in Fig. 4. These densities are used to determine the set S_+, required in (7.4), by means of (7.5).*

In this case, implementation of operator $\Psi = \Psi_\bullet \Psi_\circ$, by means of (7.9), (7.10), requires knowledge of the size density $s_N(s)$ of degradation $\mathbf{N}$, for $s \geq 0$, the size density $s_X(s)$ of shape $\mathbf{X}$, for $s \leq -1$, as well as the size density $s_{X \setminus N}(s)$ of the DRS $\mathbf{X} \setminus \mathbf{N}$, for $s \geq 0$.

EXAMPLE 7.1. We now demonstrate the use of operator $\Psi_\circ$, given by (7.4), (7.6), for restoring shapes corrupted by union noise. In this case, $\mathbf{N}_1 = 0$, a.s., and $\mathbf{Y} = \mathbf{X} \cup \mathbf{N}_2$ (see also (7.1)). The first row of Fig. 4 depicts the result of shape restoration of a random image $\mathbf{X}$ by means of (7.4) and (7.6). $\mathbf{X}$ is assumed to be a 128×128 pixel MRF with known energy function, whereas the corrupting union noise is taken to be the Ising model depicted in Fig. 1, with $T = 1.0$. The first column of Fig. 5 depicts the size densities $s_X(s)$, $s_{N_2}(s)$, $s \geq 0$, of $\mathbf{X}$ and $\mathbf{N}_2$, respectively, estimated by means of the Monte Carlo estimator (6.9), (6.11). The RHOMBUS structuring element has been employed here. These densities are then used to determine the set S_+ in (7.6) which, in turn, specifies the particular form of $\Psi_\circ$ by means of (7.4). Although 31% of the pixels in $\mathbf{Y}$ are subject to error, operator $\Psi_\circ$ was able to recover all, but 5.34%, of original pixel values.

The second row of Fig. 4 and the second column of Fig. 5 depict a similar shape restoration example with $\mathbf{X}$ being a 256×256 pixel binary Matisse image.[3] In this case, $\mathbf{N}_2$ is taken to be a MRF with known energy function that models vertical striking. The size density $s_X(s)$, $s \geq 0$, has

[3] Henri Matisse, *Woman with Amphora and Pomegranates*, 1952 – Paper on canvas.

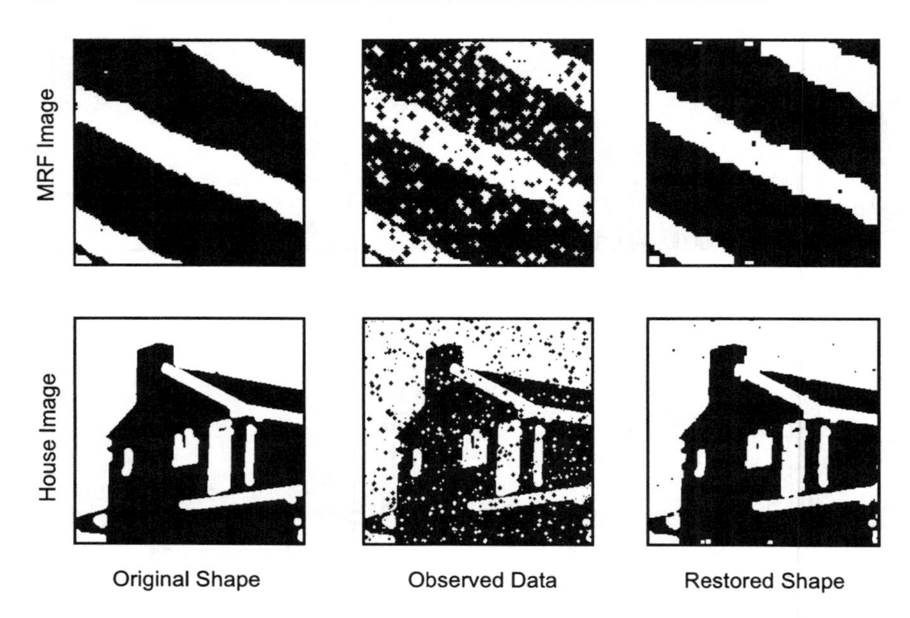

FIG. 6. *Shape restoration for a MRF image (first row) and a House image (second row) corrupted by union/intersection noise. Although the observed data experience 9.36% and 6.31% error, for the MRF and the House image, respectively, shape restoration by means of (7.9)–(7.12) produces images with only 2.19% and 1.06% error, respectively.*

been calculated by means of the empirical estimator (5.8), (5.9), whereas the size density $s_{N_2}(s)$, $s \geq 0$, has been calculated by means of the Monte Carlo estimator (6.9), (6.11). The RHOMBUS structuring element has been employed here. In this case, application of $\Psi_\circ$ on $\mathbf{Y}$ results in reducing the error from 45% down to a mere 2.28%. □

EXAMPLE 7.2. Shape restoration in the more general setting of (7.1) is clearly more complicated. In this example, we demonstrate the use of operators $\Psi_\bullet$, $\Psi_\circ$, given by (7.9) and (7.10), for restoring shapes corrupted by union/intersection noise. In this case, we take $\mathbf{Y} = (\mathbf{X} \cap \mathbf{N}_1^c) \cup \mathbf{N}_2$ (see also (7.1)), where $\mathbf{N}_1$ and $\mathbf{N}_2$ have the same distribution as a DRS $\mathbf{N}$. The first row of Fig. 6 depicts the result of shape restoration of a random image $\mathbf{X}$ by means of (7.9)–(7.12). $\mathbf{X}$ is assumed to be a 128×128 pixel MRF with known energy function, whereas the corrupting noise $\mathbf{N}$ is taken to be a *Boolean model* [2], [37]. The first column of Fig. 7 depicts the size densities $s_X(s)$, for $s \leq 0$, and $s_N(s)$, $s_{X \setminus N}(s)$, for $s \geq 0$, of $\mathbf{X}$, $\mathbf{N}$, and $\mathbf{X} \setminus \mathbf{N}$, respectively, estimated by means of the Monte Carlo estimator (6.9), (6.11). The SQUARE structuring element $B = \{(0,0),(1,0),(0,1),(1,1)\}$ has been employed here. These densities are then used to determine the sets S_-, S_+ in (7.11) and (7.12) which, in turn, specify the particular form of $\Psi_\bullet$ and $\Psi_\circ$ by means of (7.9) and (7.10). Although 9.36% of the pixels in $\mathbf{Y}$ are subject to error, operator $\Psi = \Psi_\bullet \Psi_\circ$ was able to recover all, but 2.19%, of original pixel values. The second row of Fig. 6 and the second

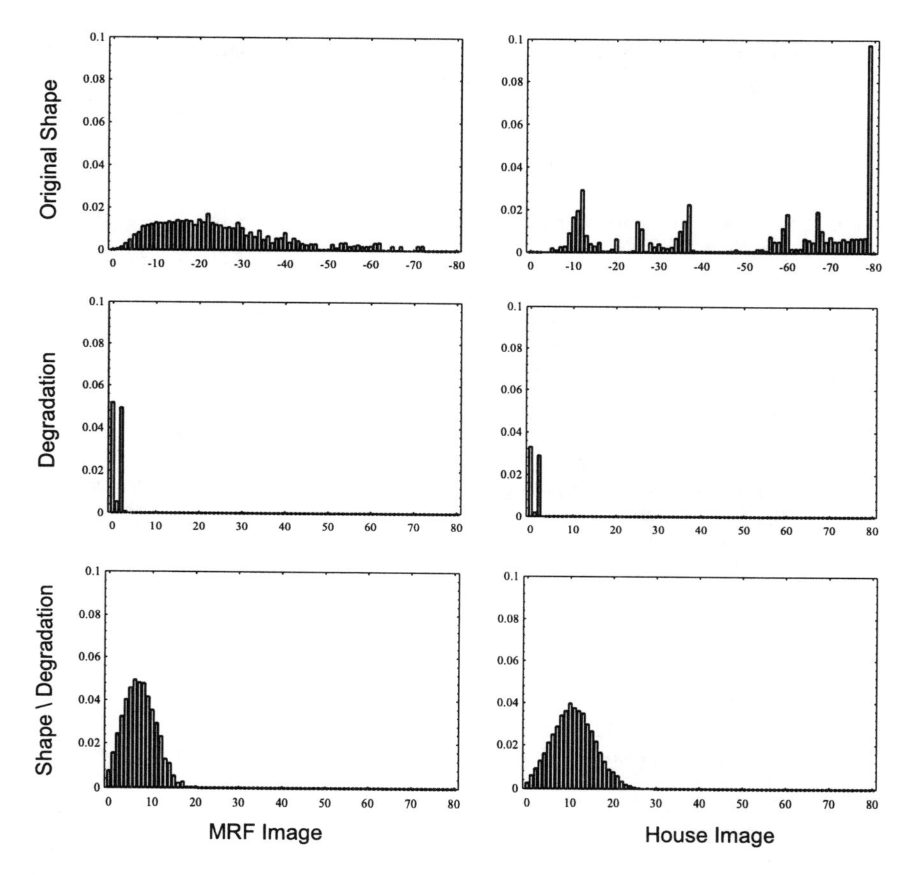

FIG. 7. *Estimated size densities of the original shape, degradation, and shape $\setminus$ degradation for the MRF (first column) and the House (second column) images depicted in Fig. 6. These densities are used to determine sets S_-, S_+, required in (7.9), (7.10) by means of (7.11) and (7.12).*

column of Fig. 7 depict a similar shape restoration example with $\mathbf{X}$ being a 256×256 pixel binary House image. $\mathbf{N}$ is taken to be a Boolean model as before. The size density $s_X(s)$, $s \leq 0$, has been calculated by means of the empirical estimator (5.8), (5.9), whereas the size densities $s_N(s)$, $s_{X \setminus N}(s)$, $s \geq 0$, have been calculated by means of the Monte Carlo estimator (6.9), (6.11). The SQUARE structuring element $B = \{(0,0),(1,0),(0,1),(1,1)\}$ has been employed. In this case, application of $\Psi = \Psi_\bullet \Psi_\circ$ on $\mathbf{Y}$ results in reducing the error from 6.31% down to a mere 1.06%. □

In order for the previously suggested shape restoration approach to be practical, we need knowledge of the associated size densities. These densities can be estimated from training data by means of either the empirical estimator discussed in Section 5 or the Monte Carlo estimator (6.1). On the other hand, if statistical models are readily available for the signal and/or degradation, then the Monte Carlo estimators discussed in Section 6 can be effectively used for size density estimation.

REFERENCES

[1] G. MATHERON, *Random Sets and Integral Geometry*, John Wiley, New York City, New York, 1975.

[2] J. SERRA, *Image Analysis and Mathematical Morphology*, Academic Press, London, England, 1982.

[3] P. MARAGOS, *Pattern spectrum and multiscale shape representation*, IEEE Transactions on Pattern Analysis and Machine Intelligence, 11 (1989), pp. 701–716.

[4] D. SCHONFELD AND J. GOUTSIAS, *Optimal morphological pattern restoration from noisy binary images*, IEEE Transactions on Pattern Analysis and Machine Intelligence, 13 (1991), pp. 14–29.

[5] E.R. DOUGHERTY, R.M. HARALICK, Y. CHEN, C. AGERSKOV, U. JACOBI, AND P.H. SLOTH, *Estimation of optimal morphological τ-opening parameters based on independent observation of signal and noise pattern spectra*, Signal Processing, 29 (1992), pp. 265–281.

[6] R.M. HARALICK, P.L. KATZ, AND E.R. DOUGHERTY, *Model-based morphology: The opening spectrum*, Graphical Models and Image Processing, 57 (1995), pp. 1–12.

[7] E.R. DOUGHERTY AND Y. CHEN, *Logical granulometric filtering in the signal-union-clutter model*, This Volume, pp. 73–95.

[8] E.R. DOUGHERTY AND J.B. PELZ, *Morphological granulometric analysis of electrophotographic images – Size distribution statistics for process control*, Optical Engineering, 30 (1991), pp. 438–445.

[9] E.R. DOUGHERTY, J.T. NEWELL, AND J.B. PELZ, *Morphological texture-based maximum-likelihood pixel classification based on local granulometric moments*, Pattern Recognition, 25 (1992), pp. 1181–1198.

[10] E.R. DOUGHERTY, J.B. PELZ, F. SAND, AND A. LENT, *Morphological image segmentation by local granulometric size distributions*, Journal of Electronic Imaging, 1 (1992), pp. 46–60.

[11] Y. CHEN AND E.R. DOUGHERTY, *Gray-scale morphological granulometric texture classification*, Optical Engineering, 33 (1994), pp. 2713–2722.

[12] E.R. DOUGHERTY AND Y. CHENG, *Morphological pattern-spectrum classification of noisy shapes: Exterior granulometries*, Pattern Recognition, 28 (1995), pp. 81–98.

[13] S. ARCHAMBAULT AND M. MOORE, *Statistiques morphologiques pour l'ajustement d'images*, International Statistical Review, 61 (1993), pp. 283–297.

[14] M. MOORE AND S. ARCHAMBAULT, *On the asymptotic behavior of some statistics based on morphological operations*, Spatial Statistics and Imaging (A. Possolo, ed.), vol. 20, Hayward, California: Institute of Mathematical Statistics, Lecture Notes, Monograph Series, 1991, pp. 258–274.

[15] K. SIVAKUMAR AND J. GOUTSIAS, *Monte Carlo estimation of morphological granulometric discrete size distributions*, Mathematical Morphology and Its Applications to Image Processing (J. Serra and P. Soille, eds.), Dordrecht, The Netherlands: Kluwer, 1994, pp. 233–240.

[16] S. GEMAN AND D. GEMAN, *Stochastic relaxation, Gibbs distributions, and the Bayesian restoration of images*, IEEE Transactions on Pattern Analysis and Machine Intelligence, 6 (1984), pp. 721–741.

[17] R.C. DUBES AND A.K. JAIN, *Random field models in image analysis*, Journal of Applied Statistics, 16 (1989), pp. 131–164.

[18] C.J. GEYER, *Practical Markov chain Monte Carlo*, Statistical Science, 7 (1992), pp. 473–511.

[19] J. BESAG, P. GREEN, D. HIGDON, AND K. MENGERSEN, *Bayesian computation and*

stochastic systems, Statistical Science, 10 (1995), pp. 3–66.

[20] B. GIDAS, *Metropolis-type Monte Carlo simulation algorithms and simulated annealing*, Topics in Contemporary Probability and its Applications (J.L. Snell, ed.), Boca Raton, Florida: CRC Press, 1995, pp. 159–232.

[21] F. SAND AND E.R. DOUGHERTY, *Asymptotic normality of the morphological pattern-spectrum moments and orthogonal granulometric generators*, Journal of Visual Communication and Image Representation, 3 (1992), pp. 203–214.

[22] F. SAND AND E.R. DOUGHERTY, *Statistics of the morphological pattern-spectrum moments for a random-grain model*, Journal of Mathematical Imaging and Vision, 1 (1992), pp. 121–135.

[23] E.R. DOUGHERTY AND F. SAND, *Representation of linear granulometric moments for deterministic and random binary Euclidean images*, Journal of Visual Communication and Image Representation, 6 (1995), pp. 69–79.

[24] B. BETTOLI AND E. R. DOUGHERTY, *Linear granulometric moments of noisy binary images*, Journal of Mathematical Imaging and Vision, 2 (1993), pp. 299–319.

[25] J. GOUTSIAS, *Morphological analysis of discrete random shapes*, Journal of Mathematical Imaging and Vision, 2 (1992), pp. 193–215.

[26] K. SIVAKUMAR AND J. GOUTSIAS, *Binary random fields, random closed sets, and morphological sampling*, IEEE Transactions on Image Processing, 5 (1996), pp. 899–912.

[27] J.J. BINNEY, N.J. DOWRICK, A.J. FISHER, AND M.E.J. NEWMAN, *The Theory of Critical Phenomena: An Introduction to the Renormalization Group*, Oxford University Press, Oxford, England, 1992.

[28] D. RUELLE, *Statistical Mechanics: Rigorous Results*, Addison-Wesley, Reading, Massachusetts, 1983.

[29] H.J.A.M. HEIJMANS, *Morphological Image Operators*, Academic Press, Boston, Massachusetts, 1994.

[30] P. DELFINER, *A generalization of the concept of size*, Journal of Microscopy, 95 (1971), pp. 203–216.

[31] J. MATTIOLI AND M. SCHMITT, *On information contained in the erosion curve*, Shape in Picture: Mathematical Description of Shape in Grey-level Images (Y.L. O, A. Toet, D. Foster, H.J.A.M. Heijmans, and P. Meer, eds.), Berlin, Germany: Springer-Verlag, 1994, pp. 177–195.

[32] J. MATTIOLI AND M. SCHMITT, *Inverse problems for granulometries by erosion*, Journal of Mathematical Imaging and Vision, 2 (1992), pp. 217–232.

[33] A. GELMAN AND D.B. RUBIN, *Inference from iterative simulation using multiple sequences*, Statistical Science, 7 (1992), pp. 457–511.

[34] G. POTAMIANOS, *Stochastic Simulation Algorithms for Partition Function Estimation of Markov Random Field Images*, PhD Thesis, Department of Electrical and Computer Engineering, The Johns Hopkins University, Baltimore, Maryland, 1994.

[35] J. GOUTSIAS, *Markov random fields: Interacting particle systems for statistical image modeling and analysis*, Tech. Rep. JHU/ECE 96-01, Department of Electrical and Computer Engineering, The Johns Hopkins University, Baltimore, Maryland 21218, 1996.

[36] D. SCHONFELD, *Optimal structuring elements for the morphological pattern restoration of binary images*, IEEE Transactions on Pattern Analysis and Machine Intelligence, 16 (1994), pp. 589–601.

[37] D. STOYAN, W.S. KENDALL, AND J. MECKE, *Stochastic Geometry and its Applications*, Second Edition, John Wiley, Chichester, England, 1995.

LOGICAL GRANULOMETRIC FILTERING IN THE SIGNAL–UNION–CLUTTER MODEL

EDWARD R. DOUGHERTY* AND YIDONG CHEN†

Abstract. A basic problem of binary morphological image filtering is to remove background clutter (noise) in order to reveal a desired target (signal). The present paper discusses the manner in which filtering can be achieved using morphological granulometric filters. Logical granulometries are unions of intersections of reconstructive openings and these use shape elements to identify image components to be passed (in full), whereas others are deleted. Assuming opening structuring elements are parameterized, the task is to find parameters that result in optimal filtering. Optimization is achieved via the notion of granulometric sizing. For situations where optimization is impractical or intractable, filter design can be achieved via adaptation. Based upon correct or incorrect decisions as to whether or not to pass a component, the filter parameter vector is adapted during training in accordance with a protocol that adapts towards correct decisions. The adaptation scheme yields a Markov chain in which the parameter space is the state space of the chain. Convergence of the adaptation procedure is characterized by the stationary distribution of the parameter vector. State-probability equations are derived via the Chapman-Kolmogorov equations and these are used to describe the steady-state distribution.

Key words. Mathematical Morphology, Logical Granulometries, Size Distribution, Optimal Morphological Filtering, Adaptive Morphological Filtering, Markov Chains.

AMS(MOS) subject classifications. 60D05, 60J10, 60J27, 68U10

1. Introduction.

A basic problem of binary filtering is to remove background clutter (noise) in order to reveal a desired target (signal). In its primary form, the problem consists of a signal random set S, a noise random set N, an observed random set $S \cup N$, and a filter Ψ for which $\Psi(S \cup N)$ provides an estimate of S, the goodness of the estimate being measured by some probabilistic error criterion $\epsilon[\Psi(S \cup N), S]$. Historically, under the assumption that signal grains are probabilistically larger than noise grains, the problem has been morphologically treated as trying to form Ψ by an opening, or a union of openings. Such an approach naturally fits into Matheron's theory of granulometries [1], and these provide the context for the present paper. The earliest paper to approach optimal statistical design assumed a very simple model in which image and granulometric generators are geometrically similar, all grains are disjoint, and optimization involves the granulometric pattern spectrum [2].

If we focus our attention on the decision as to whether to pass or not pass a grain (connected component) in the observed image, then the proper approach is to consider reconstructive granulometries, rather than

* Texas Center for Applied Technology and Department of Electrical Engineering, Texas A&M University, College Station, TX 77843-3407.

† Laboratory of Cancer Genetics, National Center for Human Genome Research, National Institutes of Health, Bethesda, MD 20892.

Euclidean granulometries. An advantage of the reconstructive approach is that the mathematical complexity of the operator effect on input grains is mitigated, thereby leading to integral representation of the error, and the possibility of finding an optimal filter when the geometries of the random sets are not complicated [3]. Even for simple models, error evaluation often requires Monte Carlo integration, so that we desire adaptive techniques that lead to good suboptimal filters. The adaptive problem has been extensively studied for single-parameter and multi-parameter reconstructive granulometries [4], [5]. The present paper introduces a structural classification of reconstructive granulometries and then discusses some of the highlights of optimal and adaptive filter design for the signal-union-noise model – as known today.

2. Convex Euclidean granulometries. Slightly changing the original terminology of Matheron, we shall say that a one-parameter family of set operators $\{\Psi_t\}$, $t > 0$, is a *granulometry* if two conditions hold: (i) for any $t > 0$, Ψ_t is a τ-*opening*, meaning it is translation invariant $[\Psi_t(S + x) = \Psi_t(S) + x]$, increasing $[S_1 \subset S_2 \Rightarrow \Psi_t(S_1) \subset \Psi_t(S_2)]$, antiextensive $[\Psi_t(S) \subset S]$, and idempotent $[\Psi_t\Psi_t = \Psi_t]$; (ii) $r \geq s > 0 \Rightarrow Inv[\Psi_r] \subset Inv[\Psi_s]$, where $Inv[\Psi_t]$ is the invariant (root) class of Ψ_t. The family is a *Euclidean granulometry* if, for any $t > 0$, Ψ_t satisfies the *Euclidean property* $[\Psi_t(S) = t\Psi_1(S/t)]$. In Matheron's original formulation, he omitted the translation invariance requirement from the definition of a granulometry and placed it with the Euclidean property in defining a Euclidean granulometry. His view is consistent with the desire to keep order properties distinct from spatial properties. Our terminology is consistent with our desire to treat increasing, translation-invariant operators.

Opening by a set (structuring element) B is defined by $S \circ B = \cup\{B + x : B + x \subset S\}$. The simplest Euclidean granulometry is a parameterized class of openings, $\{S \circ tB\}$. The most general Euclidean granulometry takes the form

$$\Psi_t(S) = \bigcup_{B \in \mathbf{B}} \bigcup_{r \geq t} S \circ rB \tag{2.1}$$

where $\mathbf{B}$ is a collection of sets and is called a *generator* of the granulometry. Assuming the sets in $\mathbf{B}$ are compact (which we assume), then a key theorem [1] states that the double union of Eq. 2.1 reduces to the single union

$$\Psi_t(S) = \bigcup_{B \in \mathbf{B}} S \circ tB \tag{2.2}$$

if and only if the sets in $\mathbf{B}$ are convex, in which case we shall say the granulometry is *convex*. The single union represents a parameterized τ-opening. If $\mathbf{B}$ consists of connected sets and $S_1, S_2, \ldots$ are mutually disjoint compact sets, then

$$\Psi_t\left(\bigcup_{i=1}^{\infty} S_i\right) = \bigcup_{i=1}^{\infty} \Psi_t(S_i) \tag{2.3}$$

A granulometry that distributes over disjoint unions of compact sets will be called *distributive.*

We restrict our attention to finite-generator convex Euclidean granulometries

$$\Psi_t(S) = \bigcup_{i=1}^{n} S \circ tB_i \tag{2.4}$$

where $\mathbf{B} = \{B_1, B_2, ..., B_n\}$ is a collection of compact, convex sets and $t > 0$, and where, for $t = 0$, we define $\Psi_0(S) = S$. For any fixed t, Ψ_t is a τ-opening and $t\mathbf{B} = \{tB_1, tB_2, ..., tB_n\}$ is a *base* for Ψ_t, meaning that set $U \in Inv[\Psi_t]$ if and only if U can be represented as a union of translates of sets in $t\mathbf{B}$. According to the size and shape of the components (grains) relative to the structuring elements, some components are eliminated, whereas others are either diminished or passed in full. The larger the value of t, the more grains are sieved from the set. It is in this sense that Matheron gave the defining granulometric conditions as an axiomatic formulation of sieving.

If a binary image is composed of signal and noise components, our goal is to filter out the noise and pass the signal. As defined, a granulometry diminishes passed grains. To correct this, we use reconstruction: The *reconstructive granulometry* $\{\Lambda_t\}$ induced by the granulometry $\{\Psi_t\}$ is defined by passing in full any component not completely eliminated by $\{\Psi_t\}$ and eliminating any component eliminated by $\{\Psi_t\}$. Some grains pass the sieve; some do not. The reconstructive granulometry is a granulometry because it satisfies the two conditions of a granulometry, but it is not Euclidean. We write $\Lambda_t = \mathcal{R}(\Psi_t)$.

Associated with every Euclidean granulometry is a size distribution. For set S, define the *size distribution* $\Omega(t) = \alpha[\Psi_t(S)]$, where α denotes Lebesgue measure. $\Omega(t)$ is a decreasing function of t. If S is compact, then $\Omega(t) = 0$ for sufficiently large t. The normalized size distribution $\Phi(t) = 1 - \Omega(t)/\Omega(0)$ increases from 0 to 1 and is continuous from the left [1], so that it is a probability distribution function. $\Phi(t)$ and $\Phi'(t) = d\Phi(t)/dt$ are often called the *pattern spectrum* of S relative to the generator $\mathbf{B}$. With S being a random set, $\Phi(t)$ is a random function and its moments, which are random variables, are used for texture classification [6]. These granulometric moments have been shown to be asymptotically normal for a fairly general class of disjoint random grain models and asymptotic moments of granulometric moments have been derived [7].

3. Logical granulometries. The representation of Eq. 2.4 can be generalized by separately parameterizing each structuring element, rather than simply scaling each by a common parameter. The result is a family $\{\Psi_{\mathbf{r}}\}$ of multiparameter τ-openings of the form

$$\Psi_{\mathbf{r}}(S) = \bigcup_{k=1}^{n} S \circ B_k[\mathbf{r}_k] \tag{3.1}$$

where $\mathbf{r}_1, \mathbf{r}_2, ..., \mathbf{r}_n$ are parameter vectors governing the convex, compact structuring elements $B_1[\mathbf{r}_1], B_2[\mathbf{r}_2], ..., B_n[\mathbf{r}_n]$ composing the base of $\Psi_{\mathbf{r}}$ and $\mathbf{r} = (\mathbf{r}_1, \mathbf{r}_2, ..., \mathbf{r}_n)$. A homothetic model arises when $\mathbf{r}_k = r_k$ is a positive scalar and there exist primitive sets $B_1, B_2, ..., B_n$ such that $B_k[\mathbf{r}_k] = r_k B_k$ for $k = 1, 2, ..., n$. To keep the notion of sizing, we require (here and subsequently) the *sizing condition* that $\mathbf{r}_k \leq \mathbf{s}_k$ implies $B_k[\mathbf{r}_k] \subset B_k[\mathbf{s}_k]$ for $k = 1, 2, ..., n$, where order in the vector lattice is defined by $(t_1, t_2, ..., t_m) \leq (v_1, v_2, ..., v_m)$ if and only if $t_j \leq v_j$ for $j = 1, 2, ..., m$.

Regarding the definition of a granulometry, two points are obvious: $\Psi_{\mathbf{r}}$ is a τ-opening because any union of openings is a τ-opening and the Euclidean condition is not satisfied. More interesting is condition (ii): $r \geq s > 0 \Rightarrow Inv[\Psi_r] \subset Inv[\Psi_s]$. Since the parameter is now a vector, this condition does not apply as stated. The obvious generalization is to order the lattice composed of vectors $\mathbf{r}$ in the usual componentwise fashion and rewrite the condition as (ii') $\mathbf{r} \geq \mathbf{s} > 0 \Rightarrow Inv[\Psi_{\mathbf{r}}] \subset Inv[\Psi_{\mathbf{s}}]$. Condition (ii') states that the mapping $\mathbf{r} \rightarrow Inv[\Psi_{\mathbf{r}}]$ is order reversing and we shall say that any family $\{\Psi_{\mathbf{r}}\}$ for which it holds is *invariance ordered.* If $\Psi_{\mathbf{r}}$ is a τ-opening for any $\mathbf{r}$ and a family $\{\Psi_{\mathbf{r}}\}$ is invariance ordered, then we call $\{\Psi_{\mathbf{r}}\}$ a granulometry. The family defined by Eq. 3.1 is not necessarily a granulometry because it need not be invariance ordered. It is not generally true that $\mathbf{r} \geq \mathbf{s} > 0 \Rightarrow Inv[\Psi_{\mathbf{r}}] \subset Inv[\Psi_{\mathbf{s}}]$. As it stands, the family $\{\Psi_{\mathbf{r}}\}$ defined by Eq. 3.1 is simply a collection of τ-openings over a parameter space. However, the induced reconstructive family $\{\Lambda_r\}$ is a granulometry (since it is invariance ordered) and we call it a *disjunctive granulometry.* Moreover, reconstruction can be performed openingwise rather than on the union:

$$\Lambda_{\mathbf{r}}(S) = \mathcal{R}\left(\bigcup_{k=1}^{n} S \circ B_k[\mathbf{r}_k]\right) = \bigcup_{k=1}^{n} \mathcal{R}(S \circ B_k[\mathbf{r}_k]) \tag{3.2}$$

Although Eq. 3.1 does not generally yield a granulometry without reconstruction, a salient special case occurs when each generator set is multiplied by a separate scalar. In this case, for any n-vector $\mathbf{t} = (t_1, t_2, ..., t_n)$, $t_i > 0$, for $i = 1, 2, ..., n$, the filter takes the form

$$\Psi_{\mathbf{t}}(S) = \bigcup_{i=1}^{n} S \circ t_i B_i \tag{3.3}$$

where, to avoid useless redundancy, we suppose that no set in $\mathbf{B}$ is open with respect to another set in $\mathbf{B}$, meaning that, for $i \neq j$, $B_i \circ B_j \neq$

B_i. For any $\mathbf{t} = (t_1, t_2, ..., t_n)$ for which there exists $t_i = 0$, we define $\Psi_{\mathbf{t}}(S) = S$. $\{\Psi_{\mathbf{t}}\}$ is a *multivariate granulometry* (even without reconstruction). The corresponding multivariate size distribution for S is defined by $\Omega(\mathbf{t}) = \alpha[\Psi_{\mathbf{t}}(S)]$ and the normalized multivariate size distribution (*multivariate pattern spectrum*) by $\Phi(\mathbf{t}) = 1 - \Omega(\mathbf{t})/\alpha[S]$. The class of finite-generator convex Euclidean granulometries is a subclass of the class of n-dimensional multivariate granulometries because, for $\mathbf{t} = (t, t, ..., t)$, $\{\Psi_{\mathbf{t}}\}$ is the Euclidean granulometry $\{\Psi_t\}$ with generator $\mathbf{B}$. The multivariate pattern spectrum is an n-dimensional probability distribution function, the asymptotic normality properties of the granulometric moments for univariate granulometries extend to multivariate granulometries, and the multivariate moments can be applied to texture classification [8].

If the union of Eq. 3.1 is changed to an intersection and all conditions qualifying Eq. 3.1 hold, the result is a family of multiparameter operators of the form

$$\Psi_{\mathbf{r}}(S) = \bigcap_{k=1}^{n} S \circ B_k[\mathbf{r}_k] \tag{3.4}$$

Each operator $\Psi_{\mathbf{r}}$ is translation invariant, increasing, and antiextensive but, unless $n = 1$, $\Psi_{\mathbf{r}}$ need not be idempotent. Hence $\Psi_{\mathbf{r}}$ is not generally a τ-opening and the family $\{\Psi_{\mathbf{r}}\}$ is not a granulometry. Each induced reconstruction $\mathcal{R}(\Psi_{\mathbf{r}})$ is a τ-opening (is idempotent) but the family $\{\mathcal{R}(\Psi_{\mathbf{r}})\}$ is not a granulometry because it is not invariance ordered. However, if reconstruction is performed openingwise, then the resulting intersection of reconstructions is invariance ordered and a granulometry. The family of operators

$$\Lambda_{\mathbf{r}}(S) = \bigcap_{k=1}^{n} \mathcal{R}(S \circ B_k[\mathbf{r}_k]) \tag{3.5}$$

is called a *conjunctive granulometry*. In the conjunctive case, the equality of Eq. 3.2 is softened to an inequality: the reconstruction of the intersection is a subset of the intersection of the reconstructions.

Conjunction and disjunction can be combined to form a more general form of reconstructive granulometry:

$$\Lambda_{\mathbf{r}}(S) = \bigcup_{k=1}^{n} \bigcap_{j=1}^{m_k} \mathcal{R}(S \circ B_{k,j}[\mathbf{r}_{k,j}]) \tag{3.6}$$

If S_i is a component of S and $x_{i,k,j}$ and y_i are the logical variables determined by the truth values of the equations $S_i \circ B_{k,j}[\mathbf{r}_{k,j}] \neq \emptyset$ and $\Lambda_{\mathbf{r}}(S_i) \neq \emptyset$ [or, equivalently, $\mathcal{R}(S_i \circ B_{k,j}[\mathbf{r}_{k,j}]) = S_i$ and $\Lambda_{\mathbf{r}}(S_i) = S_i$], respectively, then y_i possesses the logical representation

$$y_i = \sum_{k=1}^{n} \prod_{j=1}^{m_k} x_{i,k,j} \tag{3.7}$$

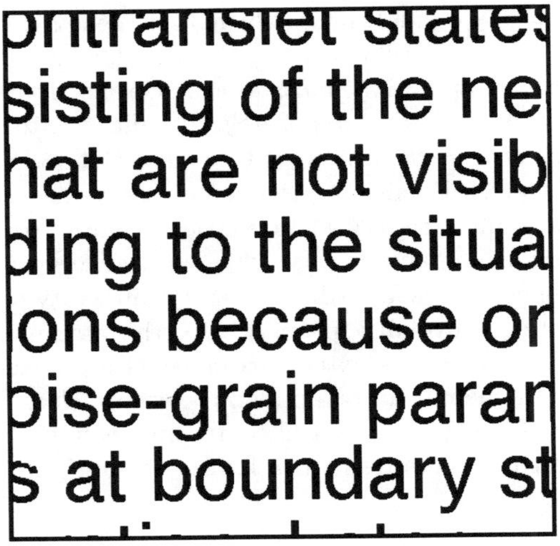

sisting of the ne
hat are not visib
ding to the situa
ons because or
oise-grain parar
s at boundary st

FIG. 1. *A sample text image.*

We call $\{\Lambda_{\mathbf{r}}\}$ a *logical granulometry.* Component S_i is passed if and only if there exists k such that, for $j = 1, 2, ..., m_k$, there exists a translate of $B_{k,j}[\mathbf{r}_{k,j}]$ that is a subset of S_i. Logical granulometries represent a class of sieving filters that locate targets among clutter based on the size and shape of the target and clutter structural components.

To illustrate the effects of conjunctive and disjunctive granulometries, defined by Eqs. 3.2 and 3.5, as well as the more general extension defined by Eq. 3.6, we apply different reconstructive openings to the sample text image shown in Fig. 1, printed with lowercase Helvetica font. The reconstructive opening by a short horizontal line, shown in the left side of Fig. 2, will find the character set {e, f, t, z} plus some other miscellaneous symbols, $\{+, -, =, \ldots\}$. Similarly, the reconstructive opening by a vertical line, shown in the right side of Fig. 2, will select {a, b, d, f, g, h, i, j, k, l, m, n, p, q, r, t, u} and some other symbols which have vertical linear components. The results of disjunctive and conjunctive granulometries are shown in the left and right sides of Fig. 3, respectively. Figure 3 clearly demonstrates that the usual disjunctive granulometry acts as a sieving process, while the conjunctive granulometry acts more like a selection process.

A more general logical granulometry

$$(3.8)\quad \Lambda(S) = [\mathcal{R}(S\circ B_{-}) \cap \mathcal{R}(S\circ B_{|})] \cup [\mathcal{R}(S\circ B_{\backslash}) \cap \mathcal{R}(S\circ B_{/})]$$

has been applied to Fig. 1 and its result is shown in Fig. 4. In Eq. 3.8, four linear structuring elements, vertical, horizontal, positive diagonal and negative diagonal, are denoted as $B_{|}$, B_{-}, $B_{/}$ and $B_{\backslash}$, respectively. Note the diagonal structuring element is at an angle slightly greater than 45° so that the character set {v, w, x, y} will be selected. We have omitted the sizing parameters from Eq. 3.8 since we assume they are known in this

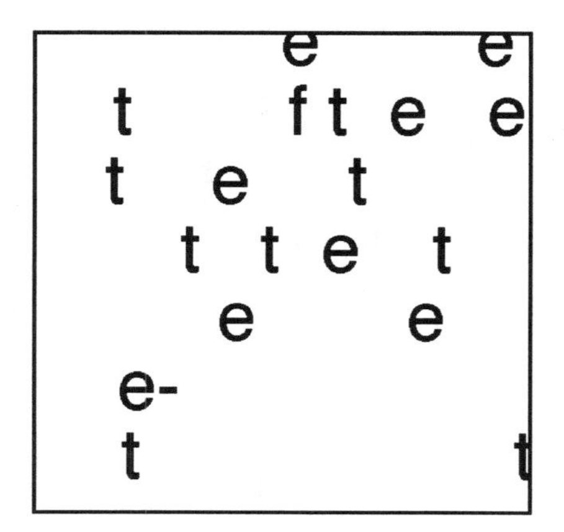
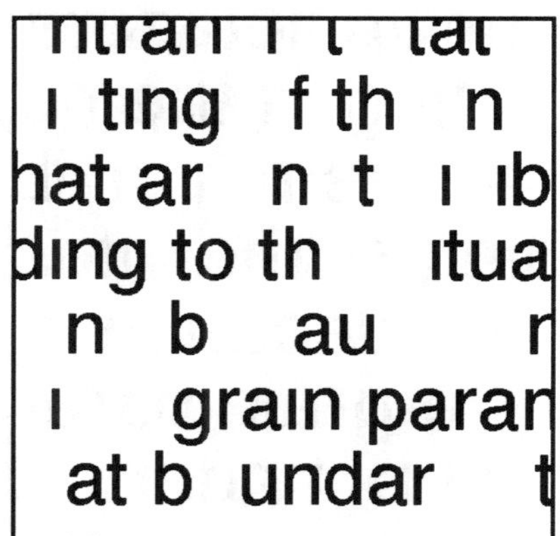

FIG. 2. *Reconstructive opening by a short horizontal line (left) and vertical line (right) of Fig. 1.*

example. In practical application, however, they need to be determined by some optimal method or adaptive approach. Similar to the result shown in Fig. 3.3 (right image), the first conjunctive granulometry of Eq. 3.8 will select only {f, t, +} and the second conjunctive granulometry of Eq. 3.8 will select only {v, w, x, y}. The final result is simply the union of these two conjunctive granulometry findings.

4. Optimal logical granulometric filters. We wish to characterize optimization of logical granulometric filters relative to a signal-union-noise random set model $S \cup N$, where

$$S = \bigcup_{i=1}^{I} C[\mathbf{s}_i] + x_i$$

$$N = \bigcup_{j=1}^{J} D[\mathbf{n}_j] + y_j \qquad (4.1)$$

I and J are random natural numbers, $\mathbf{s}_i = (s_{i1}, s_{i2}, ..., s_{im})$ and $\mathbf{n}_j = (n_{j1}, n_{j2}, ..., n_{jm})$ are independent random vectors identically distributed to random vectors $\mathbf{s}$ and $\mathbf{n}$, $C[\mathbf{s}_i]$ and $D[\mathbf{n}_j]$ are random connected, compact grains governed by $\mathbf{s}_i$ and $\mathbf{n}_j$, respectively, and x_i and y_j are random translations governing grain locations constrained by grain disjointness. Error is from signal grains erroneously removed and noise grains erroneously passed. Optimization with respect to a family $\{\Lambda_\mathbf{r}\}$ of logical granulometries is achieved by finding $\mathbf{r}$ to minimize the expected error $E[\alpha[\Lambda_\mathbf{r}(S \cup N)\Delta S]]$, where Δ denotes symmetric difference. Because $\Lambda_\mathbf{r}$ distributes over union, $\Lambda_\mathbf{r}(S \cup N) = \Lambda_\mathbf{r}(S) \cup \Lambda_\mathbf{r}(N)$.

Because $C[\mathbf{s}]$ is a random set depending on the multivariate distribution

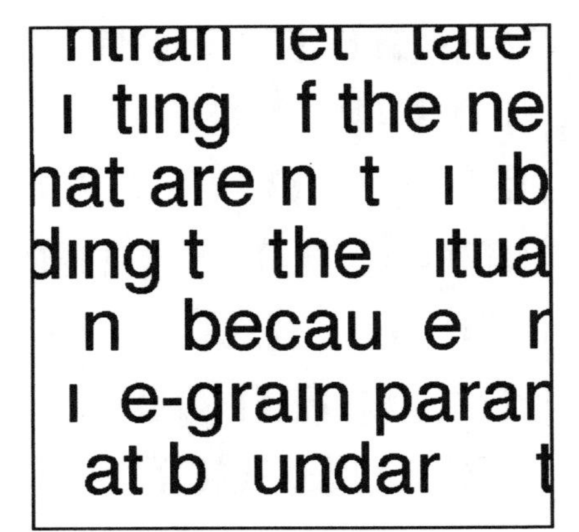

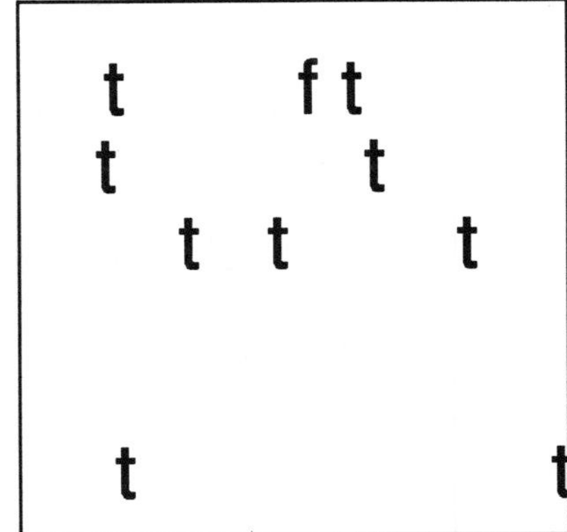

FIG. 3. *Disjunctive (left) and conjunctive (right) granulometries of Fig. 2. Notice that the image at the left side contains characters with either horizontal or vertical structuring components, while the image at the right side contains characters with both horizontal and vertical structuring components.*

of the random vector $\mathbf{s}$, the parameter set

$$\mathbf{M}_{C[\mathbf{s}]} = \{\mathbf{r} : \Lambda_{\mathbf{r}}(C[\mathbf{s}]) = C[\mathbf{s}]\} \tag{4.2}$$

is a random set composed of parameter vectors $\mathbf{r}$ for which $\Lambda_{\mathbf{r}}$ passes the random primary grain $C[\mathbf{s}]$. $\mathbf{M}_{C[\mathbf{s}]}$ and $\mathbf{M}_{D[\mathbf{n}]}$ are the regions in the parameter space where signal and noise grains, respectively, are passed. $\mathbf{M}_{C[\mathbf{s}]}$ and $\mathbf{M}_{D[\mathbf{n}]}$ are called the *signal and noise pass sets*, respectively. We often write $\mathbf{M}_{C[\mathbf{s}]}$ and $\mathbf{M}_{D[\mathbf{n}]}$ as $\mathbf{M}_S$ and $\mathbf{M}_N$, respectively. As functions of $\mathbf{s}$ and $\mathbf{n}$, $\mathbf{M}_S = \mathbf{M}_S(s_1, s_2, ..., s_m)$ and $\mathbf{M}_N = \mathbf{M}_N(n_1, n_2, ..., n_m)$. Filter error corresponding to the parameter $\mathbf{r}$ is given by

$$\begin{aligned} e[\mathbf{r}] &= E[I] \underset{\{\mathbf{s}\,:\,\mathbf{r} \notin \mathbf{M}_{C[\mathbf{s}]}\}}{\int \cdots \int} \alpha[C[\mathbf{s}]] f_S(s_1, s_2, \ldots, s_m) ds_1 ds_2 \ldots ds_m \\ &+ E[J] \underset{\{\mathbf{n}\,:\,\mathbf{r} \in \mathbf{M}_{D[\mathbf{n}]}\}}{\int \cdots \int} \alpha[D[\mathbf{n}]] f_N(n_1, n_2, \ldots, n_m) dn_1 dn_2 \ldots dn_m \end{aligned} \tag{4.3}$$

where $E[I]$ and $E[J]$ are the expected numbers of signal and noise grains, respectively, and f_S and f_N are the multivariate densities for the random vectors $\mathbf{s}$ and $\mathbf{n}$, respectively. In general, minimization of $e[\mathbf{r}]$ to find the optimal filter is mathematically prohibitive owing to the problematic nature of the domains of integration.

A special situation occurs when $\mathbf{M}_S$ and $\mathbf{M}_N$ are characterized by half-line inclusions of the component parameters: by which we mean that $\mathbf{r}$, $\mathbf{s}$, and $\mathbf{n}$ are of the same dimensions; $\mathbf{r} \in \mathbf{M}_S$ if and only if $r_1 \leq M_{S,1}(s_1)$, $r_2 \leq M_{S,2}(s_2)$, $\ldots$, $r_m \leq M_{S,m}(s_m)$; and $\mathbf{r} \in \mathbf{M}_N$ if and only if $r_1 \leq$

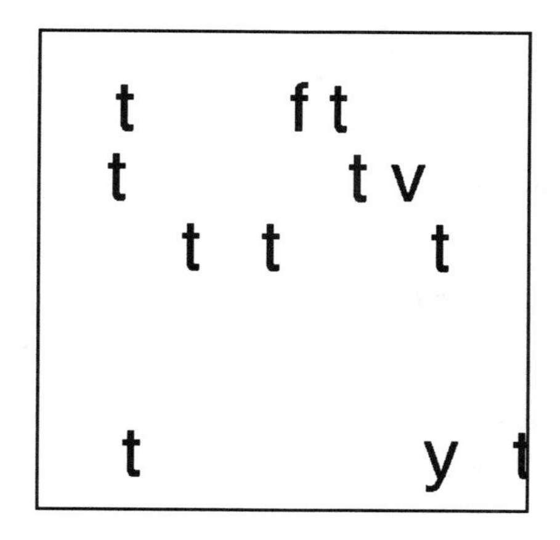

FIG. 4. *Result of logical granulometry given by Eq. 3.8 applied on Fig. 1.*

$M_{N,1}(n_1)$, $r_2 \leq M_{N,2}(n_2)$, ..., $r_m \leq M_{N,m}(n_m)$. In this case we say the model is *separable* and for $m = 2$ Eq. 4.3 reduces to

$$\begin{aligned} e[\mathbf{r}] &= E[I] \int_{-\infty}^{M_{S,1}^{-1}(r_1)} \int_{-\infty}^{M_{S,2}^{-1}(r_2)} \alpha[C[s_1, s_2]] f_S(s_1, s_2) ds_1 ds_2 \\ &+ E[J] \int_{M_{N,1}^{-1}(r_1)}^{\infty} \int_{M_{N,2}^{-1}(r_2)}^{\infty} \alpha[D[n_1, n_2]] f_N(n_1, n_2) dn_1 dn_2 \end{aligned} \tag{4.4}$$

The obvious reduction occurs for a single parameter r.

When $\{\Psi_r\}$ is a Euclidean granulometry formed according to Eq. 2.4, for any compact random set X, the $\{\Psi_r\}$–size of X is defined by $M_X = \sup\{r : \Psi_r(X) \neq \emptyset\}$. For the univariate granulometry $\{\Psi_r\} : M_S = \sup \mathbf{M}_{C[\mathbf{s}]}, M_N = \sup \mathbf{M}_{D[\mathbf{n}]}$, and the domains of integration for the first and second integrals reduce to $M_S < r$ and $M_N \geq r$, respectively. Even here the domains of integration can be very complicated and the integrals may have to be evaluated by Monte Carlo techniques.

For a mathematically straightforward example, consider the model of Eq. 4.1 and let the primary signal grain $C[\mathbf{s}]$ be a randomly rotated ellipse with random axis lengths $2u$ and $2v$, the primary noise grain $D[\mathbf{n}]$ be a randomly rotated rectangle with random sides of length $2w$ and $2z$, grain placement be constrained by disjointness, and the filter be generated by the single opening $\Psi_r(S) = S \circ rB$, where B is the unit disk. Then $M_S = \min\{u, v\}$, $M_N = \min\{w, z\}$, $\alpha[C[u, v]] = \pi uv$, and $\alpha[D[w, z]] = 4wz$. With f denoting probability densities and assuming the four sizing variables are independent,

$$e[r] = \pi E[I] \left[\int_0^r \int_0^\infty uv f(u) f(v) du dv + \int_r^\infty \int_0^r uv f(u) f(v) du dv \right]$$

$$(4.5) \qquad + \ 4E[J] \int_r^\infty \int_r^\infty wzf(w)f(z)dwdz$$

Suppose u and v are gamma distributed with parameters α and β, and w and z are exponentially distributed with parameter b. For the model parameters $\alpha = 12$, $\beta = 1$, $b = 0.2$, and $E[I] = E[J] = 20$, minimization of Eq. 4.5 occurs for $r = 5.95$ and $E[e[5.95]] = 1036.9$. Because the total expected area of the signal is $18,086.4$, the percentage of error is 5.73%.

For a conjunctive example, let the primary signal grain $C[\mathbf{s}]$ be a nonrotated cross with each bar of width 1 and random length $2w \geq 1$ and the primary noise grain $D[\mathbf{n}]$ be a nonrotated cross with each bar of width 1, one bar of length $z \geq 1$, and the other bar of length $2z$. Let grain placement be constrained by disjointness and define the filter by $\Lambda_r(S) = \mathcal{R}(S \circ rE) \cap \mathcal{R}(S \circ rF)$, where E and F are unit-length vertical and horizontal lines, respectively. Then $M_S = 2w$, $M_N = z$, $\alpha[C[w]] = 4w - 1$, $\alpha[D[z]] = 3z - 1$, and

$$(4.6) \qquad e[r] \ = \ E[I] \int_0^{r/2} (4w-1)f(w)dw + E[J] \int_r^\infty (3z-1)f(z)dz$$

Under the disjointness assumption, Eq. 4.3 is applied directly in terms of the probability models governing signal and noise. If grains are not disjoint, then segmentation is accomplished by the morphological watershed operator Ω and we need to find the probabilistic descriptions of the random outputs $\Omega(C[\mathbf{s}])$ and $\Omega(D[\mathbf{n}])$. Finding the output random-set distribution for the watershed is generally very difficult and involves statistical modeling of grain overlapping. For many granular images (gels, toner, etc.), when there is overlapping it is often very modest, with the probability of increased overlapping diminishing rapidly. The watershed produces a segmentation line between grains and its precise geometric effect depends on the random geometry of the grains and the degree of overlapping, which is itself random. Even when input grain geometry is very simple, output geometry can be very complicated (as well as dependent on overlap statistics). This problem has been addressed in the context of univariate reconstructive opening optimization [3].

5. Adaptive disjunctive granulometric filters: single parameter. Owing to the mathematical obstacles in deriving an optimal filter and the difficulty of obtaining process statistics via estimators, adaptive approaches are used to obtain a filter that is (hopefully) close to optimal. In adaptive design, a sequence of observations $T_1, T_2, T_3, \ldots$ is made and the filter is applied to each observation. Based on some criterion of goodness relating $\Lambda_{\mathbf{r}}(T_n)$ and S, the vector $\mathbf{r}$ is adapted. Adaptations yield a random-vector time series $\mathbf{r}_0, \mathbf{r}_1, \mathbf{r}_2, \mathbf{r}_3, \ldots$ resulting from transitions

$\mathbf{r}_n \to \mathbf{r}_{n+1}$, where $\mathbf{r}_n$ is the state of the process at time n and $\mathbf{r}_0$ is the initial state vector. There are various sets of conditions on the scanning process, the form of the filter, and the adaptation protocol that result in the parameter process $\mathbf{r}_n$ forming a Markov chain whose state space is the parameter space of $\mathbf{r}$. When this is so, adaptive filtering is characterized via the behavior of the Markov chain $\mathbf{r}_n$, which can be assumed to possess a single irreducible class. Convergence of the adaptive filter means existence of a steady-state distribution and characteristics of filter behavior are the stationary characteristics of the Markov chain (mean function, covariance function, etc.) in the steady state. Our adaptive estimate of the actual optimal filter depends on the steady-state distribution of $\mathbf{r}$. For instance, we might take the filter $\Lambda_{\bar{\mathbf{r}}}$, where $\bar{\mathbf{r}}$ is the mean vector of $\mathbf{r}_n$ in the steady state. The mean vector can be estimated from a single realization (from a single sequence of observations $T_1, T_2, T_3, \ldots$) owing to ergodicity in the steady state. The size of the time interval over which $\mathbf{r}_n$ needs to be averaged in the steady state for a desired degree of precision can be computed from the steady-state variance of $\mathbf{r}_n$. To date, adaptation has only been studied for disjunctive granulometric filters.

To adaptively obtain a good filter in the context of the single-parameter disjunctive granulometry

$$\Lambda_r(S) = \mathcal{R}\left(\bigcup_{i=1}^{n} S \circ rB_i\right) \tag{5.1}$$

we initialize the filter Λ_r and scan $S \cup N$ to successively encounter grains. The adaptive filter will be of the form $\Lambda_{r(n)}$, where n corresponds to the nth grain encountered. When a grain G "arrives," there are four possibilities:

$$\begin{array}{ll} \text{(a).} & G \text{ is a noise grain and } \Lambda_{r(n)}(G) = G, \\ \text{(b).} & G \text{ is a signal grain and } \Lambda_{r(n)}(G) = \emptyset, \\ \text{(c).} & G \text{ is a noise grain and } \Lambda_{r(n)}(G) = \emptyset, \\ \text{(d).} & G \text{ is a signal grain and } \Lambda_{r(n)}(G) = G. \end{array} \tag{5.2}$$

In the latter two cases, the filter has acted as desired; in either of the first two it has not. Consequently, we employ the following adaptation rule:

$$\begin{array}{lll} \text{i.} & r \to r+1 & \text{if condition (a) occurs,} \\ \text{ii.} & r \to r-1 & \text{if condition (b) occurs,} \\ \text{iii.} & r \to r & \text{if conditions (c) or (d) occur.} \end{array} \tag{5.3}$$

Each arriving grain determines a step and we treat $r(n)$ as the state of the system at step n. Since all grain sizes are independent and there is no grain overlapping, $r(n)$ determines a discrete state-space Markov chain over a discrete parameter space. Three positive stationary transition probabilities are associated with each state r:

$$(5.4) \qquad \begin{array}{ll} \text{i.} & p_{r,r+1} = P(N)P(\Lambda_r(Y) = Y) \\ \text{ii.} & p_{r,r-1} = P(S)P(\Lambda_r(X) = \emptyset) \\ \text{iii.} & p_{r,r} = P(S)P(\Lambda_r(X) = X) + P(N)P(\Lambda_r(Y) = \emptyset) \end{array}$$

where X and Y are the primary signal and noise grains, respectively, and $P(S)$ and $P(N)$ are the probabilities of a signal and noise grain arriving, respectively. $P(S)$ and $P(N)$ depend on the protocol for selecting grains in the images. A number of these, together with the corresponding probabilities are discussed in Ref. [5]: weighted random point selection, where points in the image frame are randomly selected until a point in $S \cup N$ is chosen and the grain containing the point is considered; unweighted random point selection, where each grain in $S \cup N$ is labeled and labels are uniformly randomly selected with replacement; horizontal scanning, where the image is horizontally scanned at randomly chosen points along the side of the image frame, a grain is encountered if and only if it is cut by the scan line, and the scan line traverses the entire width of the image frame.

The transition probabilities can be expressed in terms of granulometric measure:

$$(5.5) \qquad \begin{array}{ll} \text{i.} & p_{r,r+1} = P(N)P(M_N \geq r) \\ \text{ii.} & p_{r,r-1} = P(S)P(M_S < r) \\ \text{iii.} & p_{r,r} = P(S)P(M_S \geq r) + P(N)P(M_N < r) \end{array}$$

For clarity, we develop the theory with r a nonnegative integer and transitions of plus or minus one; in fact, r need not be an integer and transitions could be of the form $r \to r + m$ and $r \to r - m$, where m is some positive constant.

Equivalence classes of the Markov chain are determined by the distributions of M_S and M_N. To avoid trivial anomalies, we assume distribution supports are intervals with endpoints $a_S < b_S$ and $a_N < b_N$, where $0 \leq a_S$, $0 \leq a_N$, and it may be that $b_S = \infty$ or $b_N = \infty$. We assume $a_N \leq a_S < b_N \leq b_S$. Nonnull intersection of the supports insures that the adaptive filter does not trivially converge to an optimal filter that totally restores S. There are four cases regarding state communication.

Suppose $a_S < 1$ and $b_N = \infty$: then the Markov chain is irreducible since all states communicate (each state can be reached from every other state in a finite number of steps). Suppose $1 \leq a_S$ and $b_N = \infty$: then, for each state $r \leq a_S$, r is accessible from state s if $s < r$, but s is not accessible from r; on the other hand, all states $r \geq a_S$ communicate and form a single equivalence class. Suppose $a_S < 1$ and $b_N < \infty$: then, for each state $r \geq b_N$, r is accessible from state s if $s > r$, but s is not accessible from r; on the other hand, all states $r \leq b_N$ communicate and form a single equivalence class. Suppose $1 \leq a_S < b_N < \infty$: then states below

a_S are accessible from states below themselves, but not conversely, states above b_N are accessible from states above themselves, but not conversely, and all states r such that $a_S \leq r \leq b_N$ communicate and form a single equivalence class. In sum, the states between a_S and b_N form an irreducible equivalence class $\mathcal{C}$ of the state space and each state outside $\mathcal{C}$ is transient. With certainty, the chain will eventually enter $\mathcal{C}$ and once inside $\mathcal{C}$ will not leave. Thus, we focus our attention on $\mathcal{C}$. Within $\mathcal{C}$, the chain is irreducible and aperiodic. If it is also positive recurrent, then it will be ergodic and possess a stationary (steady-state) distribution.

Existence of a steady-state distribution is proven in Ref. [4]. Let $p_r(n)$ be the probability that the system is in state r at step n, $\lambda_k = P(N)P(M_N \geq k)$, and $\mu_k = P(S)P(M_S < k)$. The Chapman-Kolmogorov equation yields

$$\begin{aligned} p_r(n+1) - p_r(n) &= P(N)P(M_N \geq r-1)p_{r-1}(n) \\ &\quad + P(S)P(M_S < r+1)p_{r+1}(n) \\ &\quad - (P(S)P(M_S < r) + P(N)P(M_N \geq r))p_r(n) \\ &= \lambda_{r-1}p_{r-1}(n) + \mu_{r+1}p_{r+1}(n) - (\lambda_r + \mu_r)p_r(n) \end{aligned} \tag{5.6}$$

for $r \geq 1$. For $r = 0$, $p_{-1}(n) = 0$ and $\mu_0 = 0$ yield the initial state equation

$$p_0(n+1) - p_0(n) = \mu_1 p_1(n) - \lambda_0 p_0(n) \tag{5.7}$$

In the steady state, these equations form the system

$$\begin{cases} 0 = -(\lambda_k + \mu_k)p_k + \mu_{k+1}p_{k+1} + \lambda_{k-1}p_{k-1}, & k \geq 1 \\ 0 = -\lambda_0 p_0 + \mu_1 p_1 \end{cases} \tag{5.8}$$

and the solution

$$\begin{cases} p_1 = \frac{\lambda_0}{\mu_1} p_0 \\ p_r = p_0 \prod_{k=1}^{r} \frac{\lambda_{k-1}}{\mu_k}, & r \geq 1 \end{cases} \tag{5.9}$$

where

$$p_0 = \frac{1}{1 + \sum_{r=1}^{\infty} \prod_{k=1}^{r} \frac{\lambda_{k-1}}{\mu_k}} \tag{5.10}$$

Given convergence of the adaptive filter, in the sense that it reaches a steady state, key characteristics are its steady-state mean and variance. In the steady state, r is a random variable. Its mean and variance are

$$\mu_r = \sum_{r=0}^{\infty} r p_r = \sum_{r=1}^{\infty} \left[r p_0 \prod_{k=1}^{r} \frac{\lambda_{k-1}}{\mu_k} \right] \tag{5.11}$$

$$\sigma_r^2 = \sum_{r=1}^{\infty} \left[r^2 p_0 \prod_{k=1}^{r} \frac{\lambda_{k-1}}{\mu_k} \right] - \mu_r^2 \tag{5.12}$$

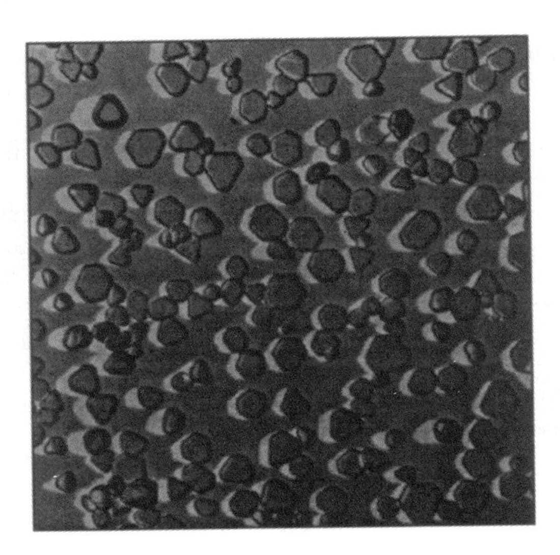

FIG. 5. *T-grain crystal image.*

Both mean and variance exist. In fact, r has finite moments of all orders.

From the standpoint of optimization, the key error measure is $E[e[r]]$. If the random primitive grains X and Y are governed by the random variables Z and W, respectively, and both M_X and M_Y are strictly increasing functions of Z and W, respectively, then

$$
\begin{aligned}
e[r] &= E[I]\int_0^{M_X^{-1}(r)} \alpha[X](z)f_Z(z)dz \\
&\quad + E[J]\int_{M_Y^{-1}(r)}^{\infty} \alpha[Y](w)f_W(w)dw
\end{aligned}
\tag{5.13}
$$

where $f_Z(z)$ and $f_W(w)$ are the probability densities for Z and W, respectively. Averaging over r in the steady state yields

$$
E[e[r]] = \sum_{r=0}^{\infty} e[r]p_r \tag{5.14}
$$

The optimal value of r, say $\hat{r}$, is found by minimizing Eq. 5.13. If, as is sometimes done, an arbitrary value of r is chosen in the steady state, then the cost of adaptivity can be measured by $E[e[r]] - e[\hat{r}]$, which must be nonnegative. If $\bar{r}$, the mean of r in the steady state, is used (as we do here), then the cost of adaptivity is $e[\bar{r}] - e[\hat{r}]$.

For an example in which there is grain overlap, consider the electron micrograph of silver-halide T-grain crystals in emulsion shown in Fig. 5. Automated crystal analysis involves removal of degenerate grains, thereby leaving well-formed crystals for measurement. A gray-scale watershed is

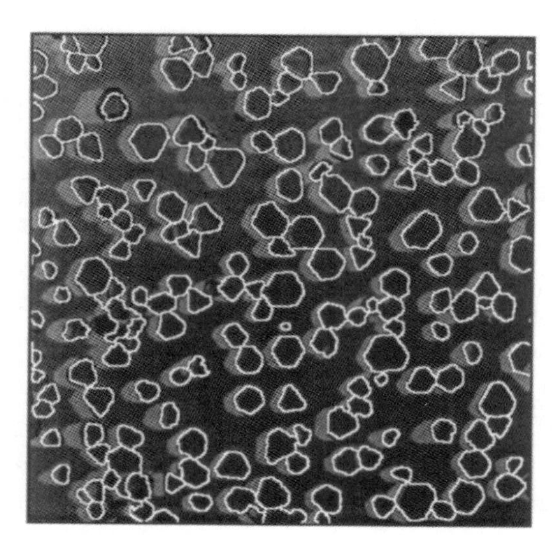

FIG. 6. *Edges from watershed segmentation superimposed on Fig. 5.*

applied to find an edge image, which is itself superimposed over the original micrograph image in Fig. 6. Each boundary formed by the watershed is filled and the crystal interiors are labeled either black (signal) or gray (noise). The resulting segmented image is shown in Fig. 7. We use an adaptive four-directional linear τ-opening, whose structuring elements are unit-length vertical, horizontal, positive-diagonal, and negative-diagonal lines. The empirical distributions of granulometric sizes of signal and noise grains are shown in Fig. 8, along with the empirical steady-state distribution of the adaptive parameter. The empirical mean and variance in the steady state are 7.781 and 0.58, respectively. Finally, choosing $t = 8$ and applying the corresponding τ-opening to the grain image of Fig. 7 yields the filtered image of Fig. 9.

6. Comparison of optimal and adaptive filters in a homothetic model. Closed-form expression of $E[e[r]]$ is rarely possible but it can be achieved when signal and noise take the forms

$$S = \bigcup_{i=1}^{I} s_i B + x_i$$

$$N = \bigcup_{j=1}^{J} n_j B + y_j \tag{6.1}$$

where the sizing parameters s_i and n_j come from known sizing distributions Π_S and Π_N, respectively, and all sizings are independent, grains are nonoverlapping, and sizing distributions Π_S and Π_N are both uniform. For unweighted random point selection and for general Π_S and Π_N (not necessarily uniform) having densities f_S and f_N, respectively, Eq. 5.13 reduces

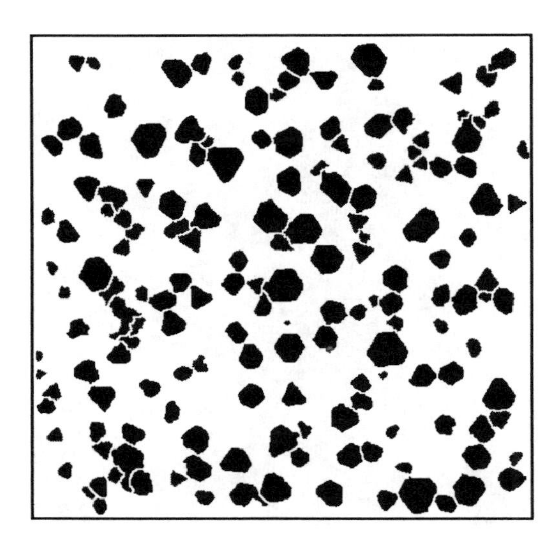

FIG. 7. *Grain Segmentation.*

to

$$(6.2)\quad e[r] = E[T]\left[\frac{P(S)}{\mu_S^{(2)}}\int_0^r z^2 f_S(z)dz + \frac{P(N)}{\mu_N^{(2)}}\int_r^\infty w^2 f_N(w)dw\right]$$

where $\mu_S^{(2)}$ and $\mu_N^{(2)}$ are the uncentered second moments of Π_S and Π_N, respectively, and where we write f_S and f_N in place of f_Z and f_W, respectively.

Let Π_S and Π_N be uniform densities over $[a,b]$ and $[c,d]$, respectively, where $a < c < b < d$ and where, for convenience, a, b, c, and d are integers, and let $m_S = (d-c)^{-1}$ and $m_N = (b-a)^{-1}$. The effective state space for the parameter r is $[c,b]$ because all other states are transient and r will move into $[c,b]$ and remain there. As shown in Ref. [4], the steady-state probabilities, mean size, size variance, and steady-state filter error are given by

$$(6.3)\quad \begin{cases} p_c = \left(\dfrac{m_S P(S)}{m_S P(S) + m_N P(N)}\right)^{b-c} \\ p_{c+i} = p_c\dbinom{b-c}{i}\left[\dfrac{m_N P(N)}{m_S P(S)}\right]^i, \quad 1 \le i \le b-c \end{cases}$$

$$(6.4)\quad \mu_r = \frac{c \times m_S P(S) + b \times m_N P(N)}{m_S P(S) + m_N P(N)}$$

$$(6.5)\quad \sigma_r^2 = \frac{m_S m_N P(S) P(N)(b-c)}{(m_S P(S) + m_N P(N))^2}$$

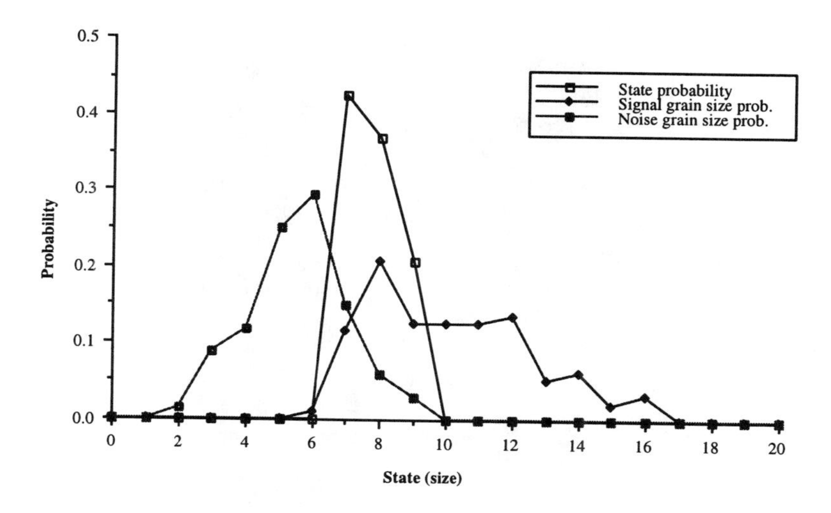

FIG. 8. *Empirical size distribution of silver grains.*

$$e[r] = E[T]\left[P(S)\frac{r^3 - c^3}{d^3 - c^3} + P(N)\frac{b^3 - r^3}{b^3 - a^3}\right] \tag{6.6}$$

Define the discriminant

$$D = \frac{P(S)}{d^3 - c^3} - \frac{P(N)}{b^3 - a^3} \tag{6.7}$$

Minimization yields the optimal solution for various conditions:

$$\min_r\{e[r]\} = \begin{cases} E[T]P(S)\dfrac{b^3 - c^3}{d^3 - c^3}, & \text{at } r = b, \quad \text{if } D < 0 \\ E[T]P(N)\dfrac{b^3 - c^3}{b^3 - a^3}, & \text{at } r = c, \quad \text{if } D > 0 \\ E[T]P(N)\dfrac{b^3 - c^3}{b^3 - a^3}, & \text{at } r \in [c, b], \quad \text{if } D = 0 \end{cases} \tag{6.8}$$

For $D < 0$, the optimum occurs at the upper endpoint b of the interval over which the states can vary; for $D > 0$, the optimum occurs at the lower endpoint c; for $D = 0$, all states in $[c, b]$ are equivalent, and hence optimal. For $D < 0$, the mean size tends towards b; for $D > 0$, the mean tends towards c; for $D = 0$, the mean is $(c + b)/2$.

The steady-state expected error is given by

$$E[e(r)] = \sum_{r=c}^{b} E[T]\left[P(S)\frac{r^3 - c^3}{d^3 - c^3} + P(N)\frac{b^3 - r^3}{b^3 - a^3}\right] p_r \tag{6.9}$$

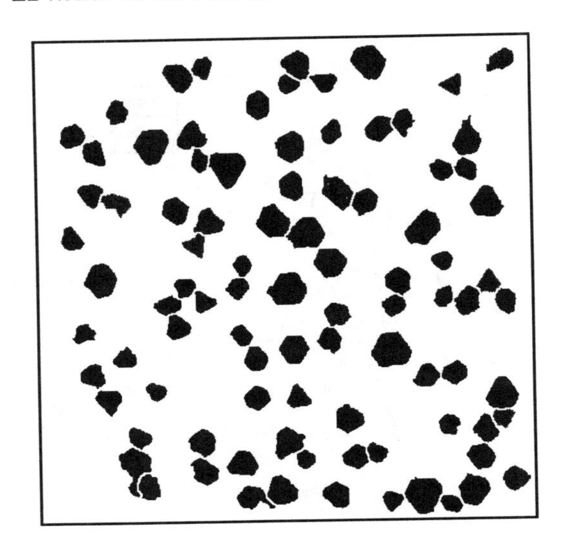

FIG. 9. *Filtered grain image at $t = 8$.*

and, as shown in Ref. [4],

$$(6.10)\quad \begin{aligned} &\min\left\{\frac{P(S)}{d^3-c^3}, \frac{P(N)}{b^3-a^3}\right\} E[T](b^3-c^3) \\ &\leq E[e[r]] \leq \max\left\{\frac{P(S)}{d^3-c^3}, \frac{P(N)}{b^3-a^3}\right\} E[T](b^3-c^3) \end{aligned}$$

The optimal filter has an error bounded by the expected steady-state error for the adaptive filter. For the special case $D = 0$, the two errors must agree because all filters whose parameters lie in the single recurrent class of the Markov chain have equal error.

7. Adaptation in a multiparameter disjunctive model. Adaptation protocols can vary extensively for logical granulometries. A number of them have been described and studied for disjunctive granulometries [5]. We restrict ourselves to a two-parameter τ-opening $\Psi_{\mathbf{r}}$, $\mathbf{r} = (r_1, r_2)$, and its reconstruction $\Lambda_{\mathbf{r}}$. The single-parameter grain-arrival possibilities of Eq. 4.5 apply again, with scalar r being replaced by the vector $\mathbf{r}$. We employ the following generic adaptation scheme:

$$(7.1)\quad \begin{array}{lll} \text{i.} & r_1 \to r_1 + 1 \text{ and/or } r_2 \to r_2 + 1 & \text{if condition (a) occurs,} \\ \text{ii.} & r_1 \to r_1 - 1 \text{ and/or } r_2 \to r_2 - 1 & \text{if condition (b) occurs,} \\ \text{iii.} & r_1 \to r_1 \text{ and } r_2 \to r_2 & \text{if conditions (c) or (d) occur.} \end{array}$$

Assuming grain arrivals and primary-grain realizations are independent, (r_1, r_2) determines a 2D discrete-state-space Markov chain. The protocol is generic because a protocol depends on interpretation of both *and/or's*, which depends on the form of the τ-opening.

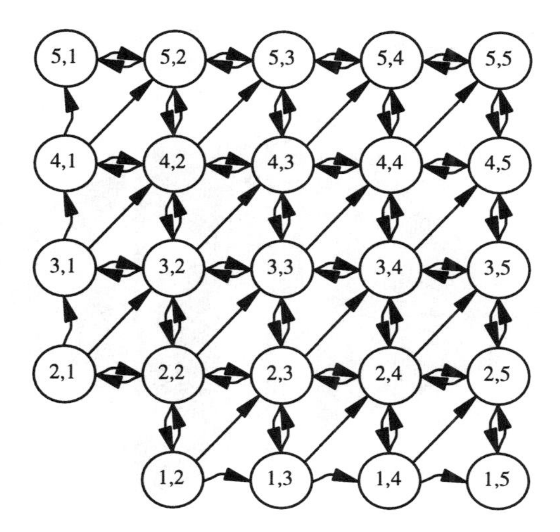

FIG. 10. *State transition diagram of Type-$[II, 1]$.*

Reference [5] discusses four two-parameter protocols. For the type-$[I, 0]$ model, $\Psi_{\mathbf{r}}$ is an opening by one two-parameter structuring element, there is no information as to which parameter causes nonfitting, and adaptation proceeds solely on the basis of whether or not a translate of the structuring element fits within the grain. For the type-$[I, 1]$ model, $\Psi_{\mathbf{r}}$ is again a two-parameter opening, but this time, if a signal grain is erroneously not passed, it is known which parameter has caused the erroneous decision. For the type-$[II, 0]$ model, $\Psi_{\mathbf{r}}$ is a τ-opening formed as a union of two openings, each by a structuring element $B_i[r_i]$ depending on a single parameter, and there is no information about which structuring element causes nonfitting. For the type-$[II, 1]$ model, $\Psi_{\mathbf{r}}$ is a τ-opening formed as a union of two openings and fitting information regarding each structuring element is fed back. Here we only consider the type-$[II, 1]$ model.

For the type-$[II, 1]$ model, if a signal grain is erroneously not passed, neither structuring element fits and hence there must be a randomization regarding the choice of parameter to decrement. A general description of the type-$[II, 1]$ model requires a general description of the transition probabilities. This can be done; however, the resulting equations are very messy. Thus, we provide the equations when the model is separable, in which case transition probabilities can be expressed in terms of granulometric measure:

$$\begin{aligned}
&\text{i.} \quad p_{(r_1,r_2),(r_1+1,r_2)} = P(N)P((M_N^{B_1} \geq r_1) \cap (M_N^{B_2} < r_2)) \\
&\text{ii.} \quad p_{(r_1,r_2),(r_1,r_2+1)} = P(N)P((M_N^{B_1} < r_1) \cap (M_N^{B_2} \geq r_2)) \\
&\text{iii.} \quad p_{(r_1,r_2),(r_1+1,r_2+1)} = P(N)P((M_N^{B_1} \geq r_1) \cap (M_N^{B_2} \geq r_2)) \\
(7.2) \quad &\text{iv.} \quad p_{(r_1,r_2),(r_1-1,r_2)} = \frac{1}{2}P(S)P((M_S^{B_1} < r_1) \cap (M_S^{B_2} < r_2))
\end{aligned}$$

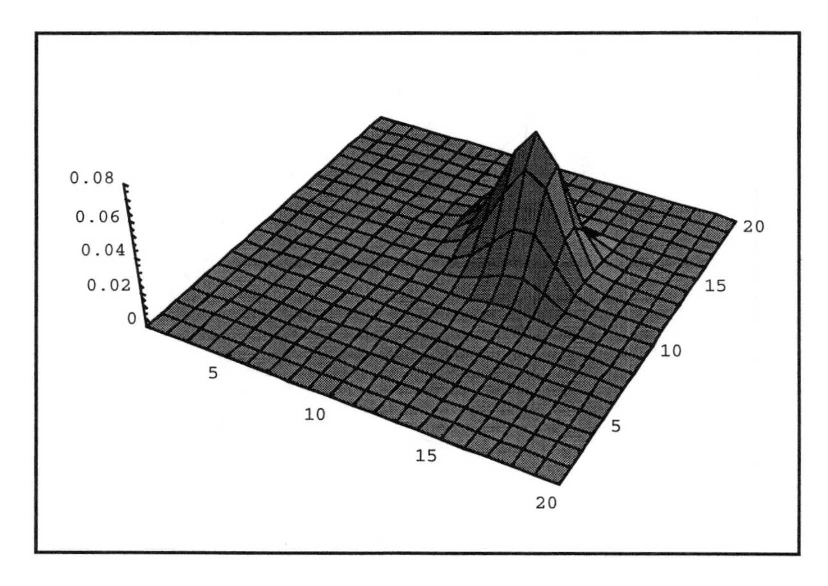

FIG. 11. *Numerical steady-state solution for Type-[II,1].*

$$\text{v.}\quad p_{(r_1,r_2),(r_1,r_2-1)} = \frac{1}{2}P(S)P((M_S^{B_1} < r_1) \cap (M_S^{B_2} < r_2))$$

$$\text{vi.}\quad p_{(r_1,r_2),(r_1,r_2)} = P(S)P((M_S^{B_1} \geq r_1) \cup (M_S^{B_2} \geq r_2)) + P(N)P((M_N^{B_1} < r_1) \cap (M_N^{B_2} < r_2))$$

A typical transition diagram is shown in Fig. 10. The state space may or may not be infinite. To avoid a trivial optimal solution we assume nonnull intersection between signal and noise pass sets. Figure 10 omits transient states and shows only the equivalence class $\mathcal{C}$ of communicating states. With certainty, the chain will enter $\mathcal{C}$ and, once inside, will not leave. Within $\mathcal{C}$ the chain is aperiodic. If it is also positive recurrent, it will be ergodic and possess a stationary distribution.

Let λ_{1,r_1,r_2}, λ_{2,r_1,r_2}, λ_{3,r_1,r_2}, μ_{1,r_1,r_2}, and μ_{2,r_1,r_2} denote the transition probabilities (i) through (v) of Eq. 7.2, respectively. The Chapman-Kolmogorov equation yields

$$\begin{aligned} &p_{r_1,r_2}(n+1) - p_{r_1,r_2}(n) \\ &\quad = \lambda_{1,r_1-1,r_2}p_{r_1-1,r_2}(n) + \lambda_{2,r_1,r_2-1}p_{r_1,r_2-1}(n) \\ &\quad + \lambda_{3,r_1-1,r_2-1}p_{r_1-1,r_2-1}(n) \\ &\quad + \mu_{1,r_1+1,r_2}p_{r_1+1,r_2}(n) + \mu_{2,r_1,r_2+1}p_{r_1,r_2+1}(n) \\ &\quad - (\lambda_{1,r_1,r_2} + \lambda_{2,r_1,r_2} + \lambda_{3,r_1,r_2} + \mu_{1,r_1,r_2} + \mu_{2,r_1,r_2})p_{r_1,r_2}(n) \end{aligned} \tag{7.3}$$

Left and bottom boundary conditions are given by three cases: $r_1 = r_2 = 2$; $r_1 = 1$ and $r_2 \geq 2$; and $r_2 = 1$ and $r_1 \geq 2$. Results of the Chapman-Kolmogorov equation for all three cases are given in Ref. [5]. Whether or

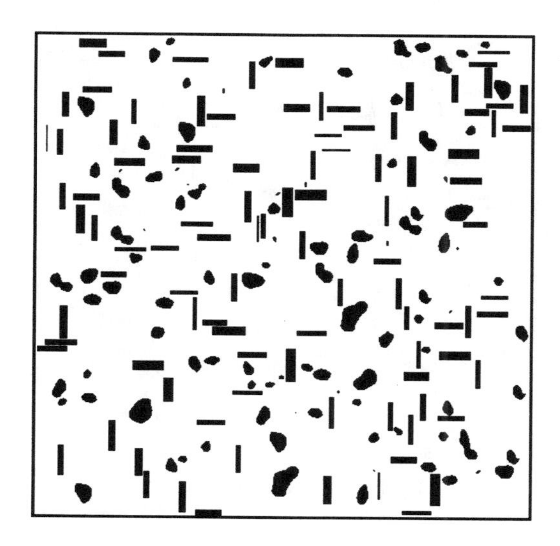

FIG. 12. *Realization of signal grains and irregular shaped noise grains.*

not there exist right and upper boundary states depends on the signal and noise parameter distributions. These can be treated similarly.

Convergence for two-parameter adaptive systems is characterized via the existence of steady-state probability distributions. If it exists, the steady-state distribution is defined by the limiting probabilities

$$p_{r_1,r_2} = \lim_{n\to\infty} p_{r_1,r_2}(n). \tag{7.4}$$

Due to complicated boundary-state conditions, it is extremely difficult to obtain general solutions for the systems discussed, although we are assured of the existence of a steady-state solution when the state space is finite. Setting $p(n+1) - p(n) = 0$ does not appear to work and this contention is supported by showing an analogy between adaptive τ-opening systems and Markovian queue networking systems [5].

We proceed numerically, assuming uniformly distributed signal and noise parameters over the square regions from $[7, 7]$ to $[16, 16]$ and from $[5, 5]$ to $[14, 14]$, respectively. We assume the arrival probabilities of signal and noise grains to be $P(S) = 2/3$ and $P(N) = 1/3$. The numerically computed steady-state distribution for the type-$[II, 1]$ model is shown in Fig. 11. The optimal filter occurs for either $\mathbf{r} = (7, 15)$ or $(15, 7)$ with minimum error $0.293E[T]$, T being total image area. Owing to directional symmetry throughout the model, existence of two optimal parameter vectors should be expected. The numerically approximated expected adaptive filter error is $E_{ss}[e[r]] = 0.327E[T]$. For the numerical steady-state distribution, the center of mass is at $(12.86, 12.86)$. At first glance, this appears to differ markedly from the two optima; however, $e[(13, 13)] = 0.327E[T]$, which is close to the optimal value (and, to three decimal places, agrees with the

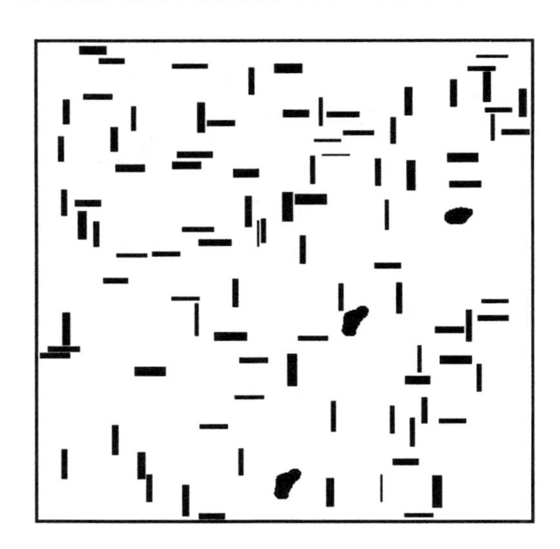

FIG. 13. *Processed image of Fig. 12 by type-$[II,1]$ filter.*

expected filter error in the steady state). Due to strongly overlapping uniform signal and noise distributions, there is a large region in the parameter plane for which $e[\mathbf{r}]$ is fairly stable.

To apply the type-$[II,1]$ adaptive model, consider the realization shown in Fig. 12. Signal grains are horizontal and vertical rectangles. The sides are normally distributed: the long with mean 30, variance 9; the short with mean 8, variance 4. The noise consists of irregularly shaped grains with comparable areas to the signal grains. There is no grain overlap. A type-$[II,1]$ adaptive reconstructive τ-opening with vertical and horizontal linear structuring elements has been applied. The steady-state distribution (50 realizations) has center of mass $(25.99, 25.45)$. Figure 13 shows the reconstructed τ-opened realization of Fig. 12 using vertical and horizontal linear structuring elements of length 26.

8. Conclusion. Logical granulometries provide a morphological means to distinguish signal from clutter in the context of Matheron's theory of granulometries. Based on signal and clutter random-set models, optimal granulometries can be determined via granulometric size in the disjoint-grain model. In situations where the error integral cannot be readily evaluated, the underlying models cannot be estimated, or the segmentation is so complicated that the output model from the segmentation cannot be described, an adaptive approach can be employed. The key to adaptation is representation of the steady-state distribution via the Chapman-Kolmogorov equation. Rarely can the state-probability equation be analytically solved, so that Monte Carlo methods need typically be employed.

REFERENCES

[1] G. Matheron, *Random Sets and Integral Geometry*, John Wiley, New York, 1975.

[2] E.R. Dougherty, R.M. Haralick, Y. Chen, C. Agerskov, U. Jacobi, and P.H. Sloth, *Estimation of optimal τ-opening parameters based on independent observation of signal and noise pattern spectra*, Signal Processing, 29 (1992) pp. 265–281.

[3] E.R. Dougherty and C. Cuciurean-Zapan, *Optimal reconstructive τ-openings for disjoint and statistically modeled nondisjoint grains*, Signal Processing, 56 (1997), pp. 45–58.

[4] Y. Chen and E.R. Dougherty, *Adaptive reconstructive τ-openings: Convergence and the steady-state distribution*, Electronic Imaging, 5 (1996), pp. 266–282.

[5] Y. Chen and E.R. Dougherty, *Markovian analysis of adaptive reconstructive multiparameter τ-openings*, Journal of Mathematical Imaging and Vision (submitted).

[6] E.R. Dougherty, J.T. Newell, and J.B. Pelz, *Morphological texture-based maximum-likelihood pixel classification based on local granulometric moments*, Pattern Recognition, 25 (1992), pp. 1181–1198.

[7] E.R. Dougherty and F. Sand, *Representation of linear granulometric moments for deterministic and random binary Euclidean images*, Journal of Visual Communication and Image Representation, 6 (1995), pp. 69–79.

[8] S. Batman and E.R. Dougherty, *Size distributions for multivariate morphological granulometries: Texture classification and statistical properties*, Optical Engineering, 36 (1997), pp. 1518–1529.

ON OPTIMAL FILTERING OF MORPHOLOGICALLY SMOOTH DISCRETE RANDOM SETS AND RELATED OPEN PROBLEMS

NIKOLAOS D. SIDIROPOULOS*

Abstract. Recently, it has been shown that morphological openings and closings can be viewed as consistent MAP estimators of (morphologically) smooth random sets immersed in clutter, or suffering from random dropouts [1]–[3]. These results hold for one-sided (union *or* intersection) noise. In the case of two-sided (union *and* intersection) noise we now know that this is no longer the case: in this more general setting, as it turns out, optimal estimators are not increasing operators [4]. Then, how may one *efficiently* compute an optimal random set estimate in this more general setting? For 1-D and certain restricted 2-D random set models the answer is provided by the Viterbi algorithm [5]. In the general case of 2-D random set models in two-sided noise the answer is unknown, and the task of finding it constitutes a challenging research problem.

Key words. Random Sets, Morphological Filtering, Opening, Closing, Dynamic Programming, Viterbi Algorithm.

AMS(MOS) subject classifications. 60D05, 49L20

1. Introduction.

The celebrated Choquet - Kendall - Matheron theorem [6]–[9] for Random Closed Sets (RACS) states that a RACS X is completely characterized by its *capacity functional*, i.e., the collection of hitting probabilities over a sufficiently rich family of so-called *test sets*. RACS theory is a mature branch of theoretical and applied probability, whose scope is the study of set-valued random variables.

This paper is concerned with some fundamental optimal filtering problems for uniformly bounded discrete random set signals embedded in noise. A uniformly bounded Discrete Random Set (DRS, for short) X may be thought of as a (finite) sampled version of an underlying RACS; cf. [10] for a rigorous analysis of a suitable sampling process. DRS's can be viewed as finite-alphabet random variables, taking values in a complete lattice with a finite least upper bound (usually, the power set, $\mathcal{P}(B)$, of some finite observation window $B \subset \mathbb{Z}^2$). Thus, the only difference with ordinary finite-alphabet random variables is that the DRS alphabet naturally possesses only a partial order relation, instead of a total order relation. Now, in the case of DRS's, there exist simple closed-form expressions that relate the capacity functional to the basic probability assignment $P_X(\cdot)$ over $\mathcal{P}(B)$. Let $p_X(\cdot)$ denote the restriction of $P_X(\cdot)$ to the atoms, i.e., the elements of $\mathcal{P}(B)$. This is the *probability mass functional* of the DRS X. The capacity functional is defined as

$$T_X(A) \triangleq P_X(X \cap A \neq \emptyset).$$

* Department of Electrical Engineering, University of Virginia, Charlottesville VA 22903. This work was supported in part by NSF CDR 8803012.

Obviously,

$$T_X(A) = 1 - \sum_{S \subseteq A^c} p_X(S),$$

where $A^c = B \setminus A$, from which it follows (cf. [11]–[15]: this is *Möbius inversion* for Boolean Algebras)

$$p_X(S) = \sum_{A \subseteq S} (-1)^{|A|}(1 - T_X(S^c \cup A)) = \sum_{A \subseteq S} (-1)^{|S|-|A|}(1 - T_X(A^c)).$$

Obviously, either one of p_X, T_X (and, in fact, other functionals as well) may be used to specify the DRS X.

This is a nice result, yet it appears that these formulas are intractable. However, as I found out during the course of this workshop, thanks to Drs. Jaffray, Nguyen, Black, and others, there exists a fast Möbius Transform [16] that allows one to move relatively efficiently between the two specifications, p_X, T_X. The interesting observation, then, is that for a certain key class of DRS's, namely discrete Boolean models for which the capacity functional T_X assumes a simple, tractable closed form, one may proceed with Bayesian inference (e.g., hypothesis testing between two Boolean DRS models) without being faced with intractability problems. This is certainly not the case for general Boolean RACS models.

Random set theory is closely related to mathematical morphology. The fundamental elements of mathematical morphology have been developed by Matheron [8], [9], Serra [17], [18], and their collaborators. Morphological filtering is one of the most popular and successful branches of this theory [19].

In this paper, we begin by reviewing some of our earlier work on statistical optimality of morphological openings and closings in the case of one-sided noise. In the case of two-sided noise, the corresponding MAP filtering problem is a hard nonlinear optimization problem. We show that for 1-D and certain restricted 2-D random set models the answer is provided by the Viterbi algorithm [5]. We also discuss future challenges, and point to some potentially fruitful approaches to solving related open problems.

Dougherty et al. [20]–[25], Schonfeld et al. [26]–[28], and Goutsias [29] have worked on several related problems, using different measures of optimality and/or families of filters.

2. Statistical optimality of openings and closings. Some key results of [3] provide our starting point.

THEOREM 2.1. (*MAP Optimality*) We observe $\mathbf{Y}^{(M)} = [Y_1, Y_2, ..., Y_M]$, where $Y_i = X \cup N_i$, $\{N_i\}_{i=1}^{M}$ is an independent *but not necessarily identically distributed* sequence of noise DRS's, which is independent of X, and each N_i is an *otherwise arbitrary* DRS taking values in some arbitrary collection, $\Psi_i(B) \subseteq \mathcal{P}(B)$, of subsets of the observation window B. Let us further assume that X is uniformly distributed over a collection, $\Phi(B) \subseteq \mathcal{P}(B)$, of all subsets K of B which are spanned by unions of translates of a family of structuring elements, W_l, $l = 1, 2, ..., L$, i.e., those $K \subseteq B$ which can be written as[1] $K = \cup_{l=1}^{L} K_l$, $K_l \in O_{W_l}(B)$, $l = 1, 2, ..., L$. Let $Y \circ W$ stand for the *opening* of Y by W. Then $\hat{X}_{MAP}(\mathbf{Y}^{(M)}) = \bigcup_{l=1}^{L} \left(\left(\cap_{i=1}^{M} Y_i \right) \circ W_l \right)$ *is a MAP estimator* of X on the basis of $\mathbf{Y}^{(M)}$.

PROOF. The proof is based on the fact that

$$G(S, j) \triangleq \{V \in \Psi_j(B) \mid Y_j \setminus S \subseteq V \subseteq Y_j\},$$

i.e., the set of all N_j-realizations which are consistent with the jth observation (i.e., Y_j) under the hypothesis that the true signal is S, is a monotone non-decreasing function of $S \in \Phi(B) \cap \mathcal{P}\left(\cap_{i=1}^{M} Y_i\right)$, for every $j = 1, 2, ..., M$. Non-uniqueness of the functional form of the MAP estimator is a direct consequence of the fact that $G(S, j)$ is generally not *strictly* increasing function of S. □

THEOREM 2.2. (*Strong Consistency*) In addition, if the following holds:

CONDITION 1. *$\forall z \in B$ there exists $0 \leq r < 1$, such that $Pr(z \in N_j) \leq r$, for infinitely many indices j.*[2] *In other words, for every $z \in B$, there exists $0 \leq r < 1$, and an infinitely long subsequence, $\mathcal{I}$, of observation indices (both possibly dependent on z), such that $Pr(z \in N_j) \leq r$, $\forall j \in \mathcal{I}$.*

then, under the foregoing assumptions,

$$\hat{X}_{MAP}(\mathbf{Y}^{(M)}) \longrightarrow X, \textit{ almost surely (a.s.) as } M \rightarrow +\infty,$$

i.e., this MAP estimator is strongly consistent.

PROOF. The proof is based on the union bound and the fact that the observation window is *finite*. □

By duality, it follows that:

THEOREM 2.3. (*MAP Optimality Dual*) Assume we observe $\mathbf{Y}^{(M)} = [Y_1, Y_2, ..., Y_M]$, where $Y_i = X \cap N_i$, $\{N_i\}_{i=1}^{M}$ is an independent *but not necessarily identically distributed* sequence of noise DRS's, which is independent of X, and each N_i is an *otherwise arbitrary* DRS taking values in some arbitrary collection, $\Psi_i(B) \subseteq \mathcal{P}(B)$, of subsets of the observation window B. Let us further assume that X is uniformly distributed over a collection, $\Phi(B) \subseteq \mathcal{P}(B)$, of all subsets K of B which can be written as $K = \cap_{l=1}^{L} K_l$, $K_l \in C_{W_l}(B)$, $l = 1, 2, ..., L$. Let $Y \bullet W$ stand for the *closing* of Y by W. Then $\hat{X}_{MAP}(\mathbf{Y}^{(M)}) = \bigcap_{l=1}^{L} \left(\left(\cup_{i=1}^{M} Y_i \right) \bullet W_l \right)$ *is a MAP estimator* of X on the basis of $\mathbf{Y}^{(M)}$.

[1] Note that one or more of the K_l's can be empty, since $\emptyset \in O_W(B)$, $\forall$ W.

[2] Observe that this is a condition on marginal noise statistics only.

THEOREM 2.4. (*Strong Consistency Dual*) In addition, if the following holds

CONDITION 2. *$\forall z \in B$ there exists $0 \leq r < 1$, such that $Pr(z \notin N_j) \leq r$, for infinitely many indices j.*[3] *In other words, for every $z \in B$, there exists $0 \leq r < 1$, and an infinitely long subsequence, $\mathcal{I}$, of observation indices (both possibly dependent on z), such that $Pr(z \notin N_j) \leq r$, $\forall j \in \mathcal{I}$.*

then, under the foregoing assumptions,

$$\hat{X}_{MAP}(\mathbf{Y}^{(M)}) \longrightarrow X, \text{ a.s. as } M \to +\infty,$$

i.e., this MAP estimator is strongly consistent.

2.1. Discussion. These theorems crucially depend on B being finite.[4] In practice, random set observations are usually discrete and finite, so we do not view this as a significant restriction, at least in practical terms. The second observation is that the results are fairly general: apart from (mild) Condition 1, which is needed for consistency, and the requirement that $\{N_j\}$ is a sequence of independent DRS's, which is independent of X, we have imposed absolutely no other restrictions on the sequence of noise DRS's $\{N_j\}$.

As a special case:

- $\{N_j\}$ is a sequence of independent Discrete Boolean Random Sets [30], which is independent of X. This particular case is of great interest, since the Boolean union noise model is arguably one of the best models for clutter. For all practical purposes (i.e., all Boolean models of practical interest), Condition 1 is satisfied, and, therefore, optimality and strong consistency can be warranted.

The final point is that these results hold under the assumption of one-sided noise. It is natural to ask what happens in the case of two-sided noise. As we will see, this transition significantly changes the ground rules.

3. The Case of two–sided noise.

3.1. The 1–D case. Let consider 1-D DRS's, i.e., random binary sequences of finite extent. For brevity and readability we switch notation to a sequence-oriented one, instead of a set-oriented one.

Consider a finite random binary sequence $\mathbf{x} = \{x(n)\}_{n=0}^{N-1}$, $x(n) \in \{0,1\}$, which is uniformly distributed over the set of all binary sequences of length N that are both open and closed (i.e., these sequences are not affected by a morphological opening or closing) with respect to a simple M point window ($M \leq N$). $\mathbf{x}$ is observed through a binary symmetric memoryless channel (BSC) of symbol inversion probability $\epsilon < \frac{1}{2}$, i.e., we observe $\mathbf{y} = \{y(n)\}_{n=0}^{N-1}$, where $y(n) \in \{0,1\}$, and $y(n)$ is equal to $1 - x(n)$

[3] Again, this is a condition on marginal noise statistics only.

[4] The size of B can be made as large as one wishes, as long as it is finite.

with probability $\epsilon \in [0, 0.5)$, or $x(n)$ with probability $1 - \epsilon$, independently of all others. Let P_M^N denote the set of all sequences of N binary digits which are piecewise constant of plateau (run) length $\geq M$. This is exactly the set of all binary sequences of length N which are both open and closed with respect to an M point window.

CLAIM 1. *It is easy to verify that:*

$$\hat{\mathbf{x}}_{MAP}(\mathbf{y}) = argmin_{\mathbf{x} \in P_M^N} \sum_{n=0}^{N-1} |y(n) - x(n)|,$$

where "argmin" should be interpreted as "an argument that minimizes," is a MAP estimator of $\mathbf{x}$ *on the basis of* $\mathbf{y}$.

The question now is, how does one *compute* such a minimizer efficiently, and what are the properties of the implicit estimator?

It has been shown in [4], [5] that a minimizer may be efficiently computed using the Viterbi algorithm [31], [32], an instance of Dynamic Programming [33], [34]. The associated computational complexity is $O(M \times N)$.

The possibility of having multiple solutions (minimizers) implies that a tie-breaking strategy is required in order to obtain an associated MAP input-output operator, i.e., a *MAP estimator.*

PROPOSITION 3.1. Regardless of the choice of the tie-breaking strategy, the MAP estimator is idempotent, meaning that $\hat{\mathbf{x}}_{MAP}(\hat{\mathbf{x}}_{MAP}(\mathbf{y})) = \hat{\mathbf{x}}_{MAP}(\mathbf{y})$.

PROPOSITION 3.2. By a special choice of the tie-breaking strategy, the MAP estimator may be designed to be self-dual, meaning that $(\hat{\mathbf{x}}_{MAP}(\mathbf{y}^c))^c = \hat{\mathbf{x}}_{MAP}(\mathbf{y})$, where ^{c} stands for the bit-wise toggle operator.

PROPOSITION 3.3. The MAP estimator *is not* an increasing operator, i.e., it does not necessarily preserve order.

The proofs can be found in [5]. A counter-example illustrating that any MAP estimator may be forced to violate the order-preservation principle can be found in [4].

The fact that the optimal MAP estimator is not increasing means that it may not be obtained as a composition of elementary openings and closings, and it is not a morphological filter in the (strict) sense of [19]. This fact changes the ground rules of the game, and forces one to opt for more elaborate inference tools, the simplest of which is the Viterbi algorithm.

Conceptually, it is not difficult to generalize this result to the case of arbitrary 1-D structuring elements, although this requires some work in setting up a suitable Viterbi trellis for each individual structuring element. It is also possible to make this work for some 2-D DRS models employing relatively small structuring elements, although we will not further pursue this direction here. A word of caution: the price that one pays in so doing is a significant increase in computational complexity and/or latency.

3.2. The 2–D case. Consider a DRS X uniformly distributed over $\Phi_W(B)$, the set of all subsets of B which are both open and closed with respect to some structuring element W. X is observed through a BSC of pixel inversion probability $\epsilon < \frac{1}{2}$, i.e., we observe Y, such that $Y(z) \in \{0, 1\}$ (the pixel binary random variable at site z) is equal to $1 - X(z)$ with probability $\epsilon \in [0, 0.5)$, or $X(z)$ with probability $1 - \epsilon$, independently of all others.

CLAIM 2. *It is easy to verify that:*

$$\hat{X}_{MAP}(Y) = argmin_{X \in \Phi_W(B)} \sum_{z \in B} |Y(z) - X(z)|,$$

where "argmin" should be interpreted as "an argument that minimizes," is a MAP estimator of X on the basis of Y.

Again the question is, how does one *compute* such a minimizer efficiently. The answer, in general, is an open problem. The Viterbi algorithm does not work (meaning: it becomes very complex) for arbitrary structuring elements (although it solves the problem nicely for some "special" ones, especially line structuring elements). For general structuring elements the above is a hard nonlinear optimization problem. One possibility is to use Graduated Non-Convexity [35] or *simulated annealing–type* algorithms [36]–[39]. However, these are iterative "relaxation" schemes that may also entail a significant computational complexity cost, and, unlike the Viterbi Algorithm, they are not guaranteed to provide an optimum solution, but rather opt for a high probability sample.

4. Discussion. An efficient solution to this latter problem will have a significant impact, not only in the area of image filtering for de-noising pattern matching and restoration, but also in the area of optimal decoder design for 2-D holographic data storage systems, where closely related problems arise.

REFERENCES

[1] N.D. SIDIROPOULOS, J.S. BARAS, AND C.A. BERENSTEIN, *Optimal Filtering of Digital Binary Images Corrupted by Union/Intersection Noise*, IEEE Trans. Image Processing, vol. 3, pp. 382–403, 1994.

[2] N.D. SIDIROPOULOS, D. MELEAS, AND T. STRAGAS, *MAP Signal Estimation in Noisy Sequences of Morphologically Smooth Images*, IEEE Trans. Image Processing, vol. 5, pp. 1088–1093, 1996.

[3] N.D. SIDIROPOULOS, J.S. BARAS, AND C.A. BERENSTEIN, *Further results on MAP Optimality and Strong Consistency of certain classes of Morphological Filters*, IEEE Trans. Image Processing, vol. 5, pp. 762–764, 1996.

[4] N.D. SIDIROPOULOS, *The Viterbi Optimal Runlength-Constrained Approximation Nonlinear Filter*, in Mathematical Morphology and its Applications to Image and Signal Processing, P. Maragos, R.W. Schafer, M. Akmal Butt (Eds), pp. 195–202, Kluwer, Boston, Massachusetts, 1996.

[5] N.D. SIDIROPOULOS, *The Viterbi Optimal Runlength-Constrained Approximation Nonlinear Filter*, IEEE Trans. Signal Processing, vol. 44, pp. 586–598, 1996.

[6] G. CHOQUET, *Theory of capacities*, Ann. Institute Fourier, vol. 5, pp. 131–295, 1953.

[7] D.G. KENDALL, *Foundations of a theory of random sets*, in Stochastic Geometry, E.F. Harding and D.G. Kendall, Eds., pp. 322–376. John Wiley, London, England, 1974.

[8] G. MATHERON, *Elements pour une theorie des Milieux Poreux*, Masson, 1967.

[9] G. MATHERON, *Random Sets and Integral Geometry*, Wiley, New York, 1975.

[10] K. SIVAKUMAR AND J. GOUTSIAS, *Binary Random Fields, Random Closed Sets, and Morphological Sampling*, IEEE Trans. Image Processing, vol. 5, pp. 899–912, 1996.

[11] L. WEISNER, *Abstract theory of inversion of finite series*, Trans. AMS, vol. 38, pp. 474–484, 1935.

[12] M. WARD, *The algebra of lattice functions*, Duke Math. J., vol. 5, pp. 357–371, 1939.

[13] GIAN-CARLO ROTA, *On the Foundations of Combinatorial Theory: I. Theory of Moebius Functions*, Z. Wahrscheinlichkeitstheorie und Verw. Gebiete, vol. 2, pp. 340–368, 1964. Also appears in Classic Papers in Combinatorics, I. Gessel and G-C. Rota (Eds.), Birkhauser, Boston,1987.

[14] H.H. CRAPO, *Moebius Inversion in Lattices*, Arch. Math., vol. 19, pp. 595–607, 1968. Also appears in Classic Papers in Combinatorics, I. Gessel and G-C. Rota (Eds.), Birkhauser, Boston,1987.

[15] M. AIGNER, *Combinatorial Theory*, Springer-Verlag, New York, 1979.

[16] H. MATHIS THOMA, *Belief function computations*, in Conditional Logic in Expert Systems, I.R. Goodman, M.M Gupta, H.T. Nguyen, and G.S. Rogers, Eds. 1991, pp. 269–308, Elsevier Science Publishers B.V. (North Holland).

[17] J. SERRA, *Image Analysis and Mathematical Morphology*, Academic Press, New York, 1982.

[18] J. SERRA, *Image Analysis and Mathematical Morphology, vol. 2, Theoretical Advances*, Academic Press, San Diego, 1988.

[19] H.J.A.M. HEIJMANS, *Morphological Image Operators*, Academic Press, Boston, 1994.

[20] E.R. DOUGHERTY, *Optimal Mean Square N-Observation Digital Morphological Filters - I. Optimal Binary Filters*, Computer Vision, Graphics, and Image Processing: Image Understanding, vol. 55, pp. 36–54, 1992.

[21] R.M. HARALICK, E.R. DOUGHERTY, AND P.L. KATZ, *Model-based morphology*, in Proc. SPIE, Vol. 1472, Orlando, Florida. Society for Optical Engineering, April 1991.

[22] E.R. DOUGHERTY, A. MATHEW, AND V. SWARNAKAR, *A conditional-expectation-based implementation of the optimal mean-square binary morphological filter*, in Proc. SPIE, Vol. 1451, San Jose, California. Society for Optical Engineering, February 1991.

[23] E.R. DOUGHERTY, ED., *Mathematical Morphology in Image Processing*, Marcel Dekker, New York, 1993.

[24] E.R. DOUGHERTY, *Optimal mean-absolute-error filtering of gray-scale signals by the morphological hit-or-miss transform*, Journal of Mathematical Imaging and Vision, vol. 4, pp. 255–271, 1994.

[25] E.R. DOUGHERTY, J. NEWELL, AND J. PELZ, *Morphological texture-based maximum-likelihood pixel classification based on local granulometric moments*, Pattern Recognition, vol. 25, pp. 1181–1198, 1992.

[26] D. SCHONFELD AND J. GOUTSIAS, *Optimal morphological pattern restoration from*

noisy binary images, IEEE Transactions on Pattern Anal. Mach. Intell., vol. 13, pp. 14–29, 1991.

[27] D. Schonfeld, *Optimal Structuring Elements for the Morphological Pattern Restoration of Binary Images*, IEEE Transactions on Pattern Anal. Mach. Intell., vol. 16, pp. 589–601, 1994.

[28] D. Schonfeld and J. Goutsias, *On the morphological representation of binary images in a noisy environment*, Journal of Visual Communication and Image Representation, vol. 2, pp. 17–30, 1991.

[29] J. Goutsias, *Morphological Analysis of Discrete Random Shapes*, Journal of Mathematical Imaging and Vision, vol. 2, pp. 193–215, 1992.

[30] N.D. Sidiropoulos, J.S. Baras, and C.A. Berenstein, *An Algebraic Analysis of the Generating Functional for Discrete Random Sets, and Statistical Inference for Intensity in the Discrete Boolean Random Set Model*, Journal of Mathematical Imaging and Vision, vol. 4, pp. 273–290, 1994.

[31] A.J. Viterbi, *Error bounds for convolutional codes and an asymptotically optimum decoding algorithm*, IEEE Trans. Information Theory, vol. 13, pp. 260–269, 1967.

[32] J.K. Omura, *On the Viterbi decoding algorithm*, IEEE Trans. Information Theory, vol. 15, pp. 177–179, 1969.

[33] R. Bellman, *Dynamic Programming*, Princeton University Press, Princeton, N.J., 1957.

[34] R. Bellman and S. Dreyfus, *Applied Dynamic Programming*, Princeton University Press, Princeton, N.J., 1962.

[35] A. Blake and A. Zisserman, *Visual Reconstruction*, MIT Press, Cambridge, Mass., 1987.

[36] S. Geman and D. Geman, *Stochastic relaxation, Gibbs distributions, and the Bayesian restoration of images*, IEEE Trans. PAMI, vol. 6, pp. 721–741, 1984.

[37] G.L. Bibro, W.E. Snyder, S.J. Garnier, and J.W. Gault, *Mean field annealing: A formalism for constructing GNC-like algorithms*, IEEE Trans. Neural Networks, vol. 3, pp. 131–138, 1992.

[38] D. Geman, S. Geman, C. Graffigne, and P. Dong, *Boundary detection by constrained optimization*, IEEE Trans. PAMI, vol. 12, pp. 609–627, 1990.

[39] D. Geman and C. Yang, *Nonlinear image recovery with half-quadratic regularization*, IEEE Trans. Image Processing, vol. 4, pp. 932–946, 1995.

PART II
Information/Data Fusion and Expert Systems

ON THE MAXIMUM OF CONDITIONAL ENTROPY FOR UPPER/LOWER PROBABILITIES GENERATED BY RANDOM SETS

JEAN-YVES JAFFRAY*

Abstract. Imprecision on probabilities is expressed through upper/lower probability intervals. The lower probability is always assumed to be a convex capacity and sometimes to be an ∞–monotone capacity. A justification of this latter assumption based on the existence of underlying random sets is given. Standard uncertainty measures of information theory, such as the Shannon entropy and other indices consistent with the Lorenz ordering, are extended to imprecise probability situations by taking their worst-case evaluation, which is shown to be achieved for a common probability. In the same spirit, the information brought by a question is evaluated by the maximum value of the associated conditional entropy. Computational aspects, which involve the resolution of decomposable convex programs, are discussed.

Key words. Ambiguity, Capacities, Conditioning, Entropy, Random Sets, Uncertainty, Upper/Lower Probabilities.

AMS(MOS) subject classifications. 62B10, 90C25, 94A15, 94A17

1. Introduction. Potential applications of information theory seem to be limited by the fact that it only applies to a probabilized environment. Clearly, in most technical or managerial problems, there is little hope that the available data provide a sufficient basis for reliable probability estimates; furthermore, experts are likely to feel more comfortable in formulating their opinions in terms of subjective probability intervals than in the form of crisp probabilities.

It is thus important to examine to what extent information theory can still be useful in situations of imprecisely probabilized uncertainty, and especially in the most frequently encountered ones. As a matter of fact, it turns out that, for the commonest form of information sources, namely data bases, the expression of either imprecision or incompleteness is perfectly captured by the formalism of random sets, which modelization automatically ensures that all the intervening lower (resp. upper) probabilities are ∞–monotone (resp. ∞–alternating) capacities. Moreover, in some domains, such as robust statistics, natural assumptions on the existing imprecision lead to lower probabilities which, although not ∞–monotone, are convex capacities. Thus the limitation of this study to convex lower probabilities and even, for some parts, to ∞–monotone lower probabilities, does not seem to restrict too much the prospective applications of its results.

We are interested in establishing results that enable one to extend questionnaire theory to the imprecise probability case. Questionnaire theory [16], a major chapter of information theory, has useful applications in,

* LIP6, Université Paris VI, 4, pl. Jussieu, F–75252 Paris Cedex 05, France. Results in Sections 6 and 7 have been obtained jointly with Serge LORIT from LIP6.

among others, classification and data mining. A questionnaire consists in a series of questions (each one being likely to be triggered by specific answers to previously asked ones) enabling one to identify any particular member of a given set (as in the game of Twenty Questions). The fundamental problem of questionnaire theory is to construct a questionnaire which is optimal in the sense of requiring a minimum expected number of questions. There exists an exact method (the Huffman–Picard algorithm), which however does not allow for the restriction of questions to a pre-selected set and, moreover, requires the complete construction of the questionnaire (which is a tree of questions). For these reasons, in practice, the following myopic heuristic is used repeatedly: "choose the most informative question, i.e., that question which gives the greatest expected decrease of the Shannon entropy;" thus only one path of the questionnaire needs be constructed for each identification performed. The price to pay in suboptimality for this computational advantage has been shown empirically to remain reasonably low.

The extension of questionnaire theory to the imprecise probability case requires a suitable adaptation of the uncertainty and information measures of the probabilistic case. Adopting a worst case point of view, we attribute to an uncertain situation the maximum value of the entropy over all probabilities consistent with the data. The same point of view is taken to evaluate the expected level of uncertainty after a question. The evaluations are the optimal values of very specific optimization problems. The greater part of the paper is devoted to their study.

The paper is organized as follows: Section 2 establishes the suitability of the random set formalism for representing imprecision in samples and data bases; Section 3 recalls the essential properties of capacities and Section 4 the relations between uncertainty measures such as the Shannon entropy with the Lorenz ordering; in Section 5 the main properties are studied and related computational aspects discussed; Section 6 studies the properties of the worst case probabilities used in evaluating questions and proposes a method for computing them; Section 7 presents an example, and is followed by the conclusion section.

2. Imprecise data and random sets. Statisticians and data base managers are inevitably confronted with the problem of missing or imprecise data: for instance, the medical files of certain patients do not mention whether or not they have been vaccinated against tetanus, or, when they do, do not precisely indicate the year of the vaccination; or, certain consumers, when asked what brand of a particular product they use, decline answering the question or only provide the information that they "do not use brand X."

A simple method for handling such imprecisions consists in edicting default rules which ascribe precise values whenever data are ambiguous. This drastic treatment of imprecision is clearly a potential source of biases and may induce overconfidence in the conclusions of subsequent analyses based on the data. For these reasons, the alternative solution, which consists in managing the original, unaltered, data directly, is worth being considered despite its greater complexity.

When imprecise, data concerning a person (or any other entity) ω of population Ω are no longer representable by a single element (e.g., a list, or a vector of attribute values) of some multi-dimensional set $\mathcal{X}$ but by a subset of this set. Mathematically speaking, data are not expressed by a point-to-point mapping, $\omega \longmapsto X(\omega) = x$, from Ω into $\mathcal{X}$, but by a multi-valued (point-to-set mapping), $\omega \longmapsto \Gamma(\omega) = B$, $B \in \mathcal{X}$, from Ω into $2^{\mathcal{X}}$.

Whereas in the precise case the variations of the values x in the population can be exactly captured by the (relative) frequencies:

$$p(x) = \frac{1}{|\Omega|}\big|\{\omega \in \Omega : X(\omega) = x\}\big|, \quad x \in \mathcal{X}, \tag{2.1}$$

and thus

$$p(A) = \sum_{x \in A} p(x), \quad A \in 2^{\mathcal{X}} \tag{2.2}$$

in the imprecise case, data only allow for evaluations of upper and lower bounds of $p(A)$:

$$F(A) = \frac{1}{|\Omega|}\big|\{\omega \in \Omega : \Gamma(\omega) \cap A \neq \emptyset\}\big| , \tag{2.3}$$

and

$$f(A) = \frac{1}{|\Omega|}\big|\{\omega \in \Omega : \Gamma(\omega) \subseteq A\}\big|. \tag{2.4}$$

Note that F and f are respectively an alternating and a monotone capacity of order infinity (see Subsection 3.1 below).

One of the main motivations for collecting data is prediction: given past data, what is to be expected in the future? The passage from observation to prediction can only be made possible by additional, (in general) probabilistic, assumptions, which depend on the context: for instance, it may be that Ω is an exhaustive sample from a finite population Ω_0 ($= \Omega$) and that its members can be assumed to be equally likely to be selected in the future; or it may be that Ω is only a subpopulation, but a representative one, so that the complete population Ω_0 can be endowed with any probability/multi-valued mapping pair (π, Γ_0) such that

$$\pi(\{\omega \in \Omega_0 : \Gamma_0(w) \cap A \neq \emptyset\}) = F(A) , \tag{2.5}$$

and

$$\pi(\{\omega \in \Omega_0 : \Gamma_0(w) \subseteq A\}) = f(A) , \tag{2.6}$$

with $F(A)$ and $f(A)$ given by (2.3) and (2.4); or, in less favorable cases, it may be that the estimation of probability π require the use of more sophisticated statistical methods.

In all these cases, the probabilistic model based on the imprecise data base initially considered consists in a probability space $(\Omega, 2^{\Omega}, \pi)$, a state space $(\mathcal{X}, 2^{\mathcal{X}})$ and a multi-valued application $\Gamma : \Omega \to 2^{\mathcal{X}}$ (where Ω and Γ may in fact be the Ω_0 and Γ_0 introduced above); such a triple constitutes, by definition, a *random set*.

3. Capacities. Capacities, which are non-additive measures, appear naturally in the presence of imprecision for expressing lower and upper probabilities.[1]

3.1. Definitions and properties. Given $\mathcal{X} = \{x_1, \dots, x_i, \dots, x_n\}$, $\mathcal{A} = 2^{\mathcal{X}}$ and $\mathcal{A}^* = \mathcal{A} \setminus \{\emptyset\}$, a mapping $f : \mathcal{A} \to \mathbb{R}$ is a *capacity* on $(\mathcal{X}, \mathcal{A})$ whenever:

$$f(\emptyset) = 0; \quad f(\mathcal{X}) = 1; \tag{3.1}$$

$$A \subseteq B \Rightarrow f(A) \leq f(B) \quad \text{for all } A, B \in \mathcal{A}. \tag{3.2}$$

A *convex* capacity is a capacity which is *supermodular*, i.e.,

$$f(A \cup B) + f(A \cap B) \geq f(A) + f(B) \quad \text{for all } A, B \in \mathcal{A}. \tag{3.3}$$

More generally, a capacity is *monotone at order k, or k–monotone* ($k \geq 2$) whenever

$$f\Big(\bigcup_{j=1}^{k} A_j\Big) \geq \sum_{L \subseteq \{1,2,\dots,k\},\, L \neq \emptyset} (-1)^{|L|+1} f\Big(\bigcap_{j \in L} A_j\Big) \quad \text{for all } A_j \in \mathcal{A}. \tag{3.4}$$

A capacity is *monotone at order infinity*, or ∞–*monotone* when it is k–monotone for all $k \geq 2$. An ∞–monotone capacity is also known as a *belief function* ([17]).

The *dual* of capacity f is the mapping $F : \mathcal{A} \to \mathbb{R}$ defined by

$$A \longmapsto F(A) = 1 - f(\mathcal{X} \setminus A); \tag{3.5}$$

F is also a capacity and is *submodular*, whenever f is supermodular, i.e.,

$$F(A \cup B) + F(A \cap B) \leq F(A) + F(B) \quad \text{for all } A, B \in \mathcal{A}. \tag{3.6}$$

The dual F of an ∞–monotone capacity or belief function f is alternating at order k (the property induced by f's k–monotonicity) for all $k \geq 2$ and called an ∞–*alternating* capacity or a *plausibility function* [17].

[1] The following exposition has been voluntarily limited to a finite setting. Much of it can be in fact extended, more or less readily, to infinite settings, see e.g. [14].

A capacity f is characterized by its *Möbius transform* $\phi : \mathcal{A} \to \mathbb{R}$, defined by

$$E \longmapsto \varphi(E) = \sum_{B \subseteq E} (-1)^{|E \setminus B|} f(B) \quad \text{for all } E \in \mathcal{A}; \tag{3.7}$$

more precisely,

$$f(A) = \sum_{E \subseteq A} \varphi(E) \quad \text{and} \quad F(A) = \sum_{E \cap A \neq \emptyset} \varphi(E) \quad \text{for all } A \in \mathcal{A}. \tag{3.8}$$

Properties (3.1) and (3.2) of capacity f are respectively equivalent to the following properties of its Möbius transform φ [2]:

$$\varphi(\emptyset) = 0; \qquad \sum_{E \in \mathcal{A}} \varphi(E) = 1; \tag{3.9}$$

$$\sum_{\{x\} \subseteq B \subseteq A} \varphi(B) \geq 0 \quad \text{for all } x \in \mathcal{X},\ A \in \mathcal{A} \text{ with } x \in A. \tag{3.10}$$

A capacity f is convex whenever

$$\sum_{\{x,y\} \subseteq B \subseteq A} \varphi(B) \geq 0 \quad \text{for all } x, y \in \mathcal{X},\ A \in \mathcal{A} \text{ with } \{x, y\} \subseteq A. \tag{3.11}$$

Belief functions (∞–monotone capacities) are characterized by Möbius transforms which are nonnegative and satisfy (3.9).

Let $\mathcal{L}$ be the set of all probabilities on $(\mathcal{X},\mathcal{A})$. Given a capacity f one defines its *core* by either one of the equivalent formulas

$$\mathit{core}\, f = \{P \in \mathcal{L} : P(A) \geq f(A) \quad \text{for all } A \in \mathcal{A}\}, \tag{3.12}$$

$$\mathit{core}\, f = \{P \in \mathcal{L} : P(A) \leq F(A) \quad \text{for all } A \in \mathcal{A}\}. \tag{3.13}$$

The core of a convex capacity (hence of an ∞–monotone capacity) f is non-empty [18] and can be described alternatively as follows: Let $\mathcal{G}$ be the support of the Möbius transform φ of f, i.e.,

$$\mathcal{G} = \{G \in \mathcal{A} : \varphi(G) \neq 0\}. \tag{3.14}$$

An *allocation* of f is a probability P which takes on singletons values of the form

$$P(\{x\}) = \sum_{G \ni x} \lambda(G, x) \varphi(G) \quad \text{for all } x \in \mathcal{X}, \tag{3.15}$$

with $\lambda(G, x) \geq 0$, and $\sum_{x \in G} \lambda(G, x) = 1$ for all $x \in G$, all $G \in \mathcal{G}$.

It is known ([2],[5]) that, when f is an ∞*–monotone* capacity:

$$P \in \textit{core } f \textit{ if and only if } P \textit{ is an allocation of } f,$$

and, moreover, the "only if" part remains valid for *any* capacity:

$$\textit{every } P \in \textit{core } f \textit{ must be an allocation of } f.$$

Finally, convex capacities f have the property that $f = \inf_{core\, f} P$, for a pointwise infimum, which is in fact achieved:

$$\textit{For every } A \in \mathcal{A}, \textit{ there exists } P \in \textit{core } f \textit{ such that } P(A) = f(A).$$

(In fact: *for any increasing sequence* (A_i), *there exists* $P \in$ *core* f *such that* $P(A_i) = f(A_i)$ *for all* i*'s*). Similarly, $F = \sup_{core\, f} P$.

3.2. Random sets and ∞–monotone capacities. Consider the random set defined by the probability space $(\Omega, 2^{\Omega}, \pi)$ and the mapping $\Gamma : \Omega \to \mathcal{A}\ (= 2^{\mathcal{X}})$. Define φ on $\mathcal{A}$ by:

$$\varphi(B) = \pi(\{\omega \in \Omega : \Gamma(\omega) = B\}) \quad \text{for all } B \in \mathcal{A}. \tag{3.16}$$

Then, for all $A \in \mathcal{A}$

$$f(A) = \pi(\{\omega \in \Omega : \Gamma(\omega) \subseteq A\}) = \sum_{B \subseteq A} \varphi(B), \tag{3.17}$$

and

$$F(A) = \pi(\{\omega \in \Omega : \Gamma(\omega) \cap A \neq \emptyset\}) = \sum_{B \cap A \neq \emptyset} \varphi(B). \tag{3.18}$$

Since $\varphi \geq 0$, f is an ∞–monotone capacity and F is the dual ∞–alternating capacity.

Consider all selections of Γ, i.e., all mappings $X : \Omega \to \mathcal{X}$ obtained by selecting arbitrarily, for every $\omega \in \Omega$, an element $X(\omega)$ in $\Gamma(\omega)$. Each X is a random variable and can be interpreted as describing a situation of uncertainty consistent with the available imprecise data. It generates a probability $P = \pi \circ X^{-1}$ on $(\mathcal{X}, \mathcal{A})$ which is an allocation of f of the particular type:[2] $\lambda(G, x) = 1$ for some $x \in G$. Although these probabilities only form a subset $\mathcal{P}_1$ of core f, it can be shown [2] that $f = \inf_{\mathcal{P}_1} P$ and $F = \sup_{\mathcal{P}_1} P$ thus justifying their denominations of lower probability and upper probability associated with the random set.

[2] The other allocations can make sense in some applications. For instance, in a survey on favorite sports, ω can be a group of people rather than a single individual, for whom $\Gamma(\omega) = \{x, y\}$, with $x =$ running and $y =$ tennis; it is possible that λ% of the people in the group prefer running and $(1 - \lambda)$% prefer playing tennis.

3.3. Other types of information and convex capacities. Data are not exclusively gathered by sampling. In particular, a common source of information is expert opinion which is likely to be expressed through assertions such as: "the probability of event A is at least $\alpha\%$ and at most $\beta\%$," or "event A is more likely than event B." Let $\mathcal{P}$ be the set of all probabilities consistent with these data.

We shall only consider situations of uncertainty where $\mathcal{P}$ has the following properties:

$$(3.19) \quad f = \inf_{\mathcal{P}} P \text{ defined by } A \mapsto f(A) = \inf\{P(A) : P \in \mathcal{P}\},\ A \in \mathcal{A},$$

is a convex capacity;

$$(3.20) \qquad \mathcal{P} = \mathit{core}\, f.$$

Although these properties do not always hold in applications, there are important cases where they necessarily do. For instance, in robust statistics, sets of prior probabilities of the form $\{P \in \mathcal{L} : |P(A) - P_0(A)| \leq \varepsilon$ for all $A \in \mathcal{A}\}$ and $\{P \in \mathcal{L} : (1-\varepsilon)P_0(A) \leq P(A) \leq (1+\varepsilon)P_0(A)$ for all $A \in \mathcal{A}\}$ (where $P_0 \in \mathcal{L}$ and $\varepsilon > 0$ are given) are frequently introduced (ε–*contamination*). In both cases, the lower probabilities are 2–monotone but not 3–monotone [3], [7]. As we shall see, many important results concerning ∞–monotone lower probabilities extend to the case where they are only convex.

4. Entropy functionals and the Lorenz ordering. The classical theory of information addresses situations of probabilistic uncertainty: in the case of a finite set of states of nature $\mathcal{X} = \{x_1, \ldots, x_i, \ldots, x_n\}$, the algebra of events $\mathcal{A} = 2^{\mathcal{X}}$ is endowed with a probability P. Since P is completely determined by its values on the elementary events $P(\{x_i\}) = p_i$ $(i = 1, \ldots, n)$, the set $\mathcal{L}$ of all probabilities on $\mathcal{A}$ can be identified (through $P(\{x_i\}) = p_i$), with the n–simplex

$$(4.1) \quad \mathcal{S}_n = \left\{ p = (p_1, \ldots, p_i, \ldots, p_n) : \sum_{i=1}^n p_i = 1;\ \ p_i \geq 0,\ i = 1, \ldots, n \right\};$$

thus $P(E) = \sum_{x_i \in E} p_i$. A standard measure of uncertainty is the (Shannon) *entropy* functional $H : \mathcal{L} \to \mathbb{R}$, given by

$$(4.2) \qquad P \longmapsto H(P) = -\sum_{i=1}^n P(\{x_i\}) \log P(\{x_i\}),$$

to which corresponds the n–variable function $h : \mathcal{S} \to \mathbb{R}$ given by

$$(4.3) \qquad P \longmapsto h(P) = -\sum_{i=1}^n p_i \log p_i$$

(note that, by convention, $p_i \log p_i = 0$ for $p_i = 0$).

Several other functionals have been proposed as uncertainty indices. They are all of the form

$$P \longmapsto V(p) = \sum_{i=1}^{n} v(p_i) \,, \tag{4.4}$$

where $v : [0,1] \to \mathbb{R}$ is concave.

Of special interest is the case of quadratic functions v_0 and V_0:

$$t \longmapsto v_0(t) = -t^2, \quad p \longmapsto V_0(p) = -\sum_{i=1}^{n} {p_i}^2. \tag{4.5}$$

This index is sometimes given under the equivalent form

$$p \longmapsto W_0(p) = \sum_{i=1}^{n} p_i(1 - p_i). \tag{4.6}$$

Despite their apparent diversity, these indices are fairly coherent, since it results directly from a famous theorem of Hardy, Littlewood and Polya [6] (quoted in, e.g., [1]) that, given $p, q \in \mathcal{S}_n$ the following three statements are equivalent:

$$V(p) \geq V(q), \text{ for all } V \in \mathcal{V}; \tag{4.7}$$

(4.8) there exists a bi-stochastic matrix S such that $p = Sq$;

$$\inf \{P(E) : |E| = \ell\} \geq \inf \{Q(E) : |E| = \ell\} \quad \text{for all } \ell \leq n. \tag{4.9}$$

By definition, (4.9) states that p *dominates* q with respect to the *Lorenz ordering*. Thus this partial ordering exactly expresses the consensus which exists between the various uncertainty indices.

Let us then turn to the problem raised by Nguyen and Walker in [15] (and later solved by Meyerowitz, Richman and Walker [12]):

> *"Find the maximum-entropy probability P^0 in core f, where f is an ∞–monotone capacity."*

Since $h(p)$ is a continuous and strictly concave function, it achieves a unique maximum in the compact subset of $\mathcal{S}_n$ identified with *core* f; thus P^0 exists and is unique.

Consider now the following problem:

> *"Find the greatest element P^* in core f for the Lorenz ordering, where f is an ∞–monotone capacity."*

Since the Lorenz ordering is neither complete nor antisymmetric the existence of a unique greatest element P^* in *core* f is not straightforward; however if P^* exists, then, according to the Hardy et al. theorem, it must coincide with the unique entropy-maximizing probability P^0 in *core* f (hence P^* is itself unique); moreover it also maximizes the other uncertainty indices V belonging to $\mathcal{V}$ (although perhaps not uniquely).

We shall in fact consider the slightly more general problems where f is only assumed to be a convex capacity. Dutta and Ray [4] have already raised and solved these problems in the different context of social choice theory; we however propose an alternative line of proof.

5. The maximum entropy probability: Construction and properties.

5.1. Method. Given a convex capacity f on $(\mathcal{X}, \mathcal{A})$ or, equivalently, the dual submodular capacity F, we shall exhibit a particular probability P^* and show successively that: P^* belongs to *core* f; P^* maximizes all V of $\mathcal{V}$ in *core* f; and P^* is the greatest element of *core* f for the Lorenz ordering, i.e.:

$$\inf \{P^*(E) : |E| = \ell\} \geq \inf \{Q(E) : |E| = \ell\} \quad \text{for all } \ell \leq n, \quad \text{for all } Q \in \mathit{core}\, f. \tag{5.1}$$

There exists a largest event A in $\mathcal{A}$ such that $F(A) = 0$, since, by (3.6),

$$F(A_1) = F(A_2) = 0 \quad \Rightarrow \quad F(A_1 \cup A_2) = 0; \tag{5.2}$$

the restriction F' of F to $\mathcal{A}' = 2^{\mathcal{X} \setminus A}$ is itself a submodular capacity; in particular,

$$1 \geq F(\mathcal{X} \setminus A) = F(A) + F(\mathcal{X} \setminus A) \geq F(\mathcal{X}) + F(\emptyset) = 1; \tag{5.3}$$

and the elements of *core* f are those of *core* f' completed by zeros. Therefore, for any $V \in \mathcal{V}$ the contribution to $V(p)$ of the partial sum $\sum_{x_i \in A} v(p_i)$ is $|A|.\ v(0)$ for all $P \in \mathit{core}\, f$ and the V–maximization problem can be reduced to a maximization problem in $\mathcal{A}'$. Similarly, the relative Lorenz ordering of any $P, Q \in \mathit{core}\, f$ is the same as that of their restrictions to $\mathcal{A}'$. Thus we can henceforth assume without restriction that

$$F \text{ is a submodular capacity and is positive on } \mathcal{A}^*. \tag{5.4}$$

5.2. Construction of candidate solution P^*. In $\mathcal{L}$ all uncertainty indices as well as the Lorenz ordering rank first the uniform distribution, which suggests that the probability we are looking for should be "closest" to uniformity in some sense. One can start by trying to make the smallest elementary probability $P^*(\{x\})$ as large as possible by allocating the whole of $F(A_1)$ to set $A_1 \in \mathcal{A}^*$ such that

$$\frac{F(A_1)}{|A_1|} = \inf \left\{ \frac{F(E)}{|E|} : E \in \mathcal{A}^* \right\}, \tag{5.5}$$

and divide it equally between its members. Then, the similarity between the remaining partial optimization problem ("allocate probability mass $[1-F(A_1)]$ in an optimal way in $\mathcal{X} \setminus A_1$") and the initial one suggests that this first allocation step should be repeated recursively, leading to the following procedure:

5.3. Procedure P^*.

Initialization: $A_0 = B_0 = \emptyset$.

Iteration: Choose A_k such that $\emptyset \neq A_k \subseteq \mathcal{X} \setminus B_{k-1}$ and

$$\frac{F(B_{k-1} \cup A_k) - F(B_{k-1})}{|A_k|} = \inf \left\{ \frac{F(B_{k-1} \cup E) - F(B_{k-1})}{|E|} : \emptyset \neq E \subseteq \mathcal{X} \setminus B_{k-1} \right\}. \tag{5.6}$$

$$\text{Set } B_k = B_{k-1} \cup A_k \text{ and } \alpha_k = \frac{F(B_{k-1} \cup A_k) - F(B_{k-1})}{|A_k|}. \tag{5.7}$$

End: Halt for $k = K$ such that $B_K = \mathcal{X}$; define P^* by

$$P^*(\{x\}) = \alpha_k \quad \text{for } x \in A_k \text{ and } k = 1, \ldots, K. \tag{5.8}$$

This procedure is equivalent to those described in [4] and [12]. Note that although the A_k's and B_k's are not uniquely determined by the procedure, P^* is in fact unique (as implied by the results below).

Let us now check that:

PROPOSITION 5.1. *$P^* \in$ core f.*

PROOF. We have to show that

$$E \in \mathcal{A} \;\Rightarrow\; P^*(E) \leq F(E). \tag{5.9}$$

However,

$$P^*(E) = \sum_{k=1}^{K} P^*(E \cap A_k)\,, \tag{5.10}$$

and, by (5.6), (5.7), (5.8), and F's submodularity,

$$\begin{aligned} P^*(E \cap A_k) &= \frac{|E \cap A_k|}{|A_k|}[F(B_k) - F(B_{k-1})] \\ &\leq F(B_{k-1} \cup (E \cap A_k)) - F(B_{k-1}) \\ &\leq F((E \cap B_{k-1}) \cup (E \cap A_k)) - F(E \cap B_{k-1}) \\ &= F(E \cap B_k) - F(E \cap B_{k-1})\,. \end{aligned} \tag{5.11}$$

Thus

$$P^*(E) \leq \sum_{k=1}^{K}[F(E \cap B_k) - F(E \cap B_{k-1})] = F(E \cap B_K) = F(E). \tag{5.12}$$

□

Before proceeding further, we need the following lemma, which simply states that the probabilities attributed by the procedure to the elementary events become progressively larger and larger.

LEMMA 5.1. $0 < \alpha_k \leq \alpha_{k+1}$ for $k = 1, \ldots, K-1$.

PROOF. $\alpha_1 = \dfrac{F(A_1)}{|A_1|} > 0$ follows from the positivity of F on $\mathcal{A}^*$ [hypothesis (5.4)]. Moreover, by taking $E = A_k \cup A_{k+1}$ in (5.6), one gets

$$\alpha_k = \frac{F(B_k) - F(B_{k-1})}{|A_k|} \leq \frac{F(B_{k+1}) - F(B_{k-1})}{|A_k \cup A_{k+1}|}, \tag{5.13}$$

from which it follows, since $|A_k \cup A_{k+1}| = |A_k| + |A_{k+1}|$, that

$$\frac{F(B_k) - F(B_{k-1})}{|A_k|} \leq \frac{F(B_{k+1}) - F(B_k)}{|A_{k+1}|} = \alpha_{k+1}. \tag{5.14}$$

□

We can now show that P^* is the unique greatest element in *core* f for the Lorenz ordering.

PROPOSITION 5.2. For all $Q \in$ core f and for all $\ell \leq n$

$$\inf\{P^*(E) : |E| = \ell\} \geq \inf\{Q(E) : |E| = \ell\}, \tag{5.15}$$

Moreover (5.15) uniquely characterizes P^* in *core* f.

PROOF. For any fixed $\ell \leq n$, it results immediately from Lemma 5.1 that the infimum of $P^*(E)$ in $\{E \in \mathcal{A} : |E| = \ell\}$ is achieved at every E in $\mathcal{E}_\ell$, where:

$$\mathcal{E}_\ell = \{E \in \mathcal{A} : E = B_k \cup G,\ G \subset A_{k+1},\ |G| = \ell - |B_k|\}, \tag{5.16}$$

and that its value is

$$\inf\{P^*(E) : |E| = \ell\} = F(B_k) + r\,\alpha_{k+1}, \quad \text{where} \quad \alpha_{k+1} = \frac{F(B_{k+1}) - F(B_k)}{|A_{k+1}|} \quad \text{and} \quad r = \ell - |B_k|. \tag{5.17}$$

Let $Q \in$ *core* f and suppose that

$$Q(E) \geq P^*(E) \quad \text{for all } E \in \mathcal{E}_\ell, \tag{5.18}$$

in which case the average value of $Q(E)$ in $\mathcal{E}_\ell$ is also bounded by $P^*(E)$. Since each x in A_{k+1} belongs to the same proportion, $r/|A_k+1|$ of the G's, this last inequality can be expressed as

$$Q(B_k) + \frac{r}{|A_{k+1}|}Q(A_{k+1}) \geq F(B_k) + r\alpha_{k+1}, \tag{5.19}$$

which is itself equivalent to

$$(|A_{k+1}| - r)Q(B_k) + rQ(B_{k+1}) \geq (|A_{k+1}| - r)F(B_k) + rF(B_{k+1}). \tag{5.20}$$

Since $Q(B_k) \leq F(B_k)$ for all k, the equality must hold in (5.20) and in (5.18) as well:

$$Q(E) = P^*(E) \quad \text{for all } E \in \mathcal{E}_\ell. \tag{5.21}$$

Thus, whether (5.18) holds or not, for $Q \in$ *core* f,

$$\begin{aligned} \inf\{Q(E) : |E| = \ell\} &\leq \inf\{Q(E) : E \in \mathcal{E}_\ell\} \\ &\leq F(B_k) + r\alpha_{k+1} = \inf\{P^*(E) : |E| = \ell\}. \end{aligned} \tag{5.22}$$

Moreover, if Q is another greatest element, (5.18) must hold; then (5.21) must hold too, which implies that Q and P^* must coincide on A_{k+1}. Since this conclusion is valid for any $\ell \leq n$, necessarily $Q = P^*$. □

The Hardy et al. theorem allows one to infer from the preceding result that P^* maximizes all $V \in \mathcal{V}$ in *core* f and, in particular, is the maximum-entropy probability in *core* f. This result can however also be achieved directly as follows.

PROPOSITION 5.3. P^* maximizes every V of $\mathcal{V}$ in[3]

$$\mathcal{Q} = \left\{ p \in \mathcal{S}_n : \sum_{x_i \in B_k} p_i \leq F(B_k),\ k = 1, \ldots, K-1 \right\}. \tag{5.23}$$

PROOF. As a concave function on [0,1], v possesses subgradients γ_k and γ_{k+1} satisfying $\gamma_k \geq \gamma_{k+1}$ at, respectively, α_k and α_{k+1}, since $0 < \alpha_k \leq \alpha_{k+1} < 1$ $(k = 1, \ldots, K-1)$. Thus, since $\alpha_k = p_i^*$ for all $x_i \in A_k$, we get

$$v(p_i) - v(p_i^*) \leq \gamma_k[p_i - p_i^*] \quad \text{for all } p_i \in [0,1] \text{ and } x_i \in A_k, \tag{5.24}$$

hence, by summation,

$$V(p) - V(p^*) \ \leq\ \sum_{k=1}^{K} \gamma_k \sum_{x_i \in A_k} (p_i - p_i^*) \quad \text{for all } p_i \in [0,1]^n, \tag{5.25}$$

[3] Under a summation sign, an expression such as "$x_i \in E_j$" indicates a summation over "all i's s.t. $x_i \in E_j$."

and, for all $p \in \mathcal{S}_n$ (thus such that $\sum_{x_i \in B_K} p_i = 1 = F(B_K)$),

$$(5.26) \qquad V(p) - V(p^*) \leq \sum_{k=1}^{K-1} (\gamma_k - \gamma_{k+1}) \left[\sum_{x_i \in B_k} p_i - F(B_k) \right],$$

from which it follows immediately that

$$(5.27) \qquad V(p) - V(p^*) \leq 0 \quad \text{for all } p \in \mathcal{Q}.$$

□

Since $core\, f \in \mathcal{Q}$ and $P^* \in core\, f$ (Proposition 5.1), we can conclude that

PROPOSITION 5.4. P^* maximizes every V of $\mathcal{V}$ in *core f*.

More generally P^* maximizes (resp. minimizes) in *core f* any function $W : p \longmapsto W(p)$ which is increasing (resp. decreasing) with respect to the Lorenz ordering, also known as a *Schur–concave* (resp. *Schur–convex*) function; the class of decreasing functions includes in particular the "suitable convex functions" of [12], such as

$$(5.28) \qquad W(p) = \sum_{i=1}^{n} \sum_{r=1}^{n} |p_i - p_r|$$

(see [8]).

5.4. Computational considerations. The procedure P^* requires at each step the comparison of as much as 2^n ratios of the form $[F(B_k \cup E) - F(B_k)]/|E|$; moreover, in general, F is not directly known and its values have to be derived from those of φ (e.g., by the *fast Möbius transform* [19], [10]). The number of comparisons may be reduced somewhat, in virtue of the following result.

LEMMA 5.2. In Procedure P^*, sets B_k, $\ell \leq k \leq K$, have complementary sets $B_k^c = \mathcal{X} \setminus B_k$ of the form

$$(5.29) \qquad B_k^c = \bigcup_{G \in \mathcal{G},\, G \cap B_k = \emptyset} G,$$

where $\mathcal{G}$ is the support of φ.

PROOF. Let B_k' be the complementary set of $\cup_{G \in \mathcal{G},\, G \cap B_k = \emptyset} G$. Clearly $B_k' \supseteq B_k$ and $F(B_k') = \sum_{G \in \mathcal{G},\, G \cap B_k \neq \emptyset} \varphi(G) = F(B_k)$. Moreover, by construction of A_k and B_k, $A_k' = B_k' \setminus B_{k-1}$ and B_k' satisfy

$$(5.30) \qquad \frac{F(B_k') - F(B_{k-1})}{|A_k'|} \geq \frac{F(B_k) - F(B_{k-1})}{|A_k|},$$

from which it follows that $|A'_k| \leq |A_k|$, hence, since $A'_k \supseteq A_k$, that $A'_k = A_k$ and $B'_k = B_k$. □

Lemma 5.2 shows that the number of ratios to be compared at each step is at most $2^{|\mathcal{G}|}$, which may still be prohibitive.

An alternative way for computing P^* is offered *in the case where f is an ∞–monotone capacity*, by the property already recalled in Section 3 that *core* f is exactly composed of the allocations of f. Thus, by indexing $\mathcal{X}$ as $\{x_i,\ i \in I\}$ and $\mathcal{G}$ as $\{G_\ell,\ \ell \in L\}$ and setting for simplicity $\varphi(G_\ell) = \varphi_\ell$ and $\lambda(G_\ell, x_i) = \lambda_{\ell i}$, our problem can be restated as the following program

$$
(5.31)\qquad
\begin{aligned}
&\min\quad -\sum_{i=1}^{n} v(p_i) \\
&\sum_{G_\ell \ni x_i} \lambda_{\ell i}\varphi_\ell - p_i = 0 \qquad (i \in I) \\
&\sum_{x_i \in G_\ell} \lambda_{\ell i} = 1 \qquad (\ell \in L) \\
&\lambda_{\ell i} \geq 0 \qquad (i \in I,\ \ell \in L)
\end{aligned}
$$

which is a convex program with linear constraints, and even a quadratic program for the particular choice of $v : t \longmapsto v(t) = -t^2$. Standard convex and quadratic algorithms can therefore be used to determine P^*.

This technique does not extend to the case where f is only a convex capacity. However, if $\mathcal{P} = \textit{core}\, f$ is directly described by a set of inequalities such as "$P(A) \in [m, M]$" or "$P(A) \geq P(B)$" which can be transcribed as linear inequalities "$a \cdot p \leq b$," the maximum-entropy probability is again the solution of a convex program with linear constraints.

Finally, for the ∞–monotone case, [12] proposes a method based on successive reallocations of the $\varphi(G)$ masses governed by a principle of reduction of a "maximum shrinkable distance" between the allocated masses. The comparison between these diverse methods is the object of work in progress.

6. Conditional entropy maximization.

6.1. Questions and conditional entropy. In questionnaire theory, information is typically provided by answers to questions such as "which event E_{j^*} of partition $\mathcal{E} = \{E_1, \ldots, E_j, \ldots, E_m\}$ is true?" which we shall call *question* $\mathcal{E}$. Since the entropy after observation of event E_{j^*} would be

$$
(6.1)\qquad H_{E_{j^*}}(P) = -\sum_{x_i \in E_{j^*}} \left[\frac{P(\{x_i\})}{P(E_{j^*})}\right] \log \left[\frac{P(\{x_i\})}{P(E_{j^*})}\right],
$$

the ex-ante expected entropy after observation, or *conditional entropy given* $\mathcal{E}$, is[4]

$$
\begin{aligned}
H_{\mathcal{E}}(P) &= \sum_{j=1}^{m} P(E_j) H_{E_j}(P) \\
&= -\sum_{i=1}^{n} P(\{x_i\}) \log P(\{x_i\}) + \sum_{j=1}^{m} P(E_j) \log P(E_j),
\end{aligned} \tag{6.2}
$$

which, again, can be identified with the n-variable function

$$
h_{\mathcal{E}}(p) = -\sum_{j=1}^{n} p_i \log p_i + \sum_{j=1}^{m} \Big(\sum_{x_i \in E_j} p_i \Big) \log \Big(\sum_{x_i \in E_j} p_i \Big); \tag{6.3}
$$

moreover, by introducing probabilities

$$
q_j = P(E_j) = \sum_{x_i \in E_j} p_i \quad (j = 1, \ldots, m), \tag{6.4}
$$

this function can be conveniently written as

$$
h_{\mathcal{E}}(p) = -\sum_{j=1}^{n} p_i \log p_i + \sum_{j=1}^{m} q_j \log q_j . \tag{6.5}
$$

By definition, question $\mathcal{E}$ is *more informative* than question $\mathcal{E}'$ when $h_{\mathcal{E}}(p) < h_{\mathcal{E}'}(p)$.

In the case of *ambiguity*, i.e., when the probability P is only known to belong to a certain subset $\mathcal{L}$ and thus p to belong to the corresponding subset $\mathcal{P}$ of $\mathcal{S}_n$, a natural, though pessimistic, extension of the preceding criterion consists in considering question $\mathcal{E}$ as *more informative* than question $\mathcal{E}'$ if and only if

$$
\sup_{\mathcal{P}} h_{\mathcal{E}}(p) < \sup_{\mathcal{P}} h_{\mathcal{E}'}(p). \tag{6.6}
$$

This leads to the consideration of the constrained optimization problem:

$$
\max_{\mathcal{P}} h_{\mathcal{E}}(p).
$$

6.2. Concavity of the conditional entropy. Consider the extension g of $h_{\mathcal{E}}$ defined on the whole of $\mathbb{R}_n^+$ by

$$
g(p) = -\sum_{i=1}^{n} p_i \log p_i + \sum_{j=1}^{m} q_j \log q_j , \tag{6.7}
$$

where $q_j = \sum_{x_i \in E_j} p_i$.

[4] Note that $H_{\mathcal{E}}(P)$ is always defined (even if some $H_{E_{j*}}(P)$ is not).

PROPOSITION 6.1. g is a concave function on $\mathbb{R}_n^+$.

PROOF. It consists in using the fact that on the interior of $\mathbb{R}_n^+$, g is differentiable, with partial derivatives

$$\frac{\partial g(p)}{\partial p_i} = -\log p_i + \log q_j \ (E_j \ni x_i, \ i = 1, \dots, n) \tag{6.8}$$

and checking the sufficient condition [11]

$$g(p+\delta) \leq g(p) + \sum_{i=1}^{n} \frac{\partial g(p)}{\partial p_i}\delta_i \quad \text{for all } p \gg 0 \text{ and } p+\delta \gg 0. \tag{6.9}$$

The continuity of g implies then its concavity on the whole of $\mathbb{R}_n^+$. □

6.3. The conditional entropy maximization problem. We henceforth assume that

$$f \text{ is an } \infty\text{–}\textit{monotone capacity.} \tag{6.10}$$

We use again indexations $\mathcal{X} = \{x_i, i \in I\}$, $\mathcal{E} = \{E_j, j \in J\}$ and $\mathcal{G} = \{G_\ell, \ell \in L\}$ and abbreviations $\varphi_\ell = \varphi(G_\ell)$ and $\lambda_{\ell i} = \lambda(G_\ell, x_i)$. Also we note $E_{j(i)}$ the element of $\mathcal{E}$ which contains x_i: thus, $x_i \in E_{j(i)}$.

As already noticed, when f is ∞–monotone, $\mathcal{P} = \mathit{core}\, f$ can be identified to the set of all p such that:

$$\begin{aligned} p_i &= \sum_{G_\ell \ni x_i} \lambda_{\ell i}\varphi_\ell; \\ \sum_{x_i \in G_\ell} \lambda_{\ell i} &= 1 \qquad (\ell \in L); \\ \lambda_{\ell i} &\geq 0 \qquad (x_i \in G_\ell, \ \ell \in L). \end{aligned} \tag{6.11}$$

The optimization problem $\max_{\mathcal{P}} h_{\mathcal{E}}(p)$ can thus be re-expressed with respect to the new variables $\lambda_{\ell i}$ as:

$$\begin{aligned} \min \sum_{i \in I} &\left[\sum_{G_\ell \ni x_i} \lambda_{\ell i}\varphi_\ell\right] \log \left[\sum_{G_\ell \ni x_i} \lambda_{\ell i}\varphi_\ell\right] - \\ &- \sum_{j \in J}\left[\sum_{x_i \in E_j}\sum_{G_\ell \ni x_i} \lambda_{\ell i}\varphi_\ell\right] \log \left[\sum_{x_i \in E_j}\sum_{G_\ell \ni x_i} \lambda_{\ell i}\varphi_\ell\right] \\ \sum_{x_i \in G_\ell} \lambda_{\ell i} = 1 &\qquad (\ell \in L) \\ \lambda_{\ell i} \geq 0 &\qquad (x_i \in G_\ell, \ \ell \in L). \end{aligned} \tag{6.12}$$

Note that this is a convex program with linear constraints: the domain of feasible solutions is the product of $r = |L|$ simplices, a convex compact subset of $\mathbb{R}^k$ for $k = \left[\sum_{\ell \in L} |G_\ell|\right] - r$; and the objective function is continuous and convex as the composition of the convex function $(-g)$ (Proposition 6.1) and a linear function.

6.4. Optimality conditions. Thus (6.12) has an optimal solution (which may not be unique), and the Kuhn and Tucker conditions are necessary and sufficient optimality conditions at points where they are defined (Mangasarian [11]): Variables $\lambda_{\ell i}$ ($x_i \in G_\ell$, $\ell \in L$), such that all the terms in (6.13) below are defined, are an optimal solution if and only if there exists Lagrange multipliers u_ℓ ($\ell \in L$) and $v_{\ell i}$ ($x_i \in G_\ell$, $\ell \in L$) such that:

$$\begin{aligned}
&\sum_{x_i \in G_\ell} \lambda_{\ell i} = 1 \ (\ell \in L); \quad u_\ell \in \mathbb{R} \ (\ell \in L);\\
&\lambda_{\ell i} \geq 0 \text{ and } v_{\ell i} \geq 0 \qquad (x_i \in G_\ell,\ \ell \in L);\\
&\varphi_\ell \left[\log \left[\sum_{G_\ell \ni x_i} \lambda_{\ell i} \varphi_\ell \right] - \log \left[\sum_{x_{i'} \in E_{j(i)}} \sum_{G_{\ell'} \ni x_{i'}} \lambda_{\ell' i'} \varphi_{\ell'} \right] \right]\\
&\qquad - u_\ell - v_{\ell i} = 0 \qquad (x_i \in G_\ell,\ \ell \in L);\\
&\lambda_{\ell i}\, v_{\ell i} = 0 \qquad (x_i \in G_\ell,\ \ell \in L).
\end{aligned} \tag{6.13}$$

Note that, since

$$\sum_{G_\ell \ni x_i} \lambda_{\ell i} = p_i = P(\{x_i\}) \ , \tag{6.14}$$

and

$$\sum_{x_{i'} \in E_{j(i)}} \sum_{G_{\ell'} \ni x_{i'}} \lambda_{\ell' i'} \varphi_{\ell'} = \sum_{x_{i'} \in E_{j(i)}} p_{i'} = P(E_{j(i)}), \tag{6.15}$$

(6.13) in fact requires that

$$\varphi_\ell \left[\log P(\{x_i\}) - \log P(E_{j(i)}) \right] - u_\ell - v_{\ell i} = 0 \ (x_i \in G_\ell,\ \ell \in L). \tag{6.16}$$

From these conditions, one can derive necessary properties of optimal allocations: Consider $G_\ell \in \mathcal{G}$ and suppose that there exists $x_i,\ x_{i'} \in G_\ell$ such that

$$\frac{P(\{x_{i'}\})}{P(E_{j(i')})} > \frac{P(\{x_i\})}{P(E_{j(i)})} > 0; \tag{6.17}$$

then

$$\log \frac{P(\{x_i\})}{P(E_{j(i)})} = \frac{u_\ell + v_{\ell i}}{\varphi_\ell} < \frac{u_\ell + v_{\ell i'}}{\varphi_\ell} = \log \frac{P(\{x_{i'}\})}{P(E_{j(i')})}. \tag{6.18}$$

Hence,

$$\frac{P(\{x_{i'}\})}{P(E_{j(i')})} > \frac{P(\{x_i\})}{P(E_{j(i)})} \Rightarrow v_{\ell i'} > 0 \Rightarrow \lambda_{\ell i'} = 0. \tag{6.19}$$

Thus, we can state:

PROPOSITION 6.2 A conditional entropy maximizing probability P is necessarily an allocation of f in which every Möbius mass $\varphi(G)$, $G \in \mathcal{G}$, is entirely allocated to the states of nature x_i in G which have the lowest positive relative probability $P(\{x_i\})/P(E_{j(i)})$.

Conversely, an allocation $\lambda_{\ell i}$ $(x_i \in G_\ell,\ \ell \in L)$ satisfying the preceding requirement can be completed by u_ℓ $(\ell \in L)$ and $v_{\ell i}$ $(x_i \in G_\ell,\ \ell \in L)$ successively determined by (6.13) to form a solution of the Kuhn-Tucker conditions: first, there is some $x_i \in G_\ell$ which receives a positive part of the Möbius mass $\varphi_\lambda : \lambda_{\ell i} > 0$ and, thus, $v_{\ell i} = 0$; thus (6.13) determines first u_ℓ second $v_{\ell i'}$ for every other $x_{i'} \in G_\ell$.

6.5. Decomposition of the optimization problem. Fix all the components $\lambda^*_{\ell i}$ of an optimal allocation except those for which $x_i \in E_j$ for a given $E_j \in \mathcal{E}$. The remaining components $\lambda^*_{\ell i}$ $(x_i \in G_\ell \cap E_j,\ \ell \in L)$ form necessarily an optimal solution of the restricted program

$$(6.20)\quad \begin{aligned} &\min \sum_{x_i \in E_j} \left[\sum_{G_\ell \ni x_i} \lambda_{\ell i}\varphi_\ell\right] \log \left[\sum_{G_\ell \ni x_i} \lambda_{\ell i}\varphi_\ell\right] \\ &\sum_{x_i \in G_\ell \cap E_j} \lambda_{\ell i} = \sum_{x_i \in G_\ell \cap E_j} \lambda^*_{\ell i} \quad (\ell \in L) \\ &\lambda_{\ell i} \geq 0 \qquad (x_i \in G_\ell \cap E_j,\ \ell \in L). \end{aligned}$$

Let $\sum_{x_i \in G_\ell \cap E_j} \lambda^*_{\ell i} = k_\ell$ and $L^0 = \{\ell \in L;\ k_\ell > 0\}$; by introducing new variables $\mu_{\ell i} = (1/k_\ell)\lambda_{\ell i}$ and coefficients $\psi_\ell = k_\ell\varphi_\ell(\ell \in L^0)$, this program becomes

$$(6.21)\quad \begin{aligned} &\min \sum_{x_i \in E_j} \left[\sum_{G_\ell \ni x_i} \mu_{\ell i}\psi_\ell\right] \log \left[\sum_{G_\ell \ni x_i} \mu_{\ell i}\psi_\ell\right] \\ &\sum_{x_i \in G_\ell \cap E_j} \mu_{\ell i} = 1 \quad (\ell \in L^0) \\ &\mu_{\ell i} \geq 0 \quad (x_i \in G_\ell \cap E_j,\ \ell \in L^0), \end{aligned}$$

the solution of which is the maximum entropy probability consistent with the ∞–monotone capacity characterized by Möbius masses ψ_ℓ on $S^0 = \{G_\ell : \ell \in L^0\}$.

Algorithms solving this type of problem have been presented and discussed in Section 5 and can be used as subprocedures in the general problem [13]. This adaptation is likely to reduce significantly the computational time of classical convex programming methods. The investigation of this question is the object of future research.

7. Example. $\mathcal{X} = \{x_i,\ i = 1,\ldots,8\}$; $\mathcal{E} = \{E_1, E_2\}$ with $E_1 = \{x_1, x_2, x_3, x_4\}$ and $E_2 = \{x_5, x_6, x_7, x_8\}$; $\mathcal{G} = \{G_\ell,\ \ell = 1,\ldots,6\}$ with:

ℓ	1	2	3	4	5	6
G_ℓ	x_1x_5	x_2x_3	x_3x_4	$x_4x_7x_8$	x_6x_7	x_5x_6
$14\ \varphi_\ell$	3	2	2	3	2	2
$14\ \varphi_{1\ell}$	1	2	2	1	×	×
$14\ \varphi_{2\ell}$	2	×	×	2	2	2

[N.B.: x_1x_5 is short for $\{x_1, x_5\}$, etc.]

It is easily checked (by finding the corresponding allocation $(\lambda_{\ell i})$) that the uniform probability $P^0(\{x\}) = 1/8$ belongs to *core* f and therefore maximizes the prior entropy.

Sub-optimization in E_1 and E_2, given initial mass allocations $\varphi_{j\ell}$ to $G_\ell \cap E_j$ $(j = 1, 2)$ leads to probability P:

i	1	2	3	4	5	6	7	8
$14\ p_i$	1	5/3	5/3	5/3	2	2	2	2
$14\ p_i/q_j(i)$	1/6	5/18	5/18	5/18	1/4	1/4	1/4	1/4

and $14\, H_{\mathcal{E}}(P) = 19,28$.

[N.B.: p_i and $p_i/q_{j(i)}$ are short for $P(\{x_i\})$ and $P(\{x_i\})/P(E_{j(i)})$]

Since $P(\{x_1\})/P(E_1) < P(\{x_5\})/P(E_2)$ and some of the mass φ_1 on $G_1 = \{x_1, x_5\}$ goes to E_2, P is not optimal. By re-allocating φ_1 so that the new allocations $(\times 14)$ to $G_1 \cap E_1$ and $G_1 \cap E_2$ become 5/3 and 4/3 respectively, one can achieve probability P^*:

i	1	2	3	4	5	6	7	8
$14\ p_i^*$	5/3	5/3	5/3	5/3	11/6	11/6	11/6	11/6

and $14\, H_{\mathcal{E}}(P^*) = 19,408$.

P^* is optimal since its restrictions to E_1 and E_2 are uniform and:

$$\frac{P^*(\{x_1\})}{P^*(E_1)} = \frac{P^*(\{x_5\})}{P^*(E_2)}; \quad \frac{P^*(\{x_4\})}{P^*(E_1)} = \frac{P^*(\{x_7\})}{P^*(E_2)} = \frac{P^*(\{x_8\})}{P^*(E_2)}.$$

8. Conclusion. The preceding results make it possible to compare the questions which are feasible at some step of the questioning process and choose the most informative one, according to the worst case evaluation. The achievement of the final goal, which is to build a good (although still suboptimal) questionnaire, may still not be easy: even in the probabilistic case, the myopic algorithm, which constructs the tree of questions step by step by selecting at each node as next question the most informative one, is just a heuristic which has proven experimentally to be fairly efficient; here, with imprecise probabilities, the gap between the myopic questionnaire and the globally optimal one is likely to widen, due to the fact that the worst evaluations of the diverse questions may well correspond to incompatible hypotheses on the true probability. This is a well known source of dynamic inconsistency (see, e.g., [9]). Further theoretical and empirical research is needed to evaluate the impact of this phenomenon and find out how to limit it.

The results on conditional entropy maximization, like those on entropy maximization, should be generalizable to the case of convex lower probabilities. This could be achieved by first managing to express *core* f as a set of linear constraints on the allocations of f or perhaps in some other way.

These complements and extensions, as well as the development of computationally efficient algorithms, is the object of future research.

REFERENCES

[1] C. Berge, *Espaces Topologiques, Fonctions Multivoques*, Dunod, Paris, 1966.

[2] A. Chateauneuf and J.Y. Jaffray, *Some characterizations of lower probabilities and other monotone capacities through the use of Möbius inversion*, Math. Soc. Sci., 17 (1989), pp. 263–283.

[3] A. Chateauneuf and J.Y. Jaffray, *Local Möbius transforms of monotone capacities*, Symbolic and Quantitative Approaches to Reasoning and Uncertainty (C. Froidevaux and J. Kohlas, eds.) Springer, Berlin, 1995, pp. 115–124.

[4] B. Dutta and D. Ray, *A concept of egalitarianism under participation constraints*, Econometrica, 57 (1989), pp. 615–635.

[5] P.L. Hammer, U.N. Peled, and S. Sorensen, *Pseudo–Boolean functions and game theory: Core elements and Shapley value*, Cahiers du CERO, 19 (1977), no. 1–2.

[6] G. Hardy, J. Littlewood, and G. Polya, *Inequalities*, Cambridge Univ. Press, 1934.

[7] P.J. Huber and V. Strassen, *Minimax tests and the Neyman–Person lemma for capacities*, Ann. Math. Stat., 1 (1973), pp. 251–263.

[8] F.K. Hwang and U.G. Rothblum, *Directional–quasi–convexity, asymmetric Schur–convexity and optimality of consecutive partitions*, Math. Oper. Res., 21 (1996), pp. 540–554.

[9] J.Y. Jaffray, *Dynamic decision making and belief functions*, Advances in the Dempster–Shafer Theory of Evidence, (R. Yager, M. Fedrizzi, and J. Kacprzyk, eds.), Wiley, 1994, pp. 331–351.

[10] Kennes, *Computational aspects of the Möbius transformation of graphs*, IEEE Transaction on Systems, Man and Cybernetics, 22 (1992), pp. 201–223.

[11] O.L. Mangasarian, *Nonlinear Programming*, Mc Graw–Hill, 1969.

[12] A. MEYEROWITZ, F. RICHMAN, AND E.A. WALKER, *Calculating maximum entropy probability densities for belief functions*, IJUFKS, 2 (1994), pp. 377–390.
[13] M. MINOUX, *Programmation Mathématique*, Dunod, 1983.
[14] H.T. NGUYEN, *On random sets and belief functions*, J. Math. Analysis and Applications, 65 (1978), pp. 531–542.
[15] H.T. NGUYEN AND E.A. WALKER, *On Decision–making using belief functions*, Advances in the Dempster–Shafer Theory of Evidence, (R. Yager, R. Fedrizzi, and J. Kacprzyk, eds.), Wiley, New-York, 1994, pp. 331–330.
[16] C. PICARD, *Graphes et Questionnaires*, Gauthier–Villars, 1972.
[17] G. SHAFER, *A Mathematical Theory of Evidence*, Princeton University Press, Princeton, New Jersey, 1976.
[18] L.S. SHAPLEY, *Cores of convex games*, Int.J. Game Theory, 1 (1971), pp. 11–22.
[19] H.M. THOMA, *Belief functions computations*, Conditional logic in expert systems, (I.R. Goodman, M. Gupta, H.T. Nguyen, and G.S. Rogers, eds.), North Holland, New York, 1991, pp. 269–308.

RANDOM SETS IN INFORMATION FUSION AN OVERVIEW

RONALD P. S. MAHLER*

Abstract. "Information fusion" refers to a range of military applications requiring the effective pooling of data concerning multiple targets derived from a broad range of evidence types generated by diverse sensors/sources. Methodologically speaking, information fusion has two major aspects: *multisource, multitarget estimation* and *inference using ambiguous observations (expert systems theory).* In recent years some researchers have begun to investigate random set theory as a foundation for information fusion. This paper offers a brief history of the application of random sets to information fusion, especially the work of Mori et. al., Washburn, Goodman, and Nguyen. It also summarizes the author's recent work suggesting that random set theory provides a systematic foundation for both multisource, multitarget estimation and expert-systems theory. The basic tool is a statistical theory of random *finite* sets which *directly* generalizes standard single-sensor, single-target statistics: density functions, a theory of differential and integral calculus for set functions, etc.

Key words. Data Fusion, Random Sets, Nonadditive Measure, Expert Systems.

AMS(MOS) subject classifications. 62N99, 62B10, 62B20, 60B05, 60G55, 04A72

1. Introduction. *Information fusion*, or *data fusion* as it is more commonly known [1] [17], [54], is a new engineering specialty which has arisen in the last two decades in response to increasing military requirements for achieving "battlefield awareness" on the basis of very diverse and often poor-quality data. The purpose of this paper is twofold. First, to provide a short history of the application of random set theory to information fusion. Second, to summarize recent work by the author which suggests that random set theory provides a unifying scientific foundation, as well as new algorithmic approaches, for much of information fusion.

1.1. What is information fusion. A common information fusion problem is as follows. A military surveillance aircraft is confronted with a number of possible threats: fighter airplanes, missiles, mobile ground-based weapons launchers, hardened bunkers, etc. The aircraft is equipped with several on-board ("organic") sensors: e.g. doppler radars for detecting motion, imaging sensors such as *Synthetic Aperture Radar* (SAR) or *Infrared Search and Track* (IRST), special receivers for detecting and identifying electromagnetic transmissions from friend or foe, and so on. In addition the surveillance aircraft may receive off-board ("nonorganic") information from ancillary platforms such as fighter aircraft, from special military communications links, from military satellites, and so on. Background or prior information is also available: terrain maps, weather reports, knowledge of

* Lockheed Martin Tactical Defense Systems, Eagan, MN 55121 USA.

enemy rules of engagement or tactical doctrine, heuristic rule bases, and so on.

The nature and quality of this data varies considerably. Some information is fairly precise and easily represented in statistical form, as in a range-azimuth-altitude "report" (i.e., observation) from a radar with known sensor-noise characteristics. Other observations ("contacts") may consist only of an estimated position of unknown certainty, or even just a bearing-angle-to-target. Some information consists of images contaminated by clutter and/or occlusions. Still other information may be available only in natural-language textual form or as rules.

1.2. The purpose of information fusion. The goal of information fusion is to take an often voluminous mass of diverse data types and assemble from them a comprehensive understanding of both the current and evolving military situation. Four major subfunctions are required for this effort to succeed. First, *multisource integration* (also called *Level 1 fusion*), is aimed at the detection, identification, localization, and tracking of targets of interest on the basis of data supplied by many sources. Targets of interest may be individual platforms or they can be "group targets" (brigades and battalions, aircraft carrier groups, etc.) consisting of many military platforms moving in concert. Secondly, *sensor management* is the term which describes the proper allocation of sensor and other information resources to resolve ambiguities. Third, *situation assessment* is the process of inferring the probable degrees of threat (the "threat state") posed by various platforms, as well as their intentions and most likely present and future actions. The final major aspect of information fusion, and in many ways the most difficult, is *response management.* This refers to the process of determining the most effective courses of action (e.g. evasive maneuvers, countermeasures, weapons delivery) based on knowledge of the current and predicted situation.

Four basic practical problems of information fusion remain essentially unresolved. The first is the *problem of incongruent data.* Despite the fact that information can be statistical (e.g. radar reports), imprecise (e.g. target features such a sonar frequency lines), fuzzy (e.g. natural language statements), or contingent (e.g. rules), one must somehow find a way to pool this information in a meaningful way so as to gain more knowledge than would be available from any one information type alone.

Second is the *problem of incongruent legacy systems.* A great deal of investment has been made in algorithms which perform various special functions. These algorithms are based on a wide variety of mathematical and/or heuristic paradigms (e.g. Bayesian probability, fuzzy logic). New approaches–e.g. "weak evidence accrual"–are introduced on a continuing basis in response to the pressure of real-world necessities. Integrating the knowledge produced by these "legacy" systems in a meaningful way is a major requirement of current large-scale fusion systems.

The third difficulty is the *level playing field problem.* Comparing the performance of two data fusion systems, or determining the performance of any individual system relative to some predetermined standard, is far more daunting a prospect than at first might seem to be the case. Real-world test data is expensive to collect and often hard to come by. Algorithms are typically compared using metrics which measure some very specific aspect of performance (e.g., miss distance, probability of correct identification). Not infrequently, however, optimization of an algorithm with respect to one set of such measures results in degradation in performance with respect to other measures. Also, there is no obvious over-all "figure of merit" which permits the direct comparison of two complex multi-function algorithms. In the case of performance evaluation of individual algorithms exercised with real data, it is often not possible to know how well one is doing against real–and therefore often high-ambiguity–data unless one also has some idea of the best performance that an ideal system could expect. This, in turn, requires the existence of theoretical best-performance bounds analogous to, for example, the Cramér-Rao bound.

The final practical problem is that of *multifunction integration.* With few exceptions, current information fusion systems are patchwork amalgams of various subsystems, each subsystem dedicated to the performance of a specific function. One algorithm may be dedicated to target detection, another to target tracking, still another to target identification, and so on. Better performance should result if these functions are tightly integrated–e.g. so that target I.D. can help resolve tracking ambiguities and vice-versa.

In the remainder of this introduction, we describe what we believe are, from a purely mathematical point of view, the two major methodological distinctions in information fusion: *point vs. set estimation*, and *indirect vs. direct estimation.*

1.3. Point estimation vs. set estimation. From a methodological point of view information fusion breaks down into two major parts: *multisource, multitarget estimation* and *expert systems theory.* The first tends to be heavily statistical and refers to the process of constructing estimates of target numbers, identities, positions, velocities, etc. from observations collected from many sensors and other sources. The second tends to mix mathematics and heuristics in varying proportion, and refers to the process of pooling observations which are inherently imprecise, vague, or contingent–e.g. natural language statements, rules, signature features, etc. These two methodological schools of information fusion exist in an often uneasy alliance. From a mathematical point of view, however, the distinction between them is essentially the same as the distinction between *point estimation* and *set estimation.* In statistics, *maximum likelihood estimation* is the most common example of point estimation whereas *interval estimation* is the most familiar example of set estimation. In information fusion, the distinction between point estimation and set estimation is what essentially separates *multisource, multitarget estimation* from *expert systems theory* (fuzzy logic, Dempster-Shafer evidential theory, rule-based inference, etc. [42]).

The goal of multisensor, multitarget estimation is, just as is the case in conventional single-sensor, single-target estimation, to construct "estimates" which are specific points in some state space, along with an estimate of the degree of inaccuracy in those estimates. The goal of expert systems theory, however, is more commonly that of using evidence to constrain target states to membership in some subset–or fuzzy subset–of possible states to some specified degree of accuracy. (For example, in response to the observation *RED* one does not try to construct a point estimate which consists of some generically *RED* target, but rather a set estimate–the set of all red targets–which limits the possible identities of the target.)

1.4. Indirect vs. direct multitarget estimation. A second important distinction in information fusion is that between *indirect* and *direct* multisensor, multitarget estimation. To illustrate this distinction, let us suppose that we have two targets on the real line which have states x_1 and x_2, and let us assume that each target generates a unique observation and that any observation is generated by only one of the two targets. Suppose that observations are collected and that they are z_1 and z_2. There are two possible ways–called "report-to-track associations"–that the targets could have generated the observed data:

$$\begin{array}{ll} z_1 \leftrightarrow x_1, & z_2 \leftrightarrow x_2 \\ z_2 \leftrightarrow x_1, & z_1 \leftrightarrow x_2 \end{array}$$

Indirect estimation approaches take the point of view that the "correct" association is an observable parameter of the system and thus can be estimated. Given that this is the case, if $z_1 \leftrightarrow x_1, z_2 \leftrightarrow x_2$ were the correct association then one could use Kalman filters to update track x_1 using report z_1 and update track x_2 using report z_2. On the other hand, if $z_2 \leftrightarrow x_1, z_1 \leftrightarrow x_2$ were the correct association then one could update x_1 using z_2 and x_2 using z_1. One could thus reduce any multitarget estimation problem to a collection of ordinary single-target estimation problems. This is the point of view adopted by the dominant indirect estimation approach, *ideal multihypothesis estimation (MHE)*. MHE is frequently described as "the optimal" solution to multitarget estimation [4], [3], [5], [47], [52].

Direct estimation techniques, on the other hand, take the view that the correct report-to-track association is an *unobservable parameter of the system* and that, consequently, if one tries to estimate it one is improperly introducing information that cannot be found in the data. That is, one is introducing an *inherent bias*. Although we cannot go into this issue in any great detail in this brief article, this bias has been demonstrated in the case of MHE [20], [22]. Counterexamples have been constructed which demonstrate that MHE is "optimal"–that is, produces the correct Bayesian posterior distribution–only under certain restrictive assumptions [51], [2].

1.5. Outline of the paper. The remainder of the paper is divided into three main sections. In Section 2 we try to answer the question, *Why are random sets necessary for information fusion?* This section includes a short history of the application of random set theory to two major aspects of information fusion: *multisensor, multitarget estimation* and *expert systems theory.* In Section 3 we summarize the mathematical basis of our approach to information fusion: a special case of random set theory which we call "finite-set statistics." Section 4 is devoted to showing how finite-set statistics can be used to integrate expert systems theory with multisensor, multitarget estimation, thus resulting in a fully "unified" approach to information fusion. Conclusions and a summary may be found in Section 5.

2. Why random sets. In this section we will summarize the major reasons why random set theory might reasonably be considered to be of interest in information fusion. We also attempt to provide a short history of the application of random set theory to information fusion problems. We begin, in Section 2.1, by showing that random set theory provides a natural way of looking at one major aspect of information fusion: *multisensor, multitarget estimation.* We describe the pioneering venture in this direction, research due to Mori, Chong, Tse, and Wishner. We also describe a closely related effort, a multitarget filter based on point process theory, due to Washburn. In Section 2.2 we describe the application of random set theory to another major aspect of information fusion: *expert systems theory.* This section describes how random set theory provides a common foundation for the Dempster-Shafer theory of imprecise evidence, for fuzzy logic, and for rule-based inference. It also includes a short description of the relationship between random set theory and a "generalized fuzzy logic" due to Li. The Section concludes with a discussion of why one might prefer random set models to vector models or point process models in information fusion applications.

2.1. Why random sets: Multitarget estimation. Let us first begin with the multitarget estimation problem and consider the situation of a single sensor collecting observations from three non–moving targets $\otimes_1, \otimes_2, \otimes_3$. At some instant the sensor interrogates the three targets and collects *four* observations $\star_1, \star_2, \star_3, \star_4$, three of which are more or less associated with specific targets, and a false alarm. The sensor interrogates once again at the same instant and collects *two* observations $\bullet_1, \bullet_2$, with one target being entirely undetected. A final interrogation is performed, resulting in *three* observations $\blacksquare_1, \blacksquare_2, \blacksquare_3$. The point to notice is that the *actual observations* of the problem are the observation-sets

$$Z_1 = \{\star_1, \star_2, \star_3, \star_4\}$$

$$Z_2 = \{\bullet_1, \bullet_2\}$$
$$Z_3 = \{\blacksquare_1, \blacksquare_2, \blacksquare_3\}$$

That is, the "observation" is a *randomly varying finite set* whose randomness comprises not just statistical variability in the specific observations themselves, but also in their *number*. That is, the observation is a *random finite set of observation space.*

Likewise, suppose that some multitarget tracking algorithm constructs estimates of the state of the system from this data. Such estimates could take the form $X_1 = \{\hat{\otimes}_1, \hat{\otimes}_2, \hat{\otimes}_3\}$, $X_2 = \{\hat{\otimes}_1, \hat{\otimes}_2\}$ or $X_3 = \{\hat{\otimes}_1, \hat{\otimes}_2, \hat{\otimes}_3, \hat{\otimes}_4\}$. That is, the *estimate* produced by the multitarget tracker can (with certain subtleties that need not be discussed in detail here, e.g. the possibility that $\hat{\otimes}_1 = \hat{\otimes}_2$ but $|X_1| = 3$) itself be regarded as a *random finite set of state space.*

The Mori–Chong–Tse–Wishner (MCTW) filter. The first information fusion researchers to apply a random set perspective to multisensor, multitarget estimation seem to have been Shozo Mori, Chee-Yee Chong, Edward Tse, and Richard Wishner of Advanced Information and Decision Systems Corp. In an unpublished 1984 report [40], they argued that the multitarget tracking problem is "non-classical" in the sense that:

> (1) The number of the objects to be estimated is, in general, random and unknown. The number of measurements in each sensor output is random and a part of observation information. (2) Generally, there is no a priori labeling of targets and the order of measurements in any sensor output does not contain any useful information. For example, a measurement couple $(\mathbf{y}_1, \mathbf{y}_2)$ from a sensor is totally equivalent to $(\mathbf{y}_2, \mathbf{y}_1)$. When a target is detected for the first time and we know it is one of n targets which have never been seen before, the probability that the measurement originat[ed] from a particular target is the same for any such target, i.e., it is $1/n$. The above properties (1) and (2) are properly reflected when both targets and sensor measurements are considered as random sets as defined in [Mathéron]....The uncertainty of the origin of each measurement in every sensor output should then be embedded in a sensor model as a stochastic mechanism which converts a random set (set of targets) into another random set (sensor outputs)...

Rather than making use of Mathéron's systematic random-set formalism [38], however, they further argued that:

> ...a random finite set X of reals can be probabilistically completely described by specifying probability $Prob.\{|X| = n\}$

> for each nonnegative n and joint probability distribution with density $p_n(\mathbf{x}_1, ..., \mathbf{x}_n)$ of elements of the set for each positive n...[which is completely symmetric in the variables $\mathbf{x}_1, ..., \mathbf{x}_n$] [40, pp. 1-2].

They then went on to construct a representation of random finite-set variates in terms of conventional stochastic quantities (i.e., discrete variates and continuous vector variates). Under Defense Advanced Research Projects Administration (DARPA) funding they used this formalism to develop a multihypothesis-type, fully integrated multisensor, multitarget detection, association, classification, and tracking algorithm [41].

Simplifying their discussion somewhat, Mori et. al. represented target states as elements of a hybrid continuous-discrete space of the form $\mathcal{R} = \mathbb{R}^n \times U$ (where U is a finite set) endowed with the hybrid (i.e., product Lebesgue-counting) measure. Finite subsets of state space are modeled as vectors in disjoint-union spaces of the form

$$\mathcal{R}^{(n)} \triangleq \bigcup_{k=0}^{\infty} \mathcal{R}^k \times \{k\}$$

Thus a typical element of $\mathcal{R}^{(n)}$ is $(\mathbf{x}, k)$ where $\mathbf{x} = (x_1, ..., x_k)$; and a random finite subset of $\mathcal{R}$ is represented as a random variable $(\mathbf{X}, N_T)$ on $\mathcal{R}^{(n)}$. A finite stochastic subset of $\mathcal{R}$ is represented as a discrete-time stochastic process $(\mathbf{X}(\alpha), N_T)$ where N_T is a random variable which is constant with respect to time. Various assumptions are made with permit the reduction of simultaneous multisensor, multitarget observations to a discrete time-series of multitarget observations. Given a sensor suite with sensor tags $U_\odot = \{1, ..., s\}$, the observation space has the general form

$$\mathcal{R}_\odot^{(m)} \triangleq \mathcal{R}^{(m)} \times \mathbb{N} \times U_\odot$$

where $\mathbb{N} = \{1, 2, ...\}$ is the set of possible discrete time tags beginning with the time tag 1 of the initial time-instant. Any given observation thus has the form $(\mathbf{y}, m, \alpha, s)$ where $\mathbf{y} = (y_1, ..., y_m)$, where α is a time tag, and where $s \in U_\odot$ is a sensor tag. The time-series of observations is a discrete-time stochastic process of the form $(\mathbf{Y}(\alpha), N_M(\alpha), \alpha, s_\alpha)$. Here, for each time tag α the random integer $s_\alpha \in U_\odot$ is the identifying sensor tag for the sensor active at that moment. The random integer $N_M(\alpha)$ is the number of measurements collected by that sensor. The vector $\mathbf{Y}(\alpha)$ represents the (random) subset of actual measurements.

The MCTW formalism is an *indirect estimation* approach in the sense described in Section 1. State estimation (i.e., determining the number, identities, and geodynamics of targets) is contingent on a *multihypothesis* track estimation scheme which necessitates the determination of correct report-to-track associations.

Washburn's point–process filter. In 1987, under Office of Naval Research funding, Robert Washburn [55] of Alphatech, Inc. proposed a theoretical approach to multitarget tracking which bears some similarities to random set theory since it is based on *point process theory* (see [7]). Washburn noted that the basis of the point process approach

> ...is to consider the observations occurring in one time period or scan of data as an image of points rather than as a list of measurements. The approach is intuitively appealing because it corresponds to one's natural idea of radar and sonar displays–devices that provide two-dimensional images of each scan of data. [55, p. 1846]

Washburn made use of the fact that randomly varying finite sets of data can be mathematically represented as *random integer-valued ("counting") measures.* (For example, if Σ is a random finite subset of measurement space then $N_\Sigma(S) \triangleq |\Sigma \cap S|$ defines a random counting measure N_Σ.) In this formalism the "measurement" collected at time-instant α by a single sensor in a multitarget environment is an integer-valued measure μ_α.

Washburn assumes that the number n of targets is known and he specifies a measurement model of the form

$$\mu_\alpha \triangleq \tau_\alpha + \nu_\alpha$$

The random measure ν_α, which models clutter, is assumed to be a Poisson measure. The random measure τ_α models detected targets:

$$\tau_\alpha \triangleq \sum_{i=1}^{n} l_{i;\alpha}\ \delta_{\mathbf{y}_{i;\alpha}}$$

Here, (1) for any measurement $\mathbf{y}$ the Dirac measure $\delta_\mathbf{y}$ is defined by $\delta_\mathbf{y}(S) = 1$ if $\mathbf{y} \in S$ and $\delta_\mathbf{y}(S) = 0$ otherwise; (2) $\mathbf{y}_{i;\alpha}$ is the (random) observation at time-step α generated by the ith target $\mathbf{x}_i$ if detected, and whose statistics are governed by a probability measure $h_\alpha(B\,|\,\mathbf{x}_i)$; (3) $l_{i;\alpha}$ is a random integer which takes only the values 1 or 0 depending on whether the ith target is detected at time-step α or not, and whose statistics are governed by the probability distribution $q_\alpha(\mathbf{x}_i)$. Various independence relationships are assumed to exist between $\nu_\alpha, l_{i;\alpha}, \mathbf{y}_{i;\alpha}, \mathbf{x}_i$.

Washburn shows how to reformulate the measurement model $\mu_\alpha = \tau_\alpha + \nu_\alpha$ as a measurement-model likelihood $f(\mu_\alpha|\mathbf{x})$ and then incorporates this into a Bayesian nonlinear filtering procedure. Unlike the MCTW filter, the Washburn filter is a *direct* multisensor, multitarget estimation approach. See [55] for more details.

2.2. Why random sets: Expert–systems theory. A second and perhaps more compelling reason why one might want to use random set theory as a scientific basis for information fusion arises from a body of recent research which shows that random set theory provides a unifying basis for much of expert systems theory. I.R. Goodman of Naval Research and Development (NRaD) is generally recognized as the pioneering advocate of random set techniques for application of expert systems in information fusion [12]. The best basic general reference concerning the relationship between random set theory and expert systems theory is [24] (see also [13], [46]). We briefly describe this research in the following paragraphs. In all cases, a *finite* universe U is assumed.

The Dempster–Shafer theory. The Dempster-Shafer theory of evidence was devised as a means of dealing with *imprecise* evidence [17], [24], [54]. Evidence concerning an unknown target is represented as a nonnegative set function $m : \mathcal{P}(U) \to [0,1]$ where $\mathcal{P}(U)$ denotes the set of subsets of the finite universe U such that $m(\emptyset) = 0$ and $\sum_{S \subseteq U} m(S) = 1$. The set function m is called a *mass assignment* and models a range of possible beliefs about propositional hypotheses of the general form $P_S \triangleq$ "target is in S," where $m(S)$ is the *weight of belief* in the hypothesis P_S. The quantity $m(S)$ is usually interpreted as the degree of belief that accrues to S but to no proper subset of S. The weight of belief $m(U)$ attached to the entire universe is called the *weight of uncertainty* and models our belief in the possibility that the evidence m in question is completely erroneous. The quantities

$$Bel_m(S) \triangleq \sum_{T \subseteq S} m(T), \qquad Pl_m(S) \triangleq \sum_{T \cap S \neq \emptyset} m(T)$$

are called the *belief* and *plausibility* of the evidence, respectively. The relationships $Bel_m(S) \leq Pl_m(S)$ and $Bel_m(S) = 1 - Pl_m(S^c)$ are true identically and the interval $[Bel_m(S), Pl_m(S)]$ is called the *interval of uncertainty*. The mass assignment can be recovered from the belief function via the *Möbius transform*:

$$m(S) = \sum_{T \subseteq S} (-1)^{|S-T|} \, Bel_m(T)$$

The quantity

$$(m * n)(S) \triangleq \frac{1}{1-K} \sum_{X \cap Y = S} m(X)n(Y)$$

is called *Dempster's rule of combination*, where $K \triangleq \sum_{X \cap Y = \emptyset} m(X)n(Y)$ is called the *conflict* between the evidence m and the evidence n.

In the finite-universe case, at least, the Dempster-Shafer theory coincides with the theory of independent, nonempty random subsets of U (see [18], [24], [43]; or for a dissenting view, see [50]). Given a mass assignment m it is always possible to find a random subset Σ of U such that $m(S) = p(\Sigma = S)$. In this case $Bel_m(S) = p(\Sigma \subseteq S) = \beta_\Sigma(S)$ and $Pl_m(S) = p(\Sigma \cap S \neq \emptyset) = \rho_\Sigma(S)$ where β_Σ and ρ_Σ are the *belief* and *plausibility* measures of Σ, respectively. Likewise, construct independent random subsets Σ, Λ of U such that $m(S) = p(\Sigma = S)$ and $n(S) = p(\Lambda = S)$ for all $S \subseteq U$. Then it is easy to show that

$$(m * n)(S) = p(\Sigma \cap \Lambda \mid \Sigma \cap \Lambda \neq \emptyset)$$

for all $S \subseteq U$.

Fuzzy logic. Recall that a fuzzy subset of the finite universe U is specified by its fuzzy membership function $f : U \to [0,1]$. Natural-language evidence is often mathematically modeled as fuzzy subsets of some universe [13], as is vague or approximate numerical data [23].

Goodman [8], Orlov [44], [45] and others have shown that fuzzy sets can be regarded as "local averages" of random subsets (see also [24]). Given a random subset Σ of U the fuzzy membership function $\mu_\Sigma(u) \triangleq p(u \in \Sigma)$ is called the *one-point covering function* of Σ [13]. Thus every random subset of Σ induces a fuzzy subset. Conversely, let f be a fuzzy membership function on U, let A be a uniformly distributed random number on $[0,1]$ and define the "canonical" random subset $\Sigma_A(f)$ of U by

$$\Sigma_A(f) \triangleq \{u \in U \mid A \leq f(u)\}$$

Then it is easily shown that

$$\Sigma_A(f) \cap \Sigma_A(g) = \Sigma_A(f \wedge g), \qquad \Sigma_A(f) \cup \Sigma_A(g) = \Sigma_A(f \vee g)$$

where $(f \wedge g)(u) = \min\{f(u), g(u)\}$ and $(f \vee g)(u) = \max\{f(u), g(u)\}$ denote the conjunction and disjunction operators of the usual Zadeh max-min fuzzy logic. However, $\Sigma_A(f)^c = \Sigma^*_{1-A}(1-f)$ where $\Sigma^*_A(f) = \{u \in U \mid A < f(u)\}$. More generally, let B be another uniformly distributed random number on $[0,1]$. Then it can be shown that

$$\begin{aligned} \mu_{\Sigma_A(f) \cap \Sigma_B(g)}(u) &= F_{A,B}(f(u), g(u)) \\ \mu_{\Sigma_A(f) \cup \Sigma_B(g)}(u) &= G_{A,B}(f(u), g(u)) \end{aligned}$$

where $F_{A,B}(a,b) \triangleq p(A \leq a, B \leq b)$ and $G_{A,B}(a,b) \triangleq 1 - F_{A,B}(1-a, 1-b)$ are called a *copula* and *co-copula*, respectively [48]. In many cases the binary operations $F_{A,B}(a,b)$ and $G_{A,B}(a,b)$ form an alternative fuzzy logic. (This is the case, for example, if A, B are independent, in which case $F_{A,B}(a,b) = ab$ and $G_{A,B}(a,b) = a + b - ab$, the so-called *prodsum logic.*) Goodman has also established more general homomorphism-like relationships between fuzzy logics and random sets than those considered here [8], [9].

These relationships suggest that many fuzzy logics correspond to different assumptions about the statistical correlations between fuzzy subsets.

This fact is implicitly recognized by many practitioners in information fusion (see, for example, [6]) who allow the choice of fuzzy logic to be dynamically determined by the current evidential situation.

Fred Kramer and I.R. Goodman of Naval Research and Development have used random set representations of fuzzy subsets to devise a multitarget tracker, PACT, which makes use of both standard statistical evidence and natural-language evidence [10].

Rule–based evidence. Knowledge-base rules have the form

$$X \Rightarrow S = \text{“\textit{If target has property } } X \text{ \textit{then it has property} } S\text{”}$$

where S, X are subsets of a (finite) universe U. For example, suppose that P is the proposition "target is a tank" and Q is the proposition "target is amphibious." Let X be the set of all targets satisfying P (i.e., all tanks) and S the set of all targets satisfying Q (i.e., all amphibious targets). Then the language rule "no tank will be found in a lake," or $P \Rightarrow Q$, can be expressed in terms of events as $X \Rightarrow S^c$ if we treat X, S as stand-ins for P, Q.

The events $S \subseteq U$ of U form a Boolean algebra. One can ask: Does there exist a Boolean algebra $\hat{U}$ which has the rules of U as its elements and which satisfies the following elementary properties:

$$(U \Rightarrow S) = S, \qquad (X \Rightarrow S) \cap X = X \cap S$$

If so, given any probability measure q on U is there a probability measure $\hat{q}$ on $\hat{U}$ such that

$$\hat{q}(X \Rightarrow S) = q(S|X) \triangleq \frac{q(S \cap X)}{q(X)}$$

for all $S, X \subseteq U$ with $q(X) \neq 0$? That is, is conditional probability a *true probability measure* on some Boolean algebra of rules? The answer to both questions turns out to be "yes."

It can be shown that the usual approach used in symbolic logic–setting $X \Rightarrow S = X \to S \triangleq (S \cap X) \cup X^c$ (i.e., the material implication)–is not consistent with probability theory in the sense that

$$q(S \cap X) < q(X \to S) < q(S|X)$$

except under trivial circumstances. Lewis [25], [14, p. 14] showed that, except under trivial circumstances, there is *no* binary operator " $\Rightarrow$ " on U which will solve the problem. Accordingly, the object $X \Rightarrow S$ must belong to some logical system which strictly extends the Boolean logic $\mathcal{P}(U)$. Any such extension–Boolean or otherwise–is called a *conditional event algebra*. Several conditional event algebras have been discovered [14]. It turns out, however, that there indeed exists a Boolean algebra of rules, first discovered by van Fraassen [53] and independently rediscovered fifteen years later by Goodman and Nguyen [11], which has the desired consistency with conditional probability.

It has been shown [34], [35] that there is at least one way to represent knowledge-base rules in random set form. This is accomplished by first representing rules as elements of the so-called Goodman-Nguyen-Walker (GNW) conditional event algebra [14], [11] and then–in analogy with the canonical representation of a fuzzy set by a random subset–representing the elements of this algebra as random subsets of the original universe. Specifically, let $(S|X)$ be a GNW conditional event in a *finite* universe U, where $S, X \subseteq U$. Let Φ be a *uniformly distributed* random subset of U–that is, a random subset of U whose probability distribution is $p(\Phi = S) = 2^{-|U|}$ for all $S \subseteq U$. Then define the random subset $\Sigma_\Phi(S|X)$ of U by

$$\Sigma_\Phi(S|X) \triangleq (S \cap X) \cup (X^c \cap \Phi)$$

Ordinary events arise by setting $X = U$, in which case we get $(S|U) = S$. Moreover, the following homomorphism-like relationships are true:

$$\begin{aligned} \Sigma_\Phi(S|X) \cap \Sigma_\Phi(T|Y) &= \Sigma_\Phi((S|X) \wedge (T|Y)) \\ \Sigma_\Phi(S|X) \cup \Sigma_\Phi(T|Y) &= \Sigma_\Phi((S|X) \vee (T|Y)) \\ \Sigma_\Phi(S|X)^c &= \Sigma_{\Phi^c}((S|X)^{c,GNW}) \end{aligned}$$

where '$\wedge$', '$\vee$', and '^{c,GNW}' represent the conjunction, disjunction, and complement operators of the GNW logic.

Li's generalized fuzzy set theory. We conclude our discussion of the relationships between expert-systems theory and random set theory with a "generalized" fuzzy set theory proposed in the Ph.D. Thesis of Ying Li. The purpose of Li's thesis was to provide a probabilistic basis for fuzzy logic. However, an adaptation of Li's basic construction (see [26]; see also the proof of Proposition 6 of [28] for a similar idea) results in a simple way of constructing examples of random subsets of a finite universe U that generalizes Goodman's "canonical" random-set representations of fuzzy sets.

Let $I = [0, 1]$ denote the unit interval and define $U^* \triangleq U \times I$. Let $\pi_U : U^* \to U$ be the projection mapping defined by $\pi_U(u, a) = u$, for all $u \in U$ and $a \in I$. The finite universe U is assumed to have an *a priori* probability measure p, I has the usual Lebesgue measure (which in this case is a probability measure), and the set $U \times I$ is given the product probability measure. Li calls the events $W \subseteq U^*$ "generalized fuzzy subsets of U." Let $W \subseteq U^*$ and for any $u \in U$ define

$$p(W|u) \triangleq p(W|u \times I) = \frac{p(W \cap (u \times I))}{p(u)}$$

Now, let A be a random number uniformly distributed on I. Define the "canonical" random subset of U generated by W, denoted by $\Sigma_A(W)$, by:

$$\Sigma_A(W)(\omega) \triangleq \pi_U(W \cap (U \times A(\omega)))$$

for all $\omega \in \Omega$. The following relationships are easily established:

$$\begin{aligned} \Sigma_A(V \cap W) &= \Sigma_A(V) \cap \Sigma_A(W), \qquad \Sigma_A(V \cup W) = \Sigma_A(V) \cup \Sigma_A(W) \\ \Sigma_A(W^c) &= \Sigma_A(W)^c, \qquad \Sigma_A(\emptyset) = \emptyset, \qquad \Sigma_A(U^*) = U \end{aligned}$$

Also, it is easy to show that the one-point covering function of $\Sigma_A(W)$ is

$$\mu_{\Sigma_A(W)}(u) \triangleq p(u \in \Sigma_A(W)) = p(W|u)$$

Let $f : U \to I$ be a fuzzy membership function on U. Define

$$W_f \triangleq \{(u, a) \in U^* |\ a \leq f(u)\}$$

Then it is easily verified that $p(W_f|u) = f(u)$, that

$$W_{f \wedge g} = W_f \cap W_g, \qquad W_{f \vee g} = W_f \cup W_g$$

(where '$\wedge$' and '$\vee$' denote the usual min/max conjunction and disjunction operators of the standard Zadeh fuzzy logic) and that

$$\Sigma_A(W_f) = \{u \in U |\ A(\omega) \leq f(u)\} = \Sigma_A(f)$$

where $\Sigma_A(f)$ is the canonical random subset induced by f as defined previously.

2.3. Why not vectors or point processes. The preceding sections show that, if the goal is to provide a unifying scientific foundation for information fusion–in particular, unification of the expert systems aspects with the multisensor, multitarget estimation aspects–then there is good reason to consider random set theory as a possible means of doing so. However, skeptics might still ask: *Why not simply use vector models* (like Mori, Chong, Tse and Wishner) *or point process models* (like Washburn)? It could be objected, for example, that what we will later call the "global density" $f_\Sigma(Z)$ of a finite random subset Σ is just a new name and notation for the so-called "Janossy densities" $j_n(\mathbf{z}_1, ..., \mathbf{z}_n)$ [7, pp. 122-123] of the corresponding simple finite point process defined by $N_\Sigma(S) \triangleq |\Sigma \cap S|$ for all measurable S.

In response to possible such objections we offer the following responses. First, vector approaches encourage carelessness in regard to basic questions. For example, to apply the theorems of conventional estimation theory one must clearly identify a measurement space and a state space and specify their topological and metrical properties. For example, the standard proof of the consistency of the maximal likelihood and maximum *a posteriori* estimators assumes that state space is a metric space. Or, as another instance, suppose that we want to perform a *sensitivity analysis* on a given information fusion algorithm–that is, determine whether it is the case that small deviations in input data can result in large deviations in output data. To answer this question one must first have some idea of what *distance* means in both measurement space and state space. The standard Euclidean metric is clearly not adequate: If we represent an observation set $\{\mathbf{z}_1, \mathbf{z}_2\}$ as a vector $(\mathbf{z}_1, \mathbf{z}_2)$ then $\|(\mathbf{z}_1, \mathbf{z}_2) - (\mathbf{z}_2, \mathbf{z}_1)\| \neq 0$ even though the order of measurements should not matter. Likewise one might ask, what is the distance between $(\mathbf{z}_1, \mathbf{z}_2)$ and $\mathbf{z}_3$? Whereas both finite set theory and point process theory have rigorous metrical concepts, attempts to define metrics for vector models can quickly degenerate into *ad hoc* invention. More generally, the use of vector models has resulted in piecemeal solutions to information fusion problems (most typically, the assumption that the number of targets is known *a priori*). Lastly, any attempt to incorporate expert systems theory into the vector approach results in extremely awkward attempts to make vectors behave as though they were finite sets.

Second, the random set approach is *explicitly geometric* in that the random variates in question are actual sets of observations–rather than, say, abstract integer-valued measures.

Third, and as we shall see shortly, systematic adherence to a random set perspective results in a series of direct parallels–most directly, the concept of the *set integral* and the *set derivative*–between single-sensor, single-target statistics and multisensor, multitarget statistics. In comparison to the point process approach, this parallelism results in a methodology for information fusion that is nearly identical in general behavior to the "Statistics 101" formalism with which engineering practitioners and theorists are already familiar. More importantly, it leads to a systematic approach to solving information fusion problems (see Section 3 below) that allows standard single-sensor, single-target statistical techniques to be directly generalized to the multisensor, multitarget case.

Fourth, because the random set approach provides a systematic foundation for both expert systems theory *and* multisensor, multitarget estimation, it permits a systematic and mathematically rigorous integration of these two quite different aspects of information fusion–a question left unaddressed by either the vector or point-process models.

Fifth, an analogous situation holds in the case of random subsets of $\mathbb{R}^n$ which are *convex* and *bounded*. Given a bounded convex subset $T \subseteq \mathbb{R}^n$, the *support function* of T is defined by $s_T(\mathbf{e}) \triangleq \sup_{\mathbf{x} \in T} \langle \mathbf{e}, \mathbf{x} \rangle$ for all vectors $\mathbf{e}$ on the unit hypersphere in $\mathbb{R}^n$, where '$\langle -, - \rangle$' denotes the inner product on $\mathbb{R}^n$. The assignment $\Sigma \to s_\Sigma$ establishes a very faithful embedding of random bounded convex sets Σ into the random functions on the unit hypersphere, in the sense that it encodes the behavior of bounded convex sets into vector mathematics [27], [39]. Nevertheless, it does not follow that the theory of random bounded convex subsets is a special case of random function theory. Rather, random functions provide a useful tool for studying the behavior of random bounded convex sets.

In like manner, finite point processes are best understood as specific–and by no means the only or the most useful–representations of random finite subsets as elements of some abstract vector space. The assignment $Z \to N_Z$ embeds finite sets into the space of signed measures, but N_Z is not a particularly faithful representation of Z. For one thing, $N_{Z \cup Z'} \neq N_Z + N_{Z'}$ unless $Z \cap Z' = \emptyset$. Random point processes are better understood as vector representations of random *multisets*–that is, as randomizations of unordered lists L. In this case $N_{L \cup L'} = N_L + N_{L'}$ identically.

3. Finite–set statistics. In this section we will summarize the process by which the basic elements of ordinary statistics–integral, derivative, densities, estimators, etc.–can be directly generalized to a statistics of finite sets. We propose a mathematical approach which is capable of providing a rigorous, fully probabilistic scientific foundation for the following aspects of information fusion:

- *Multisource integration* based on parametric estimation and Markov techniques [29], [33], [32].
- *Prior information* regarding the numbers, identities, and geokinematics of targets [31].
- *Sensor management* based on information theory and nonlinear control theory [37].
- *Performance evaluation* using information theory and nonparametric estimation [31], [37].
- *Expert-systems theory*: fuzzy logic, evidential theory, rule-based inference [36]

One of the consequences of this unification is the existence of algorithms which fully unify in *a single statistical process* the following functions of information fusion: detection, classification, and tracking; prior information with respect to detection, classification, and tracking; ambiguous evidence as well as precise data; expert systems theory; and sensor management. Included under the umbrella as well is a systematic approach to performance evaluation.

We temper these claims as follows. We will consider only sensors whose observations are of the following types: point-source, range-profile, line-of-bearing, natural-language, and rule bases. The only types of imaging sensors which can (potentially) be directly included in the approach are those whose target-images are either point "firefly" sources or relatively small clusters of point energy reflectors (so-called "extended targets"). More complex image data require prior processing by automatic target recognition algorithms to extract relevant I.D. or signature-feature information. We also restrict ourselves to report-to-track fusion, though our basic techniques can be applied in principle to distributed fusion as well. We will also ignore the communications aspects of the problem.

In Section 3.2 we describe the technical foundation for our approach to data fusion: an integral and differential calculus of set functions. We will describe the *set integral*, the *set derivative*, and show how these lead to "global" probability density functions which govern the statistical behavior of entire multisensor, multitarget problems. Then, in Section 3.3, we show how this leads to a powerful parallelism between ordinary (i.e., single-sensor, single-target) and finite-set (i.e., multisensor, multitarget) statistics. We show how this parallelism leads to fundamentally new ways of attacking information fusion problems. In particular, we will show how to measure the information of entire multisensor, multitarget systems; construct *Receiver Operating Characteristic* (ROC) curves for entire multisensor, multitarget systems; construct multisensor, multitarget analogs of the maximum likelihood and other estimators; deal with multiple dynamic targets using a generalization of Bayesian nonlinear filtering theory; and, finally, apply nonlinear control theory to the sensor management problem.

3.1. The basic approach. The basic approach can be described as follows. Assume that we have a known suite sensors which reports to a *central data fusion site*, and an unknown number of targets which have unknown identities and position, velocity, etc. Regard the sensor suite as though it were a single "global sensor" and the target set as though it were a single "global target." Observations collected by the sensor suite at approximately the same time can be regarded as a single observation set. Simplifying somewhat, each individual observation will have the form $\xi = (\mathbf{z}, u, i)$ where $\mathbf{z}$ is a continuous variable (geokinematics, signal intensity, etc.) in $\mathbb{R}^n$, u is a discrete variable (e.g. possible target I.D.s) drawn from a finite universe U of possibilities, and i is a "sensor tag" which identifies the sensor which supplied the measurement. Let us regard the total observation-set $Z = \{\xi_1, ..., \xi_k\}$ collected by the global sensor as though it were a single "global observation." This global observation is a specific realization of a *randomly varying finite observation-set* Σ. In this manner the multisensor, multitarget problem can be reformulated as a single-sensor, single-target problem. From the random observation set Σ we can construct a so-called *belief measure*, $\beta_\Sigma(S) \triangleq p(\Sigma \subseteq S)$. We then show how to construct a "global density function" $f_\Sigma(Z)$ from the belief measure, which describes the likelihood that the random set Σ takes the finite subset Z as a specific realization.

The reformulation of multisource, multitarget problems is not just a mathematical "bookkeeping" device. Generally speaking, *a group of targets observed by imperfect sensors must be analyzed as a single indivisible entity rather than as a collection of unrelated individuals.* When measurement uncertainties are large in comparison to target separations there will always be a significant likelihood that any given measurement in Z was generated by any given target. This means that *every measurement* can be associated–partially or in some degree of proportion–to *every target.* The more irresolvable the targets are, the more our estimates of them will be statistically correlated and thus the more that they will seem as though they are a *single* target. Observations can no longer be regarded as separate entities generated by individual targets but rather as collective phenomena *generated by the entire multitarget system.* The important thing to realize is that this remains true even when target separations are large in comparison to sensor uncertainties. Though in this case the likelihood is very small that a given observation was generated by any other target than the one it is intuitively associated with, nevertheless this likelihood is nonvanishing. The resolvable-target scenario is just a *limiting case* of the irresolvable-target scenario.

3.2. A calculus of set functions. If an additive measure $p_{\mathbf{Z}}$ is absolutely continuous with respect to Lebesgue measure then one can determine the *density function* $f_{\mathbf{Z}} = dp_{\mathbf{Z}}/d\lambda$ that corresponds to it. Conversely, the measure can be recovered from the density through application of the Lebesgue integral: $\int_S f_{\mathbf{Z}}(\mathbf{z})d\lambda(\mathbf{z}) = p_{\mathbf{Z}}(S)$. In this section we describe an integral and differential calculus of functions of a set variable which obeys similar properties. That is, given a vector-valued function $f(Z)$ of a *finite-set* variable Z we will define a "set integral" of the form $\int_S f(Z)\delta Z$. Conversely, given a vector-valued function $\Phi(S)$ of a *closed set* variable S, we will define a "set derivative" of the form $\delta\Phi/\delta Z$. These operations are inverse to each other in the sense that, under certain assumptions,

$$\begin{aligned} \frac{\delta}{\delta Z}\int_S f(Z)\delta Z\bigg|_{S=\emptyset} &= f(Z) \\ \int_S \frac{\delta\Phi}{\delta Z}(\emptyset)\,\delta Z &= \Phi(S) \end{aligned}$$

More importantly, if β_Σ is the belief measure of a random observation set Σ then, given certain absolute continuity assumptions, the quantity $f_\Sigma(Z) \triangleq [\delta\beta_\Sigma/\delta Z](\emptyset)$ is the *density function* of the random finite subset Σ.

It should be emphasized that this particular "calculus of set functions" was devised *specifically with application to information fusion problems in mind.* There are many ways of formulating a calculus of set functions, most notably the Huber-Strassen derivative [21] and the Graf integral [16]. The calculus described here was devised precisely because it yields a statistical theory of random finite sets which is strongly analogous to the simple "Statistics 101" formalism of conventional point-variate statistics–and therefore because it leads to the strong parallelisms between single-sensor, single-target problems and multisensor, multitarget problems summarized in Section 3.3.

Let us assume, then, that the "global observations" collected by the "global sensor" are particular realizations of a random finite set Σ. From Mathéron's random set theory we know that the class of finite subsets of measurement space has a topology, the so-called *hit-or-miss topology* [38],

[39]. If O is any Borel subset of this topology then the statistics of Σ are characterized by the associated probability measure $p_\Sigma(O) \triangleq p(\Sigma \in O)$. One consequence of the Choquet-Mathéron capacity theorem [38, p. 30] is that we are allowed to restrict ourselves to Borel sets of the specific form $O^c_{S^c}$–i.e., the class whose elements are all closed subsets C of measurement space such that $C \cap S^c = \emptyset$ (i.e., $C \subseteq S$) where S is some closed subset of measurement space. In this case $p_\Sigma(O^c_{S^c}) = p(\Sigma \subseteq S) = \beta_\Sigma(S)$ and thus we can substitute the *nonadditive* measure β_Σ in the place of the *additive* measure p_Σ.

The set function β_Σ is called the *belief measure* of Σ. Despite the fact that it is nonadditive, the belief measure β_Σ plays the *same role in multisensor, multitarget statistics that ordinary probability measures play in single-sensor, single-target statistics.* To see this, let us begin by defining the *set integral* and the *set derivative.*

The set integral. Let O be any Borel set of the relative hit-or-miss topology on the finite subsets of the product space $\mathcal{R} \triangleq \mathbb{R}^n \times U$ (which is endowed with the product measure of Lebesgue measure on $\mathbb{R}^n$ and the counting measure on U) and let $f(Z)$ be any function of a finite-set argument such that $f(Z)$ vanishes for all Z with $|Z| \geq M$ for some $M > 0$, and such that the functions defined by

$$f_k(\xi_1, ..., \xi_k) \triangleq \begin{cases} f(\{\xi_1, ..., \xi_k\}), & \text{if } \xi_1, ..., \xi_k \text{ are distinct} \\ 0, & \text{if } \xi_1, ..., \xi_k \text{ are not distinct} \end{cases}$$

for all $0 \leq k \leq M$ are integrable with respect to the product measure. Then the *set integral* concentrated on O is defined as

$$\int_O f(Z)\delta Z \triangleq f(\emptyset) + \sum_{k=1}^{M} \frac{1}{k!} \int_{\chi_k^{-1}(O \cap c_k(\mathcal{R}))} f_k(\xi_1, ..., \xi_k)\, d\xi_1 \cdots d\xi_k$$

Here, $c_k(\mathcal{R})$ denotes the class of finite k-element subsets of $\mathcal{R}$ and $\chi_k^{-1}(O \cap c_k(\mathcal{R}))$ denotes the subset of all k-tuples $(\xi_1, ..., \xi_k) \in \mathcal{R}^k$ such that $\{\xi_1, ..., \xi_k\} \in O \cap c_k(\mathcal{R})$. Also, the integrals on the right-hand side of the equation are so-called "hybrid" (i.e., continuous-discrete) integrals which arise from the product measures on $\mathcal{R}^k$. In particular, assume that $O = O^c_{S^c}$ where S is a closed subset of $\mathcal{R}$. Then $\chi_k^{-1}(O^c_{S^c} \cap c_k(\mathcal{R})) = S \times \cdots \times S = S^k$ (Cartesian product taken k times) and we write

$$\int_S f(Z)\delta Z \triangleq \sum_{k=0}^{M} \frac{1}{k!} \int_{S^k} f_k(\xi_1, ..., \xi_k)\, d\xi_1 \cdots d\xi_k$$

(Note: Though not explicitly identified as such, the set integral in this form arises frequently in statistical mechanics in connection with the theory of polyatomic fluids [19].) If $S = \mathcal{R}$ then we write $\int f(Z)\delta Z = \int_{\mathcal{R}} f(Z)\delta Z$.

The set derivative. The concept which is inverse to the set integral is the *set derivative*, which in turn requires some preliminary discussion regarding the various *constructive definitions of the Radon-Nikodým derivative.* As usual, let $\lambda(S)$ denote Lebesgue measure on $\mathbb{R}^n$ for any closed Lebesgue-measurable set S of $\mathbb{R}^n$. Let q be a nonnegative measure defined on the Lebesgue-measurable subsets of $\mathbb{R}^n$ which is absolutely continuous in the sense that $q(S) = 0$ whenever $\lambda(S) = 0$. Then by the Radon-Nikodým theorem there is an almost-everywhere unique function f such that $q(S) = \int_S f(\mathbf{z})d\lambda(\mathbf{z})$ for all measurable $S \subseteq \mathbb{R}^n$ in which case f is called the *Radon-Nikodým derivative* of q with respect to λ, denoted by $f = dq/d\lambda$. Thus $dq/d\lambda$ is defined as an *anti-integral of the Lebesgue integral.*

However, there are several ways to define the Radon-Nikodým derivative *constructively.* One form of the *Lebesgue density theorem* (see Definition 1 and Theorems 5 and 6 of [49, pp. 220-222]) states that

$$\lim_{\varepsilon \downarrow 0} \frac{1}{\lambda(E_{\mathbf{z};\varepsilon})} \int_{E_{\mathbf{z};\varepsilon}} f(\mathbf{y})d\lambda(\mathbf{y}) = f(\mathbf{z})$$

for almost all $\mathbf{z} \in \mathbb{R}^n$, where $E_{\mathbf{z};\varepsilon}$ denotes the *closed ball* of radius ε centered at $\mathbf{z}$. It thus follows from the Radon-Nikodým theorem that a constructive definition of the Radon-Nikodým derivative of q (with respect to Lebesgue measure) is:

$$\frac{dq}{d\lambda}(\mathbf{z}) = \lim_{\varepsilon \downarrow 0} \frac{q(E_{\mathbf{z};\varepsilon})}{\lambda(E_{\mathbf{z};\varepsilon})} = \lim_{\varepsilon \downarrow 0} \frac{1}{\lambda(E_{\mathbf{z};\varepsilon})} \int_{E;\varepsilon} f(\mathbf{y})d\lambda(\mathbf{y}) = f(\mathbf{z})$$

almost everywhere. An alternative approach is based on "nets." A net is a nested, countably infinite sequence of countable partitions of $\mathbb{R}^n$ by Borel sets [49, p. 208]. That is, each partition P_j is a countable sequence $Q_{j,1}, \ldots, Q_{j,k}, \ldots$ of Borel subsets of $\mathbb{R}^n$ which are mutually disjoint and whose union is all of $\mathbb{R}^n$. This sequence of partitions is nested, in the sense that given any cell $Q_{j,k}$ of the jth partition P_j, there is a subsequence of the partition $P_{j+1} : Q_{j+1,1}, \ldots, Q_{j+1,i}, \ldots$ which is a partition of $Q_{j,k}$. Given this, another constructive definition of the Radon-Nikodým derivative is, provided that the limit exists,

$$\frac{dq}{d\lambda}(\mathbf{z}) = \lim_{i \to \infty} \frac{q(E_{\mathbf{z};i})}{\lambda(E_{\mathbf{z};i})}$$

where in this case the $E_{\mathbf{z};i}$ are any sequence of sets belonging to the net which converge to the singleton set $\{\mathbf{z}\}$.

Still another approach for constructively defining the Radon-Nikodým derivative involves the use of Vitali systems instead of nets [49, pp. 209-215].

Whichever approach we use, we can rest assured that a rigorous theory of limits exists which is rich enough to generalize the Radon-Nikodým derivative in the manner we propose. (For purposes of application it is simpler to use the Lebesgue density theorem version. If the limit exists then it is enough to compute it for closed balls of radius $1/i$ as $i \to \infty$. This is what we will usually assume in what follows.)

Let $\Phi(S)$ be a vector-valued function of closed subsets S of $\mathcal{R}$ and let $\xi = (\mathbf{z}, u) \in \mathcal{R} \triangleq \mathbb{R}^n \times U$. If it exists, the *generalized Radon-Nikodým derivative* of Φ at ξ is the set function defined by:

$$\frac{\delta\Phi}{\delta\xi}(T) \triangleq \lim_{j\to\infty} \lim_{i\to\infty} \frac{\Phi((T - (F_{\mathbf{z};j} \times u)) \cup (E_{\mathbf{z};i} \times u)) - \Phi(T - (F_{\mathbf{z};j} \times u))}{\lambda(E_{\mathbf{z};i})}$$

for all closed subsets $T \subseteq \mathcal{R}$, where $E_{\mathbf{z};i}$ is a sequence of *closed* balls converging to $\{\mathbf{z}\}$; and where the $F_{\mathbf{z};j}$ is a sequence of *open* balls whose closures converge to $\{\mathbf{z}\}$. Also, we have abbreviated $T - (F_{\mathbf{z};j} \times u) = T \cap (F_{\mathbf{z};j} \times u)^c$ and $E_{\mathbf{z};i} \times u = E_{\mathbf{z};i} \times \{u\}$.

Note that if Φ is an additive measure on R^n which is absolutely continuous with respect to Lebesgue measure, then the conventional Radon-Nikodým derivative of Φ (taken with respect to Lebesgue measure) can be recovered as

$$\frac{d\Phi}{d\lambda}(\mathbf{z}) = \frac{\delta\Phi}{\delta\mathbf{z}}(T)$$

for all $\mathbf{z} \in \mathbb{R}^n$ and any closed $T \subseteq \mathbb{R}^n$.

Since the generalized Radon-Nikodým derivative of a set function is again a set function, one can take the generalized Radon-Nikodým derivative again. The *set derivative* of the set function Φ, if it exists, is defined by iterating the generalized Radon-Nikodým derivative:

$$\frac{\delta\Phi}{\delta Z}(T) \triangleq \frac{\delta^k\Phi}{\delta\xi_k \cdots \delta\xi_1}(T) = \frac{\delta}{\delta\xi_k} \frac{\delta^{k-1}\Phi}{\delta\xi_{k-1} \cdots \delta\xi_1}(T)$$

for all $Z = \{\xi_1, ..., \xi_k\} \subseteq \mathcal{R}$ with $|Z| = k$ and all closed subsets $T \subseteq \mathcal{R}$. Note that an underlying reference measure–Lebesgue measure–is always assumed and therefore does not explicitly appear in our notation.

It can be shown that the set derivative is a continuous analog of the Möbius transform of Dempster-Shafer theory. Specifically, let Φ be a set function and let $\xi_1, ..., \xi_k$ be distinct elements of $\mathcal{R}$ with $\xi_j = (\mathbf{z}_j, u_j)$ for $j = 1, ..., k$. Define the set function Φ_i by

$$\Phi_i(\{\xi_1, ..., \xi_k\}) \triangleq \Phi((E_{\mathbf{z}_1;i} \times u_1) \cup \cdots \cup (E_{\mathbf{z}_k;i} \times u_k))$$

where as usual $E_{\mathbf{z}_1;i}$ denotes a closed ball of radius $1/i$ centered at $\mathbf{z} \in \mathbb{R}^n$ and $\xi_1, ..., \xi_k$ are distinct. Assume that all iterated generalized Radon-Nikodým derivatives of Φ exist. If $Z = \{\xi_1, ..., \xi_k\}$ then:

$$\frac{\delta^k\Phi}{\delta\xi_1 \cdots \delta\xi_k}(\emptyset) = \lim_{i\to\infty} \frac{1}{\lambda(E_{\mathbf{z}_1;i}) \cdots \lambda(E_{\mathbf{z}_k;i})} \sum_{Y \subseteq Z} (-1)^{|Z-Y|} \Phi_i(Y)$$

Finally, it is worth mentioning the relationship between these concepts and allied concepts in point process theory. As already noted in Section 2.3, the quantities

$$f_\Sigma(Z) = \frac{\delta\beta_\Sigma}{\delta Z}(\emptyset)$$

for all finite $Z \subseteq \mathcal{R}$ are known in point process theory as the *Janossy densities* [7, pp. 122-123] of the point process defined by $N_\Sigma(S) \triangleq |\Sigma \cap S|$. Likewise, the quantities

$$D_\Sigma(Z) = \frac{\delta\beta_\Sigma}{\delta Z}(\mathcal{R})$$

are known as the *factorial moment densities* [7, pp. 130-150] of N_Σ. In information fusion, the graphs of the D_Σ are also known as *probability hypothesis surfaces.*

Global probability densities. Let β_Σ be the belief measure of the random finite subset Σ of observations. Then under suitable assumptions of absolute continuity (essentially this means the absolute continuity, with respect to product measure on $\mathcal{R}$, of the Janossy densities of the point process $N_\Sigma(S) = |\Sigma \cap S|$ induced by Σ) it can be shown that the quantity

$$f_\Sigma(Z) \triangleq \frac{\delta\beta_\Sigma}{\delta Z}(\emptyset)$$

exists. It is called the *global density* of the random finite subset Σ. The global density has a completely Bayesian interpretation. Suppose that $f_\Sigma(Z|X)$ is a global density with a *set parameter* $X = \{\zeta_1, ..., \zeta_t\}$. Then $f_\Sigma(Z|X)$ is the *total probability density of association* between the measurements in Z and the parameters in X.

The density function of a conventional sensor describes only the self-noise statistics of the sensor. The global density of a sensor suite differs from conventional densities in that it *encapsulates the comprehensive statistical behavior of the entire sensor suite into a single mathematical object.* That is, in its most general form a global density will have the form $f_\Sigma(Z|X;Y)$, thus including the following information:

- the observation-set $Z = \{\xi_1, ..., \xi_k\}$
- the set $X = \{\zeta_1, ..., \zeta_t\}$ of unknown parameters
- the states $Y = \{\eta_1, ..., \eta_s\}$ of the sensors (dwells, modes, etc.)
- the sensor-noise distributions of the individual sensors
- the probabilities of detection and false alarm for the individual sensors
- clutter models
- detection profiles for the individual sensors (as functions of range, aspect, etc.)

The density function corresponding to a given sensor suite can be computed explicitly by applying the generalized Radon-Nikodým derivative to the belief measure of the random observation set Σ. For example, suppose that we are given a single sensor with sensor-noise density $f(\xi|\zeta)$ with no false alarms and constant probability of detection p_D, and that observations are independent. Then the global density which specifies the multitarget motion model for the sensor is

$$f_\Sigma(\{\xi_1,...,\xi_k\}|\{\zeta_1,...,\zeta_t\}) = p_D^k(1-p_D)^{t-k} \sum_{1\leq i_1\neq\cdots\neq i_k\leq t} f(\xi_1|\zeta_{i_1})\cdots f(\xi_k|\zeta_{i_k})$$

where the summation is taken over all distinct $i_1,...,i_k$ such that $1 \leq i_1,...,i_k \leq t$. (Here, $\xi_1,...,\xi_k$ are assumed distinct, as are $\zeta_1,...,\zeta_t$.)

Properties of global densities. It can be shown that the belief measure of a random finite set can be recovered from its corresponding global density function:

$$\beta_\Sigma(S) = \int_S f_\Sigma(Z)\delta Z$$

for all closed subsets S of $\mathcal{R}$. Likewise, if $\mathbf{T}(Z)$ is a measurable vector-valued transformation of a finite-set variable, it can be shown that the expectation $E[\mathbf{T}(\Sigma)]$ of the random vector $\mathbf{T}(\Sigma)$ is just what one would expect:

$$E[\mathbf{T}(\Sigma)] = \int \mathbf{T}(Z)\ f_\Sigma(Z)\ \delta Z$$

Finally, it is amusing to note that the Choquet integral (see [15]), which is *nonlinear*, is related to the set integral, which is *linear*, as follows:

$$\int h(\mathbf{z})\ d\beta_\Sigma(\mathbf{z}) = \int \min_h(Z)\ f_\Sigma(Z)\,\delta Z$$

where $\min_h(Z) = \min_{\mathbf{z}\in Z} h(\mathbf{z})$ and h is a suitably well-behaved real-valued function on $\mathbb{R}^n$.

The parallelism between point– and finite–set statistics. Because of the set derivative and the set integral, it obviously becomes possible to compile a set of direct mathematical parallels between the world of single-sensor, single-target statistics and the world of multisensor, multitarget statistics. These parallels can be expressed as a kind of translation dictionary:

Random Vector, Z	**Finite Random Set, Σ**
sensor, $\odot$	global sensor, $\odot^*$
target, $\otimes$	global target, $\otimes^*$
observation, $\mathbf{z}$	global observation-set, Z
parameter, $\mathbf{x}$	global parameter-set, X
differentiation, $dp_{\mathbf{Z}}/d\mathbf{z}$	set differentiation, $\delta\beta_\Sigma/\delta Z$
integration, $\int_S f_{\mathbf{Z}}(\mathbf{z}\vert\mathbf{x})d\lambda(\mathbf{z})$	set integration, $\int_S f_\Sigma(Z\vert X)\delta Z$
probability measure, $p_{\mathbf{Z}}(S\vert\mathbf{x})$	belief measure, $\beta_\Sigma(S\vert X)$
density, $f_{\mathbf{Z}}(\mathbf{z}\vert\mathbf{x})$	global density, $f_\Sigma(Z\vert X)$
prior density, $f_{\mathbf{X}}(\mathbf{x})$	global prior density, $f_\Gamma(X)$
motion models, $f_{\alpha+1\vert\alpha}(\mathbf{x}_{\alpha+1}\vert\mathbf{x}_\alpha)$	global models, $f_{\alpha+1\vert\alpha}(X_{\alpha+1}\vert X_\alpha)$

The parallelism is so close that it suggests a general way of attacking information fusion problems. Any theorem or algorithm in conventional statistics can be thought of as a "sentence" in a language whose "words" and "grammar" consist of the basic concepts in the left-hand column above. The above "dictionary" establishes a direct correspondence between the words and grammar of the random-vector language and the cognate words and grammar of the finite-set language. Consequently, nearly any "sentence"–any theorem or mathematical algorithm–phrased in terms of the random vector language can, in principle, be directly translated into a corresponding "sentence"–theorem or algorithm–in the random-set language.

We say "nearly" because, as with any translation process, the correspondence between dictionaries is not precisely one-to-one. Unlike the situation in the random vector world, for example, there seems to be no natural way to add and subtract finite sets as one does vectors. Nevertheless, the parallelism is complete enough that, with the exercise of some prudence, a hundred years of conventional statistics can be *directly* brought to bear on multisensor, multitarget information fusion problems. Thus we get:

parametric estimators	global parametric estimators
information theory (entropy) metrics	global information metrics
nonlinear filtering	global nonlinear filtering
nonlinear control theory	global nonlinear control theory

In the remainder of this section we describe this relationship in somewhat greater detail.

Information metrics. Suppose that we wish to attack the problem of *performance evaluation* of information fusion algorithms in a scientifically defensible manner. In ordinary statistics one has information metrics such as the Kullback-Leibler discrimination:

$$I(f_{\mathbf{X}}; f_{\mathbf{Y}}) = \int f_{\mathbf{X}}(\mathbf{x}) \ln\left(\frac{f_{\mathbf{X}}(\mathbf{x})}{f_{\mathbf{Y}}(\mathbf{x})}\right) d\lambda(\mathbf{x})$$

where the density $f_{\mathbf{X}}$ is absolutely continuous with respect to the reference density $f_{\mathbf{Y}}$.

In like manner one can define a "global" version of this metric which is applicable to multisensor, multitarget problems:

$$I(f_{\Sigma}; f_{\Gamma}) \triangleq \int f_{\Sigma}(X) \ln\left(\frac{f_{\Sigma}(X)}{f_{\Gamma}(X)}\right) \delta X$$

where f_{Σ} is absolutely continuous with respect to f_{Γ} in a sense which will not be specified here. See [31] for more details.

Decision theory. Another example of the potential usefulness of the parallelism between point-variate and finite-set-variate statistics arises from ordinary decision theory. In single-sensor, single-target problems the *Receiver Operating Characteristic* (ROC) curve is defined as the parameterized curve $\tau \rightarrow (p_{FA}(\tau), p_D(\tau))$ where

$$\begin{aligned} p_{FA}(\tau) &= \int_{L(\mathbf{z}_1,\ldots,\mathbf{z}_k)<\tau} f_{\mathbf{Z}_1,\ldots,\mathbf{Z}_k|H_0}(\mathbf{z}_1,\ldots,\mathbf{z}_k|H_0)\, d\lambda(\mathbf{z}_1)\cdots d\lambda(\mathbf{z}_k) \\ p_D(\tau) &= 1 - \int_{L(\mathbf{z}_1,\ldots,\mathbf{z}_k)>\tau} f_{\mathbf{Z}_1,\ldots,\mathbf{Z}_k|H_1}(\mathbf{z}_1,\ldots,\mathbf{z}_k|H_1)\, d\lambda(\mathbf{z}_1)\cdots d\lambda(\mathbf{z}_k) \end{aligned}$$

for all $\tau > 0$. Here H_0, H_1 are two hypotheses and

$$L(\mathbf{z}_1,\ldots,\mathbf{z}_k) \triangleq \frac{f_{\mathbf{Z}_1,\ldots,\mathbf{Z}_k|H_0}(\mathbf{z}_1,\ldots,\mathbf{z}_k|H_0)}{f_{\mathbf{Z}_1,\ldots,\mathbf{Z}_k|H_1}(\mathbf{z}_1,\ldots,\mathbf{z}_k|H_1)}$$

is the likelihood ratio for the decision problem.

In like manner, one can in principle define a ROC curve for an *entire multisensor, multitarget problem* as follows:

$$\begin{aligned} p_{FA}(\tau) &= \int_{L(Z_1,\ldots,Z_k)<\tau} f_{\Sigma_1,\ldots,\Sigma_k|H_0}(Z_1,\ldots,Z_k|H_0)\, \delta Z_1\cdots\delta Z_k \\ p_D(\tau) &= 1 - \int_{L(Z_1,\ldots,Z_k)>\tau} f_{\Sigma_1,\ldots,\Sigma_k|H_1}(Z_1,\ldots,Z_k|H_1)\, \delta Z_1\cdots\delta Z_k \end{aligned}$$

where

$$L(Z_1,\ldots,Z_k) \triangleq \frac{f_{\Sigma_1,\ldots,\Sigma_k|H_0}(Z_1,\ldots,Z_k|H_0)}{f_{\Sigma_1,\ldots,\Sigma_k|H_1}(Z_1,\ldots,Z_k|H_1)}$$

is the "global" likelihood ratio for the problem.

Estimation theory. In conventional statistics, an *estimator* of a parameter $\mathbf{x}$ of the parameterized density $f_{\mathbf{Z}}(\mathbf{z}|\mathbf{x})$ of an unknown random vector $\mathbf{Z}$ is a function $\hat{\mathbf{x}} = \mathbf{J}(\mathbf{z}_1, ..., \mathbf{z}_m)$ of the collected measurements $\mathbf{z}_1, ..., \mathbf{z}_m$. The most familiar estimator is the *maximum likelihood estimator* (MLE), defined by

$$\mathbf{J}_{MLE}(\mathbf{z}_1, ..., \mathbf{z}_m) \triangleq \arg\sup_{\mathbf{x}} \; L(\mathbf{x}|\mathbf{z}_1, ..., \mathbf{z}_m)$$

where $L(\mathbf{x}|\mathbf{z}_1, ..., \mathbf{z}_m) = f_{\mathbf{Z}}(\mathbf{z}_1|\mathbf{x}) \cdots f_{\mathbf{Z}}(\mathbf{z}_m|\mathbf{x})$ is the likelihood function. A Bayesian version of the MLE is the *maximum a posteriori* (MAP) estimator, defined by

$$\mathbf{J}_{MAP}(\mathbf{z}_1, ..., \mathbf{z}_m) \triangleq \arg\sup_{\mathbf{x}} \; f_{\mathbf{X}|\mathbf{Z}_1,...,\mathbf{Z}_m}(\mathbf{x}|\mathbf{z}_1, ..., \mathbf{z}_m)$$

where

$$f_{\mathbf{X}|\mathbf{Z}_1,...,\mathbf{Z}_m}(\mathbf{x}|\mathbf{z}_1, ..., \mathbf{z}_m) = \frac{f_{\mathbf{Z}_1,...,\mathbf{Z}_m|\mathbf{X}}(\mathbf{z}_1, ..., \mathbf{z}_m|\mathbf{x}) \; f_{\mathbf{X}}(\mathbf{x})}{f_{\mathbf{Z}_1,...,\mathbf{Z}_m}(\mathbf{z}_1, ..., \mathbf{z}_m)}$$

is the Bayesian posterior density conditioned on the measurements $\mathbf{z}_1, ..., \mathbf{z}_m$.

In like manner, a *global* (i.e., multisensor, multitarget) *estimator* of the set parameter of a global density $f_\Sigma(Z|X)$ is a function $\hat{X} = J(Z_1, ..., Z_m)$ of global measurements $Z_1, ..., Z_m$. One can define a multisensor, multitarget version of the MLE by

$$J_{GMLE}(Z_1, ..., Z_m) \triangleq \arg\sup_{X} L(X|Z_1, ..., Z_m)$$

where $L(X|Z_1, ..., Z_m) \triangleq f_\Sigma(Z_1|X) \cdots f_\Sigma(Z_m|X)$ is the "global" likelihood function. (Notice that in determining X, one is estimating not only the geokinematics and identities of targets, but also their number as well. Thus detection, localization, and identification are unified into a single statistical operation. This operation is a *direct* multisource, multitarget estimation technique in the sense defined in Section 1.) The definition of a multisensor, multitarget analog of the MAP estimator is less straightforward since global posterior densities $f_{\Gamma|\Sigma_1,...,\Sigma_m}(X|Z_1, ..., Z_m)$ have units which vary with the cardinality of X. Nevertheless, it is possible to define a global MAP estimator, as well as prove that it is statistically consistent. The proof is a direct generalization of a standard proof of the consistency of the ML and MAP estimators.

It is also possible to generalize *nonparametric estimation*–in particular, kernel estimators–to multisensor, multitarget scenarios as well. See [37] for more details.

Cramér–Rao inequalities. One of the major achievements of conventional estimation theory is the Cramér-Rao inequality which, given knowledge of the measurement-noise distribution $f_{\mathbf{Z}|\mathbf{X}}(\mathbf{z}|\mathbf{x})$ of the sensor, sets a lower bound on the precision with which any algorithm can estimate target parameters, using only data supplied by that sensor. In its more familiar form the Cramér-Rao inequality applies only to unbiased estimators. Let $\mathbf{J} = \mathbf{J}(\mathbf{z}_1, ..., \mathbf{z}_m)$ be an estimator of data $\mathbf{z}_1, ..., \mathbf{z}_m$, let $\bar{\mathbf{X}}$ be the expected value of the random vector $\mathbf{X} = \mathbf{J}(\mathbf{Z}_1, ..., \mathbf{Z}_m)$ where $\mathbf{Z}_1, ..., \mathbf{Z}_m$ are i.i.d. with density $f_{\mathbf{Z}|\mathbf{X}}$ and let the linear transformation $C_{\mathbf{J},\mathbf{x}}$ defined by

$$\begin{aligned} C_{\mathbf{J},\mathbf{x}}(\mathbf{w}) &\triangleq E_{\mathbf{x}}[\langle \mathbf{X} - \bar{\mathbf{X}}, \mathbf{w}\rangle \; (\mathbf{X} - \bar{\mathbf{X}})] \\ &= \int \langle \mathbf{J}(\mathbf{z}_1, ..., \mathbf{z}_m) - \bar{\mathbf{X}}, \mathbf{w}\rangle \; (\mathbf{J}(\mathbf{z}_1, ..., \mathbf{z}_m) - \bar{\mathbf{X}}) \\ &\qquad \cdot f_{\mathbf{Z}_1,...,\mathbf{Z}_m|\mathbf{X}}(\mathbf{z}_1, ..., \mathbf{z}_m|\mathbf{x}) \; d\lambda(\mathbf{z}_1) \cdots d\lambda(\mathbf{z}_m) \end{aligned}$$

for all $\mathbf{w}$ be the covariance of $\mathbf{J}$ (where '$\langle -, - \rangle$' denotes the inner product.) If $\mathbf{J}$ is unbiased in the sense that $E_{\mathbf{x}}[\mathbf{X}] = \mathbf{x}$ then the Cramér-Rao inequality is

$$\langle \mathbf{v}, \; C_{\mathbf{J},\mathbf{x}} \rangle \geq \langle \mathbf{v}, \; L_{\mathbf{x}}^{-1}(\mathbf{v}) \rangle$$

for all $\mathbf{v}$, where $L_{\mathbf{x}}$ is the linear transformation defined by

$$\langle \mathbf{v}, \; L_{\mathbf{x}}(\mathbf{w}) \rangle = E_{\mathbf{x}} \left[\left(\frac{\partial \ln f}{\partial \mathbf{v}} \right) \left(\frac{\partial \ln f}{\partial \mathbf{w}} \right) \right]$$

for all $\mathbf{v}, \mathbf{w}$ and where we have abbreviated $f = f_{\mathbf{Z}_1,...,\mathbf{Z}_m|\mathbf{X}}$. In the case of *biased* estimators the Cramér-Rao inequality takes the more general form

$$\langle \mathbf{v}, \; C_{\mathbf{J},\mathbf{x}} \rangle \cdot \langle \mathbf{w}, \; L_{\mathbf{x}}(\mathbf{w}) \rangle \geq \langle \mathbf{v}, \; \frac{\partial}{\partial \mathbf{w}} E_{\mathbf{x}}[\mathbf{X}] \rangle$$

for all $\mathbf{v}, \mathbf{w}$, where the directional derivative $\partial / \partial \mathbf{w}$ is applied to the function $\mathbf{x} \longmapsto E_{\mathbf{x}}[\mathbf{X}]$.

In like manner, let $f_{\Sigma|\Gamma}(Z|X)$ be the global density of the global sensor and let $\mathbf{J}(Z_1, ..., Z_m)$ be a *vector-valued* global estimator of some *vector-valued* function $\mathbf{F}(X)$ of the set parameter X. If $\Sigma_1, ..., \Sigma_m$ are i.i.d. with global density $f_{\Sigma|\Gamma}(Z|X)$ and $\mathbf{X} = \mathbf{J}(\Sigma_1, ..., \Sigma_m)$ then define the covariance $C_{\mathbf{J},X}$ in the obvious manner. In this case it is possible to show (under assumptions analogous to those used in the proof of the conventional Cramér-Rao inequality) that

$$\langle \mathbf{v}, \; C_{\mathbf{J},X} \rangle \cdot \langle \mathbf{w}, \; L_{X,\mathbf{x}}(\mathbf{w}) \rangle \geq \langle \mathbf{v}, \; \frac{\partial}{\partial_{\mathbf{x}} \mathbf{w}} E_X[\mathbf{X}] \rangle$$

for all $\mathbf{v}, \mathbf{w}$, where $L_{X,\mathbf{x}}$ is defined by

$$\langle \mathbf{v}, \; L_{X,\mathbf{x}}(\mathbf{w}) \rangle = E_X \left[\left(\frac{\partial \ln f}{\partial_{\mathbf{x}} \mathbf{v}} \right) \left(\frac{\partial \ln f}{\partial_{\mathbf{x}} \mathbf{w}} \right) \right]$$

for all $\mathbf{v}, \mathbf{w}$, where $f = f_{\Sigma_1,\ldots,\Sigma_m|\Gamma}$, and where the directional derivative $\partial f / \partial_{\mathbf{x}} \mathbf{v}$ of the function $f(X)$ of a finite-set variable X, if it exists, is defined by

$$\begin{aligned} \frac{\partial f}{\partial_{\mathbf{x}} \mathbf{v}}(X) &\triangleq \lim_{\varepsilon \to 0} \frac{f((X - \{\mathbf{x}\}) \cup \{\mathbf{x} + \varepsilon \mathbf{v}\}) - f(X)}{\varepsilon} && (\text{if } \mathbf{x} \in X) \\ \frac{\partial f}{\partial_{\mathbf{x}} \mathbf{v}}(X) &= 0 && (\text{if } \mathbf{x} \notin X) \end{aligned}$$

For more details, see [31].

Nonlinear filtering. Provided that one makes suitable Markov and conditional-independence assumptions, in single-sensor, single-target statistics dynamic (i.e. moving) targets can be tracked using the Bayesian update equation

$$f_{\alpha+1|\alpha+1}(\mathbf{x}_{\alpha+1}|Z^{\alpha+1}) = \frac{f(\mathbf{z}_{\alpha+1}|\mathbf{x}_{\alpha+1})\, f_{\alpha+1|\alpha}(\mathbf{x}_{\alpha+1}|Z^{\alpha})}{\int f(\mathbf{z}_{\alpha+1}|\mathbf{y}_{\alpha+1})\, f_{\alpha+1|\alpha}(\mathbf{y}_{\alpha+1}|Z^{\alpha})\, d\lambda(\mathbf{y}_{\alpha+1})}$$

together with the Markov prediction integral

$$f_{\alpha+1|\alpha}(\mathbf{x}_{\alpha+1}|Z^{\alpha}) = \int f_{\alpha+1|\alpha}(\mathbf{x}_{\alpha+1}|\mathbf{x}_{\alpha})\, f_{\alpha|\alpha}(\mathbf{x}_{\alpha}|Z^{\alpha})\, d\lambda(\mathbf{x}_{\alpha})$$

Here, $f(\mathbf{z}_{\alpha+1}|\mathbf{x}_{\alpha+1})$ is the measurement model, $f_{\alpha+1|\alpha}(\mathbf{x}_{\alpha+1}|\mathbf{x}_{\alpha})$ is the Markov motion model, and $f_{\alpha|\alpha}(\mathbf{x}_{\alpha}|Z^{\alpha})$ is the Bayesian posterior conditioned upon the time-accumulated evidence $Z^{\alpha} = \{\mathbf{z}_1, \ldots, \mathbf{z}_{\alpha}\}$. These are the *Bayesian nonlinear filtering equations.*

In like manner, one has nonlinear filtering equations for multisensor, multitarget problems.

$$\begin{aligned} f_{\alpha+1|\alpha+1}(X_{\alpha+1}|Z^{(\alpha+1)}) &= \frac{f(Z_{\alpha+1}|X_{\alpha+1})\, f_{\alpha+1|\alpha}(X_{\alpha+1}|Z^{(\alpha)})}{\int f(Z_{\alpha+1}|Y_{\alpha+1})\, f_{\alpha+1|\alpha}(Y_{\alpha+1}|Z^{(\alpha)})\, \delta Y_{\alpha+1}} \\ f_{\alpha+1|\alpha}(X_{\alpha+1}|Z^{(\alpha)}) &= \int f_{\alpha+1|\alpha}(X_{\alpha+1}|X_{\alpha})\, f_{\alpha|\alpha}(X_{\alpha}|Z^{(\alpha)})\, \delta X_{\alpha} \end{aligned}$$

The global density $f(Z_{\alpha+1}|X_{\alpha+1})$ is the measurement model for the multisensor, multitarget problem, and the global density $f_{\alpha+1|\alpha}(X_{\alpha+1}|X_{\alpha})$ is the motion model *for the entire multitarget system.*

Sensor management and nonlinear control theory. Sensor management requires the control of multiple allocatable sensors so as to resolve ambiguities in our knowledge about multiple, possibly unknown targets. The parallelism between point-variate and finite-set-variate statistics suggests one way of attacking this problem, by first looking at what is done in the *single-sensor, single-target* case. Consider, for example, a single controlled sensor–e.g. a missile-tracking camera–as it attempts to follow a missile. The camera must adjust its azimuth, elevation, and focal length in such a way as to anticipate the location of the missile at the time the next image of the missile is recorded.

This is a standard problem in *optimal control theory*. The sensor as well as the target has a time-varying state vector, and the sensor as well as the target is observed (by actuator sensors) in order to determine the sensor state. The problem is solved by treating the sensor and target as a *single system* whose parameters are to be estimated simultaneously. This is accomplished by defining a *controlled vector*–associated with the camera–and a *reference vector*–associated with the target–and attempting to keep the distance between these two vectors as small as possible. (The magnitudes of the input controls $\mathbf{u}_\alpha$ are also minimized as well.)

An approach to the multisensor, multitarget *sensor management* problem becomes evident if we use the random set approach to reformulate such problems as a single-sensor, single-target problem. In this case the "global" sensor follows a "global" target (some of whose individual targets may not even be detected yet). The motion of the multitarget system is modeled using a global Markov transition density. The only undetermined aspect of the problem is how to define analogs of the controlled and reference vectors. This is done by determining the Kullback-Leibler information distance between two suitable global densities. For more details see [30].

4. Unified information fusion. In the preceding sections we have indicated how "finite-set statistics" provides a unifying framework for multisensor, multitarget estimation. Perhaps the most interesting facet of the approach, however, is that it also provides a means of integrating "ambiguous"–i.e., imprecise, vague, or contingent–information into the same framework [36]. In this section we briefly summarize how this can be done.

We begin in Section 4.1 by explaining the difference between two types of observations: "precise" observations or *data*, and "ambiguous" observations or *evidence*. We show how data should be regarded as *points* of measurement space, while evidence should be regarded as *random subsets* of measurement space. In Section 4.2, assuming that we are working in a *finite* universe, we show how to define *measurement models* for both data and evidence and how to construct recursive estimation formulas for both data and evidence. We also describe a "Bayesian" approach to defining what a *rule of evidential combination* is. Finally, in Section 4.3, we show how the reasoning used in the finite-universe case can be extended to the general case by using the differential and integral calculus of set functions.

4.1. Data vs. evidence. Conventional observations are *precise* in the sense that they provide a possibly inaccurate and/or incomplete snapshot of the target state. For example, if observations are of the form $\mathbf{z} = \mathbf{x} + \mathbf{v}$ where $\mathbf{v}$ is random noise, then $\mathbf{z}$ is a precise–but inaccurate–estimate of the actual state $\mathbf{x}$ of the target. We can say that $\mathbf{z}$ "is" $\mathbf{x}$ except for some degree of uncertainty.

More generally it is possible for evidence to be *imprecise* in that it merely *constrains* the state of the target. For example, if T is a subset of state space then the proposition "$\mathbf{x} \in T$" states, rather unequivocally, that the state $\mathbf{x}$ must be in the set T and cannot be outside of it. Imprecise evidence can be equivocal, however, in that it may consist of a range of hypotheses "$\mathbf{x} \in T_1$",...,"$\mathbf{x} \in T_m$" where $T_1, ..., T_m$ are subsets of state space and the hypothesis "$\mathbf{x} \in T_i$" is held to be true with degree of belief m_i. In this case, imprecise evidence can be represented as a random subset Γ of state space such that $p(\Gamma = T_i) = m_i$, for all $i = 1, ..., m$.

More generally, however, even precise measurement models are more ambiguous than this. For example, if H is a singular matrix then $\mathbf{z} = H\mathbf{x} + \mathbf{v}$ models measurements $\mathbf{z}$ which can be incomplete representations of the state $\mathbf{x}$. In this case the relationship between precise measurements and the state is more indirect. In like manner, imprecise evidence can exert only an *indirect constraint* on the state of a target, in that at best it constrains *data* and only thereby the state. For example, evidence could take the form of a statement "$\mathbf{z} \in R$" where R is a subset of observation space. In this case the state $\mathbf{x}$ would still be constrained by the evidence R, but only in the sense that $H\mathbf{x}+\mathbf{v} \in R$. This kind of indirect constraint can also be equivocal, in the sense that it takes the form of range of hypotheses "$\mathbf{z} \in R_1$",...,"$\mathbf{z} \in R_m$", where $R_1, ..., R_m$ are subsets of observation space and the hypothesis "$\mathbf{z} \in R_i$" is held to be true with degree of belief n_i. Accordingly, we make the following definitions:

- precise observations, or *data*, are *points* in observation space; whereas
- imprecise observations, or *evidence*, are *random subsets* of observation space.

4.2. Measurement models for evidence: Finite–universe case. In the finite-universe case, Bayesian *measurement models* for precise observations have the form

$$p_{z_1,...,z_m|x}(a_1, ..., a_m|b) \triangleq p(z_1 = a_1, ..., z_m = a_m|x = b) = \frac{p(z_1 = a_1, ..., z_m = a_m, \ x = b)}{p(x = b)}$$

where $z_1, ..., z_m$ are random variables on the measurement space such that the marginal distributions $p_{z_i|x}(a|b) = p(z_i = a|x = b)$ are identical, where x is a random variable on the state space, where $a, a_1, ..., a_m$ are elements of measurement space, and where b is an element of state space. The distribution $p_{z_1,...,z_m|x}(a_1, ..., a_m|b)$ expresses the likelihood of observing the sequence of observations $a_1, ..., a_m$ given that a target with state b is present. When we pass to the case of observations which are imprecise, vague, etc. however, the concept of a measurement model becomes considerably more complex. In fact there are many plausible such models, expressing the varying degree to which evidence can be understood to constrain data.

For example, let $Z_1, ..., Z_m$ be subsets of a finite observation space and

$\Theta_1, ..., \Theta_{m'}$ random subsets of observation space representing pieces of ambiguous evidence. Let $\Sigma_1, ..., \Sigma_{m+m'}$ be random subsets of observation space which are identically distributed in the sense that the marginal distributions $m_{\Sigma_j}(Z) = p(\Sigma_j = Z)$ are identical for all $j = 1, ..., m+m'$. A *conventional* measurement model for data alone would have the form

$$m_{\Sigma_1,...,\Sigma_m|\Gamma}(Z_1, ..., Z_m|X) \triangleq p(\Sigma_1 = Z_1, ..., \Sigma_m = Z_m|\Gamma = X)$$

An obvious extension to evidence is given by

$$\begin{aligned} & m_{\Sigma_1,...,\Sigma_{m+m'}|\Gamma}(Z_1, ..., Z_m, \Theta_1, ..., \Theta_{m'}|X) \\ \triangleq \quad & p(\Sigma_1 = Z_1, ..., \Sigma_m = Z_m, \Theta_1 \supseteq \Sigma_{m+1}, ..., \Theta_{m'} \supseteq \Sigma_{m+m'}|\Gamma = X) \end{aligned}$$

This model requires that data must be *completely consistent* with evidence—but only in an *overall, probabilistic* sense.

A measurement model in which data is far more constrained by evidence is

$$\begin{aligned} & m_{\Sigma_1,...,\Sigma_m|\Gamma}(Z_1, ..., Z_m, \Theta_1, ..., \Theta_{m'}|X) \\ \triangleq \; & p(\Sigma_1 = Z_1, ..., \Sigma_m = Z_m, \Theta_1 \supseteq Z_1, ..., \Theta_j \supseteq Z_i, ..., \Theta_{m'} \supseteq Z_m|\Gamma = X) \\ = \; & p(\Sigma_1 = Z_1, ..., \Sigma_m = Z_m, \Theta_1 \cap \cdots \cap \Theta_{m'} \supseteq Z_1 \cup \cdots \cup Z_m|\Gamma = X) \end{aligned}$$

This model stipulates that *each* data set must be *directly* (not merely probabilistically) constrained by *all* evidence at hand.

In either case, with suitable conditional independence assumptions one can derive recursive estimation formulas for posterior distributions conditioned on both data and evidence. For example, in the case of the first measurement model we can define posterior distributions in the usual way by the proportionality

$$\begin{aligned} & m_{\Gamma|\Sigma_1,...,\Sigma_{m+m'}}(X|Z_1, ..., Z_m, \Theta_1, ..., \Theta_{m'}) \\ & \qquad \propto m_{\Sigma_1,...,\Sigma_{m+m'}|\Gamma}(Z_1, ..., Z_m, \Theta_1, ..., \Theta_{m'}|X) \, m_\Gamma(X) \end{aligned}$$

Assuming conditional independence this results in *two* recursive update proportionalities, one for *data* and one for *evidence*:

$$\begin{aligned} & m_{\Gamma|\Sigma_1,...,\Sigma_{m+m'}}(X|Z_1, ..., Z_m, \Theta_1, ..., \Theta_{m'}) \\ & \propto m_{\Sigma|\Gamma}(Z_m|X) \, m_{\Gamma|\Sigma_1,...,\Sigma_{m-1},\Sigma_{m+1},...,\Sigma_{m+m'}}(X|Z_1, ..., Z_{m-1}, \Theta_1, ..., \Theta_{m'}) \end{aligned}$$

and

$$\begin{aligned} & m_{\Gamma|\Sigma_1,...,\Sigma_{m+m'}}(X|Z_1, ..., Z_m, \Theta_1, ..., \Theta_{m'}) \\ & \qquad \propto \beta_{\Sigma|\Gamma}(\Theta_{m'}|X) \, m_{\Gamma|\Sigma_1,...,\Sigma_{m+m'-1}}(X|Z_1, ..., Z_m, \Theta_1, ..., \Theta_{m'-1}) \end{aligned}$$

where

$$\beta_{\Sigma|\Gamma}(\Theta|X) \triangleq p(\Sigma \subseteq \Theta|\Gamma = X) = \sum_{T \subseteq U} p(\Sigma \subseteq T, \Theta = T|\Gamma = X)$$

Likewise, in the case of the second measurement model we can derive the recursive update equations

$$m_{\Gamma|\Sigma_1,...,\Sigma_{m+1}}(X|Z^{(m+1)}, \Theta^{(m'+1)}) \\ \propto m_\Sigma(Z_{m+1}|X)\, m_{\Gamma|\Sigma_1,...,\Sigma_m}(X|Z^{(m)}, \Theta^{(m'+1)})$$

and

$$m_{\Gamma|\Sigma_1,...,\Sigma_{m+1}}(X|Z^{(m+1)}, \Theta^{(m'+1)}) \\ \propto \delta_{\Theta_{m'+1}}(Z_1 \cup ... \cup Z_{m+1}|X)\, m_{\Gamma|\Sigma_1,...,\Sigma_{m+1}}(X|Z^{(m+1)}, \Theta^{(m')})$$

where $Z^{(i)} \triangleq \{Z_1, ..., Z_i\}$ and $\Theta^{(j)} \triangleq \Theta_1 \cap \cdots \cap \Theta_j$ and where

$$\delta_\Theta(Z|X) \triangleq p(\Theta \supseteq Z|\Gamma = X)$$

is a commonality measure.

A "Bayesian interpretation" of rules of evidential combination. If the latter measurement model is assumed, so that the effect of evidence on data will be very constraining, then posterior distributions will have the very interesting property that

$$m_{\Gamma|\Sigma_1,...,\Sigma_m}(X|Z_1, ..., Z_m, \Theta_1, ..., \Theta_{m'}) \\ = m_{\Gamma|\Sigma_1,...,\Sigma_m}(X|Z_1, ..., Z_m, \Theta_1 \cap \cdots \cap \Theta_{m'})$$

In other words, the random-set intersection operator '$\cap$' may be interpreted as a means of fusing multiple pieces of ambiguous evidence in such a way that posteriors conditioned on the fused evidence are identical to posteriors conditioned on the individual evidence and computed using Bayes' rule alone. Thus, for example, suppose that

$$\Theta_1 = \Sigma_x(f), \qquad \Theta_2 = \Sigma_x(g)$$

where f, g are two fuzzy membership functions on U and x is a uniformly distributed random number on $[0, 1]$. Let '$\wedge$' denote the Zadeh "min" fuzzy *AND* operation and *define* the posteriors $m_{\Gamma|\Sigma}(X|Z, f, g)$ and $m_{\Gamma|\Sigma}(X|Z, f \wedge g)$ by

$$\begin{aligned} m_{\Gamma|\Sigma}(X|Z, f, g) &= m_{\Gamma|\Sigma}(X|Z, \Sigma_x(f), \Sigma_x(g)) \\ m_{\Gamma|\Sigma}(X|Z, f \wedge g) &= m_{\Gamma|\Sigma}(X|Z, \Sigma_x(f \wedge g)) \end{aligned}$$

We know that $\Sigma_x(f) \cap \Sigma_x(g) = \Sigma_x(f \wedge g)$ and thus that

$$m_{\Gamma|\Sigma}(X|Z, f, g) = m_{\Gamma|\Sigma}(X|Z, \Sigma_x(f), \Sigma_x(g)) = m_{\Gamma|\Sigma}(X|Z, \Sigma_x(f) \cap \Sigma_x(g)) \\ = m_{\Gamma|\Sigma}(X|Z, \Sigma_x(f \wedge g)) = m_{\Gamma|\Sigma}(X|Z, f \wedge g)$$

and so

$$m_{\Gamma|\Sigma}(X|Z,f,g) = m_{\Gamma|\Sigma}(X|Z,f \wedge g)$$

That is: the fuzzy *AND* is a means of fusing fuzzy evidence in such a way that posteriors conditioned on the fused evidence are identical to posteriors conditioned on the fuzzy evidence individually and computed using Bayes' rule alone. Thus fuzzy logic is entirely consistent with Bayesian probability–*provided that it is first represented in random set form*, and *provided that we use a specific measurement model for ambiguous observations.*

Similar observations apply to *any* rule of evidential combination which bears a homomorphic relationship with the random set intersection operator.

4.3. Measurement models for evidence: General case. The concepts discussed in Section 4.2 were developed under the assumption of a finite universe case $\mathcal{R} = U$, but must be extended to the continuous-discrete case $\mathcal{R} \triangleq \mathbb{R}^n \times U$ if we are to apply them to information fusion problems. This can be done, provided that we use *discrete* random closed subsets to represent ambiguous observations in the continuous-discrete case.

A random closed subset Θ of observation space is *discrete* if there are closed subsets $C_1, ..., C_r$ of $\mathcal{R}$ and nonnegative numbers $q_1, ..., q_r$ such that $\sum_{i=1}^r q_i = 1$ and $p(\Theta = C_i) = q_i$, for all $i = 1, ..., r$. We abbreviate $m_\Theta(T_i) = p(\Theta = T_i)$, for all $i = 1, ..., r$ and $m_\Theta(T) = 0$ if $T \neq T_i$, for any $i = 1, ..., r$.

Provided that we accept this restriction on the form of the random subsets that we use to model ambiguous observations, it is not difficult to extend the finite-universe results just described. That is, assuming one or another measurement model for ambiguous observations, one can define *global measurement densities*

$$f_{\Sigma_1,...,\Sigma_{m+m'}|\Gamma}(Z_1, ..., Z_m, \Theta_1, ..., \Theta_{m'}|X)$$

as well as global posterior densities conditioned on both data and evidence:

$$f_{\Gamma|\Sigma_1,...,\Sigma_{m+m'}}(X|Z_1, ..., Z_m, \Theta_1, ..., \Theta_{m'})$$

Likewise, assuming one or another measurement model, one can derive recursive update equations for both data and evidence. For example, assuming one measurement model we get

$$\begin{aligned}&f_{\Gamma|\Sigma_1,...,\Sigma_{m+m'}}(X|Z_1, ..., Z_m, \Theta_1, ..., \Theta_{m'}) \\ &\quad\propto f_{\Sigma|\Gamma}(Z_m|X)\; m_{\Gamma|\Sigma_1,...,\Sigma_{m-1},\Sigma_{m+1},...,\Sigma_{m+m'}}(X|Z_1, ..., Z_{m-1}, \Theta_1, ..., \Theta_{m'})\end{aligned}$$

and

$$\begin{aligned}&f_{\Gamma|\Sigma_1,...,\Sigma_{m+m'}}(X|Z_1, ..., Z_m, \Theta_1, ..., \Theta_{m'}) \\ &\qquad\qquad\propto \beta_{\Sigma|\Gamma}(\Theta_{m'}|X)\; f_{\Gamma|\Sigma_1,...,\Sigma_{m+m'-1}}(X|Z_1, ..., Z_m, \Theta_1, ..., \Theta_{m'-1})\end{aligned}$$

where $\beta_{\Sigma|\Gamma}(\Theta|X) \triangleq p(\Sigma \subseteq \Theta|\Gamma = X) = \sum_T p(\Sigma \subseteq T, \Theta = T|\Gamma = X)$. With a different choice of measurement model one also can derive equations of the form

$$f_{\Gamma|\Sigma}(X|Z, f, g) = f_{\Gamma|\Sigma}(X|Z, f \wedge g)$$

where now f, g are *finite-level* fuzzy membership functions (i.e., the images of f, g are in some fixed finite subset of $[0, 1]$.

It thereby becomes possible to extend the Bayesian nonlinear filtering equations so that both precise and ambiguous observations can be accommodated into dynamic multisensor, multitarget estimation. For example, if one assumes one possible measurement model for evidence, one gets the following update equation:

$$f_{\alpha+1|\alpha+1}(X_{\alpha+1}|Z^{(\alpha+1)}, \Theta^{(\alpha+1)}) = \frac{\beta(\Theta_{\alpha+1}|X_{\alpha+1})\; f_{\alpha+1|\alpha}(X_{\alpha+1}|Z^{(\alpha)}, \Theta^{(\alpha)})}{\int f(Z_{\alpha+1}|Y_{\alpha+1})\; f_{\alpha+1|\alpha}(Y_{\alpha+1}|Z^{(\alpha)}, \Theta^{(\alpha)})\; \delta Y_{\alpha+1}}$$

where $\Theta^{(\alpha)} \triangleq \{\Theta_1, ..., \Theta_\alpha\}$.

5. Summary and conclusions. In this paper we began by providing a brief methodological survey of information fusion, based on the distinctions between *point estimation vs. set estimation* and between *indirect estimation and direct estimation.* We also sketched a history of the application of random set techniques in information fusion: the MCTW and Washburn filters in multisensor, multitarget estimation; and Dempster-Shafer theory, fuzzy logic, rule-based inference in expert systems theory. We then summarized our approach to information fusion, one which makes use of random set theory as a common unifying foundation for both expert systems theory and multisensor, multitarget estimation.

The application of random set theory to information fusion is an endeavor which is still in its infancy. Though the importance of the theory as a unifying foundation for expert systems theory has slowly been gaining recognition since the work of Goodman, Nguyen, and others in the late 1970s, in the case of multisensor, multitarget estimation similar efforts–those of Mori, Chong, Tse, and Wishner and, more indirectly, those of Washburn–are as recent as the mid-to-late 1980s. It is the hope of the author that the efforts of these pioneers, and the work reported in this paper, will stimulate interest in random set techniques in information fusion as well as the development of significant new information fusion algorithms.

REFERENCES

[1] R.T. ANTONY, *Principles of Data Fusion Automation*, Artech House, Dedham, Massachusetts, 1995.

[2] C.A. Barlow, L.D. Stone, and M.V. Finn, *Unified data fusion*, Proceedings of the 9th National Symposium on Sensor Fusion, vol. I (Unclassified), Naval Postgraduate School, Monterey CA, March 11–13, 1996.

[3] Y. Bar–Shalom and T.E. Fortmann, *Tracking and Data Association*, Academic Press, New York City, New York, 1988.

[4] Y. Bar–Shalom and X.–R. Li, *Estimation and Tracking: Principles, Techniques, and Software*, Artech House, Dedham, Massachusetts, 1993.

[5] S.S. Blackman, *Multiple–Target Tracking with Radar Applications*, Artech House, Dedham MA, 1986.

[6] P.P. Bonissone and N.C. Wood, *T–norm based reasoning in situation assessment applications*, Uncertainty in Artificial Intelligence (L.N. Kanal, T.S. Levitt, and J.F. Lemmer, eds.), vol. 3, New York City, New York: Elsevier Publishers, 1989, pp. 241–256.

[7] D.J. Daley and D. Vere–Jones, *An Introduction to the Theory of Point Processes*, Springer–Verlag, New York City, New York, 1988.

[8] I.R. Goodman, *Fuzzy sets as equivalence classes of random sets*, Fuzzy Sets and Possibility Theory (R. Yager, ed.), Pergamon Press, 1982, pp. 327–343.

[9] I.R. Goodman, *A new characterization of fuzzy logic operators producing homomorphic–like relations with one–point coverages of random sets*, Advances in Fuzzy Theory and Technology, vol. II, (P.P. Wang, ed.), Duke University, Durham, NC, 1994, pp. 133–159.

[10] I.R. Goodman, *Pact: An approach to combining linguistic–based and probabilistic information in correlation and tracking*, Tech. Rep. 878, Naval Ocean Command and Control Ocean Systems Center, RDT&E Division, San Diego, California, March 1986; and *A revised approach to combining linguistic and probabilistic information in correlation*, Tech. Rep. 1386, Naval Ocean Command and Control Ocean Systems Center, RDT&E Division, San Diego, California, July 1992.

[11] I.R. Goodman, *Toward a comprehensive theory of linguistic and probabilistic evidence: Two new approaches to conditional event algebra*, IEEE Transactions on Systems, Man and Cybernetics, 24 (1994), pp. 1685–1698.

[12] I.R. Goodman, *A unified approach to modeling and combining of evidence through random set theory*, Proceedings of the 6th MIT / ONR Workshop on C3 Systems, Massachusetts Institute of Technology, Cambridge, MA, pp. 42–47.

[13] I.R. Goodman and H.T. Nguyen, *Uncertainty Models for Knowledge Based Systems*, North–Holland, Amsterdam, The Netherlands, 1985.

[14] I.R. Goodman, H.T. Nguyen, and E.A. Walker, *Conditional Inference and Logic for Intelligent Systems: A Theory of Measure–Free Conditioning*, North–Holland, Amsterdam, The Netherlands, 1991.

[15] M. Grabisch, H.T. Nguyen, and E.A. Walker, *Fundamentals of Uncertainty Calculi With Applications to Fuzzy Inference*, Kluwer Academic Publishers, Dordrecht, The Netherlands, 1995.

[16] S. Graf, *A Radon–Nikodým theorem for capacities*, Journal für Reine und Angewandte Mathematik, 320 (1980), pp. 192–214.

[17] D.L. Hall, *Mathematical Techniques in Multisensor Data Fusion*, Artech House, Dedham, Massachusetts, 1992.

[18] K. Hestir, H.T. Nguyen, and G.S. Rogers, *A random set formalism for evidential reasoning*, Conditional Logic in Expert Systems (I.R. Goodman, M.M. Gupta, H.T. Nguyen and G.S. Rogers, eds.), Amsterdam, The Netherlands: North–Holland, 1991, pp. 309–344.

[19] T.L. Hill, *Statistical Mechanics: Principles and Selected Applications*, Dover Publications, New York City, New York, 1956.

[20] P.B. Kantor, *Orbit space and closely spaced targets*, Proceedings of the SDI Panels on Tracking, no. 2, 1991.

[21] P.J. Huber and V. Strassen, *Minimax tests and the Neyman–Pearson lemma for capacities*, Annals of Statistics, 1 (1973), pp. 251–263.

[22] K. KASTELLA AND C. LUTES, *Coherent maximum likelihood estimation and mean-field theory in multi-target tracking*, Proceedings of the 6th Joint Service Data Fusion Symposium, vol. I (Part 2), Johns Hopkins Applied Physics Laboratory, Laurel, MD, June 14–18, 1993, pp. 971–982.

[23] R. KRUSE AND K.D. MEYER, *Statistics with Vague Data*, D. Reidel/Kluwer Academic Publishers, Dordrecht, The Netherlands, 1987.

[24] R. KRUSE, E. SCHWENCKE, AND J. HEINSOHN, *Uncertainty and Vagueness in Knowledge-Based Systems*, Springer-Verlag, New York City, New York, 1991.

[25] D. LEWIS, *Probabilities of conditionals and conditional probabilities*, Philosophical Review, 85 (1976), pp. 297–315.

[26] Y. LI, *Probabilistic Interpretations of Fuzzy Sets and Systems*, Ph.D. Thesis, Department of Electrical Engineering and Computer Science, Massachusetts Institute of Technology, Cambridge, MA, 1994.

[27] N.N. LYSHENKO, *Statistics of random compact sets in Euclidean space*, Journal of Soviet Mathematics, 21 (1983), pp. 76–92.

[28] R.P.S. MAHLER, *Combining ambiguous evidence with respect to ambiguous a priori knowledge. Part II: Fuzzy logic*, Fuzzy Sets and Systems, 75 (1995), pp. 319–354.

[29] R.P.S. MAHLER, *Global integrated data fusion*, Proceedings of the 7th National Symposium on Sensor Fusion, vol. I (Unclassified), March 16–18, 1994, Sandia National Laboratories, Albuquerque, NM, pp. 187–199.

[30] R.P.S. MAHLER, *Global optimal sensor allocation*, Proceedings of the 9th National Symposium on Sensor Fusion, vol. I (Unclassified), Naval Postgraduate School, Monterey CA, March 11–13, 1996.

[31] R.P.S. MAHLER, *Information theory and data fusion*, Proceedings of the 8th National Symposium on Sensor Fusion, vol. I (Unclassified), Texas Instruments, Dallas, TX, March 17–19, 1995, pp. 279–292.

[32] R.P.S. MAHLER, *Nonadditive probability, finite-set statistics, and information fusion*, Proceedings of the 34th IEEE Conference on Decision and Control, New Orleans, LA, December 1995, pp. 1947–1952.

[33] R.P.S. MAHLER, *The random-set approach to data fusion*, SPIE Proceedings, vol. 2234, 1994, pp. 287–295.

[34] R.P.S. MAHLER, *Representing rules as random sets. I: Statistical correlations between rules*, Information Sciences, 88 (1996), pp. 47–68.

[35] R.P.S. MAHLER, *Representing rules as random sets. II: Iterated rules*, International Journal of Intelligent Systems, 11 (1996), pp. 583–610.

[36] R.P.S. MAHLER, *Unified data fusion: Fuzzy logic, evidence, and rules*, SPIE Proceedings, 2755 (1996), pp. 226–237.

[37] R.P.S. MAHLER, *Unified nonparametric data fusion*, SPIE Proceedings, vol. 2484, 1995, pp. 66–74.

[38] G. MATHÉRON, *Random Sets and Integral Geometry*, John Wiley, New York City, New York, 1975.

[39] I.S. MOLCHANOV, *Limit Theorems for Unions of Random Closed Sets*, Springer-Verlag Lecture Notes in Mathematics, vol. 1561, Springer-Verlag, Berlin, Germany, 1993.

[40] S. MORI, C.-Y. CHONG, E. TSE, AND R.P. WISHNER, *Multitarget multisensor tracking problems. Part I: A general solution and a unified view on Bayesian approaches*, Revised Version, Tech. Rep. TR-1048-01, Advanced Information and Decision Systems, Inc., Mountain View CA, August 1984. My thanks to Dr. Mori for making this report available to me (Dr. Shozo Mori, Personal Communication, February 28, 1995).

[41] S. MORI, C.-Y. CHONG, E. TSE, AND R.P. WISHNER, *Tracking and classifying multiple targets without a priori identification*, IEEE Transactions on Automatic Control, 31 (1986), pp. 401–409.

[42] R.E. NEAPOLITAN, *A survey of uncertain and approximate inference*, Fuzzy Logic for the Management of Uncertainty (L. Zadeh and J. Kocprzyk, eds.), New

York City, New York: John Wiley, 1992.

[43] H.T. NGUYEN, *On random sets and belief functions*, Journal of Mathematical Analysis and Applications, 65 (1978), pp. 531–542.

[44] A.I. ORLOV, *Relationships between fuzzy and random sets: Fuzzy tolerances*, Issledovania po Veroyatnostnostatishesk. Modelirovaniu Realnikh System, 1977, Moscow, Union of Soviet Socialist Republics.

[45] A.I. ORLOV, *Fuzzy and random sets*, Prikladnoi Mnogomerni Statisticheskii Analys, 1978, Moscow, Union of Soviet Socialist Republics.

[46] P. QUINIO AND T. MATSUYAMA, *Random closed sets: A unified approach to the representation of imprecision and uncertainty*, Symbolic and Quantitative Approaches to Uncertainty (R. Kruse and P. Siegel, eds.), New York City, New York: Springer–Verlag, 1991, pp. 282–286.

[47] D.B. REID, *An algorithm for tracking multiple targets*, IEEE Transactions on Automatic Control, 24 (1979), pp. 843–854.

[48] B. SCHWEIZER AND A.SKLAR, *Probabilistic Metric Spaces*, North–Holland, Amsterdam, The Netherlands, 1983.

[49] G.E. SHILOV AND B.L. GUREVICH, *Integral, Measure, and Derivative: A Unified Approach*, Prentice–Hall, New York City, New York, 1966.

[50] *P. Smets, The transferable belief model and random sets*, International Journal of Intelligent Systems, 7 (1992), pp. 37–46.

[51] L.D. STONE, M.V. FINN, AND C.A. BARLOW, *Unified data fusion*, Tech. Rep., Metron Corp., January 26, 1996.

[52] J.K. UHLMANN, *Algorithms for multiple–target tracking*, American Scientist, 80 (1992), pp. 128–141.

[53] B.C. VAN FRAASSEN, *Probabilities of conditionals*, Foundations of Probability Theory, Statistical Inference, and Statistical Theories of Science (W.L. Harper and E.A. Hooker, eds.), vol. I, Dordrecht, The Netherlands: D. Reidel, 1976, pp. 261–308.

[54] E. WALTZ AND J. LLINAS, *Multisensor Data Fusion*, Artech House, Dedham, Massachusetts, 1990.

[55] R.B. WASHBURN, *A random point process approach to multiobject tracking*, Proceedings of the American Control Conference, vol. 3, June 10–12, 1987, Minneapolis, Minnesota, pp. 1846–1852.

CRAMÉR–RAO TYPE BOUNDS FOR RANDOM SET PROBLEMS

FRED E. DAUM*

Abstract. Two lower bounds on the error covariance matrix are described for tracking in a dense multiple target environment. The first bound uses Bayesian theory and equivalence classes of random sets. The second bound, however, does not use random sets, but rather it is based on symmetric polynomials. An interesting and previously unexplored connection between random sets and symmetric polynomials at an abstract level is suggested. Apparently, the shortest path between random sets and symmetric polynomials is through a Banach space.

Key words. Bounds on Performance, Cramér–Rao Bound, Estimation, Fuzzy Logic, Fuzzy Sets, Multiple Target Tracking, Nonlinear Filters, Random Sets, Symmetric Polynomials.

AMS(MOS) subject classifications. 12Y05, 60D05, 60G35, 60J70, 93E11

1. Introduction. A good way to ruin the performance of a Kalman filter is to put the wrong data into it. This is the basic problem of tracking in a dense multiple-target environment in many systems using radar, sonar, infrared, and other sensors. The effects of clutter, jamming, measurement noise, false alarms, missed detections, unresolved measurements, and target maneuvers make the problem very challenging. The problem is extremely difficult in terms of performance as well as computational complexity. Despite the plethora of multiple target tracking algorithms that have been developed, no nontrivial theoretical bound on performance was published prior to [1]. Table 1 lists some of the algorithms for multiple target tracking. At a recent meeting of experts on multiple target tracking, the lack of theoretical performance bounds was identified as a critical issue. Indeed, one expert (Oliver Drummond) noted that: "There is no Cramér–Rao bound for multiple target tracking." The theory in [1] provides a bound on the error covariance matrix similar to the Cramér–Rao bound. On the other hand, the bound in [1] requires Monte Carlo simulation, whereas a true Cramér–Rao bound, such as the one reported in [31], does not. The relative merits of the two bounds in [1] and [31] depend on the computational resources available and the specific parameters of a given problem. Both [1] and [31] are lower bounds on the error covariance matrix, and therefore a tighter lower bound could be computed by taking the larger lower bound of the two. In general, it is not obvious a priori whether [1] or [31] would result in a larger lower bound.

More generally, the lack of Cramér–Rao bounds is also evident in random set problems. In particular, one of the plenary speakers at this IMA workshop on random sets (Ilya Molchanov) noted that "There is no

* Raytheon Company, 1001 Boston Post Road, Marlborough, MA 01752.

TABLE 1
Comparison of multiple target tracking algorithms.

Formalism or Algorithm	Time Horizon Considered (No. of Samples)	Number of Data Association Hypotheses	Unresolved Data Modeled in Algorithm	Relative Performance in Dense Multiple Target Environments		Computational Complexity	
				Unresolved Data	Resolved Data	Exact Solution	Approximate Solution
Nearest Neighbor	1	1	No	Poor	Poor	Low	Low
Nearest Neighbor-M	1	1	Yes	Fair	Poor	Low	Low
Probabilistic Data Association (PDA)	1	1	No	Poor	Fair	Low	Low
Joint Probabilistic Data Association (JPDA)	1	1	No	Fair	Good	Exp	Medium
JPDAM	1	1	Yes	Good	Good	Exp	Medium
Nearest Neighbor JPDA	1	1	No	Fair	Good to Excellent	Poly	Low
Assignment	1	1	No	Fair	Good to Excellent	Poly	Medium
Dynamic Programming (Viterbi)	Many	1	No	Poor	Good	Poly	Medium
Hough Transform	Many	1	No	Fair	Good	Poly	Medium
Multiple Hypothesis Tracking (MHT)	Many	Many	No	Good	Optimal	Exp	High
MHT-M	Many	Many	Yes	Best	Excellent	Exp	High
Morefield	Many	Many	No	Fair	Excellent	Exp	High
Symmetric Measurements EKF	Many	Many	No	?	Good	Poly	High
Symmetric Measurements Non-recursive	Many	Many	No	?	Excellent	Exp	High
Branching	Many	Many	No	Fair	Excellent	Bounded	Med/High
Branching-M	Many	Many	Yes	Good	Excellent	Bounded	Med/High
Multidimensional Assignment	Many	Many	No	Good	Excellent	–	High
Multidimensional Assignment-M	Many	Many	Yes	Excellent	Excellent	–	High
Exact N-Best Hypotheses	Many	N	No	Good	Excellent	–	Medium

Cramér–Rao bound for random sets." Although this paper is written in the context of multiple target tracking problems, it is clear that the two performance bounds described here can be applied to more general random set problems.

Bayesian performance bounds for tracking in a dense multiple target environment can also be obtained without using random sets. Such bounds are derived using symmetric polynomials [31]. The formulation of *multiple target tracking* (MTT) problems using symmetric polynomials is due to Kamen [32]; this approach removes the combinatorial complexity of this problem and replaces it with a nonlinear filtering problem. That is, a discrete problem is replaced by a continuous problem using Kamen's idea. This is analogous to the proof of the prime number theorem using complex contour integrals of the Riemann zeta-function in analytical number theory; other examples of the interplay between discrete and continuous

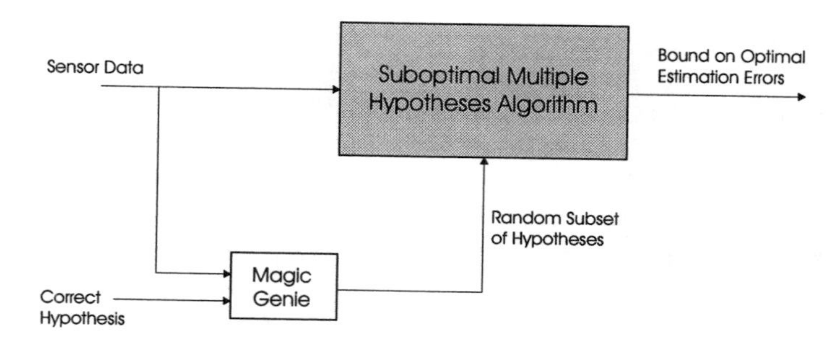

FIG. 1. *Bound on estimation error using random subsets.*

problems are discussed in Section 4. As shown in this paper, there is an obvious connection between random sets and symmetric polynomials in the context of MTT. This suggests that random sets and symmetric polynomials should be connected at an abstract level as well. Such a connection has not been noted before, and it would be useful to develop the abstract theory of random sets and symmetric polynomials. Apparently the shortest path between random sets and symmetric polynomials is through a Banach space.

Further background on multiple target tracking is given in [1] and [12], as well as in the papers by R. Mahler and S. Mori in this volume.

2. Bounds using random sets. Figure 1 shows the basic intuitive idea of the error bound in [1]. The magic genie provides a random subset of hypotheses to a data association algorithm. Each "hypothesis" is a full explanation about the association of sensor measurements to physical objects for a given data set. The magic genie knows the correct data association hypothesis, but it is important for the magic genie to keep it hidden, so that the algorithm can't cheat; otherwise the bound would not be tight. Therefore, a *random* subset of hypotheses is required. The genie gives the algorithm a *subset* of hypotheses in order to reduce computational complexity. This algorithm, aided by the genie, will produce better estimation accuracy (on average) than the optimal algorithm without help from a magic genie. The mathematical details of this bound are purely Bayesian in [1]. Figure 1 and the intuitive sketch of the bound given above show how *random sets* occur in a natural way. Further details on the bound itself are given below.

The Bayesian theory in [1] provides a lower bound on the estimation error covariance matrix:

$$E(C) \geq E(C^*)$$

where

C = Estimation error covariance matrix.
C^* = Covariance matrix computed as shown in Figure 1.
$E(\cdot)$ = Expected value of $(\cdot)$ with respect to sensor measurements.

A precise mathematical statement of this result, along with a proof, is given in [1].

The matrix C^* is computed with the help of a magic genie as shown in Figure 1. The *multiple hypothesis tracking* (MHT) algorithm shown in Figure 1 is a slight modification of a standard MHT algorithm such as the one described in [1]. The standard MHT algorithm is modified to accept helpful hints from the magic genie; otherwise, the MHT algorithm is unchanged. As noted in Figure 1, the MHT algorithm is "suboptimal" in the sense that the total number of hypotheses is limited by standard pruning and combining heuristics. The alternative, an optimal MHT algorithm, which must consider every feasible hypothesis, is completely out of the question owing to its enormously high computational complexity.

The suboptimal MHT algorithm is given simulated sensor measurements or real sensor measurements that have been recorded. The MHT algorithm runs off-line, and it does not run in real time. Its purpose is to evaluate the best possible system performance, rather than produce such performance.

The magic genie has access to the sensor data, as well as to the "correct hypothesis" about measurement-to-measurement association. The genie knows exactly which measurements arise from which targets; the genie also knows which measurements are noise or clutter and which measurements are unresolved. In short, the genie knows everything, but the genie is not allowed to tell the MHT algorithm the whole truth. In particular, the genie supplies the MHT algorithm with a random subset of hypotheses about measurement-to-measurement association, which measurements are due to noise or clutter, and which are unresolved. This random subset of hypotheses must contain the correct hypothesis, but the genie is not allowed to tell the MHT algorithm which hypothesis is correct. If the genie divulged this correct information, then the lower bound on the covariance matrix would degenerate to the trivial bound corresponding to perfectly known origin of measurements. The genie must hide the correct hypothesis within the random subset of other (wrong) hypotheses. This can be done by randomly permuting the order of hypotheses within the subset and by making sure that the number of hypotheses does not give any clue about which one is correct.

Figure 2 shows a typical output from the block diagram in Figure 1. The lower bound corresponds to $E(C^*)$, which is computed by the MHT algorithm using help from the magic genie. A real-time, on-line MHT algorithm cannot possibly do better than this, because it does not have any help from the magic genie. This is the intuitive "proof" of the inequality $E(C) \geq E(C^*)$. For example, an on-line, real-time MHT algorithm might not consider the correct hypothesis at all. Moreover, even if it did, the correct hypothesis would be competing with an enormous number of other hypotheses, and it would not be obvious which one actually is correct. The genie tells the MHT algorithm which hypotheses might be correct, and

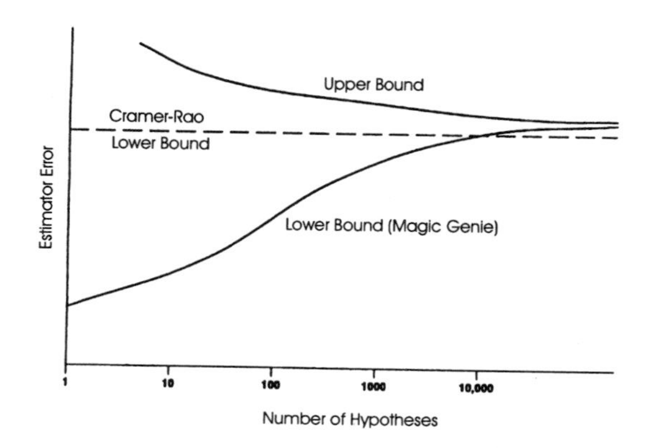

FIG. 2. *Performance bounds for multiple target tracking.*

thereby limits the total number of hypotheses that must be considered.

The upper bound in Figure 2 can be produced by an MHT algorithm that considers a limited number of hypotheses but without any help from the genie. This is exactly what a practical on-line, real-time MHT algorithm would do. The performance of the optimal MHT algorithm, by definition, is no worse than any practical MHT algorithm, hence the upper bound in Figure 2.

As shown in Figure 2, the upper and lower bounds should converge as the number of hypotheses is increased. The upper bound is decreasing because the suboptimal MHT algorithm is getting better and better as the number of hypotheses is increased. Likewise, the lower bound is increasing, because the magic genie is providing less and less help to the MHT algorithm. If the genie gave all hypotheses to the MHT algorithm, this would be no help at all! The rate at which the upper and lower bounds converge depends on the specific application. We implicitly assume in Figure 2 that the hypotheses are selected in a way to speed convergence. One good approach is to use the same pruning and combining heuristics used in standard MHT algorithms; discard the unlikely hypotheses and keep the most likely ones. This strategy speeds convergence for both the upper bound and lower bound.

3. Bounds using nonlinear filters. The Cramér–Rao bound derived in [31] is based on a clever formulation of the data association problem using nonlinear *symmetric measurement equations* (SME) due to Kamen [32]. The basic idea of this bound is shown in Figure 3. Kamen's SME formulation removes the combinatorial nature of the data association problem, and replaces it with a nonlinear filter problem. In particular, for two targets, the symmetric measurement equations would be:

$$z_1 = y_1 + y_2 \ ,$$

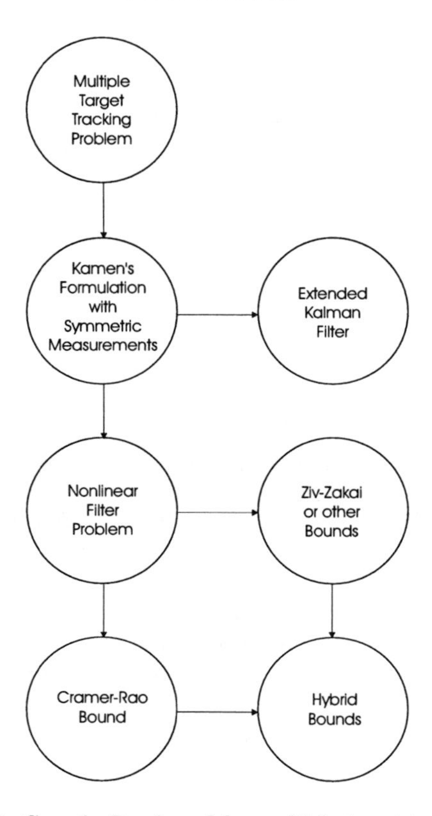

FIG. 3. *Cramér–Rao bound for multiple target tracking.*

$$z_2 = y_1 y_2 \ ,$$

in which

y_1 = Physical measurement from target no. 1.
y_2 = Physical measurement from target no. 2.
z_1, z_2 = Synthetic measurements.

The measurement equations are "symmetric" in the sense that permuting y_1 and y_2 does not change the values of z_1 or z_2. Obviously this can be generalized to any number of targets. For example, with three targets:

$$z_1 = y_1 + y_2 + y_3 \ ,$$

$$z_2 = y_1 y_2 + y_1 y_3 + y_2 y_3 \ ,$$

$$z_3 = y_1 y_2 y_3 \ ,$$

and so on .... More generally, the SME's could be written as:

$$z_k(y_1, y_2, \ldots, y_N) = \sum y_{i_1} y_{i_2} \cdots y_{i_k} \ ,$$

for $k = 1, 2, \ldots, N$; in which the summation is over certain sets of indices such that $1 \leq i_1 < i_2 < \cdots < i_k \leq N$. On the other hand, the two examples for $N = 2$ and $N = 3$ given above were easier for me to write, easier for my secretary to type, and probably easier for the reader to understand. So much for generality! Note that these equations are nonlinear in the physical measurements. Therefore, the resulting filtering problem is nonlinear, but random sets are not used at all. It is interesting to ask: "Where did the random sets go to in the SME formulation?" The answer is that the random sets disappeared into a Banach space; see Section 4 for an elaboration of this answer.

Note that these particular SME's correspond to the elementary symmetric polynomials, which occur in Newton's work on the binomial theorem, binomial coefficients and symmetric polynomials. Other symmetric measurement equations are possible; for example, with two targets:

$$z_1 = y_1 + y_2 \,,$$

$$z_2 = y_1^2 + y_2^2 \,.$$

It turns out that different choices of symmetric measurement equations do not affect the Cramér–Rao bound in [31], and neither does this choice affect the optimal filter performance. However, the performance of a suboptimal nonlinear filter, such as the *extended Kalman filter* (EKF), does depend on the specific SME formulation.

The above SME formulation is for scalar-valued measurements from any given target; however, in practical applications, we are typically interested in vector-valued measurements. At the IMA workshop, R. Mahler asked me how to handle this case without incurring a computational complexity that is exponential in the dimension of the measurement vector (m). The answer is to use a well known formulation in Kalman filtering called "Battin's trick," whereby a vector valued measurement can be processed as a sequence of m scalar-valued measurement updates of the Kalman filter. It is intuitively obvious that this can be done if the measurement error covariance matrix is diagonal, which is typically the case in practical applications. In any case, the measurement error covariance matrix can always be diagonalized, because it is symmetric and positive definite. Generally, this would be done by selecting the measurement coordinate system corresponding to the principal axes of the measurement error ellipsoid; this is an off-line procedure (usually by inspection) that incurs no additional computational complexity in real-time. However, if the measurement error covariance matrix must be diagonalized in real-time (for some reason or another), then the computational complexity is bounded by m^3. On the other hand, if real time diagonalization of the measurement error covariance matrix is not required, then the increase in computational complexity is obviously bounded by a factor of m. That is, the computational complexity is linear in m, rather than being exponential in m.

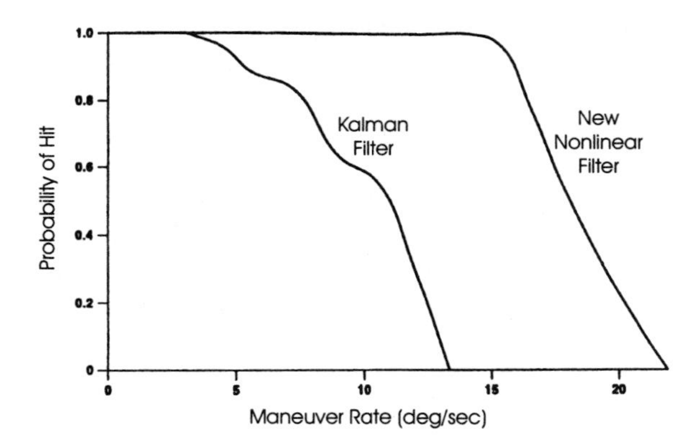

FIG. 4. *New nonlinear filters vs. Kalman filter (see [27]).*

In Kamen's work, the goal is an algorithm for data association, that uses an EKF to approximate the solution to the nonlinear filtering problem. In contrast, the error bound in [31] does not use an EKF, but rather it computes the covariance matrix bound directly from the nonlinear filter problem. More generally, there is no need to use an EKF for the algorithm using Kamen's SME formulation, but rather one could solve the nonlinear filter problem exactly or else use some approximations other than the EKF. Such an alternative to the EKF was recently reported in [33], and it shows a significant improvement over the EKF, but the SME performance is still suboptimal, owing to the approximate solution of the nonlinear filter problem used in [33]. In particular, the JPDA filter was superior to the *new SME filter* (NIF) in the Monte Carlo runs reported in [33].

The poor performance of the EKF reported in [33] is not surprising in general. There is a vast literature on nonlinear filters, both exact and approximate, reporting algorithms with superior performance to the EKF; for example, see Figure 4. Other references that discuss nonlinear filters include [17]–[30]. Table 2 lists several exact nonlinear filters that have been developed recently. Details of these exact filters are given in [20]–[25].

A fundamental flaw in the EKF is that it can only represent unimodal probability densities. More specifically, the EKF is based on a Gaussian approximation of the exact conditional probability density. The multiple target tracking problem, however, is characterized by multimodal densities, owing to the ambiguity in data association. Therefore, it is not surprising that the EKF has poor performance for such nonlinear filtering problems, as shown in [33]. In contrast, the new exact nonlinear filter in [25] can represent multimodal probability densities from the exponential family. This suggests that the SME filter performance can be improved by using this new exact nonlinear filter, rather than the EKF or NIF.

The advantage of the SME formulation of MTT is a reduction in computational complexity for exact implementations. In particular, as shown

TABLE 2
Exact nonlinear recursive filters.

Filter	Conditional Density $p(x,t \mid Z_k)$	Class of Dynamics	Propagation Equations
1. Kalman (1960)	η = Gaussian	$\frac{\partial f}{\partial x} = A(t)$	$\dot{m} = Am$ $\dot{P} = AP + PA^T + GG^T$
2. Benes (1981)	$\eta \exp[\int f(x)dx]$	$\frac{\partial f}{\partial x} = (\frac{\partial f}{\partial x})^T$ and $\lVert f(x) \rVert^2 + \mathrm{tr}(\frac{\partial f}{\partial x}) = x^T Ax + b^T x + c$	$\dot{m} = -PAm - \frac{1}{2}Pb$ $\dot{P} = I - PAP$
3. Daum (1986)	$\eta P_{ss}^a(x)$	$f - aQr^T = Dx + E$ and $\mathrm{tr}(\frac{\partial f}{\partial x}) + \frac{a}{2}rQr^T = x^T Ax + b^T x + c$ where $r = \frac{\partial}{\partial x}\log P_{ss}(x)$	$\dot{m} = 2(a-1)PAm + Dm + (a-1)Pb + E$ $\dot{P} = 2(a-1)PAP + DP + PD^T + Q$
4. Daum (1986)	$\eta q^a(x,t)$	Same as filter 3, but with $r = \frac{\partial}{\partial x}\log q(x,t)$	Same as filter 3.
5. Daum (1986)	$\eta Q(x,t)$	$\frac{\partial f}{\partial x} - (\frac{\partial f}{\partial x})^T = D^T - D$ and $\frac{\partial f}{\partial t} + \dot{D}x + \dot{E} = -(\frac{\partial f}{\partial t})^T f - \frac{1}{2}[\frac{\partial}{\partial x}\mathrm{tr}(\frac{\partial f}{\partial x})^T]$ $+(2A + D^T D)x + D^T E + b$	$\dot{m} = -(2PA + D)m - E - Pb$ $\dot{P} = -2PAP - PD^T - DP + I$
6. New Nonlinear Filter [25]	$p(x,t)\exp[\theta^T(x,t)\psi(Z_k,t)]$	Solution of PDE for $\theta(x,t)$	$\dot{\psi} = A^T\psi + \Gamma$ where $\Gamma = (\Gamma_1, \Gamma_2, \ldots, \Gamma_M)^T$ with $\Gamma_j = \psi^T B_j \psi$

in Table 1, the computational complexity of exact MHT is exponential, whereas the SME approach is polynomial. More specifically, the computational complexity for our Cramér–Rao bound [31] is:

$$cc = \begin{cases} (nN)^3 \,, & \text{for standard matrix multiplies} \\ M(nN) \,, & \text{for more efficient matrix multiplies} \end{cases}$$

in which

N = Number of targets.
n = Dimension of state vector for each target.
$M(q)$ = Computational complexity to multiply two matrices of size $q \times q$.

Standard matrix multiplication for unstructured dense matrices requires $M(q) = q^3$, whereas the best current estimate of computational complexity, due to Coopersmith and Winograd [51] is theoretically:

$$M(q) = k\, q^{2.38} \,.$$

Unfortunately, k is rather large [51]. Matrix multiplications of size $q = nN$ determines the computational complexity for our Cramér–Rao bound, owing to the use of EKFs, using a result due to Taylor (see [31]). Other bounds that may be tighter than the Cramér–Rao bound could also be used; see [31] and [70] for a discussion of such alternative bounds.

In the above, we have implicitly assumed that the number of targets (N) is known *a priori* exactly, and furthermore that the number of detections (D) satisfies the condition $D = N$ for all time. Obviously these assumptions are extremely unrealistic in practical applications; these assumptions are essentially never satisfied in the real world. Nevertheless, the SME theory can be generalized, with the result that the computational complexity for the Cramér–Rao bound is still polynomial in n, N and D (see [31] for details). In contrast, MHT has exponential computational complexity.

Kamen's MTT algorithm using SME with EKF's also has polynomial complexity. Moreover, the use of the exact nonlinear filters in Table 2 preserves this complexity. On the other hand, if the exact conditional density is not from an exponential family, then the complexity is, in general, much higher. Remember: the use of an EKF is approximate for an MTT algorithm, but it is exact for the Cramér–Rao bound!

4. Connection between random sets and symmetric measurement equations. As shown in Figure 5, random sets can be used to develop MHT algorithms and performance bounds, and Kamen's nonlinear SMEs can be used for the same purposes. This suggests that there is a connection between random sets and the SME formulation. Figure 5 was drawn in the limited context of MTT applications; however, it seems to me that the connection between random sets and SME could be developed in a much broader (and more abstract) context. Apparently this connection has not been noted before.

I am not suggesting that random sets are isomorphic to SMEs or that one approach is better than the other. In fact, the natural formulation of MTT problems using SME appears to be somewhat less general than the random sets formulation. In particular, the natural SME formulation of MTT assumes that the number of true targets is known exactly *a priori* [32], whereas random sets can easily model an unknown number of targets. In this sense, as a rough analogy, random sets are like a general graph, whereas SME is like a tree (i.e., a special kind of graph). Trees are very nice to use in those applications where they are appropriate. For example, queuing theory analysis for data networks modelled as trees is often much more tractable than for general graphs. Obviously, the down side of trees is that they cannot model the most general network. On the other hand, a general graph can be decomposed into a set of trees, which is often useful for analysis and/or synthesis. Likewise, it is possible to combine several SME structures for MTT to handle the case of an uncertain number of real

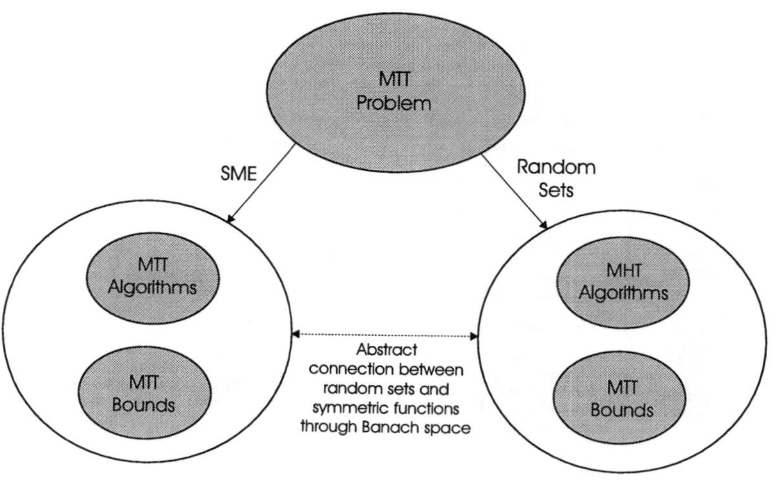

FIG. 5. *Connection between random sets and nonlinear symmetric measurement formulation.*

targets (see [31]).

The connection between SME and trees noted above is actually more than just an analogy. In particular, there is an intimate connection between polynomials and trees going back to Cayley [52]. Moreover, the symmetric polynomials used in SME are an algebraic representation of the symmetric group, which is generated by transpositions, which can be arranged in a tree, as the graphical representation of permutations. In particular, it is well known that a collection of $n-1$ transpositions on M objects generates the symmetric group $\mathcal{S}_n$ if and only if the graph with n vertices and $n-1$ edges (each edge corresponding to one transposition) is a tree [62]. Of course, there are many other representations of the symmetric group that do not correspond to trees [63]. In fact, the representation theory of the symmetric group is still an active area of research. Further details on the connection between trees, symmetric polynomials and the symmetric group are given in [59]–[66]. An elementary introduction to this subject is given in [72], [73], with the latter reference having been written ostensibly for high school students. More generally, the complexity of an unstructured graph can be measured by counting the number of trees into which it can be decomposed, by computing the so-called "tree-generating determinant" [53]. From this perspective, trees are the simplest graphs! Moreover, the sum of the weights of all the trees of a graph is an invariant of the graph, which can be computed as the Whitney–Tutte polynomial, whose original application was to determine the chromatic polynomial of a graph [54]. As an aside, polynomials also play a key role in the topology of knots: Alexander polynomials, Conway polynomials, Laurent Polynomials, Jones polynomials, generalized polynomials, and the tree-polynomial (which lists all the maximal trees in a graph) [58]. Of course, the use of such polynomi-

TABLE 3
Comparison of random sets and symmetric polynomials for MTT performance bounds.

ITEM	Random Sets	Symmetric Polynomials	Hybrid A	Hybrid B
1. Method	Magic Genie	Nonlinear Filter	Nonlinear Filter	Nonlinear Filter
2. Computational Complexity	Exponential	Polynomial	Polynomial	Polynomial
3. Graph	Unstructured	Tree	Forest	Forest
4. Assumptions for MTT Problem	Targets can be created and annihilated at arbitrary times	Continuity of Targets in Time	Continuity of Targets in Time	Bounded Complexity of Creation and Annihilation of Targets
5. Assumptions about number of Detections (D) vs. number of Targets (N)	D and N are arbitrary and unknown *a priori*	D = N Also, D and N are known *a priori*	D and N are arbitrary (within bounds) and unknown *a priori*	D and N are arbitrary (within bounds) and unknown *a priori*

als in topology and graph theory are not unrelated (see Chapter 2 in [57]).

Table 3 summarizes the comparison between SME and random sets used for performance bounds in MTT. It is clear that SME has achieved a dramatic reduction in computational complexity by exploiting a special case of MTT, rather than using a general formulation (as in MHT using random sets) where targets can be created and annihilated at arbitrary times. In general, real targets do not appear and disappear with such rapidity, but rather real targets are characterized by continuity in time, which is built into Kamen's SME formulation of MTT, but which is lacking in the standard formulation of MHT using unstructured random sets.

More generally, and beyond the context of MTT, the use of structured random sets would seem to be advantageous for analysis and synthesis, analogous to the use of trees in graph theory. Of course, the exploitation of such structures, especially trees, is standard in random graph theory [55] and the theory of random mappings [56].

There appears to be an interesting connection between random sets and the SME method, which is obvious for MTT applications, but which has not been studied at an abstract level. Apparently the shortest path between random sets and symmetric polynomials is through a Banach space. In particular, the SME formulation creates a nonlinear filtering problem, the solution of which is the conditional probability density $p(x|Z_k)$, which is a Banach space valued random variable. From another viewpoint, it is well known that random set problems can be embedded in the framework of Banach space valued random variables, within which the random sets disappear. Moreover, H. Nguyen noted in his lecture at this IMA workshop that this is perhaps the basic reason that random sets are not studied more widely as a subject in pure mathematics: they are subsumed within the more general framework of Banach space valued random variables. The basic tool in this area is the Radström embedding theorem [67]; see R. Taylor's paper in this volume for a discussion of this Banach space viewpoint, including [67]–[69].

Kamen's SME formulation of MTT reminds me of several other methods that transform a discrete (or combinatorial) problem into a continuous (or analytic) problem that can be solved using calculus. For example, Brockett uses ODE's to sort lists and solve other manifestly discrete problems [39]. The assignment problem solved using relaxation methods is another example [43], and Karmarkar's solution of linear programming problems is a third example [40]. The first proofs of the prime number theorem, by Hadamard and de la Vallée Poussin, a century ago, use complex contour integrals of the Riemann zeta-function; in contrast, the so-called "elementary" proof by Selberg and Erdös (1949) does not use complex analysis. The sentiment appears to be unanimous that the old analytic proofs are simpler, in some sense, than the modern proof based on Selberg's formula. To quote Tom Apostol, the Selberg–Erdös proof is "quite intricate," whereas the analytic proof is "more transparent" (p. 74 in [47]). Of course, Tom Apostol is biased, being the author of a book on analytic number theory [47], and hence we should consult another authority. According to Hardy and Wright, the Selberg–Erdös proof "is not easy" (p. 9 in [49]); that's good enough for me. But seriously, whether a proof is simpler or not depends on how comfortable you are with the tools involved: complex contour integrals of the Riemann zeta-function in Chapter 13 of [47] vs. the Selberg formula and the Möbius inversion formula in Chapter XXII of [49].

More generally, the application of analysis (e.g., limits and continuity) to number theory is an entire branch of mathematics, called "analytic number theory." The proof of the prime number theorem using complex function theory is but one example among many [47].

At a more elementary level, the solution of combinatorial problems using calculus is the theme of [45]. Moreover, M. Kac has emphasized the connection between number theory and randomness in [37], [38]. Recently, G. Chaitin has revealed a deep relationship between arithmetic and randomness. The Möbius function is another powerful analytic tool for solving combinatorial problems. The interplay between discrete and continuous mathematics is developed at length in [71]. The book by Schroeder contains a wealth of such examples in which calculus is used to solve discrete problems [41]. More generally, Bezout's theorem, homotopy methods and the Atiyah–Singer index theorem also come to mind in this context. Finally, one might add the curious, but well established, utility of elliptic curves in number theory, including Wiles' proof of Fermat's last theorem [35], [36].

TABLE 4
Interplay between the discrete and continuous.

Mathematics	Physics	Algorithms
1. Proof of the prime number theorem (1896 versions)	14. Wave-particle duality of light in classical physics (i.e., Newton vs. Huygens)	19. Brockett sorting lists by solution of ODEs
2. Discrete spectrum of locally compact linear operators	15. Wave-particle duality in quantum mechanics (i.e., Schrödinger vs. Heisenberg)	20. Assignment problem solved by relaxation algorithm
3. Isomorphism of $L^2 \ell^2$ and as Hilbert spaces	16. Fokker-Planck equation	21. Karmarkar's solution of linear programing
4. Atiyah-Singer index theorem	17. Boltzmann transport equation	22. Homotopy methods
5. Bezout's theorem	18. Navier-Stokes equation	23. Interpolation and sampling
6. Generating functions, Möbius function, z-transform		24. Shannon's information theory
7. Dimension of a topological space		25. Numerical solution of PDE's
8. Smale's "topology of algorithms"		
9. M. Kac's use of statistical independence in number theory		
10. Analytic number theory		
11. Proof of Fermat's last theorem		
12. Knot theory		
13. Graph theory		

As shown by the list of examples in Table 4, the use of continuous mathematics (e.g., calculus or complex analysis) to solve what appears to be a discrete problem is not so rare after all! Table 4 is a list which includes: algorithms, formulas, proofs of theorems, physics and other examples. Perhaps, algebraic topology and algebraic geometry should be added to the list.

5. Ironical connection with fuzzy sets. It is well known (by some) that fuzzy sets are really "equivalence classes of random sets" (ECORS) in disguise. Irwin R. Goodman deserves the credit for developing this connection with mathematical rigor [2]–[4]. Apparently this connection between fuzzy sets and ECORS is not altogether welcome by fuzzy researchers; in particular, in Bart Kosko's recent book [6] he says:

> "There was one probability criticism that shook Zadeh. I think this is why he stayed out of probability fights. He and I have often talked about it but he does not write about it or tend to discuss it in public talks. He said it showed up as early as 1966 in the Soviet Union. The criticism says you can view a fuzzy set as a random set.

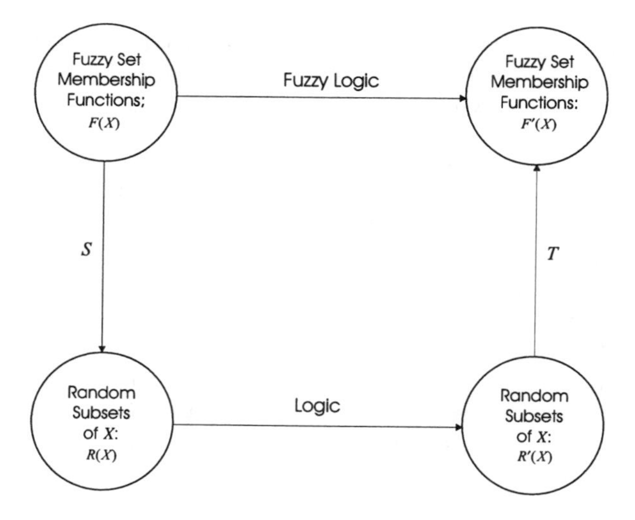

FIG. 6. *Fuzzy logic as logic over equivalence classes of random sets (Goodman).*

> Most probabilists have never heard of random sets .... In the next chapter I mention how I have used this probability view when I wanted to prove a probability theorem about a fuzzy system. It's just a complicated way of looking at things."

Bart Kosko's assertion that "most probabilists have never heard of random sets," is clearly an overstatement considering the growing literature on random sets going back several decades [7]–[10] and continuing with this volume itself on random sets!

Goodman's basic idea is shown in Figure 6, which shows that standard logic (Boolean algebra) can be used to obtain the same results as fuzzy logic, where the standard logic operates on random subsets, whereas fuzzy logic operates on fuzzy set membership functions.

This connection between fuzzy logic and Bayesian probability is rather ironical. The irony derives from the fact that there is an on-going debate between fuzzy advocates and Bayesians. In particular, a recent special issue of the *IEEE Transactions on Fuzzy Systems* is devoted entirely to this debate [13]. For example, one Bayesian, D. V. Lindley [14], writes:

> "I repeat the challenge that I have made elsewhere — that anything that can be done by alternatives to probability can be better done by probability. No one has provided me with an example of a problem that concerns personal uncertainty and has a more satisfactory solution outside the probability calculus than within it. If it is true that Japan is using fuzzy systems in their transport, then they would do better to use probability there. The improvement the

> Japanese notice most likely comes from the mere recognition of some uncertainty, rather than from the arbitrary calculations of fuzzy logic ...."

Professor Lindley's paper [14] is remarkably short (one-third page), and it is one of only two Bayesian papers, compared with seven fuzzy papers in [13]. A further irony is that Goodman's work is not mentioned at all in the entire special issue [13] on this debate! As far as I know, Goodman is the only person who has formulated the appropriate question in mathematically precise terms, rather than merely philosophize about it. The word "philosophy" is interesting in this context, as Professor Mendel points out in his tutorial on fuzzy logic [15]. In particular, Professor Mendel emphasizes the *philosophical* difference between fuzzy logic and probability, but he is rather careful not to assert that there is a rigorous mathematical theorem that proves that fuzzy logic is more general than probability (see p. 361 of [15]). Indeed, such a theorem would contradict both Goodman's results, as well as Professor Kosko's statement quoted earlier.

REFERENCES

[1] F.E. DAUM, *Bounds on performance for multiple target tracking*, IEEE Transactions on Automatic Control, 35 (1990), pp. 443–446.

[2] I.R. GOODMAN, *Fuzzy sets as equivalence classes of random sets*, Fuzzy Sets and Possibility Theory (R.R. Yager, ed.), pp. 327–343, Pergamon Press, 1982.

[3] I.R. GOODMAN AND H.T. NGUYEN, *Uncertainty Models For Knowledge-Based Systems*, North–Holland Publishing, 1985.

[4] I.R. GOODMAN, *Algebraic and probabilistic bases for fuzzy sets and the development of fuzzy conditioning*, Conditional Logic in Expert Systems (I.R. Goodman, M.M. Gupta, H.T. Nguyen, and G.S. Rogers, eds.), North–Holland Publishing, 1991.

[5] K. HESTIR, H.T. NGUYEN, AND G.S. ROGERS, *A random set formalism for evidential reasoning*, Conditional Logic in Expert Systems (I.R. Goodman, M.M. Gupta, H.T. Nguyen, and G.S. Rogers, eds.), pp. 309–344, North–Holland Publishing, 1991.

[6] B. KOSKO, *Fuzzy Thinking*, Hyperion Publishers, 1993.

[7] G. MATHERON, *Random Sets and Integral Geometry*, John Wiley & Sons, 1975.

[8] D.G. KENDALL, *Foundations of a theory of random sets*, Stochastic Geometry (E.F. Harding and D.G. Kendall, eds.), pp. 322–376, John Wiley & Sons, 1974.

[9] T. NORBERG, *Convergence and existence of random set distributions*, Annals Probability, 32 (1984), pp. 331–349.

[10] H.T. NGUYEN, *On random sets and belief functions*, Journal of Mathematical Analysis and Applications, 65 (1978), pp. 531–542.

[11] R.B. WASHBURN, *Review of bounds on performance for multiple target tracking*, (unpublished), March 27, 1989.

[12] F.E. DAUM, *A system approach to multiple target tracking*, Multitarget–Multisensor Tracking, Volume II (Y. Bar–Shalom, ed.), pp. 149–181, Artech House, 1992.

[13] *Fuzziness vs. Probability –* N^{th} *Round*, Special Issue of IEEE Transactions on Fuzzy Systems, 2 (1994).

[14] D.V. LINDLEY, *Comments on 'The efficacy of fuzzy representations of uncertainty'*,

Special Issue of IEEE Transactions on Fuzzy Systems, 2 (1994), p. 37.

[15] J.M. MENDEL, *Fuzzy logic systems for engineering: A tutorial*, Proceedings of the IEEE, 83 (1995), pp. 345-377.

[16] N. WIENER AND A. WINTNER, *Certain invariant characterizations of the empty set*, J. T. Math, (New Series), II (1940), pp. 20–35.

[17] V.E. BENEŠ, *Exact finite–dimensional filters for certain diffusions with nonlinear drift*, Stochastics, 5 (1981), pp. 65–92.

[18] V.E. BENEŠ, *New exact nonlinear filters with large Lie algebras*, Systems and Control Letters, 5 (1985), pp. 217–221.

[19] V.E. BENEŠ, *Nonlinear filtering: Problems, examples, applications*, Advances in Statistical Signal Processing, Vol. 1, pp. 1–14, JAI Press, 1987.

[20] F.E. DAUM, *Exact finite dimensional nonlinear filters*, *IEEE Trans. Automatic Control*, 31 (1986), pp. 616–622.

[21] F.E. DAUM, *Exact nonlinear recursive filters*, Proceedings of the 20th Conference on Information Sciences and Systems, pp. 516–519, Princeton University, March 1986.

[22] F.E. DAUM, *New exact nonlinear filters*, Bayesian Analysis of Time Series and Dynamic Models (J.C. Spall, ed.), pp. 199–226, Marcel Dekker, New York, 1988.

[23] F.E. DAUM, *Solution of the Zakai equation by separation of variables*, IEEE Trans. Automatic Control, 32 (1987), pp. 941–943.

[24] F.E. DAUM, *New exact nonlinear filters: Theory and applications*, Proceedings of the SPIE Conference on Signal and Data Processing of Small Targets, pp. 636–649, Orlando, Florida, April 1994.

[25] F.E. DAUM, *Beyond Kalman filters: Practical design of nonlinear filters*, Proceedings of the SPIE Conference on Signal and Data Processing of Small Targets, pp. 252–262, Orlando, Florida, April 1995.

[26] A.H. JAZWINSKI, *Stochastic Processes and Filtering Theory*, Academic Press, New York, 1970.

[27] G.C. SCHMIDT, *Designing nonlinear filters based on Daum's Theory*, AIAA Journal of Guidance, Control and Dynamics, 16 (1993), pp. 371–376.

[28] H.W. SORENSON, *On the development of practical nonlinear filters*, Information Science, 7 (1974), pp. 253–270.

[29] H.W. SORENSON, *Recursive estimation for nonlinear dynamic systems*, Bayesian Analysis of Time Series and Dynamic Models (J.C. Spall, ed.), pp. 127–165, Marcel Dekker, New York, 1988.

[30] L.F. TAM, W.S. WONG, AND S.S.T. YAU, *On a necessary and sufficient condition for finite dimensionality of estimation algebras*, SIAM J. Control and Optimization, 28 (1990), pp. 173–185.

[31] F.E. DAUM, *Cramér–Rao bound for multiple target tracking*, Proceedings of the SPIE Conference on Signal and Data Processing of Small Targets, Vol. 1481, pp. 582–590, 1991.

[32] E. KAMEN, *Multiple target tracking based on symmetric measurement equations*, Proceedings of the American Control Conference, pp. 263–268, 1989.

[33] R.L. BELLAIRE AND E.W. KAMEN, *A new implementation of the SME filter*, Proceedings of the SPIE Conference on Signal and Data Processing of Small Targets, Vol. 2759, pp. 477–487, 1996.

[34] S. SMALE, *The topology of algorithms*, Journal of Complexity, 5 (1986), pp. 149–171.

[35] A. WILES, *Modular elliptic curves and Fermat's last theorem*, Annals of Mathematics, 141 (1995), pp. 443–551.

[36] *Seminar on FLT*, edited by V.K. Murty, CMS Conf. Proc. Vol. 17, AMS 1995.

[37] M. KAC, *Statistical Independence in Probability, Analysis, and Number Theory*, Mathematical Association of America, 1959.

[38] M. KAC, *Probability, Number Theory, and Statistical Physics: Selected Papers*, MIT Press, 1979.

[39] R.W. Brockett, *Dynamical systems that sort lists, diagonalize matrices and solve linear programming problems*, Proceedings of the IEEE Conf. on Decision and Control, pp. 799–803, 1988.
[40] N. Karmarkar, *A new polynomial time algorithm for linear programming*, Combinatorica, 4 (1984), pp. 373–395.
[41] M.R. Schroeder, *Number Theory in Science and Communication*, Springer–Verlag, 1986.
[42] A. Weil, *Number Theory*, Birkhäuser, 1984.
[43] D.P. Bertsekas, *The auction algorithm: A distributed relaxation method for the assignment problem*, Annals of Operations Research, 14 (1988), pp. 105–123.
[44] M. Waldschmidt, P. Moussa, J.–M. Luck, and C. Itzykson (Eds.), *From Number Theory to Physics*, Springer–Verlag, 1992.
[45] R.M. Young, *Excursions in Calculus: An Interplay of the Continuous and the Discrete*, Mathematical Association of America, 1992.
[46] B. Booss and D.D. Bleecker, *Topology and Analysis: The Atiyah–Singer Index Formula and Gauge–Theoretic Physics*, Springer–Verlag, 1985.
[47] T.M. Apostol, *Introduction to Analytic Number Theory*, Springer–Verlag, 1976.
[48] I. MacDonald, *Symmetric Functions and Hall Polynomials*, 2nd Edition, Oxford Univ. Press, 1995.
[49] G.H. Hardy and E.M. Wright, *An Introduction to the Theory of Numbers*, 5th Edition, Oxford Univ. Press, 1979.
[50] M.R. Schroeder, *Number Theory in Science and Communication*, Springer–Verlag, New York, 1990.
[51] D. Coppersmith and S. Winograd, *Matrix multiplication via arithmetic progressions*, Proceedings of the 19th ACM Symposium on Theory of Computing, pp. 472–492, ACM 1987.
[52] A. Cayley, *On the theory of the analytical forms called trees*, Philosophical Magazine, 4 (1857), pp. 172–176.
[53] H.N.V. Temperley, *Graph Theory*, John Wiley & Sons, 1981.
[54] Béla Bollobás, *Graph Theory*, Springer–Verlag, 1979.
[55] Béla Bollobás, *Random Graphs*, Academic Press, 1985.
[56] V.F. Kolchin, *Random Mappings*, Optimization Software Inc., 1986.
[57] J. Stillwell, *Classical Topology and Combinatorial Group Theory*, 2nd Edition, Springer–Verlag, 1993.
[58] L.H. Kauffman, *On Knots*, Princeton University Press, 1987.
[59] H. Wielandt, *Finite Permutation Groups*, Academic Press, 1964.
[60] D. Passman, *Permutation Groups*, Benjamin, 1968.
[61] O. Ore, *Theory of Graphs*, AMS, 1962.
[62] G. Polya, *"Kombinatorische anzahlbstimmungen für gruppen, graphen und chemische verbindungen*, Acta Math. 68 (1937), pp. 145–254.
[63] G. de B. Robinson, *Representation Theory of the Symmetric Group*, University of Toronto Press, 1961.
[64] J.D. Dixon and B. Mortimer, *Permutation Groups*, Springer–Verlag, 1996.
[65] W. Magnus, A. Karrass, and D. Solitar, *Combinatorial Group Theory*, Dover Books Inc., 1976.
[66] J.–P. Serre, *Trees*, Springer–Verlag, 1980.
[67] H. Radström, *An embedding theorem for spaces of convex sets*, Proceedings of the AMS, pp. 165–169, 1952.
[68] M. Puri and D. Ralescu, *Limit theorems for random compact sets in Banach spaces*, Math. Proc. Cambridge Phil. Soc., 97 (1985), pp. 151–158.
[69] E. Gine, M. Hahn, and J. Zinn, *Limit Theorems for Random Sets*, Probability in Banach Space, pp. 112–135, Springer–Verlag, 1984.
[70] K.L. Bell, Y. Steinberg, Y. Ephraim, and H.L. Van Trees, *Extended Ziv–Zakai lower bound for vector parameter estimation*, IEEE Trans. Information Theory, 43 (1997), pp. 624–637.
[71] *Relations between combinatorics and other parts of mathematics*, Proceedings of

Symposia in Pure Mathematics, Volume XXXIV, American Math. Society, 1979.

[72] H.W. LEVINSON, *Cayley diagrams*, Mathematical Vistas, Annals of the NY Academy of Sciences, Volume 607, pp. 62–88, 1990.

[73] I. GROSSMAN AND W. MAGNUS, *Groups and their Graphs*, Mathematical Association of America, 1964.

RANDOM SETS IN DATA FUSION MULTI–OBJECT STATE–ESTIMATION AS A FOUNDATION OF DATA FUSION THEORY

SHOZO MORI*

Abstract. A general class of multi-object state-estimation problems is defined as the problems for estimating random sets based on information given as collections of random sets. A general solution is described in several forms, including the one using the conditional Choquet's capacity functional. This paper explores the possibility of the abstract formulation of multi-object state-estimation problems becoming a foundation of data fusion theory. Distributed data processing among multiple "intelligent" processing nodes will also be briefly discussed as an important aspect of data fusion theory.

Key words. Data Fusion, Distributed Data Processing, Multi-Object State-Estimation, Multi-Target/Multi-Sensor Target-Tracking.

AMS(MOS) subject classifications. 60D05, 60G35, 60G55, 62F03, 62F15

1. Introduction. In my rather subjective view, data-fusion problems, otherwise referred to as information integration problems, provide us – engineers, mathematicians, and statisticians, alike – with an exciting opportunity to explore a new breed of uniquely modern problems. This is so mainly because such problems require us to shift our focus from *single-object* to *multiple-objects*, as entities in estimation problems, which is probably a significant evolutionary milestone of estimation theory. Around the middle of this century, the concept of random variables was extended spatially to random vectors[1] and temporally to stochastic processes. Even with these extensions, however, the objects to be estimated were traditionally "singular." In data fusion problems, multiple objects must be treated individually as well as collectively.

What Is "Data Fusion"?: There is an "official" definition:

> *Definition: Data fusion is a process dealing with association, correlation, and combination of data and information from single and multiple sources to achieve refined position and identity estimation, and complete the timely assessments of situation and threats, and their significance.*

This definition [1] was one agreed upon by a panel, the Joint Directors of Laboratories, composed of fusion suppliers, i.e., the developers of data

* Texas Instruments, Inc., Advanced C^3I Systems, 1290 Parkmoor Avenue, San José, CA 95126. Earlier work related to this study was sponsored by the Defense Advanced Research Projects Agency. Recent work, including participation in the workshop, was supported by Texas Instruments' internal research and promotional resources.

[1] "Multi-variate" rather-than scalar parameters, and even into infinite-dimensional abstract linear vector spaces.

fusion software, and of its users, i.e., various U.S. military organizations. It is interesting to compare the above definition with the following quotation taken from one of the most recent books on stochastic geometry [2]:

> *Object recognition is the task of interpreting a noisy image to identify certain geometrical features. An object recognition algorithm must decide where there are any objects of a specified kind in the scene, and if so, determine the number of objects and their locations, shapes, sizes and spatial relationship.*

We may immediately observe similar or parallel concepts in the above two quoted statements. In each process, determining the existence of certain objects, i.e., object detection, is an immediate goal. A secondary goal may be to determine the relationship between hypothesized detected objects, referred to as correlation or association problems in data fusion. The final goal may be to estimate the "states" of individual objects, such as location, velocity, size, shape parameters, emission parameters, etc.

Multi-Object State-Estimation: A main objective of this paper is to explore the possibility of an abstract multi-object state-estimation theory becoming a foundation of these new kinds of processes for data fusion in military and civilian applications, as well as object recognition in image analysis. Multi-object state-estimation, also known as multi-target tracking,[2] is a natural extension of "single-object" state-estimation, i.e., tracking of a physical moving object, such as an aircraft, a ship, a car, etc. Indeed, tracking in that sense was a significant motivation for establishing filtering theory, such as the classical work [3] during World War II. We will use the term "object" instead of "target" to refer to an abstract entity rather than a particular physical existence, and also, "state-estimation" instead of "tracking" to indicate possible inclusion of various kinds of state variables. Including a pioneer work [4] written in the 1960's, the major works in multi-target tracking are well documented in the technical surveys,[3] [5] and [6], and the books, [7] and [8].

It is my opinion that the state-of-the-art of multi-target tracking had been a chaotic collection of various algorithms, many of which carried snappy acronyms, until the works by R.D. Reid were made public in the late 70's. I believe that one of his papers, [10], is the most significant work in the history of multi-target tracking. The paper provides us with the first almost completely comprehensive description of the multiple-hypothesis filtering approach to data fusion problems, although a clear definition of

---

[2] In multi-target terminology, a "target" means an object to be tracked, not necessarily to be shot down or aimed at.

[3] [11] also contains an extensive list of the literature on the subject, while the reference list in [10] is concise but contains all the essential works.

data-to-data association hypotheses and tracks appeared earlier in a paper by C.L. Morefield [9]. Like many other papers on multi-target tracking algorithms, however, Reid's paper [10] does not clearly mention exactly what mathematical problem is solved by his multi-hypothesis algorithm. It was proven later in [12] that his algorithm is indeed *optimal* in a strict Bayesian sense or a reasonable approximation to an optimal solution as an estimate of a set of "targets" given as a random set. At least partly because of my special recognition of this work [10] by R.D. Reid, I believe any new development in multi-object state-estimation should be compared with Reid's multi-hypothesis filtering algorithm, one way or another. It is an objective of this paper to describe a mathematical model, a problem statement, and an optimal solution, totally in random set formalism, while proving the optimality of Reid's algorithm.

Development of General Theory: The problem statement and the theoretical development in the report [11] by I.R. Goodman motivated me to formulate the multi-object tracking problem in abstract terms in [12] and [13], which were the first serious attempts to generalize Reid's algorithm. Although the "general" model in [11] clearly defines data association hypotheses as a random partition of the cumulative collection of data sets, the connection of results in [11] to Reid's algorithm is not clear. One difficulty may originate from the fact that the model in [11] allows the set of objects as a random set to be a countably infinite set with a positive probability. Although [11] through [13] were attempts to establish an abstract theory of multi-target tracking, a random-set formalism was not used explicitly.

In the recent works [14] and [15], the concept of random sets is used explicitly. In the work [14] by R.B. Washburn, the objects are modeled by a fixed ordered tuple, i.e., a "multi-variate" model, but the measurements are explicitly modeled as random sets represented by random binary fields and their Laplace transformations. In [15], R. Mahler describes a formulation of multi-object tracking problems totally in a random-set formalism. As suggested in [15], this formalism leads to a new kind of algorithms, which we may call *correlation-free* algorithms, like the ones described in [16] and [17]. In my view, however, the relationship of the theory and the algorithms in [14]–[17] to Reid's algorithm has not been made clear. One difficulty may be due to the fact that explicit expressions of conditional and unconditional probability density functions for random sets are not very well understood.

In this paper, and in order to specify objects and measurements as random sets, Choquet's capacity functionals as described in [18] will be considered, and their relationship to probability density functions will be discussed. This will make a bridge between *correlation-based* algorithms such as Reid's algorithm [10] and its generalization described in [12] and [13], and *correlation-free* algorithms described in [15] through [17]. It will be shown that the essence of a correlation-based algorithm is to produce "labeled" representation maintaining *a posteriori* object-wise independence, while a correlation-free algorithm is to produce a mixture of "label-less" representation that is obtained by "scrambling" a correlation-based representation.

As previously stated, the primary objective of this paper is to provide a random-set based abstract multi-object state-estimation formalism as a possible theoretical foundation of data fusion technology. In this regard, the distributed processing aspect is another important dimension of data fusion. In distributed processing environments, there are multiple information processing nodes or agents, each of which has its own set of sensors or information gathering devices, and those nodes exchange information among themselves as necessary. Informational exchanges and resulting informational gains are very important elements in any large-scale system involving many decision makers as well as many powerful distributed information processing machines. Naturally, we will propose the distributed version of multi-object state-estimation formalism as a theoretical foundation in such environments.

2. Preliminaries. In the remainder of this paper, we consider the problem of "estimating" a random set of *objects* based on information given in terms of random sets. The set of objects to be estimated is given as a finite random set $\mathbf{X}$ in a space E, or a point process ([18] and [19]) in E. Let us assume that the space E, which we call the *state space*, is a locally compact Hausdorff space satisfying the second axiom of countability so that we can use conditioning relatively freely. Let $\mathcal{F}$, $\mathcal{K}$, and $\mathcal{I}$ be the collections of all closed sets, compact sets, and finite sets in E, resp. We may use the so-called *hit-or-miss* topology for $\mathcal{F}$, or the *myopic* topology for $\mathcal{K}$, ([18]) as needed. A random set[4] in E is a measurable mapping from the underlying probability space to the measurable space $\mathcal{F}$ with the Borel field on it. By saying that the random set $\mathbf{X}$ is finite, we mean that[5] Prob.$\{\#(\mathbf{X}) < \infty\} = 1$, i.e., Prob.$\{\mathbf{X} \in \mathcal{I}\} = 1$. Whenever necessary, we assume that the state space E, as a measurable space with its Borel field, has a σ-finite measure μ, and assume further, if necessary, that the topology on E is metrizable and equivalent to a metric d. In most practical applications, however, we only need the state space E *to be* a so-called *hybrid space*,[6] i.e., the product space of a Euclidean space (or its subset) and a finite (or possibly countably infinite) set, to allow the state of each object to have continuous components such as a geolocational state (position, velocity, acceleration, etc.) and a parametric state (parameters character-

[4] We only need to consider random closed sets.

[5] By $\#(A)$, we mean the cardinality of a set A. We write a set of elements x in a set X satisfying a condition C as $\{x \in X|C\}$. When the condition C is for points in the underlying probability space, by $\{C\}$, we mean an event in which the condition C holds, instead of writing $\{\omega \in \text{ prob.}|C \text{ holds}\}$. Prob.$(\cdot)$ is the probability measure of the underlying probability space that we implicitly assume. Prob.$\{C\}$ is shorthand of Prob.$(\{C\})$.

[6] In the sense that it is a *hybrid* of continuous and discrete spaces.

izing size, shape, emission, etc.), and discrete components, such as kinds, types, classifications, identifications, functional states, operational modes, etc. Such a space can be given the direct-product measure composed by a Lebesgue measure and a counting measure, and the direct-product metric composed by a Euclidean metric and a discrete metric. We will not distinguish an object from its state in the space E.

Finite Random Sequence Model: In a pre-random-set setting such as in [12] and [13], the objects are modeled as a finite random sequence $\mathbf{x}$ in the space E, or in other words, a random element $\mathbf{x}$ in the space of the direct sum[7] $\mathcal{E} = \bigcup_{n=1}^{\infty} E^n$ of the direct products $E^n = \overbrace{E \times \cdots \times E}^{n}$. Each element x in $\mathcal{E}$ is a finite sequence in E, i.e., $x = (x_i)_{i=1}^n \stackrel{\text{def}}{=} (x_1, \ldots, x_n) \in E^n$ for some n. For any element x in $\mathcal{E}$, let the length of x be denoted by $\ell(x)$, i.e., $\ell(x) = n \Leftrightarrow x \in E^n$. Whenever the state space E has a measure μ or a metric d, it can be naturally mapped[8] into space $\mathcal{E}$. Let us consider a mapping φ from the space $\mathcal{E}$ of finite sequences in E to the space $\mathcal{I}$ of finite sets in E, by $\varphi\left((x_i)_{i=1}^n\right) = \{x_i\}_{i=1}^n \stackrel{\text{def}}{=} \bigcup_{i=1}^n \{x_i\}$ for every $(x_i)_{i=1}^n \in \mathcal{E}$. Every connected component E^n of $\mathcal{E}$ maps into a connected component $\mathcal{I}_n \stackrel{\text{def}}{=} \{X \in \mathcal{I} | \#(X) = n\}$ of $\mathcal{I}$, if we can ignore sequences with repeated elements. Indeed, ignoring repeated elements, each $\mathcal{I}_n$ is homeomorphic to a quotient space obtained from E^n by the equivalence relation defined by permutations of finite sequences. Because of this relationship between $\mathcal{E}$ and $\mathcal{I}$, often in random-set literature such as [19], a random element $\mathbf{x}$ in $\mathcal{E}$ is described as being indistinguishable from the random set $\mathbf{X} = \varphi(\mathbf{x})$ it generates. It would be easier, however, to define a finite random sequence $\mathbf{x}$ than a random set $\mathbf{X}$. Indeed, we can specify the finite random sequence $\mathbf{x}$ completely by giving, for each n, Prob.$\{\ell(\mathbf{x}) = n\}$ and a joint probability distribution, Prob.$\{\mathbf{x} \in \prod_{i=1}^{n} K_i | \ell(\mathbf{x}) = n\}$, for every $(K_1, \ldots, K_n) \in \mathcal{K}^n$, which was exactly the mathematical model used in [12] and [13].

In this way, we can define a class of finite random sequences, what we may call an *i.i.d. model*,[9] by letting Prob.$\{\mathbf{x} \in \prod_{i=1}^{n} K_i | \ell(\mathbf{x}) = n\}$

---

[7] In [12] and [13], the space $\mathcal{E}$ was defined as $\mathcal{E} = \bigcup_{n=1}^{\infty} E^n \times \{n\}$ instead. The explicit inclusion of the number of objects, or the length of the sequence, was only for clarity, to avoid any unnecessary confusion. Of course, we clearly have $E^n \neq E^m$ if $n \neq m$, unless $E = \emptyset$, which we have excluded implicitly. As is customary, E^0 should be interpreted as a singleton of any "symbol" that is used only to represent the 0–length sequence. This extended space $\mathcal{E}$ can be considered as a free monoid generated by E as its alphabets.

[8] For example, a direct sum metric can be constructed, e.g., as $d(x, y) = 1$ if $x \in E^n$ and $y \in E^m$ with $n \neq m$, otherwise an appropriate direct-product metric d_n, e.g., $d_n\left((x_i)_{i=1}^n, (y_i)_{i=1}^n\right) = \sum_{i=1}^{n} d(x_i, y_i)$.

[9] *i.i.d.* = independent, identically distributed.

$= \prod_{i=1}^{n} F(K_i)$ for each $(K_1, \ldots, K_n) \in \mathcal{K}^n$ with a probability distribution $F(\cdot)$ on the state space E. If we assume that the distribution of $\ell(\mathbf{x})$ is Poisson with mean ν, then we have a *Poisson–i.i.d.* model, which is commonly referred to as a *Poisson point process*, with a finite intensity measure $\gamma(B) = \nu F(B)$. By letting Prob.$\{\ell(\mathbf{x}) = n\} = 1$ for some n, we have a *binomial process*. As in many other engineering applications, however, it is necessary to assume that probability distributions are absolutely continuous (i.e., have density functions) with respect to appropriate measures in order to develop practical algorithms. Hence, it is quite natural to consider the probability density function for a random set. However, there seems to be a certain peculiarity associated with probability density functions for random sets, as seen below:

Probability Density Function for a Finite Random Set: For the moment, let us assume that Prob.$\{\ell(\mathbf{x}) = n\} = 1$ for a fixed $n > 0$. Applying a change-of-variable theorem[10] to the natural mapping φ from $\mathcal{E}$ to E^n, we have

$$\int_{\mathcal{I}_n} \phi(X) M_n(dX) = \int_{E^n} \phi(\varphi(x)) \mu^n(dx) , \tag{2.1}$$

for any non-negative real-valued measurable function ϕ defined on $\mathcal{I}_n$, where μ^n is the direct-product measure on E^n and $M_n(\cdot)$ is the measure induced on $\mathcal{I}_n$ by φ as $M_n(\mathcal{B}) \stackrel{\text{def}}{=} \mu^n\left(\varphi^{-1}(\mathcal{B})\right)$ for every measurable set $\mathcal{B}$ in $\mathcal{I}_n$. Consider a finite random sequence $\mathbf{x}$ which has a probability density function $f_n(\cdot)$, i.e., Prob.$\{\mathbf{x} \in dx\} = f_n(x)\mu^n(dx)$, assuming Prob.$\{\ell(\mathbf{x}) = n\} = 1$. It follows from (2.1) that the probability density $p_n(\cdot)$ of the random set $\mathbf{X} = \varphi(\mathbf{x})$ can be expressed as[11]

$$p_n\left(\{x_i\}_{i=1}^n\right) = \sum_{\pi \in \Pi_n} f_n\left((x_{\pi(i)})_{i=1}^n\right) , \tag{2.2}$$

where Π_n is the set of all permutations on $\{1, \ldots, n\}$. All the repeated elements are ignored, assuming that the set of all sequences with repeated elements has zero μ^n–measure. We may call this summation over all permutations in (2.2) *scrambling*. "Scrambling" is necessary because, when φ maps points in E^n to $\mathcal{I}_n$, the ordering is lost, and hence its inverse φ^{-1} generates equivalence up to all possible permutations. If function $f_n(\cdot)$ is permutable, we have $p_n\left(\{x_i\}_{i=1}^n\right) = n! f_n\left((x_i)_{i=1}^n\right)$, and in particular, if objects are i.i.d. with a common probability density $f_1(\cdot)$, we have that

---

[10] For example, see [20].

[11] When the right hand side of equation (2.2) is interpreted as a function of the ordered n–tuple $(x_i)_{i=1}^n$ and is multiplied by the probability Prob.$\{\ell(\mathbf{n}) = n\}$, it is known as Janossy density [27]. It should be noted that a Janossy density is not necessarily a density of a probability measure.

$p_n(\{x_i\}_{i=1}^n) = n! \prod_{i=1}^n f_1(x_i)$. Equation (2.2) might have some appearance of triviality. However, as we shall see later, this scrambling is in fact the essence of what we call a data association problem, or what data-fusionists call a *correlation problem*.

Choquet's Capacity Functional: An alternative way to characterize a random set is to use a standard tool of random-set theory, which is *Choquet's capacity functional*. For a random set $\mathbf{X}$ in E, let

$$T(K) = \text{Prob.}\{\mathbf{X} \cap K \neq \emptyset\}\,, \tag{2.3}$$

for all $K \in \mathcal{K}$. Then, (i) $0 \leq T(K) \leq 1$, (ii) $T(\cdot)$ is upper semi-continuous in $\mathcal{K}$, and (iii) $T(\cdot)$ satisfies

$$\begin{aligned} S_1(K;K_1) &\stackrel{\text{def}}{=} T(K \cup K_1) - T(K) \geq 0\,, \\ S_n(K;K_1,\ldots,K_n) &\stackrel{\text{def}}{=} S_{n-1}(K;K_1,\ldots,K_{n-1}) \\ &\quad - S_{n-1}(K \cup K_n;K_1,\ldots,K_{n-1}) \geq 0\,, \quad \text{for } n > 1\,, \end{aligned} \tag{2.4}$$

where[12] K and all K_i's are arbitrary compact sets in E. Conversely, if a functional T on $\mathcal{K}$ satisfies the above three conditions, (i)–(iii), then Choquet's theorem ([18]) guarantees that there is a uniquely defined probability measure on the space $\mathcal{F}$ of closed sets in E, from which we can define a random set $\mathbf{X}$ such that (2.3) holds.

By $T_{\mathbf{X}}$, let us denote Choquet's capacity functional that defines a random set $\mathbf{X}$. If two random sets, $\mathbf{X}_1$ and $\mathbf{X}_2$, are independent, then we have that $T_{\mathbf{X}_1 \cup \mathbf{X}_2}(K) = 1 - (1 - T_{\mathbf{X}_1}(K))(1 - T_{\mathbf{X}_2}(K))$. Thus we can define a finite random set $\mathbf{X}$ with independent n objects ($\#(\mathbf{X}) = n$ with prob. one), each having a probability distribution, $F_1, \ldots, F_n$, by Choquet's capacity functional $T_{\mathbf{X}}(K) = 1 - \prod_{i=1}^n (1 - F_i(K))$. The i.i.d. model with common probability distribution F can be rewritten as $T_{\mathbf{X}}(K) = 1 - \sum_{n=0}^{\infty} p_n(1 - F(K))^n$ with $p_n \geq 0$ and $\sum_{n=0}^{\infty} p_n = 1$. For the Poisson process, we have that $T_{\mathbf{X}}(K) = 1 - \exp(-\gamma(K))$. For the binomial process, $T_{\mathbf{X}}(K) = 1 - (1 - F(K))^n$. It may not be so easy to express a random set with dependent objects. For example, suppose $\mathbf{x}$ is a random vector in E^n with joint probability density F that induces a random finite set $\mathbf{X}$ as $\mathbf{X} = \varphi(\mathbf{x})$. Then we have $T_{\mathbf{X}}(K) = \sum_{i=1}^n F_i(K) - \sum_{i=1}^{n-1} \sum_{j=i+1}^n F_{ij}(K^2) \cdots + (-1)^{n-1} F_{1\cdots n}(K^n)$, with $F_{i_1 \cdots i_m}(\cdot)$ being the par-

[12] A functional T on $\mathcal{K}$ satisfying (i)–(iii) is said to be an *alternating Choquet capacity of infinite order* [18].

tial marginals[13] of a probability distribution F on E^n. Moreover, with $T = T_{\mathbf{X}}$ in (2.4), we have $S_n(K; K_1, \ldots, K_n) = \sum_{\pi \in \Pi_n} F\left(\prod_{i=1}^{n} K_{\pi(i)}\right)$ and $S_{n+1}(K; K_1, \ldots, K_n, K_{n+1}) \equiv 0$, whenever $K, K_1, \ldots, K_n, K_{n+1}$ are disjoint compact sets in E, which we can relate to the "scrambling" (2.2).

Data Sets: As mentioned before, one unique aspect of data fusion is information given in terms of random sets. A finite random set $\mathbf{Y}$ in a measurement space[14] E_M, a locally compact Hausdorff space satisfying the second axiom of countability, is called a *data set* if it is considered as a set of observables in E_M taken by a sensor at the same time or almost the same time. Each data set $\mathbf{Y}$ is modeled as a union $\mathbf{Y} = \mathbf{Y}_{DT} \cup \mathbf{Y}_{FA}$ of the set $\mathbf{Y}_{DT}$ of measurements originating from the detected objects in $\mathbf{X}$, and the set $\mathbf{Y}_{FA}$ of *false alarms* that are extraneous measurements not originating from any object in $\mathbf{X}$. Assuming conditional independence of $\mathbf{Y}_{DT}$ and $\mathbf{Y}_{FA}$, the conditional capacity functional of $\mathbf{Y}$ given $\mathbf{X}$ is

$$
\begin{aligned}
T_{\mathbf{Y}}(H|\mathbf{X}) &= \text{Prob.}\left(\{\mathbf{Y} \cap H \neq \emptyset\}|\mathbf{X}\right) \\
&= 1 - (1 - T_{DT}(H|\mathbf{X}))\,(1 - T_{FA}(H))\,,
\end{aligned} \tag{2.5}
$$

for each compact set H in E_M. Assuming object-wise independent but state-dependent detection, the random set $\mathbf{Y}_{DT}$ of "real" measurements can be modeled as[15]

$$
\begin{aligned}
&T_{DT}(H|\mathbf{X}) = \\
&1 - \sum_{\delta \in \Delta(\mathbf{X})} \prod_{x \in \mathbf{X}} \left(P_D(x)\,(1 - F_M(H|x))\right)^{\delta(x)} (1 - P_D(x))^{1-\delta(x)}\,,
\end{aligned} \tag{2.6}
$$

where $\Delta(A) \stackrel{\text{def}}{=} \{0,1\}^A$ for any set A, $P_D(x)$ is the probability of each object at the state x being detected and $F_M(\cdot|\cdot)$ is the state-to-measurement transition probability. The random set $\mathbf{Y}_{FA}$ of "false" measurements can be any reasonable finite random set, e.g., a Poisson point process as $T_{FA}(H) = 1 - \exp(-\gamma_{FA}(H))$ with an intensity measure γ_{FA}.

3. A pure assignment problem. Before stating a class of general multi-object state-estimation problems using the capacity functional formalism, let us consider a simple but illustrative special case. Consider a data set $\mathbf{Y}$ for which there is no false alarm, $T_{FA}(H) \equiv 0$, and no missing

---

[13] By which we mean that $F_{i_1 \cdots i_m}\left(\prod_{j=1}^{m} A_j\right) = F\left(\prod_{i=1}^{n} B_i\right)$ with $B_{i_j} = A_j$ for every j and $B_i = E$ for each i for which there is no j such that $i = i_j$.

[14] Again, a practical example for the measurement space is a hybrid space to accommodate continuous measurements as well as any discrete-valued observations.

[15] By using this model, we exclude the possibility of *split* or *merged* measurements.

object, $P_D(x) \equiv 1$ so that we have $T_Y(H|\mathbf{X}) = 1 - \prod_{x \in \mathbf{X}} (1 - F_M(H|x))$. Assume Prob.$\{\#(\mathbf{X}) = n\} = 1$ for some $n > 1$ and n independent objects with probability distributions, $F_1, \ldots, F_n$. Probability distributions F_i's and $F_M(\cdot|x)$ (for every $x \in E$) are assumed to have densities with respect to μ of E and an appropriate measure μ_M on the measurement space E_M, resp., as, $F_i(dx) = f_i(x)\mu(dx)$ and $F_M(dy|x) = f_M(y|x)\mu_M(dy)$.

Problems with A Priori Identifications: First assume that, for each $i \in \{1, \ldots, n\}$, the probability distribution of the ith object $\mathbf{x}_i$ is F_i, and consider the problem of estimating the random vector $(\mathbf{x}_i)_{i=1}^n$ rather than the random set $\mathbf{X}$. In other words, $(\mathbf{x}_i)_{i=1}^n$ is a particular enumeration of $\mathbf{X}$. In [13], such enumeration is called *a priori identification*. Suppose $(\mathbf{y}_i)_{i=1}^n$ is an arbitrary enumeration of the random set $\mathbf{Y}$. Then, abusing the notation $p(\cdot|\cdot)$ as a generic conditional probability density function notation, we have

$$
\begin{aligned}
p\left((\mathbf{y}_i)_{i=1}^n | (\mathbf{x}_i)_{i=1}^n\right) &= \sum_{\pi \in \Pi_n} p\left((\mathbf{y}_i)_{i=1}^n | \pi, (\mathbf{x}_i)_{i=1}^n\right) p\left(\pi | (\mathbf{x}_i)_{i=1}^n\right) \\
&= \frac{1}{n!} \sum_{\pi \in \Pi_n} \prod_{i=1}^n f_M\left(\mathbf{y}_{\pi(i)} | \mathbf{x}_i\right) ,
\end{aligned} \tag{3.1}
$$

where π is a random permutation[16] on $\{1, \ldots, n\}$ to model the arbitrariness of the enumeration. Since the enumeration is totally arbitrary, we should have $p\left(\pi | (\mathbf{x}_i)_{i=1}^n\right) = 1/n!$. This permutation π assigns each probability distribution F_i to a measurement $\mathbf{y}_{\pi(i)}$. Using multi-target tracking terminology, each object distribution F_i which actually enumerates the object set $\mathbf{X}$ is called a *track* and each realization of π is called a *track-to-measurement association hypothesis*.[17]

Correlation-Based Algorithm: Using these hypotheses π, the joint *a posteriori* probability distribution of the random vector $(\mathbf{x}_i)_{i=1}^n$ can be described as

$$
p\left((\mathbf{x}_i)_{i=1}^n | (\mathbf{y}_i)_{i=1}^n\right) = \sum_{\pi \in \Pi_n} W(\pi) \prod_{i=1}^n \hat{p}_i\left(\mathbf{x}_i | \mathbf{y}_{\pi(i)}\right) , \tag{3.2}
$$

with $W(\cdot)$ and $\hat{p}(\cdot|\cdot)$ being defined right below. Evaluation of each hypothesis π is done by normalizing the *association likelihood function* $L(\pi)$, i.e., $W(\pi) = p\left(\pi | (\mathbf{y}_i)_{i=1}^n\right) = L(\pi)/\text{const.}$, where $L(\pi) = \prod_{i=1}^n L_i\left(\mathbf{y}_{\pi(i)}\right)$ with

[16] It is important to define the random permutation π independently of the value of measurements, $(y_i)_{i=1}^n$. Otherwise, there may be some difficulty in evaluating the conditioning by π.

[17] In [11], it is called a *labeled partition*.

$L_i(y) \stackrel{\text{def}}{=} \int_E f_M(y|x) f_i(x) \mu(dx)$ being the individual *track-to-report likelihood function*. Thus the problem for finding a best track-to-report correlation hypothesis π to maximize $L(\pi)$ becomes a classical n–task-n–resource assignment problem. The *a posteriori* distribution of $(\mathbf{x}_i)_{i=1}^n$ is then given as a mixture of the joint conditional object state distributions

$$p\left((\mathbf{x}_i)_{i=1}^n | \pi, (\mathbf{y}_i)_{i=1}^n\right) = \prod_{i=1}^n \hat{p}_i(\mathbf{x}_i | \mathbf{y}_{\pi(i)}) ,$$

given each hypothesis π with weights $W(\pi) = p\left(\pi | (\mathbf{y}_i)_{i=1}^n\right)$, where $\hat{p}_i(\cdot|y)$ is the density of the conditional probability of the ith object $\mathbf{x}_i$ given $\mathbf{y}_{\pi(i)} = y$ assuming that the $\pi(i)^{\text{th}}$ measurement originates from $\mathbf{x}_i$, i.e., $\hat{p}_i(x|y) = L_i(y)^{-1} f_M(y|x) f_i(x)$. The *correlation-based* algorithms maintain the joint probability distribution of the random vector $(\mathbf{x}_i)_{i=1}^n$ as the weighted mixture of the generally object-wise independent object state distributions, $p\left((\mathbf{x}_i)_{i=1}^n | \pi, (\mathbf{y}_i)_{i=1}^n\right)$, given data association hypotheses π, as seen in (3.2).

Correlation–Free Algorithm: It is easy to see that it follows from (3.1) and (3.2) that

$$\begin{aligned} p\left((\mathbf{x}_i)_{i=1}^n | (\mathbf{y}_i)_{i=1}^n\right) &= \frac{p\left((\mathbf{y}_i)_{i=1}^n | (\mathbf{x}_i)_{i=1}^n\right) p\left((\mathbf{x}_i)_{i=1}^n\right)}{p\left((\mathbf{y}_i)_{i=1}^n\right)} \\ &= \frac{\sum_{\pi\in\Pi_n} \prod_{i=1}^n f_M(\mathbf{y}_{\pi(i)} | \mathbf{x}_i) f_i(\mathbf{x}_i)}{\sum_{x\in\Pi_n} \prod_{i=1}^n \int_E f_M(\mathbf{y}_{\pi(i)} | x) f_i(x) \mu(dx)} , \end{aligned} \tag{3.3}$$

where the prior is given as $p\left((\mathbf{x}_i)_{i=1}^n\right) = \prod_{i=1}^n f_i(\mathbf{x}_i)$. In other words, equation (3.2) can be obtained by mechanically applying Bayes' rule using the last expression of (3.1). Moreover, since $p\left((\mathbf{y}_i)_{i=1}^n | (\mathbf{x}_i)_{i=1}^n\right)$ is permutable with respect to $\mathbf{y}_i$'s, we have $p\left(\{\mathbf{y}_i\}_{i=1}^n | (\mathbf{x}_i)_{i=1}^n\right) = n!\ p\left((\mathbf{y}_i)_{i=1}^n | (\mathbf{x}_i)_{i=1}^n\right)$, and hence, we have $p\left((\mathbf{x}_i)_{i=1}^n | \{\mathbf{y}_i\}_{i=1}^n\right) = p\left((\mathbf{x}_i)_{i=1}^n | (\mathbf{y}_i)_{i=1}^n\right)$. Thus equation (3.3) absorbs all the correlation hypotheses into the cross-object correlation in the joint state distributions, which we may call a *correlation-free* algorithm. Equations (3.1)–(3.3) show us that the correlation-based and the correlation-free approaches are equivalent to each other. A typical example of correlation-based algorithms is Reid's multi-hypothesis algorithm or its generalization adopted to the objects with *a priori* identification, while a typical example of correlation-free algorithms may be the *joint probabilistic data association* (JPDA) algorithm ([21], [22]) that uses a Gaussian approximation and ignores the cross-object correlation in (3.3). A correlation-free algorithm maintaining the cross-object correlation of (3.3) by using a totally quantized state space has been recently reported in [17].

Problems Without A Priori Identification: We can eliminate the *a priori* identification from equation (3.2) by scrambling it as

$$
\begin{aligned}
P\left(\{\mathbf{x}_i\}_{i=1}^n | \{\mathbf{y}_i\}_{i=1}^n\right) &= \sum_{\hat{\pi}\in\Pi_n} p\left(\left(\mathbf{x}_{\hat{\pi}(i)}\right)_{i=1}^n | \{\mathbf{y}_i\}_{i=1}^n\right) \\
&= \sum_{\hat{\pi}\in\Pi_n} \sum_{\pi\in\Pi_n} W(\pi) \prod_{i=1}^n \hat{p}_i\left(\mathbf{x}_{\hat{\pi}(i)} | \mathbf{y}_{\pi(i)}\right) ,
\end{aligned} \tag{3.4}
$$

where every realization of the random permutation $\hat{\pi}$ in (3.4) is called an *object-to-data association hypothesis*, which was first introduced in [12] and is indeed a necessary element in proving the optimality of Reid's algorithm. Since this scrambling is to eliminate the *a priori* identification, we can achieve the same effect by scrambling the *a priori* distribution first. Indeed, it is easy to see that equation (3.4) can be obtained by applying Bayes' rule

$$
p\left(\{\mathbf{x}_i\}_{i=1}^n | \{\mathbf{y}_i\}_{i=1}^n\right) = \frac{p\left(\{\mathbf{y}_i\}_{i=1}^n | \{\mathbf{x}_i\}_{i=1}^n\right) p\left(\{\mathbf{x}_i\}_{i=1}^n\right)}{p\left(\{\mathbf{y}_i\}_{i=1}^n\right)} \tag{3.5}
$$

rather mechanically, by using the scrambled prior

$$
p\left(\{\mathbf{x}_i\}_{i=1}^n\right) = \sum_{\hat{\pi}\in\Pi_n} \prod_{i=1}^n f_i(x_{\hat{\pi}(i)}) . \tag{3.6}
$$

Indeed, equation (3.5) is the essence of the correlation-free algorithms for objects without *a priori* identification, as described in [15] and [16]. The fact that equation (3.4) (post-scrambling) is obtained from (3.5) and (3.6) (pre-scrambling) proves the equivalence between the correlation-based and the correlation-free algorithms, for objects without *a priori* identification.

When we consider $(\mathbf{x}_i)_{i=1}^n$ in (3.1)–(3.3) as a totally arbitrary enumeration of the random set $\mathbf{X}$, equations (3.2)–(3.5) are all equivalent to each other and can be rewritten in terms of the conditional Choquet's capacity functional as

$$
T_{\mathbf{X}}(K|\mathbf{Y}) = 1 - \frac{\sum_{\lambda\in\Lambda(\mathbf{Y})} \prod_{i=1}^n \int_{E\setminus K} f_M(\lambda(i)|x) F_i(dx)}{\sum_{\lambda\in\Lambda(Y)} \prod_{i=1}^n \int_E f_M(\lambda(i)|x) F_i(dx)} , \tag{3.7}
$$

for[18] every compact set K in E, where $\Lambda(\mathbf{Y})$ is the set of all enumerations of set $\mathbf{Y}$, i.e., all one-to-one mappings from $\{1,\ldots,n\}$ to $\mathbf{Y}$.

[18] "$\setminus$" is the set subtraction operator, i.e., $A\setminus B = \{a \in A | a \notin B\}$ for any pair of sets, A and B.

4. General Poisson–i.i.d. problem. Finally, we are ready to make a formal problem statement. Although some of the conditions can be relaxed without significant difficulty, we will let ourselves be restricted by the following two assumptions:

ASSUMPTION A1. The random set $\mathbf{X}$ of objects is a Poisson point process in the state space E with a finite intensity measure $\bar{\gamma}$, i.e., $T_{\mathbf{X}}(K) = 1 - \exp(-\bar{\gamma}(K))$ for every $K \in \mathcal{K}$.

ASSUMPTION A2. We are given N conditionally independent random sets, $\mathbf{Y}_1, \ldots, \mathbf{Y}_N$, in the measurement spaces, $E_{N_1}, \ldots, E_{M_N}$, resp., i.e., $T_{\mathbf{Y}_1 \times \cdots \times \mathbf{Y}_N}(H_1 \times \cdots \times H_N|\mathbf{X}) = \prod_{k=1}^{N} T_{\mathbf{Y}_k}(H_k|\mathbf{X})$, for any compact sets, $H_1, \ldots, H_N$, in $E_{N_1}, \ldots, E_{M_N}$, resp., such that, for each k,

$$T_{\mathbf{Y}_k}(H_k|\mathbf{X}) = 1 - e^{-\gamma_{FA_k}(H_k)} \times \sum_{\delta \in \Delta(\mathbf{X})} \prod_{x \in \mathbf{X}} \left(P_{D_k}(x)(1 - F_{M_k}(H_k|x))\right)^{\delta(x)} (1 - P_{D_k}(x))^{1-\delta(x)}, \tag{4.1}$$

where γ_{FA_k} is a finite intensity measure of the false alarm set (Poisson processes), $P_{D_k}(\cdot)$ is a $[0,1]$–valued measurable function modeling the state-dependent detection probability, and $F_{M_k}(\cdot|\cdot)$ is the state-to-measurement transition probability.[19]

With these two assumptions, our problem can be stated as follows:

PROBLEM. Express the conditional Choquet's capacity functional $T_{\mathbf{X}}(\cdot|\mathbf{Y}_1, \ldots, \mathbf{Y}_N)$ in terms of $\bar{\gamma}$ and $(P_{D_k}, F_{M_k}, \gamma_{FA_k})_{k=1}^{N}$.

This problem is to search for an optimal solution, in a strictly Bayesian sense,[20] to the problem of estimating the random set $\mathbf{X}$ based on the information given in terms of random sets $\mathbf{Y}_1, \ldots, \mathbf{Y}_N$.

In order to describe a solution, we need to refine the definition of the data association hypothesis that we discussed in the previous section. Let[21] $\mathbf{Z} = \bigcup_{k=1}^{N} \mathbf{Y}_k \times \{k\}$, and $\tau_{|k} = \{y \in \mathbf{Y}_k | (y,k) \in \tau\}$ for every $\tau \subseteq \mathbf{Z}$. Then any subcollection τ of the cumulative (tagged) data sets is called a *track* on $\mathbf{Z}$ if $\#(\tau_{|k}) \leq 1$ for every k. Any collection of non-empty, non-overlapping

---

[19] Assumption A2 excludes the possibility of split or merged measurements.

[20] By the "strict Bayesian sense," we mean ability to define the probability distribution of the random set $\mathbf{X}$ conditioned by the σ–algebra of events generated by information, i.e., the random sets, $\mathbf{Y}_1, \ldots, \mathbf{Y}_N$, so that any kind of *a posteriori* statistics concerning $\mathbf{X}$ can be calculated as required.

[21] $\mathbf{Z}$ is equivalent to $(\mathbf{Y}_k)_{k=1}^{N}$ in the sense that they induce the same conditioning σ–subalgebra of events. Each element in $\mathbf{Z}$ is a pair $(\mathbf{y}, k)$ of a member $\mathbf{y}$ of $\mathbf{Y}_k$ and the index or tag k to identify the data set $\mathbf{Y}_k$.

tracks on $\mathbf{Z}$ is called a *data association hypothesis*[22] on $\mathbf{Z}$. Let the set of all tracks on $\mathbf{Z}$ be denoted by $\mathcal{T}(\mathbf{Z})$ and the set of all data association hypotheses on $\mathbf{Z}$ by $\Lambda(\mathbf{Z})$. Using these concepts,[23] we can now state a solution in the form of the following theorem:

THEOREM 4.1. Under Assumptions A1 and A2, suppose that every false alarm intensity measure γ_{FA_k} and every state-to-measurement transition probability F_{M_k} in (4.1) have density functions, as $\gamma_{FA_k}(dy) = \beta_{FA_k}(y)$ $\mu_{M_k}(dy)$ and $F_{M_k}(dy|x) = f_{M_k}(y|x)\mu_{M_k}(dy)$ with respect to an appropriate measure μ_{M_k} on the measurement space E_{M_k}. Then, there exists a finite measure $\hat{\gamma}$ on E, a probability distribution $(\hat{p}(\lambda))_{\lambda\in\Lambda(\mathbf{Z})}$ on $\Lambda(\mathbf{Z})$, and a probability distribution $\hat{F}_\tau$ on E for each $\tau \in \mathcal{T}(Z) \setminus \{\emptyset\}$, such that

$$T_{\mathbf{X}}(K|\mathbf{Z}) = 1 - e^{-\hat{\gamma}(K)} \sum_{\lambda\in\Lambda(\mathbf{Z})} \hat{p}(\lambda) \prod_{\tau\in\lambda} \left(1 - \hat{F}_\tau(K)\right), \tag{4.2}$$

for each $K \in \mathcal{K}$.

PROOF. The above theorem is actually a re-statement of one of the results shown in [12] and [13], and hence, we will only outline the proof. Given each data association hypothesis $\lambda \in \Lambda(\mathbf{Z})$, the *a posteriori* estimation of the random set $\mathbf{X}$ of objects can be done object-wise independently, as

$$\hat{F}_\tau(K) = \frac{\int_K \prod_{k=1}^N h_k(x;\tau,\mathbf{Z})\bar{\gamma}(dx)}{\int_E \prod_{k=1}^N h_k(x;\tau,\mathbf{Z})\bar{\gamma}(dx)}, \tag{4.3}$$

for every $\tau \in \lambda$ and every $K \in \mathcal{K}$, with

$$h_k(x;\tau,\mathbf{Z}) = \begin{cases} f_{M_k}(y|x)P_{D_k}(x), & \text{if } (y,k) \in \tau \\ 1 - P_{D_k}(x), & \text{otherwise (i.e., if } \tau_{|k} = \emptyset) \end{cases}. \tag{4.4}$$

Each data association hypothesis $\lambda \in \Lambda(\mathbf{Z})$ is "evaluated" as

$$\hat{p}(\lambda) = C(\mathbf{Z})^{-1} \left(\prod_{\tau\in\lambda} \int_E \prod_{k=1}^N h_k(x;\tau,\mathbf{Z})\bar{\gamma}(dx) \right) \times \left(\prod \{\beta_{FA_k}(y) | (y,k) \in \mathbf{Z} \setminus \cup\lambda\} \right), \tag{4.5}$$

with a normalizing constant $C(\mathbf{Z})$. The problem for selecting a best hypothesis to maximize (4.5) can be formalized as a $0-1$ integer programming

[22] Called a labeled partition in [11].

[23] Formally, $\mathcal{T}(\mathbf{Z}) = \{\tau \subseteq \mathbf{Z} | \#(\tau_{|k}) \leq 1 \text{ for all } k\}$ and $\Lambda(\mathbf{Z}) = \{\lambda \subseteq \mathcal{T}(\mathbf{Z}) \setminus \{\emptyset\} |_{\tau_1\cap\tau_2} = \emptyset \text{ for all } (\tau_1,\tau_2) \in \lambda^2 \text{ such that } \tau_1 \neq \tau_2\}$.

problem ([9]), or a linear programming problem ([23]). Finally, $\hat{\gamma}$ is the finite intensity measure on E that defines the random set of all objects which remained undetected through all detection attempts represented by the data sets, $\mathbf{Y}_1, \ldots, \mathbf{Y}_N$, and is expressed as

$$\hat{\gamma}(K) = \int_K \prod_{k=1}^{N} (1 - P_{D_k}(x))\, \bar{\gamma}(dx) \,, \tag{4.6}$$

for each $K \in \mathcal{K}$.

Given the number $\#(\mathbf{X}) = n$ of objects in the random set $\mathbf{X}$, let $(\mathbf{x}_i)_{i=1}^n$ be an arbitrary enumeration of X. Then, equation (4.2) can be rewritten as

$$\text{Prob.}\{(\mathbf{x}_i)_{i=1}^n \in \prod_{i=1}^{n} K_i | \#(\mathbf{X}) = n, \mathbf{Z}\} \text{Prob.}\{\#(\mathbf{X}) = n | \mathbf{Z}\} =$$

$$\frac{e^{-\hat{\gamma}(E)}}{n!} \sum_{\substack{\lambda \in \Lambda(\mathbf{Z}) \\ \#(\lambda) \le n}} \hat{p}(\lambda) \sum_{\pi \in \Pi(\lambda, \{1,\ldots,n\})} \left(\prod_{\substack{i=1 \\ i \notin \mathrm{Im}(\pi)}}^{n} \hat{\gamma}(K_i) \right) \left(\prod_{\tau \in \lambda} \hat{F}_\tau \left(K_{\pi(\tau)} \right) \right) , \tag{4.7}$$

for[24] every $(K_i)_{i=1}^n \in \mathcal{K}^n$. For a given hypothesis $\lambda \in \Lambda(\mathbf{Z})$ and $\#(\mathbf{X}) = n$, every $\pi \in \Pi(\lambda, \{1, \ldots, n\})$ can be viewed as a *data-to-object assignment hypothesis* which assigns each track τ to an object index i, completely randomly, i.e., with each realization having the probability $(n - \#(\lambda))!/n!$. We can consider equation (4.7) as an optimal solution when our problem is formulated in terms of a random-finite-sequence model rather than a random-set model. The proof of (4.7) is given in [12] and [13], although the evaluation formulae (4.3)–(4.6) are in so-called batch-processing form while those in [12] and [13] are in a recursive form. The equivalence between the two forms is rather obvious. Thus equation (4.2) is proved as a re-statement of (4.7). This theorem, in essence, proves the optimality of Reid's algorithm. Equations (4.2) and (4.7) show that a choice of sufficient statistics of the multi-object state-estimation problem is the triple $\left((\hat{p}(\lambda))_{\lambda \in \Lambda(\mathbf{Z})}, (\hat{F}_\tau)_{\tau \in \mathcal{T}(\mathbf{Z}) \setminus \{\emptyset\}}, \hat{\gamma} \right)$ of the *hypothesis evaluation*, the *track evaluation*, and the *undetected object intensity*. □

Probability Density: Let us now consider a random-set probability density expression equivalent to (4.2) or (4.7). To do this, we need to assume that each of the *a posteriori* object state distributions $\hat{F}_\tau$ and the undetected object intensity $\hat{\gamma}$ has a density with respect to the measure μ of the state space E, as $\hat{F}_\tau(dx) = \hat{f}_\tau(x)\mu(dx)$ and $\hat{\gamma}(dx) = \hat{\beta}(x)\mu(dx)$. Then

[24] $\Pi(A, B)$ is the set of all one-to-one functions defined on an arbitrary set A taking value in another arbitrary set B. $\mathrm{Dom}(f)$ and $\mathrm{Im}(f)$ are the domain and range of any function f, resp.

we can rewrite (4.7), again abusing the notation p for probability density functions, as

$$p\left((\mathbf{x}_i)_{i=1}^{\mathbf{n}}|\mathbf{Z}\right) = (1/\mathbf{n}!) \quad e^{-\int_E \hat{\beta}(x)\mu(dx)} \times$$

$$\sum_{\substack{\lambda\in\Lambda(\mathbf{Z})\\ \#(\lambda)\leq n}} \hat{p}(\lambda) \sum_{\pi\in\Pi(\lambda(\mathbf{Z}),\{1,\ldots,\mathbf{n}\})} \left(\prod_{\substack{i=1\\ i\notin \mathrm{Im}(\pi)}}^{n} \hat{\beta}(\mathbf{x}_i)\right)\left(\prod_{\tau\in\lambda} \hat{f}_\tau\left(\mathbf{x}_{\pi(\tau)}\right)\right), \tag{4.8}$$

where $\mathbf{x} = (\mathbf{x}_i)_{i=1}^{\mathbf{n}}$ is an arbitrarily chosen enumeration of the random set $\mathbf{X}$. In order to show that the number of objects $\#(\mathbf{X}) = \ell(\mathbf{x})$ is random, $\mathbf{n}$ is used instead of n. As discussed in the previous section, given $\#(\mathbf{X}) = n$ for some n, the random-set probability density expression can be obtained by "scrambling" (4.8). Since the right hand side of (4.8) is permutable with respect to $(\mathbf{x}_i)_{i=1}^{\mathbf{n}}$, we have

$$p(\mathbf{X}|\mathbf{Z}) = \sum_{\substack{\lambda\in\Lambda(\mathbf{Z})\\ \#(\lambda)\leq\#(X)}} \hat{p}(\lambda) \sum_{\pi\in\Pi(\lambda(\mathbf{Z}),\mathbf{X})} p_{ND}\left(\mathbf{X}\setminus \mathrm{Im}(\pi)\right)\prod_{\tau\in\lambda} \hat{f}_\tau(\pi(\tau)), \tag{4.9}$$

where $p_{ND}(\cdot)$ is the probability density of the set of undetected objects in $\mathbf{Z}$ that is a Poisson process with the intensity measure $\hat{\gamma}$ having density $\hat{\beta}$ with respect to μ, i.e., $\hat{p}_{ND}(X) = e^{-\hat{\gamma}(E)} \prod_{x\in X} \hat{\beta}(x)$, for every finite set X in E, i.e., $X \in \mathcal{I}$. The *a posterior* distribution of the number of objects is then written as

$$\text{Prob.}\left\{\#(\mathbf{X}) = n|\mathbf{Z}\right\} = e^{-\hat{\gamma}(E)} \sum_{\substack{\lambda\in\Lambda\\ \#(\lambda)\leq n}} \hat{p}(\lambda)\frac{\hat{\gamma}(E)^{(n-\#(\lambda))}}{(n-\#(\lambda))!}. \tag{4.10}$$

Thus, the expected number of objects that remain undetected in the cumulative data sets $\mathbf{Z}$ is $\hat{\gamma}(E)$, independent of data association hypotheses λ.

Correlation-Free Algorithms: The scrambled version, i.e., the random-set probability density version (4.1) of the data set model, can be written as

$$p(\mathbf{Y}_k|\mathbf{X}) = \sum_{a\in\mathcal{A}(X,Y_k)} P_{FA_k}\left(\mathbf{Y}_k \setminus \mathrm{Im}(a)\right)\times$$

$$\left(\prod_{x\in \mathrm{Dom}(a)} f_{M_k}(a(x)|x)P_{D_k}(x)\right)\left(\prod_{x\in\mathbf{X}\setminus \mathrm{Dom}(a)} (1-P_{D_k}(x))\right), \tag{4.11}$$

where $\mathcal{A}(\mathbf{X},\mathbf{Y}_k) \stackrel{\text{def}}{=} \bigcup_{X_D\subseteq\mathbf{X}} \{a \in (\mathbf{Y}_k)^{X_D} | a \text{ is one-to-one}\}$, and $P_{FA_k}(\cdot)$ is the probability density expression of the Poisson process of the false alarms in $\mathbf{Y}_k$, i.e.,

$$p_{FA_k}(Y) = e^{-\gamma_{FA_k}(E_{M_k})} \prod_{y \in Y} \beta_{FA_k}(y) \tag{4.12}$$

$$= e^{-\int_{E_{M_k}} \beta_{FA_k}(y)\mu_{M_k}(dy)} \prod_{y \in Y} \beta_{FA_k}(y) ,$$

for every finite set Y in the measurement space E_{M_k}.

It can then be shown by recursion that we can obtain the random-set probability density expression (4.9) by mechanically applying Bayes' rule as

$$p(\mathbf{X}|\mathbf{Z}) = p(\mathbf{X}|\mathbf{Y}_1, \ldots, \mathbf{Y}_N) = \frac{\prod_{k=1}^{N} p(\mathbf{Y}_k|\mathbf{X})p(\mathbf{X})}{p(\mathbf{Y}_1, \ldots, \mathbf{Y}_N)} , \tag{4.13}$$

where each $p(\mathbf{Y}_k|\mathbf{X})$ is given by (4.11) and $p(\mathbf{X})$ is the "scrambled" version of the *a priori* distribution, $T_{\mathbf{X}}(K) = 1 - \exp(-\bar{\gamma}(K))$; i.e.,

$$p(\mathbf{X}) = e^{-\bar{\gamma}(E)} \prod_{x \in \mathbf{X}} \bar{\beta}(x) = e^{-\int_{E} \bar{\beta}(x)\mu(dx)} \prod_{x \in \mathbf{X}} \bar{\beta}(x) , \tag{4.14}$$

assuming that $\bar{\gamma}$ is absolutely continuous with respect to μ as $\bar{\gamma}(dx) = \bar{\beta}(x)\mu(dx)$. We may call algorithms obtained by applying the random-set observation equation (4.11) mechanically to the Bayes' rule (4.13) *correlation-free* algorithms because data association hypotheses are not explicitly used. In such algorithms, data association hypotheses are replaced by explicit cross-object correlation in the probability density function $p(\mathbf{X}|\mathbf{Z})$.

Correlation-free algorithms might have certain superficial attraction because of the appearance of simplicity in expression (4.13). There are, however, several hidden problems in applying this correlation-free approach. First of all, when using an object model without an *a priori* limit on the number of objects such as a Poisson model (Assumption A1), it is not possible to calculate $p(\mathbf{X}|\mathbf{Z})$ mechanically through (4.13) since we cannot use non-finite sufficient statistics. Even when we use a clever way to separate out the state distributions of undetected objects, the necessary computational resource would grow as the data sets are accumulated. Such growth might be equivalent to the growth of data association hypotheses in correlation-based algorithms. That may force us to use a model with an upper-bound on $\#(\mathbf{X})$ or even with $\#(\mathbf{X}) = n$ for some n with probability one, which itself may be a significant limitation. In fact, the algorithm described in [16] is for a model with an upper bound on $\#(\mathbf{X})$ and can be considered a "scrambling" version of a multi-hypothesis algorithm for the non-Poisson-i.i.d. models described in [24]. On the other hand, the algorithm described in [17] is for a model with a fixed number of objects, which we can consider as a non-Gaussian, non-combining extension of the JPDA algorithm described in [21] and [22]. Both algorithms in [16] and [17] use quantized object state spaces.

Since both the multi-hypothesis approach (4.3)–(4.6) and the correlation-free approach (4.13) produce the *optimal solution*, either (4.2) or (4.7)–(4.9), the immediate question is which approach may yield computationally more efficient practical algorithms. Unfortunately, it is probably very difficult to carry on any kind of complexity analysis to measure computational efficiency. We cannot focus only on the combinatoric part of the problem expressed by equation (4.5), since in most cases, evaluation of the *a posteriori* distributions (4.3) is generally much more computationally intensive. The computational requirements are generally significantly more for any correlation-free algorithm since it requires evaluation of probability distributions of much higher dimensions than the object-wise independent distributions (4.3). On the other hand, the multi-hypothesis approach is criticized, often excessively, for suffering from growing combinatorial complexity. There is, however, a collection of techniques known as *hypothesis management* for keeping the combinatoric growth under control, compromising optimality in accordance with available computational resources. The essence of those techniques was also discussed in [10].

One such technique is known as *probabilistic data clustering*,[25] which may be illustrated by the estimate $T_{\mathbf{X}}(K|\mathbf{Y}_1)$ of the object set $\mathbf{X}$ conditioned only by the first data set $\mathbf{Y}_1$,

$$T_{\mathbf{X}}(K|\mathbf{Y}_1) = 1 - e^{-\int_K (1-P_{D_1}(x))\bar{\gamma}(dx)} \times$$

$$\prod_{y\in Y_1} \frac{\beta_{FA_1}(y) + \int_{E\setminus K} p_{M_1}(y|x)P_{D_1}(x)\bar{\gamma}(dx)}{\beta_{FA_1}(y) + \int_E p_{M_1}(y|x)P_{D_1}(x)\bar{\gamma}(dx)}. \tag{4.15}$$

In other words, we can represent $2^{\mathbf{m}_1}$ hypotheses by a collection of $\mathbf{m}_1$ independent sets of two hypotheses; i.e., in effect, $2\mathbf{m}_1$ hypotheses. Using the correlation-free approach, it would be very difficult to exploit any kind of statistical independence, as shown in (4.15), because of "scrambling," either explicitly or implicitly. In general, however, computational complexity depends on many factors, such as the dimension of the state space, the number of quantization cells, the object and false alarm density with respect to the measurement error sizes, etc. Hence, it may be virtually impossible to quantify it as a general discussion.

Dynamical Cases: In most practical cases, it is necessary to model the random set $\mathbf{X}$ of objects that are dynamically changing. Dynamic but

[25] As opposed to the spatial clustering discussed in [2].

deterministic cases can be, at least theoretically, treated as static cases by imbedding dynamics into observation equations. Non-deterministic dynamics may be treated by considering a random-set-valued time-series, for example, a Markovian process $(\mathbf{X}_t)_{t\in[0,\infty)}$ with a state transition probability $p(\mathbf{X}_{t+\Delta t}|\mathbf{X}_t)$, or alternatively, by considering the random set $\mathbf{X}$ as an element of an appropriate infinite-dimensional space such as $E^{[0,\infty)}$. There might be, however, some serious problems in such an approach. One potential problem might be the "scrambling" that a random-set-value, time-series model may cause, which may conflict with underlying physical reality. Another worry may be that we may not maintain the local compactness of the state space. For this reason, we probably would prefer another approach in which we simply expand the state space from E to E^N, i.e., consider each object as $(\mathbf{x}_i(t_k))_{k=1}^N \in E^N$, rather than $\mathbf{x}_i \in E$. by taking this approach, it is easy to use an object-wise Markovian model with a state transition probability on the state space E such as Prob. $\{\mathbf{x}_i(t+\Delta t) \in dx | \mathbf{x}_i(t) = x\} = \Phi_{\Delta t}(dx|x)$.

A mechanical application of such an extended model is hardly practical because of the high dimensionality due to the expansion of the state space from E to E^N. In many cases, however, by arranging the estimation process in an appropriate way, we can minimize the increase in computational requirements. For example, if we can process data sets in time-order, with an object-wise Markovian model with state transition probability $\Phi_{\Delta t}(\cdot|\cdot)$, we can solve any non-deterministic version of our problem by including the *extrapolation process*, defined as a transformation $\mathcal{L}_{\Delta t}$ from a probability distribution F to $\mathcal{L}_{\Delta t}F$, defined as $(\mathcal{L}_{\Delta t}f)(K) = \int_E \Phi_{\Delta t}(K|x)\mu(dx)$ for $K \in \mathcal{K}$. Models with varying cardinality of a random set, or models with some sort of birth-death process, can also be handled smoothly with this approach, by including a discrete-state Markovian process with a set of states such as {unborn, alive, dead}.

5. Distributed multi–object estimation. In a distributed data processing system, we assume multiple data processing agents, each of which processes data from its own set of information sources, i.e., sensors, and communicates with other agents to exchange information. The spatial and temporal changes in information, held by each processing agent, can be modeled by a directed graph, which we may call an *information graph*, as defined in [25]. We may have as many *information state nodes* as necessary in this graph to mark the changes in the information state of each agent, either by receiving the data from one of the "local" sensors, or by receiving the data from other agents through some means of communication. In this way, we can model arbitrary information flow patters as "who talks what," but we assume that the processing agents communicate with sufficient statistics. In the context of multi-object state-estimation, this means the set of evaluated tracks and data-association hypotheses, as defined in the previous section.

Information Arithmetic: As in the previous section, we consider a collection of "tagged" data sets, i.e., $\mathbf{Z} = \{(\mathbf{Y}_1, k_1), (\mathbf{Y}_2, k_2), \ldots\}$, with indices k_j's that identify the time and the information source (sensor). Each information state node i in an information graph can be associated with a collection of tagged data sets, $\mathbf{Z}_i$, which we call information at i. Assuming perfect propagation of information and optimal data processing, for any information node i, the information at i, $\mathbf{Z}_i$, can be written as $\mathbf{Z}_i = \bigcup\{\mathbf{Z}_{i'}|i' \leq i\} = \bigcup\{\mathbf{Z}_{i'}|i' \mapsto i\}$ where $i' \leq i$ means the partial order of the information graph defined by the precedence, and $i' \mapsto i$ means i' is an immediate predecessor of i. We need to consider the role of the *a priori* information that we assume is common among all processing agents. For this purpose, we must add an artificial minimum element to the information graph. Let i_0 be that minimum element, so that we have $i_0 \leq i$ for all other nodes i in the information graph. We should have $\mathbf{Z}_{i_0} = \{(\emptyset, k_0)\}$ with some unique index k_0 for the artificial origin. One essential distributed multi-object state-estimation problem can be defined as the problem for calculating the sufficient statistics of a given information node i from a collection of sufficient statistics contained at a selected set of predecessors i' of node i. Thus, consider a node i and assume the problem of *fusing* information from its immediate predecessors. We can show [25] that there exists a set $\bar{I}$ of predecessors of the node i and an integer-valued function a defined on the set $\bar{I}$ such that $\sum_{\bar{i}\in\bar{I}} \alpha(\bar{i})\#(\mathbf{Z}_{\bar{i}} \cap \{z\}) = 1$ for all $z \in \mathbf{Z}_i$.

Assuming stationarity of the random set $\mathbf{X}$ of objects and using the probability-density expression, it can be shown [25] that

$$
\begin{aligned}
p(\mathbf{X}|\mathbf{Z}_i) &= p\left(\mathbf{X}|\bigcup_{i'\mapsto i}\mathbf{Z}_{i'}\right) = p\left(\mathbf{X}|\bigcup_{i'\leq i}\mathbf{Z}_{i'}\right) \\
&= p(\mathbf{Z}_i)^{-1}\prod_{\bar{i}\in\bar{I}}\left(p(\mathbf{X}|\mathbf{Z}_{\bar{i}})p(\mathbf{Z}_{\bar{i}})\right)^{\alpha(\bar{i})},
\end{aligned}
\tag{5.1}
$$

with $p(\mathbf{X}|\mathbf{Z}_{i_0}) = p(\mathbf{X})$ and $p(\mathbf{Z}_{i_0}) = 1$. Let us consider a simple example with two agents, 1 and 2, with a mutual broadcasting information exchange pattern. Suppose that, at one point, the two agents exchange information so that each agent n has the informational node i_{n1}. Then each agent n accumulates the information from its own set of sensors and updates its informational state from i_{n1} to i_{n2}, after which the two exchange information again to get the information node i_{n3}. Thus we have $\mathbf{Z}_{i_{11}} = \mathbf{Z}_{i_{21}}$, $\mathbf{Z}_{i_{11}} \subseteq \mathbf{Z}_{i_{12}}$, $\mathbf{Z}_{i_{21}} \subseteq \mathbf{Z}_{i_{22}}$, $(\mathbf{Z}_{i_{12}} \setminus \mathbf{Z}_{i_{11}}) \cap (\mathbf{Z}_{i_{22}} \setminus \mathbf{Z}_{i_{21}}) = \emptyset$, $\mathbf{Z}_{i_{13}} = \mathbf{Z}_{i_{23}} = \mathbf{Z}_{i_{12}} \cup \mathbf{Z}_{i_{22}}$, and $\mathbf{Z}_{i_{11}} = \mathbf{Z}_{i_{21}} = \mathbf{Z}_{i_{12}} \cap \mathbf{Z}_{i_{22}}$. We can define the pair $(\bar{I}, \alpha)$ as $\bar{I} = \{i_{12}, i_{22}, i_{11}\}$, $\alpha(i_{12}) = \alpha(i_{22}) = 1$ and $\alpha(i_{11}) = -1$. In other words, we have

$$
p(\mathbf{X}|\mathbf{Z}_{i_{13}}) = \frac{p(\mathbf{Z}_{i_{12}})p(\mathbf{Z}_{i_{22}})}{p(\mathbf{Z}_{i_{12}} \cup \mathbf{Z}_{i_{22}})p(\mathbf{Z}_{i_{11}})}\,\frac{p(\mathbf{X}|Z_{i_{12}})p(\mathbf{X}|Z_{i_{22}})}{p(\mathbf{X}|\mathbf{Z}_{i_{11}})},
\tag{5.2}
$$

in which we can clearly see the role of the index function α which gathers necessary information sets and eliminates information redundancy to avoid informational double counting.

Solution to Distributed Multi-Object Estimation: To translate the results in [25] into the conditional capacity version of (5.1), in addition to the assumptions described in Section 3, we need to assume that all object state probability distributions $\hat{F}(\cdot|\tau, \mathbf{Z})$, conditioned by tracks τ and cumulative data sets $\mathbf{Z}$, have probability density functions $\hat{F}(dx|\tau, \mathbf{Z}) = \hat{f}(x|\tau, \mathbf{Z})\mu(dx)$, as well as intensity measures of the undetected objects $\hat{\gamma}(dx|\mathbf{Z}) = \hat{\beta}(dx|\mathbf{Z})\mu(dx)$. Then, we can write

$$(5.3) \quad T_{\mathbf{X}}(K|\mathbf{Z}_i) = 1 - e^{-\hat{\gamma}(K)} \sum_{\gamma \in \Lambda(\mathbf{Z}_i)} \hat{p}(\lambda|\mathbf{Z}_i) \prod_{\tau \in \lambda} (1 - \hat{F}(K|\tau, \mathbf{Z}))\,,$$

for each $K \in \mathcal{K}$, where each "fused" data association hypothesis λ is evaluated as

$$(5.4) \quad p(\lambda|\mathbf{Z}_i) = c \sum_{\bar{i} \in \bar{I}} p(\lambda_i|\mathbf{Z}_i)^{\alpha(\bar{i})} \prod_{\tau \in \lambda} \int_E \prod_{\bar{i} \in \bar{I}} \hat{g}(x|\tau \cap \mathbf{Z}_{\bar{i}}, \mathbf{Z}_{\bar{i}})^{\alpha(\bar{i})} \mu(dx)\,,$$

with each $\lambda_{\bar{i}}$ being the unique predecessor, on $\mathbf{Z}_{\bar{i}}$, of each "fused" hypothesis λ, where

$$(5.5) \qquad \hat{g}\,(x|\tau_{\bar{i}}, \mathbf{Z}_{\bar{i}}) = \begin{cases} \hat{f}(x|\tau_{\bar{i}}, \mathbf{Z}_{\bar{i}}), & \text{if } \tau_i \neq \emptyset \\ \hat{\beta}(x|\mathbf{Z}_{\bar{i}}), & \text{otherwise (if } \tau_{\bar{i}} = \emptyset) \end{cases}.$$

Each "fused" track $\tau \in \lambda$ is evaluated as

$$(5.6) \quad \hat{F}(K|\tau, \mathbf{Z}_i) = \int_K \hat{f}(x|\tau, \mathbf{Z}_i)\mu(dx) = \frac{\int_K \prod_{\bar{i} \in \bar{I}} \hat{f}(x|\tau \cap \mathbf{Z}_{\bar{i}}, \mathbf{Z}_{\bar{i}})^{\alpha(\bar{i})} \mu(dx)}{\int_E \prod_{\bar{i} \in \bar{I}} \hat{f}(x|\tau \cap \mathbf{Z}_{\bar{i}}, \mathbf{Z}_{\bar{i}})^{\alpha(\bar{i})} \mu(dx)}\,,$$

while the intensities of the undetected objects are fused as

$$(5.7) \qquad \hat{\gamma}(K|\mathbf{Z}_i) = \int_K \hat{\beta}(x|\mathbf{Z}_i)\mu(dx) = \int_K \prod_{\bar{i} \in \bar{I}} \hat{\beta}(x|\mathbf{Z}_{\bar{i}})^{a(\bar{i})} \mu(dx)\,,$$

for each $K \in \mathcal{K}$.

Non-Deterministic Cases: To obtain the above results, we need to assume that each individual object is stationary or at least has deterministic dynamics. However, we can expand the object state space to account for non-deterministic object state dynamics, as discussed in the previous section. Again, it would hardly be practical to apply the theory mechanically to the expanded state space because of the resulting high dimensionality in any practical case. It is possible, however, to break down high-dimensional integration into some algorithm with additional assumptions such as the Markovian property. For example, it is shown in [26] that we can calculate the track-to-track association likelihood function that appears in (5.4) by a simple algorithm that resembles filtering processes. In many non-deterministic cases, however, the track evaluation through (5.6) may not be efficient at all, compared with simply reprocessing all the measurements in a given track by an appropriate filtering algorithm. Nevertheless, multi-object state-estimation using distributed data processing, as described above, is still meaningful in most cases, because each local agent can process local information on its own and is often capable of establishing a few "strong" data association hypotheses.

6. Conclusion. A random-set formalism of a general class of multi-object state-estimation problems was presented in an effort to explore the possibility of such a formalism becoming a foundation of data fusion theory. My initial intention was to derive an optimal solution, in a strictly Bayesian point of view, or to prove the optimality of a multi-hypothesis (correlation-based) algorithm, using the standard tool of random-set theory, i.e., conditional and unconditional Choquet's capacity functionals. Unfortunately, I was not able to do so, and in this paper, I have only re-stated the known general result in terms of conditional Choquet's capacity functionals. If the theorem stated in Section 4 had been proved by standard techniques used in a random-set formalism, it would have provided an important step towards application of random-set theory to further development in data fusion problems, as well as related problems such as image analysis, object recognition, etc.

Nonetheless, I believe several new observations have been made in this paper. For example, the relationship between order set (sequence or vector) representation and unordered set representation through "scrambling" operations is new, to the best of my knowledge, as is the description of the equivalence of the correlation-based algorithms with the correlation-free counterparts that have recently been developed. Several practical aspects of these two approaches have also been discussed. I am inclined to think that some mixture of these two apparently quite different approaches may produce a new set of algorithms in areas where the environments are too difficult for any single approach to work adequately. In my opinion, applications of random-set theory to practical problems contain significant potential for providing a theoretical foundation and practical algorithms for data fusion theory.

Acknowledgment. Earlier work related to the topics discussed in this paper was supported by the Defense Advanced Research Projects Agency. I would like to express my sincere thanks to the management of Texas Instruments for allowing me to participate in this exciting workshop, and to the workshop's organizers for having invited me. I am especially grateful to Dr. R.P.S Mahler of Lockheed-Martin Corporation, who offered me the opportunity to give a talk at the workshop. I am also indebted to Professor John Goutsias of The Johns Hopkins University, Dr. Lawrence Stone of Metron, Inc., and Dr. Fredrick Daum of Raytheon Company, for their valuable comments.

REFERENCES

[1] F.E. WHITE, JR., *Data fusion subpanel report*, Technical Proceedings of the 1991 Joint Service Data Fusion Symposium, Vol. 1, Laurel, MD, (1991), p. 342.

[2] M.N.M. VAN LIESHOUT, *Stochastic Geometry Models in Image Analysis and Spatial Statistics*, CWI Tracts, (1991), p. 11.

[3] N. WIENER, *The Extrapolation, Interpolation, and Smoothing of Stationary Time Series*, John Wiley & Sons, New York, 1949.

[4] R.W. SITTLER, *An optimal data association problem in surveillance theory*, IEEE Transactions on Military Electronics, 8 (1964), pp. 125–139.

[5] Y. BAR–SHALOM, *Tracking methods in a multitarget environment*, IEEE Transactions on Automatic Control, 23 (1978), pp. 618–626.

[6] H.L. WIENER, W.W. WILLMAN, J.H. KULLBACK, AND I.R. GOODMAN, *Naval ocean surveillance correlation handbook, 1978*, NRL Report 8340, Naval Research Laboratory, Washington D.C., 1979.

[7] S.S. BLACKMAN, *Multiple–Target Tracking with Radar Applications*, Artech House, Norwood, MA, 1986.

[8] Y. BAR–SHALOM AND T.E. FORTMANN, *Tracking and Data Association*, Academic Press, San Diego, CA, 1988.

[9] C.L. MOREFIELD, *Application of 0–1 integer programming to multitarget tracking problems*, IEEE Transactions on Automatic Control, 22 (1977), pp. 302–312.

[10] D.B. REID, *An algorithm for tracking multiple targets*, IEEE Transactions on Automatic Control, 24 (1979), pp. 843–854.

[11] I.R. GOODMAN, *A general model for the contact correlation problem*, NRL Report 8417, Naval Research Laboratory, Washington D.C., 1983.

[12] S. MORI, C.-Y. CHONG, E. TSE, AND R.P. WISHNER, *Multitarget multisensor tracking problems – Part I: A general solution and a unified view on Bayesian approaches*, A.I.&D.S. Technical Report TR–1048–01, Advanced Information and Decision Systems, Mountain View, CA, 1983, Revised 1984.

[13] S. MORI, C.-Y. CHONG, E. TSE, AND R.P. WISHNER, *Tracking and classifying multiple targets without a priori identification*, IEEE Transactions on Automatic Control, 31 (1986), pp. 401–409.

[14] R.B. WASHBURN, *A random point process approach to multiobject tracking*, Proceedings of the 1987 American Control Conference, Minneapolis, MN, (1987), pp. 1846–1852.

[15] R. MAHLER, *Random sets as a foundation for general data fusion*, Proceeding of the Sixth Joint Service Data Fusion Symposium, Vol. I, Part 1, The Johns Hopkins University, Applied Physics Laboratory, Laurel, MD, (1993), pp. 357–394.

[16] K. KASTELLA, *Discrimination gain for sensor management in multitarget detection and tracking*, Proceedings of IEEE–SMC and IMACS Multiconference CESA'96, Lille, France, July 1996.

[17] L.D. STONE, M.V. FINN, AND C.A. BARLOW, *Unified data fusion*, Report to Office of Navel Research, Metron Inc. (1996).

[18] G. MATHERON, *Random Sets and Integral Geometry*, John Wiley & Sons, New York, 1974.

[19] D. STOYAN, W.S. KENDALL, AND J. MECKE, *Stochastic Geometry and its Applications*, Second Edition, John Wiley & Sons, Chichester, England, 1995.

[20] H.L. ROYDEN, *Real Analysis*, (1968), p. 318.
[21] Y. BAR–SHALOM, T.E. FORTMANN, AND M. SCHEFFÉ, *Joint probabilistic data association for multiple targets in clutter*, Proceedings of the 1980 Conference on Information Sciences and Systems, Princeton University, Princeton, NJ, 1980.
[22] T.E. FORTMANN, Y. BAR–SHALOM, AND M. SCHEFFÉ, *Multi–target tracking using joint probabilistic data association*, Proceedings of the 19th IEEE Conference on Decision Control, Albuquerque, NM, (1980), pp. 807–812.
[23] K.R. PATTIPATI, S. DEB, Y. BAR–SHALOM, AND R.B. WASHBURN, *Passive multisensor data association using a new relaxation algorithm*, Multitarget–Multisensor Tracking: Advanced Applications (Y. Bar–Shalom, ed.), Artech House, 1990.
[24] S. MORI AND C.–Y. CHONG, *A multitarget tracking algorithm – independent but non–Poisson cases*, Proceedings of the 1985 American Control Conference, Boston, MA, (1985), pp. 1054–1055.
[25] C.–Y. CHONG, S. MORI, AND K.–C. CHANG, *Distributed multitarget multisensor tracking*, Multitarget–Multisensor Tracking: Advanced Applications (Y. Bar–Shalom, ed.), Chap. 8, Artech House, (1990), pp. 247–295.
[26] S. MORI, K.A. DEMETRI, W.H. BARKER, AND R.N. LINEBACK, *A Theoretical foundation of data fusion – generic track association metric*, Technical Proceedings of the 7th Joint Service Data Fusion Symposium, Laurel, MD, (1994), pp. 585–594.
[27] D.J. DALEY AND D. VERE–JONES, *An Introduction to the Theory of Point Processes*, Springer–Verlag, 1988.

EXTENSION OF RELATIONAL AND CONDITIONAL EVENT ALGEBRA TO RANDOM SETS WITH APPLICATIONS TO DATA FUSION

I.R. GOODMAN AND G.F. KRAMER*

Abstract. Conditional event algebra (CEA) was developed in order to represent conditional probabilities with differing antecedents by the probability evaluation of well-defined individual "conditional" events in a single larger space extending the original unconditional one. These conditional events can then be combined logically before being evaluated. A major application of CEA is to data fusion problems, especially the testing of hypotheses concerning the similarity or redundancy among inference rules through use of probabilistic distance functions which critically require probabilistic conjunctions of conditional events. Relational event algebra (REA) is a further extension of CEA, whereby functions of probabilities formally representing single event probabilities — not just divisions as in the case of CEA — are shown to represent actual "relational" events relative to appropriately determined larger probability spaces. Analogously, utilizing the logical combinations of such relational events allows for testing of hypotheses of similarity between data fusion models represented by functions of probabilities. Independent of, and prior to this work, it was proven that a major portion of fuzzy logic — a basic tool for treating natural language descriptions — can be directly related to probability theory via the use of one point random set coverage functions. In this paper, it is demonstrated that a natural extension of the one point coverage link between fuzzy logic and random set theory can be used in conjunction with CEA and REA to test for similarity of natural language descriptions.

Key words. Conditional Event Algebra, Conditional Probability, Data Fusion, Functions of Probabilities, Fuzzy Logic, One Point Coverage Functions, Probabilistic Distances, Random Sets, Relational Event Algebra.

AMS(MOS) subject classifications. 03B48, 60A05, 60A99, 03B52, 60D05, 52A22

1. Introduction.

The work carried out here is motivated directly by the basic data fusion problem: Consider a collection of multi-source information, which, for convenience, we call "models," in the form of descriptions and/or rules of inference concerning a given situation. The sources may be expert-based or sensor system-based, utilizing the medium of natural language or probability, or a mixture of both. The main task, then, is to determine:

GOAL 1. Which models can be considered similar enough to be combined or reduced in some way and which models are dissimilar so as to be considered inconsistent or contradictory and kept apart, possibly until further arriving evidence resolves the issue.

GOAL 2. Combine models declared as pertaining to the same situation.

* NCCOSC RDTE DIV (NRaD), Codes D4223, D422, respectively, Seaside, San Diego, CA 92152-7446. The work of both authors is supported by the NRaD Independent Research Program as Project ZU07.

Goals 1 and 2 are actually extensions of classical hypotheses testing and estimation, respectively, applied to those situations where the relevant probability distributions are either not available from standard procedures or involve, at least initially, non-probabilistic concepts such as natural language.

1.1. Statement of the problem. In this paper, we consider only Goal 1, with future efforts to be directed toward Goal 2. Furthermore, this is restricted to pairwise testing of hypotheses for sameness, when the two models in question are given in the formal truth-functional forms

$$\text{(1.1)} \qquad \begin{cases} \text{Model 1: } P(a) = f(P(c), P(d), P(e), P(h), \ldots) \\ \text{Model 2: } P(b) = g(P(c), P(d), P(e), P(h), \ldots)\ , \end{cases}$$

where $f, g : [0,1]^m \to [0,1]$ are known functions and contributing events $c, d, e, h, \ldots$ all belong to probability space (Ω, B, P), i.e., $c, d, e, h, \ldots \in B$, a Boolean or sigma-algebra over Ω associated with probability measure P. In general, probability involves *non*-truth-functional relations, such as the joint or conjunctive probability of two events *not* being a function of the individual marginal probability evaluations of the events. In light of this fact, in most situations, the "events" a and b in the left-hand side of eq. (1.1) are only formalisms representing overall "probabilities" for Models 1 and 2, respectively. However, if events $c, d, e, h, \ldots$ are all infinite left rays of a real one-dimensional space, and $f = g$ is any copula and P on the left-hand side of eq. (1.1) is replaced by the joint probability measure P_0 determined by f (see, e.g., Sklar's copula theorem [28]), then a and b actually exist in the form of corresponding infinite left rays in a multidimensional real space and eq. (1.1) reduces to an actual truth-functional form.

Returning back to the general case, *suppose* legitimate events a and b could be explicitly obtained in eq. (1.1) lying in some appropriate Boolean or sigma-algebra B_0 extending B in the sense of containing an isomorphic imbedding of all of the contributing events $c, d, e, h, \ldots$, independent of the choice of P, but so that for each P, the probability space (Ω_0, B_0, P_0) extends (Ω, B, P), and *suppose* $P_0(a\&b)$ could be evaluated explicitly, then one could compute any of several natural *probability distance functions* $M_{P_0}(a,b)$. Examples of such include: $D_{P_0}(a,b), R_{P_0}(a,b), E_{P_0}(a,b)$; where $D_{P_0}(a,b)$ is the *absolute probability distance* between a and b (actually a pseudometric over (Ω, B, P), see Kappos [21]); $R_{P_0}(a,b)$ is the *relative probability distance* between a and b; and $E_{P_0}(a,b)$ is a symmetrized log-conditional probability form between a and b. (See Section 1.4 or [13], [14].) For example, using the usual Boolean notation (& for conjunction, v for disjunction, $(\ )'$ for complement, $+$ for symmetric difference or sum, $\leq$ for subevent of, etc.),

$$
\begin{aligned}
D_{P_0}(a,b) &= P_0(a+b) \\
&= P_0(a'\&b) + P_0(a\&b') \qquad (1.2)\\
&= P_0(a) + P_0(b) - 2P_0(a\&b).
\end{aligned}
$$

Then, by replacing the formalisms $P(a), P(b)$ in eq. (1.1), by the probabilities $P_0(a) = f(P(c), P(d), P(e), \ldots)$ and $P_0(b) = g(P(c), P(d), P(e), \ldots)$, together with the assumption that the evaluation $P_0(a\&b)$ is meaningful and can be explicitly determined, the full evaluation of $D_{P_0}(a,b)$ can be obtained via eq. (1.2). In turn, assuming for simplicity that the higher order probability distribution of the relative atom evaluations $P_0(a\&b), P_0(a'\&b)$, $P_0(a\&b')$ are — as P and a and b are allowed to vary — uniformly distributed over the natural simplex of possible values, the *cumulative distribution function* (cdf) F_D of $D_{P_0}(a,b)$ under the hypothesis of being not identical can be ascertained. Thus, using F_D and the "statistic" $D_{P_0}(a,b)$ one can test the hypotheses that a and b are different or the same, up to P_0–measure zero. (See Sections 1.4, 1.5 for more details.)

Indeed, it has been shown that two relatively new mathematical tools — *conditional event algebra* (CEA) and the more general *relational event algebra* (REA) — can be used to legitimize a and b in eq. (1.1): CEA, developed prior to REA, yields the existence and construction of such a and b when functions f and g are identical to ordinary arithmetic division for two arguments $P(c), P(d)$ in Model 1 with $c \leq d$, and $P(e), P(h)$ in Model 2 with $e \leq h$, i.e., for conditional probabilities [18]. REA extends this to other classes of functions of probabilities, including, weighted linear combinations, weighted exponentials, and polynomials and series, among other functions [13].

Finally, it is of interest to be able to apply the above procedure when the models in question are provided through natural language descriptions. In this case, we first convert the natural language descriptions to a corresponding fuzzy logic one. Though any choice of fuzzy logic still yields a truth-functional logic, while probability logic is non-truth functional in general, it is interesting to note that the now well-developed one point random set coverage function representation of various types of fuzzy logic [10] bridges this gap: the logic of one point coverages is truth-functional (thanks to the ability to use Sklar's copula theorem here [28]). In turn, this structure fits the format of eq. (1.1) and one can then apply CEA and/or REA, as before, to obtain full evaluations of the natural event probabilistic distance-related functions and thus test hypotheses for similarity.

1.2. Overview of effort. Preliminary aspects of this work have been published in [13], [14]. But, this paper provides for the first time a unified, cohesive approach to the problem as stated in Section 1.1 for both direct probabilistic and natural language-based formulations. Section 1.3

provides some specific examples of models which can be tested for similarity. Section 1.4 gives some additional details on the computation and tightest bounds with respect to individual event probabilities of some basic probability distance functions when the conjunctive probability is not available. Section 1.5 summarizes the distributions of these probability distance functions as test statistics and the associated tests of hypotheses. Sections 2 and 3 give summaries of CEA and REA, respectively, while Section 4 provides the background for one point random set coverage representations of fuzzy logic. Section 5 shows how CEA can be combined with one point coverage theory to yield a sound and computable extension of CEA and (unconditional) fuzzy logic to conditional fuzzy logic. Finally, Section 6 reconsiders the examples presented in Section 1.3 and sketches implementation for testing similarity hypotheses.

1.3. Some examples of models. The following three examples provide some particular illustrations of the fundamental problem.

EXAMPLE 1.1 (MODELS AS WEIGHTED LINEAR FUNCTIONS OF PROBABILITIES OF POSSIBLY OVERLAPPING EVENTS). Consider the estimation of the probability of enemy attack tomorrow at the shore from two different experts who take into account the contributing probabilities of good weather holding, a calm sea state, and the enemy having an adequate supply of type 1 weapons. In the simplest kind of modeling, the experts may provide their respective probabilities as weighted sums of contributing probabilities of possibly non-overlapping events

$$(1.3)\quad \begin{cases} \text{Model 1: } P(a) = (w_{11}P(c)) + (w_{12}P(d)) + (w_{13}P(e)) \\ \text{vs.} \\ \text{Model 2: } P(b) = (w_{21}P(c)) + (w_{22}P(d)) + (w_{23}P(e)) \,, \end{cases}$$

where $0 \leq w_{ij} \leq 1, w_{i1} + w_{i2} + w_{i3} = 1, i = 1,2$; $a =$ enemy attacks tomorrow, according to Expert 1; $b =$ enemy attacks tomorrow, according to Expert 2; $c =$ good weather will hold; $d =$ calm sea state will hold; $e =$ enemy has adequate supply of type 1 weapons. Note again that c, d, e in general are not disjoint events so that the total probability theorem is not applicable here. It can be readily shown no solution exists independent of all choices of P in eq. (1.3) when a, b, c, d, e all belong literally to the *same* probability space. □

EXAMPLE 1.2 (MODELS AS CONDITIONAL PROBABILITIES). Here,

$$(1.4)\quad \begin{cases} \text{Model 1: } P(a) = P(c|d) \ (= P(c\&d)/P(d)) \\ \text{vs.} \\ \text{Model 2: } P(b) = P(c|e) \ (= P(c\&e)/P(e)) \,. \end{cases}$$

Models 1 and 2 could represent, e.g., two inference rules "if d, then c" "if e, then c" or two posterior descriptions of parameter c via different data sources corresponding to events d, e. Lewis' Theorem ([22] – see also comments in Section 2.1 here) directly shows that, in general, no possible a, b, c, d, e can exist in the same probability space, independent of the choice of probability measure. □

EXAMPLE 1.3 (MODELS AS NATURAL LANGUAGE DESCRIPTIONS). In this case, two experts independently provide their opinions concerning the same situation of interest: namely the description of an enemy ship relative to length and visible weaponry of a certain type.

(1.5)
- Model 1: Ship A is very long, or has a large number of q–type weapons on deck

  vs.

- Model 2: Ship A is fairly long, or if intelligence source 1 is reasonably accurate, it has a medium quantity of q–type weapons on deck

Translating the natural language form in eq. (1.5) to fuzzy logic form [4]:

$$(1.6)\begin{cases} \text{Model 1: } t(a) = (f_{long}(lngth(A)))^2 \vee_1 f_{large}(\#(Q)) \\ \text{vs.} \\ \text{Model 2: } t(b) = (f_{long}(lngth(A)))^{1.5} \vee_2 (f_{medium}|f_{accurate})(\#(Q), L) \end{cases}$$

where $\vee_1$ and $\vee_2$ are appropriately chosen fuzzy logic disjunction operators over $[0,1]^2$. Also, $f_c : D \to [0,1]$ denotes a fuzzy set membership function corresponding to attribute c; $f_{long} : \mathbb{R}^+ \to [0,1]$, $f_{large} : \mathbb{R}^+ \to [0,1]$, $f_{medium} : \{0,1,2,3,\ldots\} \to [0,1]$, $f_{accurate}$: class of intelligence sources $\to [0,1]$ are appropriately determined fuzzy set membership functions *representing* the attributes "long," "large," "medium," and "accurate," respectively; $(f_{medium}|f_{accurate}) : \{0,1,2,3,\ldots\}\times$ class of intelligence sources $\to [0,1]$ is a *conditional* fuzzy set membership function (to be considered in more detail in Section 5) representing the "if-then" statement. Also, $t(a) =$ truth or possibility of the description of ship A using Model 1; $t(b) =$ truth or possibility of the description of ship A using Model 2; where $A =$ ship A, $Q =$ collection of q–type weapons on deck of A, $L =$ intelligence source l; and measurement functions $lngth(\ ) =$ length of ( ) in feet, $\#(\ ) =$ no. of ( ).

In this example the issue of determining whether one could find actual events a, b, such that they and all contributing events lie in the same probability space requires first the conversion of the fuzzy logic models in eq. (1.6) to probability form. This is seen to be possible for both the unconditional and conditional cases via the one point random set coverage representation of fuzzy sets and certain fuzzy logics. (For the unconditional case, see Section 4; for the conditional case, see Section 5.) Thus, Lewis' result is applicable, showing a negative answer to the above question. □

Hence, all three examples again point up the need — if such constructions can be accomplished — to obtain an appropriate probability space *properly extending* the original given one where events a, b can be found, as well as the isomorphic imbedding of the contributing events (but, not the original events!) in eq. (1.1), independent of the choice of the given probability measure.

1.4. Probability distance functions. We summarize here some basic candidate probability distance functions $M_P(a, b)$ for any events a, b belonging to a probability space (Ω, B, P). The absolute distance $D_{P_0}(a, b)$ has already been defined in eq. (1.2) for a, b assumed to belong to a probability space (Ω_0, B_0, P_0) extending (Ω, B, P). For simplicity here, we consider any a, b belonging to probability space (Ω, B, P). First, the *naïve distance* $N_P(a, b)$ is given by the absolute difference of probabilities

$$N_P(a,b) = |P(a) - P(b)| = |P(a'\&b) - P(a\&b')|. \tag{1.7}$$

Here, there is no need to determine $P(a\&b)$ and clearly N_P is a pseudometric relative to probability space (Ω, B, P). On the other hand, a chief drawback is that there can be many events a, b which have probabilities near a half and are either disjoint or close to being disjoint, yet $N_P(a, b)$ is small, while $D_P(a, b)$ for such cases remains appropriately high. However, a drawback for the latter occurs when, in addition to a, b being nearly disjoint, both events have low probabilities, in which case $D_P(a, b)$ remains small, not reflecting the distinctness between the events. A perhaps more satisfactory distance function for this situation is the relative probability distance $R_P(a, b)$ given as

$$\begin{aligned} R_P(a,b) &= d_P(a,b)/P(a \vee b) \\ &= (P(a) + P(b) - 2P(a\&b))/(P(a) + P(b) - P(a\&b)) \\ &= (P(a'\&b) + P(a\&b'))/(P(a'\&b) + P(a\&b') + P(a\&b))\ , \end{aligned} \tag{1.8}$$

noting that the last example of near-disjoint small probability events yields a value of $R_P(a, b)$ close to unity, not zero as for $D_P(a, b)$. Note also in eqs. (1.7), (1.8) the existence of both the relative atom and the marginal-conjunction forms. It can be shown (using a tedious relative atom argument) that R_P is also a pseudometric relative to probability space (Ω, B, P), just as D_P is. (Another probability "distance" function is a symmetrization of conditional probability [13].)

Various tradeoffs for the use of each of the above functions can be compiled. In addition, one can pose a number of questions concerning the characterization of these and possibly other probability distance functions ([13], Sections 1, 2).

1.5. Additional properties of probability distance functions and tests of hypotheses. First, it should be remarked that eqs. (1.2), (1.8), again point out that full computations of $D_P(a,b), R_P(a,b), E_P(a,b)$ (but, of course not $N_P(a,b)$) require knowledge of the two marginal probabilities $P(a), P(b)$, *as well as the conjunctive probability* $P(a\&b)$. When the latter is missing, we can consider the well-known extended Fréchet–Hailperin tightest bounds [20], [3] in terms of the marginal probabilities for $P(a\&b)$ and $P(a \text{ v } b)$:

$$
\begin{aligned}
max(P(a)+P(b)-1,0) &\leq P(a\&b)\\
&\leq min(P(a),P(b))\\
&\leq wP(a)+(1-w)P(b)\\
&\leq max(P(a),P(b))\\
&\leq P(a \,\text{v}\, b)\\
&\leq min(P(a)+P(b),1),
\end{aligned}
\tag{1.9}
$$

for any weight w, $0 \leq w \leq 1$. In turn, applying inequality (1.9) to eqs. (1.2), (1.8) yields the corresponding tightest bounds on the computations of the probability distance functions as

$$N_P(a,b) \leq D_P(a,b) \leq min(P(a)+P(b), 2-P(a)-P(b)), \tag{1.10}$$

$$1-min\left(\frac{P(a)}{P(b)}, \frac{P(b)}{P(a)}\right) \leq R_P(a,b) \leq min(1, 2-P(a)-P(b)). \tag{1.11}$$

Inspection of inequalities (1.10) and (1.11) shows that considerable errors can be made in estimating the probability distance functions when $P(a\&b)$ is not obtainable. In effect, one of the roles played by CEA and REA is to address this issue through the determination of such conjunctive probabilities when a and b represent complex models as in Examples 1.1–1.3. (See Section 6.)

Eqs. (1.2), (1.7), and (1.8) also show that these probability distance functions can be expressed as functions of the relative atomic forms $P(a\&b)$, $P(a'\&b), P(a\&b')$. Then, making a basic higher order probability assumption that these three quantities can be considered also with respect to different choices of P, a, and b as random variables are jointly uniformly distributed over the natural simplex of values $\{(s,t,u) : 0 \leq s,t,u \leq 1, s+t+u \leq 1\} \subseteq [0,1]^3$, when $a \neq b$, one can then readily derive by standard transformation of probability techniques the corresponding cdf F_M for each function $M = N, D, R, E$. Thus,

$$F_N(t) = 1-(1-t)^3, \quad F_D(t) = t^2(3-2t), \quad F_R(t) = t^2, \tag{1.12}$$

for all $0 \leq t \leq 1$. To apply the above to testing hypotheses, we simply proceed in the usual way, where the null hypothesis is $H_0 : a \neq b$ and the alternative is $H_1 : a = b$. Here, for any observed (i.e., fully computable) probability distance function $M_P(a,b)$, we

$$\text{(1.13)} \qquad \begin{cases} \text{accept } H_0 \text{ (and reject } H_1\text{) iff } M_p(a,b) > C_\alpha \\ \text{accept } H_1 \text{ (and reject } H_0\text{) iff } M_p(a,b) \le C_\alpha \text{ ,} \end{cases}$$

where threshold C_α is pre-determined by the significance (or type-one error) level

$$\text{(1.14)} \qquad \alpha = P(\text{reject } H_0 \,|\, H_0 \text{ true}) = F_{M_P}(C_\alpha).$$

Thus, for all similar tests using the same statistic outcome $M_P(a,b)$, but possibly differing significance levels α (and hence thresholds C_α), considering the fixed significant level

$$\text{(1.15)} \quad \alpha = F_m(M_P(a,b)) = P(\text{reject } H_0 \text{ using } M_P(a,b)|H_0 \text{ holds}),$$

$$\text{(1.16)} \qquad \begin{cases} \text{If significance level } \alpha < \alpha_0, \text{ then accept } H_0 \\ \text{If significance level } \alpha > \alpha_0, \text{ then accept } H_1. \end{cases}$$

2. Conditional event algebra. Conditional event algebra is concerned with the following problem: Given any probability space (Ω, B, P), find a space (B_0, P_0) and an associated mapping $\psi : B^2 \to B_0$ — with B_0 (and well-defined operations over it representing conjunction, disjunction, and negation in some sense) and ψ not dependent on any particular choice of P — such that $\psi(\cdot, \Omega) : B \to B_0$ is an isomorphic imbedding and the following compatibility condition holds with respect to conditional probability:

$$\text{(2.1)} \qquad P_0(\psi(a,b)) = P(a|b), \qquad \text{for all } a,b \text{ in } B, \text{ with } P(b) > 0.$$

When $(B_0, P_0; \psi)$ exists, call it a *conditional event algebra* (CEA) extending (Ω, B, P) and call each $\psi(a,b)$ a *conditional event.* For convenience, use the notation $(a|b)$ for $\psi(a,b)$. When Ω_0 exists such that (Ω_0, B_0, P_0) is a probability space extending (Ω, B, P), call $(\Omega_0, B_0, P_0; \psi)$ a *Boolean* CEA extension. Finally, call any Boolean CEA extension with $(\Omega_0, B_0, P_0) = (\Omega, B, P)$ a *trivializing* extension.

A basic motivation for developing CEA is as follows: Note first that a natural numerical assignment of uncertainty to an inference rule or conditional statement in the form "if b, then a" (or "a, given b" or perhaps even "a is partially caused by b") when a, b are events (denoted usually as the consequent, antecedent, respectively) belonging to a probability space (Ω, B, P) is the conditional probability $P(a|b)$, i.e., formally (noting a can be replaced by $a\&b$)

$$\text{(2.2)} \qquad P(\text{if } b, \text{ then } a) = P(a|b).$$

(A number of individuals have considered this assignment as a natural one. See, e.g., Stalnaker [29], Adams [1], Rowe [26].) Then, analogous to the use of ordinary unconditional probability logic, which employs probability assignments for any well-defined logical/Boolean operations on events, it is also natural to inquire if a "conditional probability logic" (or CEA), can be derived, based on sound principles, which is applicable to inference rules.

2.1. Additional general comments. For the following special cases, the problem of constructing a specific CEA extension of the original probability space can actually be avoided:

1. *All antecedents are identical, with consequents possibly varying.* In this case, the traditional development of conditional probability theory is adequate to handle computations such as the conjunctions, disjunctions and negations applied to the statements "if b, then a," "if b, then c," where, similar to the interpretation in eq. (2.2), assuming $P(b) > 0$

$$P(\text{if } b, \text{ then } a) = P(a|b) = P_b(a),\ P(\text{if } b, \text{ then } c) = P(c|b) = P_b(c); \tag{2.3}$$

$$P((\text{if } b, \text{ then } a)\&(\text{if } b, \text{ then } c)) = P(a\&c|b) = P_b(a\&c), \tag{2.4}$$

$$P((\text{if } b, \text{ then } a) \text{ v } (\text{if } b, \text{ then } c)) = P(a \text{ v } c|b) = P_b(a \text{ v } c), \tag{2.5}$$

$$P(\text{not}(\text{if } b, \text{ then } a)) = P(a'|b) = P_b(a'), \tag{2.6}$$

where P_b is the standard notation for the conditional probability operator $P(\cdot\,|b)$, a legitimate probability measure over all of B, but also restricted, without loss of generality, to the trace Boolean (or sigma-) algebra $b\&B = \{b\&d : d \text{ in } B\}$. A further special case of this situation is when all antecedents are identical to Ω, so that all unconditional statements or events such as a, c can be interpreted as the special conditionals "if Ω, then a," "if Ω, then c," respectively.

2. *Conditional statements are assumed statistically independent, with differing antecedents.* When it seems intuitively obvious that the conditional expressions in question should not be considered dependent on each other, one can make the reasonable assumption that, with the analogue of eq. (2.3) holding,

$$P(\text{if } b, \text{ then } a) = P(a|b), \quad P(\text{if } d, \text{ then } c) = P(c|d), \tag{2.7}$$

the usual laws of probability are applicable and

$$P((\text{if } b, \text{ then } a)\&(\text{if } d, \text{ then } c)) = P(a|b)P(c|d), \tag{2.8}$$

$$P((\text{if } b, \text{ then } a) \text{ v } (\text{if } d, \text{ then } c)) = P(a|b) + P(c|d) - (P(a|b)P(c|d)). \tag{2.9}$$

A reasonable sufficient condition to assume independence of the conditional expressions is that $a\&b$, b are P–independent of $c\&d$, d (for each of four possible combinations of pairs).

3. *While the actual structure of a specific CEA may not be known, it is reasonable to assume that a Boolean one exists, so that all laws of probability are applicable.* Rowe tacitly makes this assumption in his work ([26], Chapter 8) in applying parts of the Fréchet–Hailperin bounds — as well as further P–independence reductions in the spirit of Comment 2 above. Thus, inequality (1.9) applied formally to conditionals "if b, then a," "if d, then c" with compatibility relation (2.7) yields the following bounds in terms of the marginal conditional probabilities $P(a|b), P(c|d)$, assuming $P(b), P(d) > 0$:

$$
\begin{aligned}
\max(P(a|b) + P(c|d) - 1, 0) &\leq P((\text{if } b, \text{then } a)\&(\text{if } d, \text{then } c)) \\
&\leq \min(P(a|b), P(c|d)) \\
&\leq (wP(a|b)) + ((1-w)P(c|d)) \\
&\leq \max(P(a|b), P(c|d)) \\
&\leq P((\text{if } b, \text{then } a) \text{ v } (\text{if } d, \text{then } c)) \\
&\leq \min(P(a|b) + P(c|d), 1)\ .
\end{aligned} \tag{2.10}
$$

Again, apropos to earlier comments, inspection of eq. (2.10) shows that considerable errors can arise in not being able to determine specific probabilistic conjunctions and disjunctions via some CEA.

At first glance one may propose that there already exists a candidate within classical logic which can generate a trivializing CEA: the material conditional operator $\Rightarrow$ where, as usual, for any two events a, b belonging to probability space (Ω, B, P), $b \Rightarrow a = b' \text{ v } a = b' \text{ v } (a\&b)$. However, note that [7]

$$
P(b \Rightarrow a) = 1 - P(b) + P(a\&b) = P(a|b) + (P(a'|b)P(b'))P(a|b), \tag{2.11}
$$

with strict inequality holding in general, unless $P(b) = 1$ or $P(a|b) = 1$. In fact, Lewis proved the fundamental negative result: (See also the recent work [5].)

THEOREM 2.1 (D. LEWIS [22]). In general, there does not exist any trivializing CEA.

Nevertheless, this result does not preclude non-trivializing Boolean and other CEAs from existing. Despite many positive properties [19], the chief drawback of previously proposed CEA (all using three-valued logic approaches) is their non-Boolean structure, and consequent incompatibility with many of the standard laws and extensive results of probability theory. For a history of the development of non-Boolean CEA up to five years ago, again see Goodman [7].

2.2. PSCEA: A non–trivializing Boolean CEA. Three groups independently derived a similar non-trivializing Boolean CEA: Van Fraasen, utilizing "Stalnaker Bernoulli conditionals" [30]; McGee, motivated by utility/rational betting considerations [23]; and Goodman and Nguyen [17], utilizing an algebraic analogue with arithmetic division as an infinite series and following up a comment of Bamber [2] concerning the representation of conditional probabilities as unconditional infinite trial branching processes. In short, this CEA, which we call the product space CEA (or PSCEA, for short), is constructed as follows (see [18] for further details and proofs): Let (Ω, B, P) be a given probability space. Then, form its extension (Ω_0, B_0, P_0) and mapping ψ: $B^2 \to B_0$ by defining (Ω_0, B_0, P_0) as that product probability space formed out of a countable infinity of copies of (Ω, B, P) (as its identical marginal). Hence, $\Omega_0 = \Omega \times \Omega \times \Omega \times \cdots$; $B_0 =$ sigma-algebra generated by $(B \times B \times B \times \cdots)$, etc. Define also, for any a, b in B, the conditional event $(a|b)$ as:

$$(2.12)\quad (a|b) = (a\&b|b) = V_{\infty}^{j=0}((b')^j \times (a\&b) \times \Omega_0) \qquad \text{(direct form)}$$

$$(2.13)\quad = V_{j=0}^{k-1}((b')^j \times (a\&b) \times \Omega_0) \text{ v } ((b')^k \times (a|b)), \quad k = 1, 2, 3, \ldots \qquad \text{(recursive form)},$$

where the exponential-Cartesian product notation holds for any $c, d \in B$

$$c^j \times d^k = \begin{cases} \underbrace{c \times c \times \cdots \times c}_{j \text{ factors}} \times \underbrace{d \times d \times \cdots \times d}_{k \text{ factors}}, & \text{if } j, k \text{ are positive integers} \\ \underbrace{c \times c \times \cdots \times c}_{j \text{ factors}}, & \text{if } j \text{ is a positive integer and } k = 0 \\ \underbrace{d \times d \times \cdots \times d}_{k \text{ factors}}, & \text{if } j = 0 \text{ and } k \text{ is a positive integer.} \end{cases} \tag{2.14}$$

It follows from eq. (2.12) that the ordinary membership function $\phi(a|b) : \Omega \to 0, 1$ (unlike the three-valued ones corresponding to previously proposed CEA) corresponding to $(a|b)$ is given for any $\underline{\omega} = (\omega_1, \omega_2, \omega_3, \ldots)$ in Ω_0, where for any positive integer j, eq. (2.15) implies eq. (2.16):

$$(2.15)\quad \phi(b)(\omega_1) = \phi(b)(\omega_2) = \cdots = \phi(b)(\omega_{j-1}) = 0 < \phi(b)(\omega_j) \ (= 1),$$

$$(2.16)\quad \phi(a|b)(\underline{\omega}) = \phi(a|b)(\omega_j) \ (= \phi(a)(\omega_j)).$$

The natural isomorphic imbedding here of B into B_0 is simply:

$$(2.17)\quad a \leftrightarrow (a|\Omega) = a \times \Omega_0, \text{ for all } a \in B,$$

and, indeed, for all $a, b \in B$ with $P(b) > 0$, $P_0((a|b)) = P(a|b)$.

2.3. Basic properties of PSCEA. A brief listing of properties of PSCEA is provided below, valid for any given probability space (Ω, B, P) and any $a, b, c, d \in B$, all derivable from use of the basic recursive definition for $k = 1$ in eq. (2.13) and the structure of (Ω_0, B_0, P_0) (again, see [18]):

(i) *Fixed antecedent combinations compatible with Comment 1 of Section 2.1.*

$$(a|b)\&(c|b) = (a\&c|b),\ (a|b) \vee (c|b) = (a \vee c|b),\ (a|b)' = (a'|b), \tag{2.18}$$

$$P_0((a|b)\&(c|b)) = P(a\&c|b),\ P_0((a|b) \vee (c|b)) = P(a \vee c|b), \tag{2.19}$$

$$P_0((a|b)') = P(a'|b) = 1 - P(a|b) = 1 - P_0((a|b)). \tag{2.20}$$

(ii) *Binary logical combinations.* The following are all extendible to any number of arguments where

$$(a|b)\&(c|d) = (A|b \vee d),\ (a|b) \vee (c|d) = (B|b \vee d) \text{ (formalisms)}, \tag{2.21}$$

$$A = (a\&b\&c\&d) \vee ((a\&b\&d') \times (c|d)) \vee ((b'\&c\&d) \times (a|b)), \tag{2.22}$$

$$B = (a\&b) \vee (c\&d) \vee ((a'\&b\&d') \times (c|d)) \vee ((b'\&c'\&d) \times (a|b)), \tag{2.23}$$

$$P_0((a|b)\&(c|d)) = P_0(A)/P(b \vee d), \tag{2.24}$$

$$\begin{aligned} P_0((a|b) \vee (c|d)) &= P_0(B)/P(b \vee d) \\ &= P(a|b) + P(c|d - P_0((a|b)\&(c|d)), \end{aligned} \tag{2.25}$$

$$P_0(A) = P(a\&b\&c\&d) + (P(a\&b\&d')P(c|d)) + (P(b'\&c\&d)P(a|b)), \tag{2.26}$$

$$\begin{aligned} P_0(B) = P((a\&b) \vee (c\&d)) &+ (P(a'\&b\&d')P(c|d)) \\ &+ (P(b'\&c'\&d)P(a|b)). \end{aligned} \tag{2.27}$$

In particular, note the combinations of an unconditional and conditional

$$P_0((a|b)\&(c|\Omega)) = P(a\&b\&c) + (P(b'\&c)P(a|b)), \tag{2.28}$$

$$(a|b)\&(b|\Omega) = (a\&b|\Omega) \qquad \text{(modus ponens)}, \tag{2.29}$$

whence $(a|b)$ and $(b|\Omega)$ are necessarily always P_0–independent.

(iii) *Other properties including: Higher order conditioning; partial ordering, extending unconditional event ordering; compatibility with probability ordering.*

2.4. Additional key properties of PSCEA. (Once more, see [18] for other results and all proofs.) In the following, $a, b, c, d, a_j, b_j \in B$, $j = 1, \ldots, n$, $n = 1, 2, 3, \ldots$. Apropos to Comment 2 in Section 2.1 concerning the sufficiency assumption of independence of two conditional events), PSCEA satisfies:

(iv) *Sufficiency for P_0–independence.* If $a\&b, b$ are (four-way) P–independent of $c\&d, d$, then $(a|b)$ is P_0–independent of $(c|d)$.

(v) *General independence property.* $(a_1|b_1), \ldots, (a_n|b_n)$ is always P_0–independent of $(b_1 \text{ v} \cdots \text{ v } b_n|\Omega)$. (This can be extended to show $(a_1|b_1), \ldots,$ $(a_n|b_n)$ is always P_0–independent of any $(c|\Omega)$, where $c \geq b_1 \text{ v} \cdots \text{ v } b_n$ or $c \leq (b_1)'\& \cdots \&(b_n)'$.)

(vi) *Characterization of PSCEA among all Boolean CEA:*

THEOREM 2.2 (GOODMAN AND NGUYEN [18], PP. 296–301). Any Boolean CEA which satisfies modus ponens and the general independence property must coincide with PSCEA, up to probability evaluations of all well-defined finite logical combinations (under &, v, ( )′) of conditional events.

(vii) *Compatibility between conditioning of measurable mappings and PSCEA conditional events.* Let $(\Omega_1, B_1, P_1) \xrightarrow{Z} (\Omega_2, B_2, P_1 \circ Z^{-1})$ indicate that $Z : \Omega_1 \to \Omega_2$ is a (B_1, B_2)–measurable mapping which induces probability space $(\Omega_2, B_2, P_1 \circ Z^{-1})$ from probability space (Ω_1, B_1, P_1). When some P_2 is used in place of $P_1 \circ Z^{-1}$, it is understood that $P_2 = P_1 \circ Z^{-1}$. Also, still using the notation $(\Omega_0, B_0, P_0; (\cdot \mid \cdot\cdot))$ to indicate the PSCEA extension of probability space (Ω, B, P), define the PSCEA *extension* of Z to be the mapping $Z_0 : (\Omega_1)_0 \to (\Omega_2)_0$, where

$$Z_0(\underline{\omega}) = (Z(\omega_1), Z(\omega_2), Z(\omega_3), \ldots), \quad \text{for all } \underline{\omega} = (\omega_1, \omega_2, \omega_3, \ldots) \in (\Omega_1)_0 \,. \tag{2.30}$$

LEMMA 2.1 (RESTATEMENT OF GOODMAN AND NGUYEN [18], SECTIONS 3.1, 3.4). If $(\Omega_1, B_1, P_1) \xrightarrow{Z} (\Omega_2, B_2, P_1 \circ Z^{-1})$ holds, then does $((\Omega_1)_0,$ $(B_1)_0, (P_1)_0) \xrightarrow{Z_0} ((\Omega_2)_0, (B_2)_0, (P_1 \circ Z^{-1})_0)$ hold (where we can naturally identify $(P_1)_0 \circ Z_0^{-1}$ with $(P_1 \circ Z^{-1})_0$) and say that $(\ )_0$ lifts Z to Z_0.

Next, replace Z by a joint measurable mapping X, Y, in eq. (2.30), where (Ω_1, B_1, P_1) is simply (Ω, B, P), Ω_2 is replaced by $\Omega_1 \times \Omega_2, B_2$ by the sigma-algebra generated by $(B_1 \times B_2)$, and define

$$(X, Y)(\omega) = (X(\omega), Y(\omega)), \quad \text{for any } \omega \in \Omega. \tag{2.31}$$

Thus, we have the following commutative diagram:

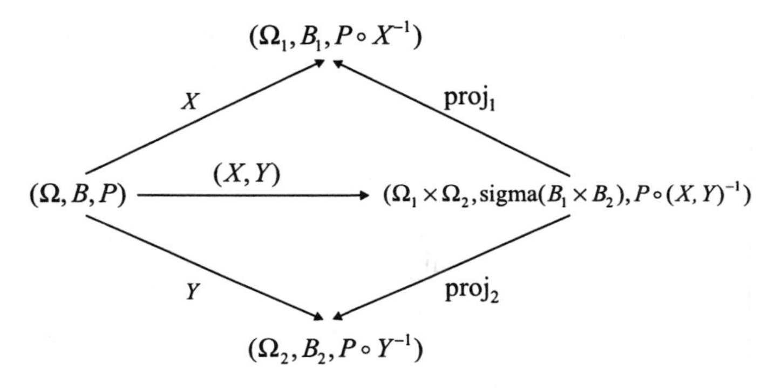

(2.32)

Finally, consider any $a \in B_1$ and $b \in B_2$ and the corresponding conditional event in the form $(a \times b|\Omega_1 \times b)$. Then, the following holds, using Lemma 2.1 and the basic structure of PSCEA:

THEOREM 2.3 (CLARIFICATION OF GOODMAN AND NGUYEN [18], SECTIONS 3.1, 3.4). The commutative diagram of arbitrary joint measurable mappings in eq. (2.32) lifts to the commutative diagram:

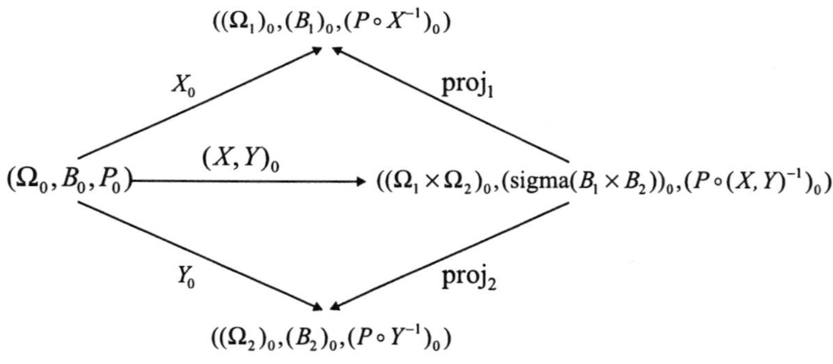

(2.33)

where, we can naturally identify $(\Omega_1 \times \Omega_2)_0$ with $(\Omega_1)_0 \times (\Omega_2)_0$, $(\text{sigma}(B_1 \times B_2))_0$ with $\text{sigma}((B_1)_0 \times (B_2)_0)$, $(P \circ X^{-1})_0$ with $P_0 \circ X_0^{-1}$, $(P \circ Y^{-1})_0$ with $P_0 \circ Y_0^{-1}$, and $(P \circ (X,Y)^{-1})_0$ with $P_0 \circ (X_0, Y_0)^{-1}$. Moreover, the basic compatibility relations always hold between conditioning of measurable mappings in unconditional events and joint measurable mappings in conditional events:

$$
\begin{aligned}
&P(X \in a|Y \in b) \ \ (= P(X^{-1}(a)|Y^{-1}(b))) = \\
(2.34)\quad &= P_0((X,Y)_0 \in (a \times b|\Omega_1 \times b)) \ \ (= P_0((X,Y)_0^{-1}((a \times b|\Omega_1 \times b)))),
\end{aligned}
$$

for all $a \in B_1$, $b \in B_2$.

3. Relational event algebra. The relational event algebra (REA) problem was stated informally in Section 1.1. More rigorously, given any two functions $f, g : [0,1]^m \to [0,1]$, and any probability space (Ω, B, P), find a probability space (Ω_0, B_0, P_0) extending (Ω, B, P) isomorphically (with B_0 not dependent on P) and find mappings $a(f), b(g) : B^m \to B_0$ such that the formal relations in eq. (1.1) are solvable where on the left-hand side P is replaced by P_0, a by $a(f)(c, d, e, h, \ldots)$, b by $b(g)(c, d, e, h, \ldots)$, with possibly some constraint on the $c, d, e, h, \ldots$ in B and the class of probability functions P. Solving the REA problem can be succinctly put as determining the commutative diagram:

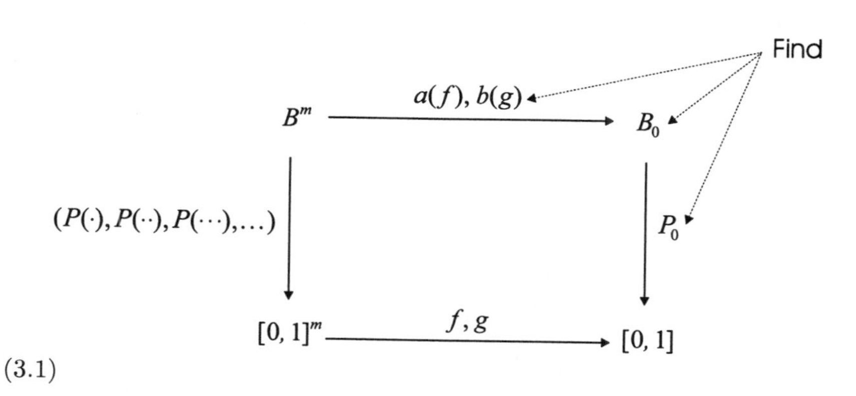

(3.1)

As mentioned before, the CEA problem is that special case of the REA problem, where f and g are each ordinary division of two probabilities with the restriction that each pair of probabilities corresponds to the first event being a subevent of the second. (See beginning of Section 2 and eq. (2.1).) Other special cases of the REA problem that have been treated include f, g being: constant-valued; weighted linear functions in multiple (common) arguments; polynomials or infinite series in one common argument; exponentials in one common argument; min and max. The first, second, and last cases are considered here; further details for all but the last case can be found in [13].

3.1. REA problem for constant–valued functions. To begin with, it is obvious that for any given measurable space (Ω, B), other than the events $\emptyset$ and Ω, there do not exist any other constant-probability-valued events belonging to probability space (Ω, B, P), independent of all possible choices of P. However, by considering the extensions (Ω_0, B_0, P_0), in a modified sense such events can be constructed. Consider first eq. (1.1) where f and g are any constants. Let probability space (Ω, B, P) be given as before and consider any real numbers s, t in [0,1]. Next, independent of the choice of s, t, pick a fixed event, say c in B, with $0 < P(c) < 1$, and define for any integers $1 \leq j \leq k \leq n$,

$$\mu(j,n;c) = c^{j-1} \times c' \times c^{n-j}, \tag{3.2}$$

utilizing notation similar to that in eq. (2.14). Note that all $\mu(j,n;c)$, as j varies, are mutually disjoint with the identical product probability evaluation

$$P_n(\mu(j,n;c)) = (P(c))^{n-1}P(c'), \tag{3.3}$$

where the product probability space $(\Omega^n, \mathrm{sigma}(B^n), P_n)$ has n marginal spaces, each identical to (Ω, B, P) with PSCEA extension $((\Omega^n)_0$, $(\mathrm{sigma}(B_n))_0$, $(P_n)_0)$. Then, consider the following conditional event with evaluation due to eq. (3.3):

$$\begin{aligned} \theta(j,k,n;c) &= \Big(\bigvee_{i=j+1}^{k} \mu(i,n;c) \big| \bigvee_{i=1}^{n} \mu(i,n;c)\Big); \\ (P_n)_0(\theta(j,k,n;c)) &= \frac{(k-j)}{n}. \end{aligned} \tag{3.4}$$

In addition, it is readily shown that for any fixed n, as j, k vary freely, $1 \le j \le k \le n$, the set of all finite disjoint unions of *constant-probability* events $\theta(j,k,n;c)$ is closed with respect to all well-defined logical combinations for $((\Omega^n)_0, (\mathrm{sigma}(B^n))_0, (P_n)_0)$ and is not dependent upon the particular choice of integer n, event $c \in B$, nor probability P, except for the assumption $0 < P(c) < 1$. In order for constant-probability events which act in a universal way to exist — analogous to the role that the boundary constant-probability events $\emptyset$ and Ω_0 play — to accommodate all possible values of s,t simultaneously, we must let $n \to \infty$ (or be sufficiently large). One way of accomplishing this is to first form for each value of n, a new product probability space with the first factor being (Ω_0, B_0, P_0) and the second factor being $((\Omega^n)_0, (\mathrm{sigma}(B^n))_0, (P_n)_0)$ and then formally allow n to approach ∞. By a slight abuse of notation, we will formally identify this space and limiting process with (Ω_0, B_0, P_0). Finally, with all of what has been stated above, we can choose any sequences of rationals converging to s,t, such as $(j_n/n)_{n=1,2,3,\ldots} \to s$, $(k_n/n)_{n=1,2,3,\ldots} \to t$ and define

$$\theta(s,t) = \lim_{n\to\infty} \theta(j_n, k_n, n; c), \quad \theta(t) = \theta(0,t), \tag{3.5}$$

with the convention, boundary values, and evaluations

$$\theta(s,t) = \emptyset, \text{ if } s \ge t,\ \theta(0) = \emptyset,\ \theta(1) = \Omega_0, \tag{3.6}$$

$$P_0(\theta(s,t)) = \max(t-s, 0),\ P_0(\theta(t)) = t, \text{ all } 0 \le s,t \le 1. \tag{3.7}$$

Hence, an REA solution to $f = s$ and $g = t$ constants in [0,1], is simply to choose $a(f) = \theta(s)$, $b(g) = \theta(t)$. A summary of logical combination properties of such "constant-probability" events $\theta(s,t)$ and $\theta(t)$ are given below: For all $0 \le s_j \le t_j \le 1$, $0 \le s \le t \le 1$, all c_i, d_j in B, $i = 1,\dots,m$, $j = 1,\dots,n$, $m \le n$,

$$\theta(s_1,t_1) \,\&\, \theta(s_2,t_2) = \theta(\max(s_1,s_2),\ \min(t_1,t_2)), \tag{3.8}$$

$$\theta(t_1) \,\&\, \theta(t_2) = \theta(\min(t_1,t_2)),\ \theta(t_1) \text{ v } \theta(t_2) = \theta(\max(t_1,t_2)), \tag{3.9}$$

$$(\theta(t))' = \theta(t,1),\ (\theta(s))' \,\&\, \theta(t) = \theta(s,t), \tag{3.10}$$

$$\begin{aligned} &(c_1 \times \cdots \times c_m \times \theta(s_1,t_1)) \,\&\, (d_1 \times \cdots \times d_n \times \theta(s_2,t_2)) = \\ &= (c_1 \& d_1) \times \cdots \times (c_m \& d_m) \times d_{m+1} \times \cdots \times d_n \times (\theta(s_1,t_1) \& \theta(s_2,t_2)), \end{aligned} \tag{3.11}$$

with all obvious corresponding probability evaluations by P_0.

3.2. REA problem for weighted linear functions. Consider the REA problem where, as before, (Ω, B, P) is a given probability space with $c_j \in B$, $j = 1,\dots,m$. For all $\underline{t} = (t_1,\dots,t_m)$ in $[0,1]^m$, now define

$$f(\underline{t}) = w_{11}t_1 + \cdots + w_{1m}t_m,\ g(\underline{t}) = w_{21}t_1 + \cdots + w_{2m}t_m\ , \tag{3.12}$$

$$0 \le w_{ij} \le 1,\ w_{i1} + \cdots + w_{im} = 1,\ i = 1,2,\ j = 1,\dots,m. \tag{3.13}$$

The following holds for any real w_j, with disjoint $c_{\underline{q}}$ replacing non-disjoint c_j:

$$\begin{aligned} &\sum_{j=1}^{m} P(c_j) \cdot w_j = \sum_{\underline{q} \in J_m} P(c_{\underline{q}}) \cdot w_{\underline{q}}; \\ &J_m = \{\emptyset, \Omega\}^m \setminus \{(\Omega,\dots,\Omega)\},\ \underline{q} = (q_1,\dots,q_m), \\ &w_{\underline{q}} = \sum_{\{j:\ 1 \le j \le m \text{ and } q_j = \emptyset\}} w_j\ ,\ c_{\underline{q}} = (c_1 + q_1) \,\&\cdots\&\, (c_m + q_m). \end{aligned} \tag{3.14}$$

Then, the REA solution for this case using eq. (3.14) and constant-probability events as constructed in the last section is seen to consist of the following disjoint disjunctions of Cartesian products

$$\begin{aligned} &a(f)(\underline{c}) = \bigvee_{\underline{q} \in J_m} c_{\underline{q}} \times \theta(w_{\underline{q}}), \qquad b(g)(\underline{c}) = \bigvee_{\underline{q} \in J_m} c_{\underline{q}} \times \theta(w_{2\underline{q}}); \\ &c_{\underline{q}} = (c_1,\dots,c_m),\ w_{i\underline{q}} = \sum_{\{j:\ 1 \le j \le m \text{ and } q_j = \emptyset\}} w_{ij}. \end{aligned} \tag{3.15}$$

Some logical combinations of REA solutions here:

$$a(f)(\underline{c}) \,\&\, b(g)(\underline{c}) = \bigvee_{q \in J_m} c_q \times \theta(\min(w_{1\underline{q}}, w_{2\underline{q}})), \tag{3.16}$$

$$a(f)(\underline{c}) \vee b(g)(\underline{c}) = \bigvee_{q \in J_m} c_q \times \theta(\max(w_{1\underline{q}}, w_{2\underline{q}})), \tag{3.17}$$

$$(a(f)(\underline{c}))' = \bigvee_{q \in J_m} c_q \times \theta(w_{1\underline{q}}, 1) \vee (c_1' \,\&\cdots\&\, c_m'). \tag{3.18}$$

A typical example of the corresponding probability evaluations for (Ω_0, B_0, P_0) is

$$P_0[a(f)(\underline{c})\&b(g)(\underline{c})] = \sum_{\underline{q} \in J_m} c_{\underline{q}} \min(w_{1\underline{q}}, w_{2\underline{q}}). \tag{3.19}$$

Applications of the above results can be made to weighted coefficient polynomials and series in one variable (replacing in eq. (3.15) c_j by c^{j-1}), as well as for weighted combinations of exponentials in one or many arguments; but computational problems arise for the latter unless special cases are considered, such as independence of terms, etc. (see [15]).

3.3. REA problem for min, max. In this case, we consider in eq. (1.1) REA solutions when one or both functions f, g involve minimum or maximum operations. For simplicity, consider f by itself in the form

$$f(s,t) = \max(s,t), \quad \text{for all } s,t \text{ in } [0,1], \tag{3.20}$$

and seek for all events c, d belonging to probability space (Ω, B, P), a relational event $a(f)(c,d)$ belonging to probability space (Ω_0, B_0, P_0) where for all P

$$P_0(a(f)(c,d)) = \max(P(c), P(d)). \tag{3.21}$$

First, it should be remarked that it can be proven that we cannot apply any techniques related to Section 3.2 where the weights are *not* dependent on P. However, a reasonable *modified* REA solution is possible based on the idea of choosing weights dependent upon P resulting in the form d, when $P(c) < P(d)$, and c, when $P(d) < P(c)$, etc. Using the REA solution in Section 3.2, we have:

$$a(f)(c,d) = (c\&d) \vee ((c\&d') \times \theta(w_{P,1})) \vee ((c'\&d) \times \theta(w_{P,2})), \tag{3.22}$$

$$w_{P,1} = \begin{cases} 1, & \text{if } P(d) < P(c) \\ 0, & \text{if } P(c) < P(d), \\ w, & \text{if } P(c) = P(d) \end{cases} \quad w_{P,2} = \begin{cases} 0, & \text{if } P(d) < P(c) \\ 1, & \text{if } P(c) < P(d). \\ 1-w, & \text{if } P(c) = P(d) \end{cases} \tag{3.23}$$

Dually, the case for $f = \min$ is also solvable by this approach.

4. One point random set coverages and coverage functions. One point random set coverages (or hittings) and their associated probabilities, called one point random set coverage functions, are the weakest way to specify random sets, analogous to the role measures of central tendency play with respect to probability distributions. In this section, we show that there is a class of fuzzy logics and a corresponding class of joint random sets which possess homomorphic-like properties with respect to probabilities of certain logical combinations of one point random set coverages, thereby identifying a part of fuzzy logic as a weakened form of probability. In turn, this motivates the proposed definition for conditional fuzzy sets in Section 5. Previous work in this area can be found in [16], [9], [10].

4.1. Preliminaries. In the following development, we assume all sets D_j finite, $j \in J$, any finite index set, and repeatedly use the measurable map notation introduced in Section 2.4 (vii) applied to random sets and 0–1–valued random variables. As usual, the distinctness of a measurable map is up to the probability measure it induces. For any $x \in D_j$, denote the filter class on x with respect to D_j as $F_x(D_j) = \{c : x \in c \subseteq D_j\}$ and, as before, denote the ordinary set membership (or indicator) functional as ϕ. Denote the power class of any D_j as $\rho(D_j)$ and the double power class of D_j as $\rho\rho(D_j)$. If S_j is any random subset of D_j, written as $(\Omega, B, P) \xrightarrow{S_j} (\rho(D), \rho\rho(D), P \circ S_j^{-1})$, use the multivariable notation $S_J = (S_j)_{j \in J}$ to indicate a collection of joint random subsets S_j of $D_j, j \in J$. Similarly, define the corresponding collection of joint 0–1 random variables $\phi(S_J) = (\phi(S_j))_{j \in J}$, noting the (x, j)–marginal random variable here is $\phi(S_j)(x)$, for any $x \in D_j$, $j \in J$ yielding the relation $S_j \leftrightarrow \phi(S_J) = (\phi(S_j)(x))_{x \in D_j, j \in J}$. Let $\dagger$ denote the separating union:

$$\text{(4.1)} \qquad \dagger(D_J) = \bigcup_{j \in J} (D_j \times \{j\}), \qquad \dagger(\rho(D_J)) = \bigcup_{j \in J} (\rho(D_J) \times \{j\}).$$

The following commutative diagram summarizes the above relations for all $x \in D_j$, $j \in J$:

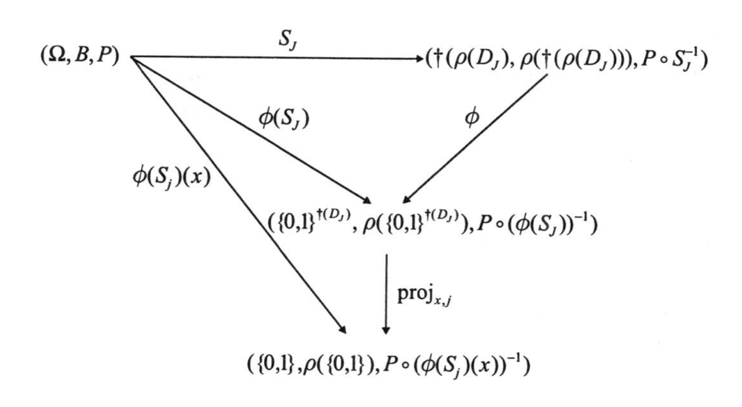

(4.2)

For each $x \in D_j$, note the equivalences of *one point random set coverages*:

$$(4.3)\quad \begin{cases} x \in S_j \text{ iff } S_j^{-1}(F_x(D_j)) \text{ occurs iff } \phi(S_j)(x) = 1 \\ x \notin S_j \text{ iff } S_j^{-1}(F_x(D_j)) \text{ does not occur iff } \phi(S_j)(x) = 0, \end{cases}$$

and for each S_j define its *one point coverage function* $f_j : D_j \to [0,1]$, a fuzzy set membership function, by

$$(4.4)\quad f_j(x) = P(x \in S_j) = P(S_j^{-1}(F_x(D_j))) = P((\phi(S_j))(x) = 1).$$

The induced probability measure $P_0 S_J^{-1}$ through $P_0\phi(S_j)^{-1}$ is completely determined by its corresponding joint probability function g_{f_J}, or equivalently, by its corresponding joint cdf F_{f_J} where $f_J = (f_j)_{j\in J} = (f_j(x))_{x\in D_j, j\in J}$. Also, use the multivariable notation $D_j = (D_j)_{j\in J}$, $f_j(x_j) = (f_j(x_j))_{j\in J}$, $x_J = (x_j)_{j\in J}$, $c_J = (c_j)_{j\in J}$, $D_j - c_j = (D_j - c_j)_{j\in J}$, etc. A copula, written cop: $[0,1]^n \to [0,1]$, is any joint cdf of one-dimensional marginal cdfs corresponding to the uniform distribution over [0,1], compatibly defined for $n = 1, 2, \ldots$ (see [3]). It will also be useful to consider the cocopula or DeMorgan transform of cop, i.e., $\text{cocop}(t_J) = 1 - \text{cop}(1_J - t_J)$, for all $t_J \in [0,1]^J$. By Sklar's copula theorem [28],

$$(4.5)\quad \begin{aligned} F_{f_J}(\underline{t}) &= \text{cop}((F_{f_j(x)}(t_{x,j}))_{x\in D_j, j\in J}), \\ \underline{t} &= (t_{x,j})_{x\in D_j, j\in J} \in [0,1]^{\dagger(D_J)} \end{aligned}$$

$$F_{f_j(x)}(t) = \begin{cases} 0, & \text{if } t < 0 \\ 1 - f_j(x), & \text{if } 0 \le t \le 1, \\ 1, & \text{if } 1 \le t \end{cases} \quad g_{f_j(x)}(t) = \begin{cases} f_j(x), & \text{if } t = 1 \\ 1 - f_j(x), & \text{if } t = 0, \end{cases}$$

(4.6)

where, in particular,

$$(4.7)\quad F_{f_j(x)}(0) = g_{f_j(x)}(0) = 1 - f_j(x), \quad \text{all } x \in D_j,\ j \in J.$$

The following is needed :

LEMMA 4.1 (A VERSION OF THE MÖBIUS INVERSION FORMULA [27]). Given any finite nonempty set J and a collection of joint 0–1–valued random variables $T_J = (T_j)_{j\in J}$, $(\Omega, B, P) \xrightarrow{T_j} (\{0,1\}, \rho(\{0,1\}), P \circ T_j^{-1})$, with $P(T_j = 1) = \lambda_j$, $P(T_j = 0) = 1 - \lambda_j, j \in J$, and given any set $L \subseteq J$, with the convention of letting $P(\&_{j\in\emptyset}(T_j = 0)) = \text{cop}((\lambda_j)_{j\in\emptyset}) = 1$ $\text{cocop}((\lambda_j)_{j\in\emptyset}) = 0$, we have that

$$(4.8)\quad \begin{aligned} &P\Big(\underset{j\in L}{\&}(T_j = 1)\ \&\ \underset{j\in J\smallsetminus L}{\&}(T_j = 0)\Big) = \\ &= \sum_{K\subseteq L} (-1)^{\text{card}(K)} P\Big(\underset{j\in K\cup(J\smallsetminus L)}{\&}(T_j = 0)\Big) = \overline{\text{cop}}\,(\lambda_L; \lambda_{J\smallsetminus L}), \end{aligned}$$

where we define

$$\overline{\mathrm{cop}}(\lambda_L;\lambda_{J\setminus L}) =$$

$$(4.9)\qquad \begin{aligned} &= \sum_{K\subseteq L} (-1)^{\mathrm{card}(K)} \ \mathrm{cop}((1-\lambda_j)_{j\in K\cup(J\setminus L)}) \\ &= \delta(L=\emptyset) + \sum_{K\subseteq L} (-1)^{\mathrm{card}(K)+1} \ \mathrm{cocop}((\lambda_j)_{j\in K\cup(J\setminus L)}), \end{aligned}$$

with δ being the Krönecker delta function. Note the special cases

$$(4.10)\quad \overline{\mathrm{cop}}(\lambda_J) = \overline{\mathrm{cop}}(\lambda_J;\lambda_\emptyset) = \sum_{\emptyset\neq K\subseteq J} (-1)^{\mathrm{card}(K)+1} \mathrm{cocop}((\lambda_j)_{j\in K}),$$

$$(4.11)\qquad \overline{\mathrm{cop}}(\lambda_\emptyset;\lambda_J) = 1 - \mathrm{cocop}((\lambda_j)_{j\in J} = \mathrm{cop}((1-\lambda_j)_{j\in J}.$$

4.2. Solution class of joint random sets one point coverage equivalent to given fuzzy sets. Next, given any collection of fuzzy set membership functions f_J, $f_j : D_j \to [0,1], j \in J$, consider $\mathcal{S}(f_J)$, the class of all collections of joint random subsets $S(f_J)$ of D_j which are *one point coverage equivalent to* f_J, i.e., each $S(f_j)$ is any random subset of D_j, which is *one point coverage equivalent to* f_j, $j \in J$, i.e.,

$$(4.12)\qquad P(x \in S(f_j)) = f_j(x), \quad \text{for all } x \in D_j, \quad j \in J\ ,$$

compatible with the converse relation in eq. (4.4). It can be shown that, when $f_j = \phi(a_j)$, $a_j \in B$, $j \in J$, then necessarily $\mathcal{S}(f_J) = \{a_J\}$, $a_J = (a_j)_{j\in J}$.

THEOREM 4.1 (GOODMAN [10]). Given any collection of fuzzy sets f_J, as above, $\mathcal{S}(f_J)$ is bijective to $\phi(\mathcal{S}(f_J)) = \{\phi(S(f_J))) : S(f_J) \in \mathcal{S}(f_J)\}$, which is bijective to $\{F_{f_J} = \mathrm{cop}((F_{f_{j(x)}})_{x\in D_j, j\in J}) : \mathrm{cop}\colon [0,1]^{\dagger D_J} \to [0,1]$ is arbitrary$\}$.

PROOF. This follows by noting that any choice of cop always makes F_{f_J} a legitimate cdf corresponding to fixed one point coverage functions f_J. □

By applying Lemma 4.1 and eq. (4.7), the explicit relation between each $S(f_J)$ and the choice of cop generating them is given, for any $c_J = (c_j)_{j\in J}$, $c_j \in B$, by

$$P(S(f_J) = c_J) =$$

$$(4.13)\qquad \begin{aligned} &= P(\mathop{\&}_{j\in J}(S(f_j) = c_j)) \\ &= P(\mathop{\&}_{x\in j, j\in J}(\phi(S(f_j))(x) = 1) \mathbin{\&} \mathop{\&}_{x\in D_j - c_j, j\in J}(\phi(S(f_j))(x)=0)) \\ &= \overline{\mathrm{cop}}\big(f_{\dagger(c_J)}(x_{\dagger(c_J)});\ f_{\dagger(D_J-c_J)}(x_{\dagger(D_J-c_J)})\big). \end{aligned}$$

In particular, when cop = min, one can easily show (appealing, e.g., to the unique determination of cdfs for 0–1–valued random variables by all joint evaluations at 0 — see Lemma 4.1 — and by eq. (4.5)) this is equivalent to choosing the *nested* random sets $S(f_j) = f_j^{-1}[U,1] = \{x : x \in D_j,\ f_j(x) \geq U\}$ as the one point coverage equivalent random sets for the *same fixed* U, a uniformly distributed random variable over [0,1]. When cop = prod (arithmetic product) the corresponding collection of joint random sets is such that $S(f_J)$ corresponds to $\phi(S(f_J))$ being a collection of independent 0–1 random variables, and hence $S(f_J)$ also corresponds to the maximal entropy solution in $\mathcal{S}(f_J)$.

4.3. Homomorphic–like relations between fuzzy logic and one point coverages. Call a pair of operators $\&_1, \mathrm{v}_1 : [0,1]^n \to [0,1]$, well-defined for $n = 1,2,3,\ldots$, a *fuzzy logic conjunction, disjunction pair*, if both operators are pointwise nondecreasing with $\&_1 \leq \mathrm{v}_1$, and for any $0 \leq t_j \leq 1$, $j = 1,\ldots,n$, letting $r = t_1 \&_1 \cdots \&_1 t_n$, $s = t_1 \mathrm{v}_1 \cdots \mathrm{v}_1 t_n$, if for any $t_j = 0$, then $r = 0$ and $s = t_1 \mathrm{v}_1 \cdots \mathrm{v}_1 t_{j-1} \mathrm{v}_1 t_{j+1} \mathrm{v}_1 \cdots \mathrm{v}_1 t_n$, and if any $t_j = 1$, then $s = 1$ and $r = t_1 \&_1 \cdots \&_1 t_{j-1} \&_1 t_{j+1} \&_1 \cdots \&_1 t_n$. Note that any cop, cocop pair qualifies as a fuzzy logic conjunction, disjunction pair, as does any *t–norm, t–conorm* pair (certain associative, commutative functions, see [16]). For example, (min, max), (prod, probsum) are two fuzzy conjunction, disjunction pairs which are also both cop, cocop and t–norm, t–conorm pairs, where probsum is defined as the DeMorgan transform of prod.

THEOREM 4.2. Let (cop, cocop) be arbitrary. Then (referring to eq. (4.10)):

(i) ($\overline{\text{cop}}$, cocop) is a conjunctive disjunctive fuzzy logic operation pair which in general is non–DeMorgan.

(ii) For any choice of fuzzy set membership functions symbolically, $f_J : D_J \to [0,1]^J$ and any $S(f_J) \in \mathcal{S}(f_J)$ determined through cop (as in Theorem 4.1), and any $x_J \in D_J$,

$$\begin{aligned} \overline{\text{cop}}(f_J(x_J)) &= P(\mathop{\&}_{j\in J}(x_j \in S(f_j))), \\ \text{cocop}(f_J(x_J)) &= P(\bigvee_{j\in J}(x_j \in S(f_j))). \end{aligned} \tag{4.14}$$

(iii) If (cop, cocop) is any continuous t–norm, t–conorm pair, then ($\overline{\text{cop}}$ = cop iff (cop, cocop) is either (min, max), (prod, probsum), or any ordinal sum of (prod, probsum).

PROOF. Part (i) follows from (ii). Part (ii) left-hand side follows from Lemma 4.1 with $T_j = \phi(S(f_j))$, $L = J$. (ii) right-hand side follows from the DeMorgan expansion of $P(\bigvee_{j\in J}(x_j \in S(f_j))) = 1 - P(\mathop{\&}_{j\in J}(\phi(S(f_j))(x_j) = 0))$ and the last result. (iii) follows from [10], Corollary 2.1. □

The validity for $\overline{\text{cop}}$ = cop without using the sufficiency conditions in (iii) above are apparently not known at present. Call the family of all (cop, cocop) pairs listed in Theorem 4.2 (iii), the *semi-distributive* family (see [10]), because of additional properties possessed by them. Call the much larger family of all ($\overline{\text{cop}}$, cocop) pairs the *alternating signed sum* family. $\overline{\text{cop}}$ is a more restricted function of cocop than a modular transform (i.e., when $\overline{\text{cop}}$ is evaluated at two arguments). In fact, Frank has found a non-trivial family characterizing all pairs of t–norms and t–conorms which are modular to each other, a proper subfamily of which consists of also copula, cocopula pairs which, in turn, properly contains the semi-distributive family [6]. When we choose as a fuzzy logic conjunction disjunction pair $(\&_1, \text{v}_1) = (\overline{\text{cop}}, \text{cocop})$, eq. (4.14) shows that there is a "homomorphic-like" relation for conjunctions and disjunctions separately between fuzzy logic and corresponding probabilities of conjunction and disjunctions of one point random set coverages. It is natural to inquire what other classes of fuzzy logic operators produce homomorphic-like relations between various well-defined *combinations* of conjunctions and disjunctions with corresponding probabilities of combinations of one point random set coverages. One such has been determined:

THEOREM 4.3 (GOODMAN [10]). Suppose $(\&_1, \text{v}_1)$ is any continuous conjunction, disjunction fuzzy logic pair. Then, the following are equivalent:

(i) For any choice of $f_{ij} : D_{ij} \to [0,1]$, $i = 1, \ldots, m$, $j = 1, \ldots, n$, m, $n \geq 1$, there is a collection of joint one point coverage equivalent random sets $S(f_{ij})$, $i = 1, \ldots, m$, $j = 1, \ldots, n$, such that, for all $x_{ij} \in D_{ij}$, the homomorphic-like relation holds:

$$\underset{i=1}{\overset{m}{\&_1}} \Big(\bigvee_{j=1}^{n}{}_1 (f_{ij}(x_{ij})) \Big) = P\Big(\underset{i=1}{\overset{m}{\&}} \Big(\bigvee_{j=1}^{n} (x_{ij} \in S(f_{ij})) \Big) \Big). \tag{4.15}$$

(ii) Same statement as (i), but with v_1 over $\&_1$ and v over &.

(iii) $(\&_1, \text{v}_1)$ is any member of the semi-distributive family.

By inspection, it is easily verified (using, e.g., the nested random set forms corresponding to min and the mutual independence property corresponding to prod) that the fuzzy logic operator pairs $(\&_1, \text{v}_1)$ = (min, max) and (prod, probsum) produce homomorphic-like relations between *all* well-defined finite combinations of these operators applied to fuzzy set membership values and probabilities of corresponding combinations of one point coverages. (The issue of whether ordinal sums also enjoy this property remains open.) However, it is also quickly shown for the case $D_j = D$, all $j \in J$, the pair (min, max) actually produces full conjunction-disjunction homomorphisms between arbitrary combinations of fuzzy set membership functions and corresponding combinations of one point coverage equivalent random sets. This fact was discovered in a different form many years ago by Negoita and Ralescu [24] in terms of non-random level sets, corresponding to the nested random sets discussed above. On the other hand, the pair (prod, probsum) can be ruled out of producing actual conjunction-disjunction homomorphisms by noting that the collection of all jointly independent $\phi(S(f))$ indexed by D, as $f : D \to [0,1]$ varies arbitrarily, is not even closed with respect to min, max (corresponding to set intersection and union). For example, note that, in general, and for any four 0–1 random variables T_j, $j = 1,2,3,4$, $P(min(T_1,T_2,T_3,T_4) = 0) \neq P(min(T_1,T_2) = 0)P(min(T_3,T_4) = 0)$.

4.4. Some applications of homomorphic–like representation to fuzzy logic concepts. The following concepts, in one related form or another, can be found in any comprehensive treatise on fuzzy logic [4], where ($\&_1$,v_1) is some chosen fuzzy logic conjunction, disjunction pair. However, in order to apply the homomorphic-like relations in Theorems 4.2 and 4.3, we now assume ($\&_1$,v_1) is any member of the alternating signed sum family.

(i) *Fuzzy negation.* Here, $1 - f$ is called the fuzzy negation of $f : D \to [0,1]$. By noting the almost trivial relation, $x \in S(1-f)$ iff not $(x \in (S(1-f))')$ the homomorphic relations presented in Theorems 4.2 and 4.3 can be reinterpreted to include negations. However, it is not true in general that $S(f) = (S(1-f))'$, even when they are generated by copulas in the semi-distributive family, compatible with the fact that fuzzy logic as a truth functional logic cannot be Boolean.

(ii) *Fuzzy logic projection.* For any $f : D_1 \times D_2 \to [0,1]$, $x \in D_1$, the fuzzy logic projection of f at x is

$$\text{(4.16)}\quad \text{fuzzy proj}_1(f)(x) = \bigvee_{x\in D_2}{}_1 \Big(f(x,y)\Big) = P\Big(\bigvee_{y\in D_2} \Big(x \in S(f(\cdot\,,y))\Big)\Big),$$

a probability projection of a corresponding random set.

(iii) *Fuzzy logic modifiers.* These correspond to "very," "more or less," "most," etc., where for any $f : D \to [0,1]$, modifier $h : [0,1] \to [0,1]$ is applied compositionally as $h \circ f : D \to [0,1]$. Then, for any $x \in D$

$$\text{(4.17)}\quad h(f(x)) = P(x \in S(h \circ f)) = P(f(x) \in S(h)) = P(x \in f^{-1}(S(h)))$$

also with obvious probability interpretations.

(iv) *Fuzzy extension principle.* In its simplest form, let f: $D \to [0,1]$, $g : D \to E$. Then the f–fuzzification of g is $g[f] : E \to [0,1]$ where at any $y \in E$

$$\text{(4.18)}\qquad g[f](y) = \bigvee_{x_J\in g^{-1}(y)}{}_1 (f(x)) = P(y \in g(S(f))),$$

the one point coverage function of the g function of a random set representing f. A more complicated situation occurs when f above is defined as the fuzzy Cartesian product $\times_1 f_J : \times D_J \to [0,1]$ of $f_j : D_j \to [0,1]$, $j \in J$, given at any $x_J \in [0,1]^J$ as $\times_1 f_J(x_J) = \&_1(f_J(x_J))$. Then, restricting $(\&_1, \mathrm{v}_1)$ to only be in the semi-distributive family, we have in place of (4.18)

$$
\begin{aligned}
g[f](y) &= \bigvee_{x_J \in g^{-1}(y)}{}_1 (\&_1(f_J(x_J)) \\
&= P\Big(\bigvee_{x_J \in g^{-1}(y)} \big(\mathop{\&}_{j \in J}(x_j \in S(f_j))\big)\Big) \\
&= P(y \in g(S(f_J))),
\end{aligned} \tag{4.19}
$$

a natural generalization of (4.18).

(v) *Fuzzy weighted averages.* For any $f_1 : D_1 \to [0,1]$, $f_2 : D_2 \to [0,1]$ and weight w, $0 \le w \le 1$, the fuzzy weighted average of f_1, f_2 at any $x \in D_1$, $y \in D_2$, is

$$
(wf_1(x)) + ((1-w)f_2(y)) = P_0(a), \tag{4.20}
$$

$$
\begin{aligned}
a = [(x \in S(f_1))\&(y \in S(f_2))] \ \mathrm{v} \ [(x \in S(f_1))\&(y \notin S(f_2)) \times \theta(w)] \\
\mathrm{v} \ [(x \notin S(f_1))\&(y \in S(f_2)) \times \theta(1-w)].
\end{aligned} \tag{4.21}
$$

This is achieved by application of the REA solution to weighted linear functions of probabilities (Section 3.2); in this case the latter are the one point coverage ones.

(vi) *Fuzzy membership functions of overall populations.* This is inspired by the well-known example in approximate reasoning: "Most blond Swedes are tall," where the membership functions corresponding to "blond & tall" and to "blond" are averaged over the population of Swedes and then one divides the first by the second to obtain the ratio of overall tallness to blondness, to which the fuzzy set corresponding to modifier "most" is then applied ([4], pp. 173–185). More generally, let $f_j : D \to [0,1]$, $j = 1,2$, and consider two measurable mappings

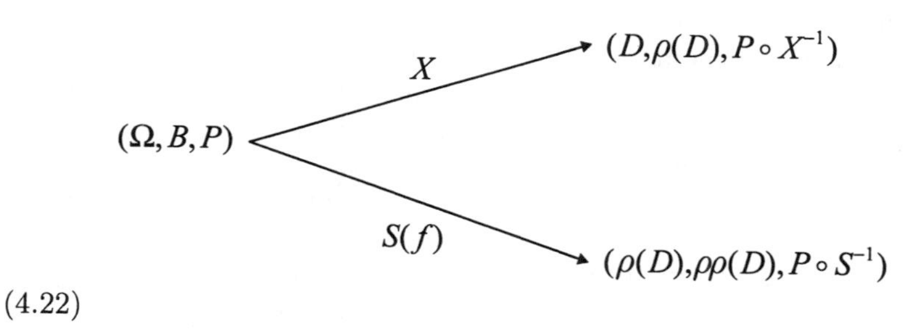

(4.22)

where X is a designated population weighting random variable. Then, the relevant fuzzy logic computations are, noting the similarity to Robbin's original application of the Fubini iterated integral theorem [25] and denoting expectation by $E_X(\ )$,

$$
\begin{aligned}
E_X(f_2(X)) &= \int_{x \in D} f_2(x)\, dP(X^{-1}(x)) \\
&= \int_{x \in D} \int_{\omega \in \Omega} \phi((S(f_2)(\omega))(x)\, dP(\omega)\, dP(X^{-1}(x)) \\
&= \int_{\omega \in \Omega} \int_{x \in D} \phi((S(f_2)(\omega))(x)\, dP(X^{-1}(x))\, dP(\omega) \qquad (4.23)\\
&= \int_{\omega \in \Omega} P(X \in S(f_2)(\omega))\, dP(\omega) \\
&= P(X \in S(f_2)),
\end{aligned}
$$

and similarly,

$$E_X(f_1(X)\&_1 f_2(X)) = P(X \in S(f_1 \&_1 f_2)) = P(X \in (S(f_1) \cap S(f_2))). \qquad (4.24)$$

Hence, the overall population tendency to attribute 1, given attribute 2, is, by using eq. (4.23) and eq. (4.24),

$$\frac{E_X(f_1(X)\&_1 f_2(X))}{E_X(f_2(X))} = P(X \in S(f_1) | X \in S(f_2)), \qquad (4.25)$$

showing that this proposed fuzzy logic quantity is the same as the conditional probability of one point coverage of random weighting with respect to $S(f_1)$, given a one point coverage of the random weighting with respect to $S(f_2)$.

5. Use of one point coverage functions to define conditional fuzzy logic. This work extends earlier ideas in [11]. (For a detailed history of the problem of attempting a sound definition for conditional fuzzy sets see Goodman [8].) Even for the special case of ordinary events, until the recent fuller development of PSCEA, many basic difficulties arose with the use of other CEA. It seems reasonable that, whatever definitions we settle upon for conditional fuzzy sets and logic, they should generalize conditional events and logic of PSCEA. In addition, we have seen in the last sections that homomorphic-like relations can be established between aspects of fuzzy logic and a truth-functional-related part of probability theory, namely the probability evaluations of logical combinations of one point coverages for random sets corresponding to given fuzzy sets. Thus, it is also reasonable to expect that the definition of conditional fuzzy sets and associated logic should tie-in with homomorphic-like relations with one point coverage equivalent random sets. This connection becomes more evident by considering the fundamental lifting and compatibility relations for joint measurable mappings provided in Theorem 2.3: Let $f_j : D_j \to [0,1]$, $j = 1,2$, be any two fuzzy set membership functions. In Theorem 2.3 and commutative diagrams (2.32), (2.33), let (Ω, B, P) be as before, any given probability space, but now replace X by $S(f_1)$, Y by $S(f_2)$, for any joint pair of one point coverage equivalent random sets $S(f_j) \in \mathcal{S}(f_j)$, $j = 1,2$. Also, replace Ω by $\rho(D_j), B_j$ by $\rho\rho(D_j)$, and in eq. (2.34), event $a \in B_1$ by filter class $F_x(D_1) \in \rho\rho(D_1)$ and event $b \in B_2$, by filter class $F_y(D_2) \in \rho\rho(D_2)$, for any choice of $x \in D_1$, $y \in D_2$. Temporarily, we assume $f_2(y) > 0$. Then, in addition to the result that diagram (2.33) lifts (2.32) via $(\ )_0$ with the corresponding joint random set and one point coverage interpretation, eq. (2.34) now becomes (recalling eq. (4.3))

$$
\begin{aligned}
(5.1)\qquad & P(x \in S(f_1) | y \in S(f_2)) = \\
& = P_0((S(f_1), S(f_2))_0 \in (F_x(D_1) \times f_y(D_2) | \rho(D_1) \times F_y(D_2))).
\end{aligned}
$$

Note the distinction between the expression in eq. (5.1) and the one point coverage function of the *conditional random set* $(S(f_1) \times S(f_2) \,|\, D_1 \times S(f_2))$, (where as usual for any, $\omega \in \Omega$, $(S(f_1) \times S(f_2) \,|\, D_1 \times S(f_2))(\omega) = (S(f_1)(\omega) \times S(f_2)(\omega) \,|\, D_1 \times S(f_2)(\omega))$. Both $(S(f_1), S(f_2))_0$ and $(S(f_1) \times S(f_2) \,|\, D_1 \times S(f_2))$ are based on the same measurable spaces, but differ on the induced probability measures (see eq. (5.3)). The latter in general produces complex infinite series evaluations for its one point coverage function (as originally attested to in [12]): For any $u = (x_1, y_1, x_2, y_2, \ldots)$ in $(D_1 \times D_2)_0$, using eq. (2.12) and eq. (4.8),

$$
\begin{aligned}
& P(u \in (S(f_1) \times S(f_2) | D_1 \times S(f_2))) = \\
(5.2)\qquad & = \sum_{j=0}^{\infty} P((x_j \in S(f_1))\, \&\, (y_j \in S(f_2))\, \&\, \mathop{\&}_{i=1}^{j-1} (y_i \notin S(f_2))) \\
& = \sum_{j=0}^{\infty} \overline{\mathrm{cop}}(f_1(x_j), f_2(y_j); f_2(y_1), \ldots, f_2(y_{j-1})).
\end{aligned}
$$

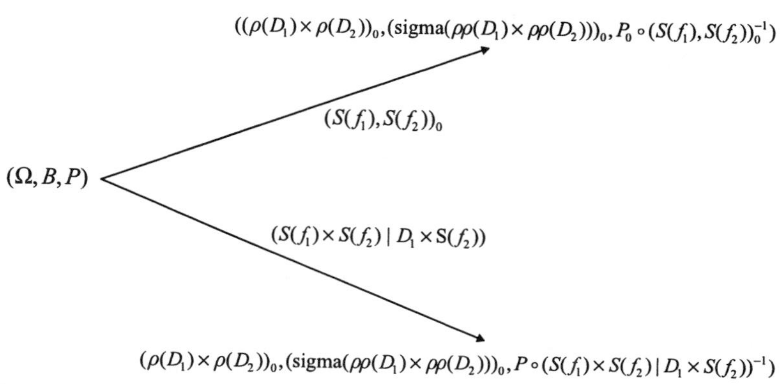

(5.3)

On the other hand, the expression in eq. (5.1) quickly simplifies to the form

$$P(x \in S(f_1)|y \in S(f_2)) = \frac{P((\phi(S(f_1))(x) = 1)\&(\phi(S(f_2))(y) = 1))}{P(\phi(S(f_2))(y) = 1)}$$

$$(5.4) \qquad = \frac{\overline{\text{cop}}(f_1(x), f_2(y))}{f_2(y)},$$

by using eq. (4.14) and the assumption that $f_2(y) > 0$. Thus, we are led in this case to define the conditional fuzzy set $(f_1|f_2)$ as

$$(5.5) \qquad (f_1|f_2)(x, y) = P(x \in S(f_1)|y \in S(f_2)) = \frac{\overline{\text{cop}}(f_1(x), f_2(y))}{f_2(y))}.$$

The case of $f_2(y) = 0$, is treated as a natural extension of the situation for the PSCEA conditional event membership function $\phi(a|b)$ in eqs. (2.15), (2.16), namely, for any $u = (x_1, y_1, x_2, y_2, \ldots)$, with x_j in D_1, y_j in D_2:

$$(5.6) \qquad \begin{array}{c} \text{If } f_2(y_1) = f_2(y_2) = \cdots = f_2(y_{j-1}) = 0 < f_2(y_j), \\ \text{then, by definition } (f_1|f_2)(u) = (f_1|f_2)(x_j, y_j), \end{array}$$

noting that the case for vacuous zero values in eq. (5.6) occurs for $j = 1$.

$$(5.7) \qquad \begin{array}{c} \text{If } f_2(y_1) = f_2(y_2) = \cdots = f_2(y_j) = \cdots = 0, \\ \text{then, by definition } (f_1|f_2)(u) = 0. \end{array}$$

More concisely, eqs. (5.6), (5.7) are equivalent to

$$(5.8) \qquad (f_1|f_2)(u) = \sum_{j=1}^{\infty} \prod_{i=1}^{j-1} \delta(f_2(y_j) = 0) \cdot ((f_1|f_2)(x_j, y_j)),$$

where $(f_1|f_2)(x_j, y_j)$ is as in eq. (5.5).

Furthermore, let us define logical operations between such conditional fuzzy membership motivated by the homomorphic-like relations discussed earlier and compatible with the definition in eqs. (5.5)–(5.8). For example, binary conjunction here between $(f_1|f_2)$ and $(f_3|f_4)$ for any arguments $u = (x_1, y_1, x_2, y_2, \ldots)$, $v = (w_1, z_1, w_2, z_2, \ldots)$ is defined by first determining that j and k so that $f_2(y_1) = f_2(y_2) = \cdots = f_2(y_{j-1}) = 0 < f_2(y_j)$, $f_4(z_1) = f_4(z_2) = \cdots = f_4(z_{k-1}) = 0 < f_4(z_k)$, assuming the two trivial cases of 0–value in eq. (5.7) are avoided here. Thus, $f_2(y_j), f_4(z_k) > 0$. Next, choose any copula and corresponding pair ($\&_1$,$\vee_1$) (= ($\overline{\text{cop}}$, cocop)) in the alternating sum family and consider the joint one point coverage equivalent random subsets of D_j, $S(f_j)$, $j = 1, 2, 3, 4$, determined through cop (as in Theorem 4.1, e.g.). Denote one point coverage events $a = (x_j \in S(f_1))$ (= $(\phi(S(f_1)) = 1) = (S(f_1))^{-1}(F_{x_j}(D_1))$, etc.), $b = (y_j \in$

$S(f_2))$, $c = (w_k \in S(f_3))$, $d = (z_k \in S(f_4))$ (all belonging to the probability space (Ω, B, P)). Finally, define a conditional fuzzy logic operator $\&_1$ as

$$(5.9)\quad \begin{aligned} ((f_1|f_2)\&_1(f_3|f_4))(u,v) &= (f_1|f_2)(u)\&_1(f_3|f_4)(v) \\ &= P_0((a|b)\&(c|d)) = P_0(A)/P(b \vee d), \end{aligned}$$

$$(5.10)\quad P_0(A) = P(a\&b\&c\&d) + (P(a\&b\&d')P(c|d)) + (P(b'\&c\&d)P(a|b)),$$

using eqs. (2.24), (2.26). In turn, each of the components needed for the full evaluation of eq. (5.9) are readily obtainable using, e.g., the bottom parts of eq. (4.13) and/or eq. (4.14). Specifically, we have:

$$(5.11)\quad P(a\&b\&c\&d) = \overline{\text{cop}}(f_1(x_j), f_2(y_j), f_3(w_k), f_4(z_k)),$$

$$(5.12)\quad P(a\&b\&d') = \overline{\text{cop}}(f_1(x_j), f_2(y_j); f_4(z_k)),$$

$$(5.13)\quad P(b'\&c\&d) = \overline{\text{cop}}(f_3(w_k), f_4(z_k); f_2(y_j)),$$

$$(5.14)\quad P(a|b) = (f_1|f_2)(x_j, y_j) = (\overline{\text{cop}}(f_1(x_j), f_2(y_j))/f_2(y_j),$$

$$(5.15)\quad P(c|d) = (f_3|f_4)(w_k, z_k) = (\overline{\text{cop}}(f_3(w_k), f_4(z_k))/f_4(z_k),$$

$$(5.16)\quad P(b \vee d) = \text{cocop}(f_2(y_j), f_4(z_k)).$$

Also, using eqs. (2.25), (2.27), we obtain similarly,

$$(5.17)\quad \begin{aligned} ((f_1|f_2)\vee_1(f_3|f_4))(u,v) &= (f_1|f_2)(u)\vee_1(f_3|f_4)(v) \\ &= P_0((a|b) \vee (c|d)) \\ &= P(a|b) + P(c|d) - P_0((a|b)\&(c|d)), \end{aligned}$$

all obviously obtainable from eqs. (5.9)-(5.16). For fuzzy complementation,

$$(5.18)\quad \begin{aligned} (f_1|f_2)'(u) &= ((f_1|f_2)(u))' = P((a|b)') = 1 - P(a|b) \\ &= P(a'|b) = 1 - (f_1|f_2)(u) = 1 - (f_1|f_2)(x_j, y_j) \\ &= (f_2(y_j) - \overline{\text{cop}}(f_1(x_j), f_2(y_j)))/f_2(y_j) \\ &= (\overline{\text{cop}} f_2(y_j); f_1(x_j))/f_2(y_j), \end{aligned}$$

all consistent. Combinations of conjunctions and disjunctions for multiple arguments follow a similar pattern of definition. The following summarizes some of the basic properties of fuzzy conditionals, sets, and logic, recalling here the multivariable notation introduced earlier:

THEOREM 5.1. Consider any collection of fuzzy set membership functions $f_J : D_J \to [0,1]^J$, any choice of cop producing joint one point coverage equivalent random subsets $S(f_J)$ of D_J with respect to probability space (Ω, B, P), and define fuzzy conditional membership functions and logic as in eqs. (5.5)–(5.18):

(i) When, any two fuzzy set membership functions reduce to ordinary set membership functions such as $f_1 = \phi(a_1), f_2 = \phi(a_2)$, $a_1, a_2 \in B$, then

$$(f_1|f_2) = \phi(a_1 \times a_2|D_1 \times a_2),$$

the ordinary membership function of a PSCEA conditional event in product form.

(ii) All well-defined combinations of fuzzy logical operations over $(f_1|f_2), (f_3|f_4), \ldots$, when $f_J = \phi(a_J)$, reduce to their PSCEA counterparts.

(iii) When $f_2 = 1$ identically, we have the natural identification

$$(f_1|f_2) = f_1 \ .$$

(iv) The following is extendible to any number of arguments: When $f_2 = f_4 = f$,

$$(f_1|f)\&_1(f_3|f) = (f_1\&_1 f_3|f),\ (f_1|f)\,\mathrm{v}_1\,(f_3|f) = (f_1\mathrm{v}_1 f_3|f),$$

where for the unconditional computations $f_1\&_1 f_3 = \overline{\mathrm{cop}}(f_1, f_3),\ f_1\mathrm{v}_1 f_3 = \overline{\mathrm{cop}}(f_1, f_3)$, etc.

(v) Modus ponens holds for conditional fuzzy logic as constructed here: $(f_1|f_2)\&_1 f_2 = f_1\&_1 f_2$.

PROOF. Straightforward, from the construction. □

Thus, a reasonable basis has been established for conditional fuzzy logic extending both PSCEA for ordinary events and unconditional fuzzy logic. The resulting conditional fuzzy logic is not significantly more difficult to compute than is PSCEA itself.

6. Three examples reconsidered. We briefly show here how the ideas developed in the previous sections can be applied to the examples of Section 1.3.

EXAMPLE 6.1. (See eq. (1.3).) Applying the REA solution for weighted linear functions from Section 3.2, we obtain

$$\begin{aligned} P_0(a\&b) = {} & P(c\&d\&e) + (\min(w_{11}+w_{12}, w_{21}+w_{22}))P(c\&d\&e') \\ & + (\min(w_{11}+w_{13}, w_{21}+w_{23}))P(c\&d'\&e) \\ & + (\min(w_{12}+w_{13}, w_{22}+w_{23}))P(c'\&d\&e) \\ & + (\min(w_{13}, w_{23}))P(c'\&d'\&e) + (\min(w_{12}, w_{22}))P(c'\&d\&e') \\ & + (\min(w_{11}, w_{21}))P(c\&d'\&e'). \end{aligned} \tag{6.1}$$

Then, choosing, e.g., the absolute probability distance function, eq. (6.1) yields

$$(6.2)\quad \begin{aligned} D_{P_0}(a,b) &= P_0(a) + P_0(b) - 2P_0(a\&b) \\ &= |(w_{11}+w_{12}) - (w_{21}+w_{22})|P(c\&d\&e') \\ &+ |(w_{11}+w_{13}) - (w_{21}+w_{23})|P(c\&d'\&e) \\ &+ |(w_{12}+w_{13}) - (w_{22}+w_{23})|P(c'\&d\&e) \\ &+ |w_{13} - w_{23}|P(c'\&d'\&e) \\ &+ |w_{12} - w_{22}|P(c'\&d\&e') + |w_{11} - w_{21}|P(c\&d'\&e'). \end{aligned}$$

In turn, use the above expression to test hypotheses $a \neq b$ vs. $a = b$, following the procedure in eqs. (1.13)–(1.16), where, e.g., the fixed significance level is

$$(6.3)\quad \alpha_0 = F_D(D_{P_0}(a,b)) = (D_{P_0}(a,b))^2\,(3 - 2D_{P_0}(a,b))\,,$$

by using eq. (6.2) for full evaluation. □

EXAMPLE 6.2. Specializing the conjunctive probability formula in eqs. (2.24), (2.26),

$$(6.4)\quad P_0(a\&b) = P_0((c|d)\&(c|e)) = P_0(A)/P(d \vee e),$$

$$(6.5)\quad P(A) = P(c\&d\&e) + (P(c\&d\&e')P(c|e)) + (P(c\&d'\&e)P(c|d)).$$

Using, e.g., the relative distance to test the hypotheses $a \neq b$ vs. $a = b$, eq. (6.4) allows us to calculate

$$(6.6)\quad R_{P_0}(a,b) = \frac{P_0(a+b)}{P_0(a \vee b)} = \frac{P(c|d) + P(c|e) - 2P_0(a\&b)}{P(c|d) + P(c|e) - P_0(a\&b)}$$

Then, we can test hypotheses $a \neq b$ vs. $a = b$, by using eqs. (1.13)–(1.16), where now the fixed significance level is

$$(6.7)\quad \alpha_0 = F_R(R_{P_0}(a,b)) = (R_{P_0}(a,b))^2$$

by using eq. (6.6) for full evaluation. □

EXAMPLE 6.3. We now suppose that in the fuzzy logic interpretation in eq. (1.6) for Model 1, v_1 is max, while for Model 2, v_2 is chosen compatible with conditional fuzzy logic, as outlined in Section 5. We also simplify the models further: The attribute "fairly long" and "long" are identified, thus avoiding exponentiation (though this can also be treated — but to avoid complications, we make the assumption). We also identify "large" with "medium." Model 2 can be evaluated from eqs. (5.9)–(5.17), where

$$c = (x_1 \in S(f_1)),\ d = (w_1 \in S(f_3)),\ e = (z_1 \in S(f_4)),$$

$$(6.8)\quad f_1(x_1) = f_{long}(\text{lngth}(A)), f_2(y_1) = 1, f_3(w_1) = f_{medium}(\#(Q)),$$
$$f_4(z_1) = f_{accurate}(L), x_1 = \text{lngth}(A), y_1 \text{ arbitrary}, w_1 = \#(Q), z_1 = L,$$

On the other hand, we are here interested in *comparing* the two models via probability representations and use of probabilistic distance functions. In summary, all of the above simplifies to

$$(6.9)\qquad \begin{cases} \text{Model 1:}\quad t(a) = P_0(a) = \max(P(c)^2,\ P(d)) \\ \text{vs.} \\ \text{Model 2:}\quad t(b) = P_0(b) = P_0((c|\Omega) \vee (d|e)) \end{cases} .$$

In turn, applying the REA approach to max in Section 3.5 and using PSCEA via eq. (5.17) with $d = \Omega$, we obtain the description events according to each expert as

$$(6.10)\quad a = (c^2\&d) \vee ((c^2\&d') \times \theta(w_{P,1})) \vee (((c^2)'\&d) \times \theta(w_{P,2})),$$

$$(6.11)\quad c^2 = c \times c,\ (c^2)' = (c \times c') \vee c',$$

$$(6.12)\quad b = c \vee (d|e) = c \vee (d\&e) \vee ((c'\&e') \times (d|e)).$$

Then, using again PSCEA and simplifying,

$$(6.13)\quad \begin{aligned} P_0(a\&b) &= P(c\&d)P(c) + P(c\&d')P(c)w_{P,1} + P(c\&d)P(c')w_{P,2} \\ &\quad + P(c'\&d\&e)w_{P,2} + P(c'\&d\&e')P(d|e)w_{P,2}. \end{aligned}$$

In turn, each expression in eq. (6.13) can be again fully evaluated via the use of eq. (4.13):

$$(6.14)\quad \begin{aligned} &P(c\&d) = \overline{\text{cop}}(f_1(x_1), f_3(w_3)),\ P(c'\&d\&e) = \overline{\text{cop}}(f_3(w_1), f_4(z_1); f_1(x_1)), \\ &P(c'\&d\&e') = \overline{\text{cop}}(f_3(w_1); f_1(x_1), f_4(z_1)),\ P(c) = f_1(x_1), \\ &P(c\&d') = \overline{\text{cop}}(f_1(x_1); f_3(w_1)) = P(c) - P(c\&d), \\ &P(d|e) = (f_3|f_4)(w_1, z_1), \end{aligned}$$

etc., with all interpretations in terms of the original corresponding attributes given in eq. (6.8). Finally, eq. (6.13) can be used, together with the evaluations of the marginal probabilities $P_0(a), P_0(b)$ in eq. (6.9) to obtain once more any of the probability distance measures used in Section 1.4 for testing the hypotheses of $a \neq b$ vs. $a = b$. □

REFERENCES

[1] E. ADAMS, *The Logic of Conditionals*, D. Reidel, Dordrecht, Holland, 1975.

[2] D.E. BAMBER, *Personal communications*, Naval Command Control Ocean Systems Center, San Diego, CA, 1992.

[3] G. DALL'AGLIO, S. KOTZ, AND G. SALINETTI (eds.), *Advances in Probability Distributions with Given Marginals*, Kluwer Academic Publishers, Dordrecht, Holland, 1991.

[4] D. DUBOIS AND H. PRADE, *Fuzzy Sets and Systems*, Academic Press, New York, 1980.

[5] E. EELLS AND B. SKYRMS (eds.), *Probability and Conditionals*, Cambridge University Press, Cambridge, U.K., 1994.

[6] M.J. FRANK, *On the simultaneous associativity of $F(x, y)$ and $x + y - F(x, y)$*, Aequationes Math., 19 (1979), pp. 194–226.

[7] I.R. GOODMAN, *Evaluation of combinations of conditioned information: A history*, Information Sciences, 57–58 (1991), pp. 79–110.

[8] I.R. GOODMAN, *Algebraic and probabilistic bases for fuzzy sets and the development of fuzzy conditioning*, Conditional Logic in Expert Systems (I.R. Goodman, M.M. Gupta, H.T. Nguyen, and G.S. Rogers, eds.), North–Holland, Amsterdam (1991), pp. 1–69.

[9] I.R. GOODMAN, *Development of a new approach to conditional event algebra with application to operator requirements in a C3 setting*, Proceedings of the 1993 Symposium on Command and Control Research, National Defense University, Washington, DC, June 28–29, 1993, pp. 144–153.

[10] I.R. GOODMAN, *A new characterization of fuzzy logic operators producing homomorphic-like relations with one-point coverages of random sets*, Advances in Fuzzy Theory and Technology, Vol. II (P. P.Wang, ed.), Duke University, Durham, NC, 1994, pp. 133–159.

[11] I.R. GOODMAN, *A new approach to conditional fuzzy sets*, Proceedings of the Second Annual Joint Conference on Information Sciences, Wrightsville Beach, NC, September 28 – October 1, 1995, pp. 229–232.

[12] I.R. GOODMAN, *Similarity measures of events, relational event algebra, and extensions to fuzzy logic*, Proceedings of the 1996 Biennial Conference of North American Fuzzy Information Processing Society–NAFIPS, University of California at Berkeley, Berkeley, CA, June 19–22, 1996, pp. 187–191.

[13] I.R. GOODMAN AND G.F. KRAMER, *Applications of relational event algebra to the development of a decision aid in command and control*, Proceedings of the 1996 Command and Control Research and Technology Symposium, Naval Postgraduate School, Monterey, CA, June 25–28, 1996, pp. 415–435.

[14] I.R. GOODMAN AND G.F. KRAMER, *Extension of relational event algebra to a general decision making setting*, Proceedings of the Conference on Intelligent Systems: A Semiotic Perspective, Vol. I, National Institute of Standards and Technology, Gaithersberg, MD, October 20–23, 1996, pp. 103–108.

[15] I.R. GOODMAN AND G.F. KRAMER, *Comparison of incompletely specified models in C4I and data fusion using relational and conditional event algebra*, Proceedings of the 3rd International Command and Control Research and Technology Symposium, National Defense University, Washington, D.C., June 17–20, 1997.

[16] I.R. GOODMAN AND H.T. NGUYEN, *Uncertainty Models for Knowledge–Based Systems*, North–Holland, Amsterdam, 1985.

[17] I.R. GOODMAN AND H.T. NGUYEN, *A theory of conditional information for probabilistic inference in intelligent systems: II product space approach; III mathematical appendix*, Information Sciences, 76 (1994), pp. 13–42; 75 (1993), pp. 253–277.

[18] I.R. GOODMAN AND H.T. NGUYEN, *Mathematical foundations of conditionals and their probabilistic assignments*, International Journal of Uncertainty, Fuzziness

and Knowledge–Based Systems, 3 (1995), pp. 247–339.

[19] I.R. GOODMAN, H.T. NGUYEN, AND E.A. WALKER, *Conditional Inference and Logic for Intelligent Systems*, North–Holland, Amsterdam, 1991.

[20] T. HAILPERIN, *Probability logic*, Notre Dame Journal of Formal Logic, 25 (1984), pp. 198–212.

[21] D.A. KAPPOS, *Probability Algebra and Stochastic Processes*, Academic Press, New York, 1969, pp. 16–17 et passim.

[22] D. LEWIS, *Probabilities of conditionals and conditional probabilities*, Philosophical Review, 85 (1976), pp. 297–315.

[23] V. MCGEE, *Conditional probabilities and compounds of conditionals*, Philosophical Review, 98 (1989), pp. 485–541.

[24] C.V. NEGOITA AND D.A. RALESCU, *Representation theorems for fuzzy concepts*, Kybernetes, 4 (1975), pp. 169–174.

[25] H.E. ROBBINS, *On the measure of a random set*, Annals of Mathematical Statistics, 15 (1944), pp. 70–74.

[26] N. ROWE, *Artificial Intelligence through PROLOG*, Prentice–Hall, Englewood Cliffs, NJ, 1988 (especially, Chapter 8).

[27] G. SHAFER, *A Mathematical Theory of Evidence*, Princeton University Press, Princeton, NJ, 1976, p. 48 et passim.

[28] A. SKLAR, *Random variables, joint distribution functions, and copulas*, Kybernetika, 9 (1973), pp. 449–460.

[29] R. STALNAKER, *Probability and conditionals*, Philosophy of Science, 37 (1970), pp. 64–80.

[30] B. VAN FRAASEN, *Probabilities of conditionals*, Foundations of Probability Theory, Statistical Inference, and Statistical Theories of Science, (W.L. Harper and C.A. Hooker, eds.), D. Reidel, Dordrecht, Holland (1976), pp. 261–300.

BELIEF FUNCTIONS AND RANDOM SETS

HUNG T. NGUYEN AND TONGHUI WANG*

Abstract. This is a tutorial about a formal connection between belief functions and random sets. It brings out the role played by random sets in the so-called theory of evidence in artificial intelligence.

Key words. Belief Function, Choquet Capacity, Plausibility Function, Random Set.

AMS(MOS) subject classifications. 60D05, 28E05

1. Introduction. The theory of evidence [14] is based upon the concept of belief functions which seem useful in modeling incomplete knowledge in various situations of artificial intelligence. As general degrees of beliefs are viewed as generalizations of probabilities, their axiomatic definitions should be some weakened forms of probability measures. It turns out that they are precisely Choquet capacities of some special kind, namely monotone of infinite order. Regardless of how philosophical objections, a random-set interpretation exists. Viewing belief functions as distributions of random sets, one can use the rigorous calculus of random sets within probability theory to derive inference procedures as well as to provide a probabilistic meaning for the notion of independent pieces of evidence in the problem of evidence fusion.

2. General degrees of belief. We denote by $\mathbb{R}$ the set of reals, $\emptyset$ the empty set, $\subseteq$ the set inclusion, $\cup$ the set union, and $\cap$ the set intersection. The power set (collection of all subsets) of a set Θ is denoted by 2^Θ. For A, $B \in 2^\Theta$, A^c denotes the set complement of A, $|A|$ denotes the cardinality of A, and $A \setminus B = A \cap B^c$.

A context in which general degrees of belief are generalizations of probability is the following:

EXAMPLE 2.1. Let $\Theta_1, \Theta_2, \ldots, \Theta_k$ be a partition of a finite set Θ, $\mathbb{P}$ be the class of all probability measures on Θ, and

$$\mathcal{P} = \{P : P \in \mathbb{P},\, P(\Theta_i) = \alpha_i,\, i = 1, 2, \ldots, k\},$$

where the given α_i's are positive and $\sum_{i=1}^k \alpha_i = 1$. Let $F : 2^\Theta \longrightarrow [0,1]$ be the "lower probability" defined by

$$F(A) = \inf_{P \in \mathcal{P}} P(A), \qquad A \in 2^\Theta.$$

□

* Department of Mathematical Sciences, New Mexico State University, Las Cruces, New Mexico 88003-8001.

Then the set function F is no longer additive. But F satisfies the following two conditions:

(i) $F(\emptyset) = 0, \quad F(\Theta) = 1.$
(ii) For any $n \geq 2$ and $A_1, A_2, \ldots, A_n$ in 2^Θ,

$$F\left(\bigcup_{i=1}^{n} A_i\right) \geq \sum_{\emptyset \neq I \subseteq \{1,2,\ldots,n\}} (-1)^{|I|+1} F\left(\bigcap_I A_i\right).$$

Note that (ii) is simply a weakening form of Poincare's equality for probability measures (i.e., if F is a probability measure, then (ii) becomes an equality). Condition (ii) together with (i) is stronger than the monotonicity of F. We recognize (ii) as the definition of a set function, *monotone of infinite order* (or *totally monotone*) in the theory of capacities [1], [13].

The proof of (ii) is simple. For $A \subseteq \Theta$, define $A_* = \cup \Theta_i$, where the union is over the Θ_i's such that $\Theta_i \subseteq A$. Define

$$H : 2^\Theta \longrightarrow [0,1] \quad \text{by} \quad H(A) = P(A_*),$$

where P is any element in $\mathcal{P}$ (note that, for any $P \in \mathcal{P}$, $P(A_*)$ is the same). Then we have

$$H(\emptyset) = P(\emptyset_*) = P(\emptyset) = 0, \quad H(\Theta) = P(\Theta_*) = P(\Theta) = 1,$$

$H(A) \leq F(A)$ for every $A \in 2^\Theta$.

Also, for any subset $\mathcal{S} \subseteq 2^\Theta$,

$$\bigcap_{A \in \mathcal{S}} A_* = \left(\bigcap_{A \in \mathcal{S}} A\right)_*, \qquad \bigcup_{A \in \mathcal{S}} A_* \subseteq \left(\bigcup_{A \in \mathcal{S}} A\right)_*.$$

Hence

$$\begin{aligned}
H\left(\bigcup_{i=1}^{n} A_i\right) &= P\left(\left(\bigcup_{i=1}^{n} A_i\right)_*\right) \geq P\left(\bigcup_{i=1}^{n} (A_i)_*\right) \\
&= \sum_{\emptyset \neq I \subseteq \{1,2,\ldots,n\}} (-1)^{|I|+1} P\left(\bigcap_I (A_i)_*\right) \\
&= \sum_{\emptyset \neq I \subseteq \{1,2,\ldots,n\}} (-1)^{|I|+1} P\left(\left(\bigcap_I A_i\right)_*\right) \\
&= \sum_{\emptyset \neq I \subseteq \{1,2,\ldots,n\}} (-1)^{|I|+1} H\left(\bigcap_I A_i\right).
\end{aligned}$$

But for each $A \in 2^\Theta$, there is a $P_A \in \mathcal{P}$ such that $P_A(A) = P_A(A_*)$, and thus

$$F(A) = \inf_{P \in \mathcal{P}} P(A) \leq P_A(A) = P_A(A_*) = H(A),$$

which together with the fact that $H(A) \leq F(A)$ shows that $F(A) = H(A)$. (Note that $\mathcal{P} = \{P \in \mathbb{P} : F(\cdot) \leq P(\cdot)\}$).

A *belief function* on a finite set Θ is defined axiomatically as a set-function F, from 2^Θ to $[0,1]$, satisfying (i) and (ii) above [14]. Clearly, any probability measure is a belief function.

REMARK 2.1. In the above example, the lower probability of the class $\mathcal{P}$ happens to be a belief function, and moreover, F generates $\mathcal{P}$, i.e.,

$$\mathcal{P} = \{P \in \mathbb{P} : F \leq P\}.$$

□

If a belief function F is given axiomatically on Θ, then there is a non-empty class $\mathcal{P}$ of probability measures on Θ for which F is precisely its lower envelope. The class $\mathcal{P}$, called the class of *compatible* probability measures (see, e.g. [2], [18]), is $\mathcal{P} = \{P \in \mathbb{P} : F \leq P\}$. The fact that $\mathcal{P} \neq \emptyset$ follows from a general result in game theory [15]. A simple proof is this.

Let $\Theta = \{\theta_1, \theta_2, \ldots, \theta_n\}$. Define $g : \Theta \to \mathbb{R}^+$ by

$$g(\theta_i) = F(\{\theta_i, \theta_{i+1}, \ldots, \theta_n\}) - F(\{\theta_{i+1}, \theta_{i+2}, \ldots, \theta_n\}),$$

for $i = 1, 2, \ldots, n-1$, and $g(\theta_n) = F(\{\theta_n\})$. Then

$$g(\theta) \geq 0, \quad \text{and} \quad \sum_{\theta \in \Theta} g(\theta) = 1.$$

That is, g is a probability density function on Θ. Let

$$P_g(A) = \sum_{\theta \in A} g(\theta) \qquad \text{for} \quad A \subseteq \Theta,$$

and $A_i = \{\theta_i, \theta_{i+1}, \ldots, \theta_n\}$. Then, we have

$$\begin{aligned} g(\theta_i) &= F(A_i) - F(A_i \setminus \{\theta_i\}) \\ &= \sum_{B \subseteq A_i} f(B) - \sum_{B \subseteq A_i \setminus \{\theta_i\}} f(B) = \sum_{\theta_i \in B \subseteq A_i} f(B), \end{aligned}$$

where f is the Möbius inverse of F. Since $f \geq 0$,

$$P_g(A_i) = \sum_{\theta \in A_i} g(\theta) = \sum_{\theta \in A_i} \sum_{\theta \in B \subseteq A_i} f(B) \geq \sum_{B \subseteq A_i} f(B) = F(A_i).$$

Now renumbering the θ_i's, the A_i above becomes an arbitrary subset of Θ, and therefore $F(A) = \inf\{P(A) : F \leq P\}$.

For an approach using multi-valued mappings see [2], [18].

As a special class of (nonnegative) measures, namely those μ such that $\mu(\Theta) = 1$, is used to model probabilities, a special class of (Choquet) *capacities*, namely those F such that $F(\Theta) = 1$ and infinitely monotone, is used to model general degrees of belief. An objection to the use of non-additive set-functions to model uncertainty is given in [8]. However, the argument of Lindley has been countered on mathematical grounds in [5].

Using the discrete counter-part of the derivative (in Combinatorics), one can introduce belief functions F on a finite set Θ as follows. The *Möbius inverse* of F is the set-function f defined on 2^Θ by

$$f(A) = \sum_{B \subseteq A} (-1)^{|A \setminus B|} F(B).$$

It can be verified that (see e.g. [14]) $f \geq 0$ if and only if F is a belief function. Thus, to define a belief function F, it suffices to specify

$$f : 2^\Theta \longrightarrow [0, 1], \quad f(\emptyset) = 0, \quad \sum_{A \subseteq \Theta} f(A) = 1,$$

and take

$$F(A) = \sum_{B \subseteq A} f(B).$$

For a recent publication of research papers on the theory of evidence (also called the Dempster-Shafer theory of evidence) in the context of artificial intelligence, see [19].

EXAMPLE 2.2. Let $\Theta = \{\theta_1, \theta_2, \theta_3, \theta_4\}$ and X be a random variable with values in Θ. Suppose that the density, g, of X can be only specified as

$$g(\theta_1) \geq 0.4, \quad g(\theta_2) \geq 0.2, \quad g(\theta_3) \geq 0.2, \quad g(\theta_4) \geq 0.1. \tag{2.1}$$

This imprecise knowledge about g forces us to consider the class $\mathcal{P}$ of probability measures on Θ whose density satisfies (2.1).

Define $f : 2^\Theta \longrightarrow [0, 1]$ by

$$f(\{\theta_1\}) = 0.4, \qquad f(\{\theta_2\}) = f(\{\theta_3\}) = 0.2,$$

$$f(\{\theta_4\}) = 0.1, \qquad f(\{\theta_1, \theta_2, \theta_3, \theta_4\}) = 0.1,$$

and $f(A) = 0$ for any other subset A of Θ. The assignment $f(\{\theta_1, \theta_2, \theta_3, \theta_4\}) = 0.1$ means that the mass 0.1 is given globally to Θ, and not to any specific element of it. Then $\sum_{A \subseteq \Theta} f(A) = 1$, and hence $F(A) = \sum_{B \subseteq A} f(B)$ is a belief function. It can be easily verified that $F = \inf \mathcal{P}$. □

For *infinite* Θ, the situation is more complicated. Fortunately, this is nothing else than the problem of construction of capacities. As in the case of measures, we would like to specify the values of capacity F only on some "small" class of subsets of Θ and yet sufficient for determining its values on 2^Θ.

In view of the connection with distributions of random sets in the next section, it is convenient to consider the *dual* of the belief function F, called a *plausibility* function:

$$G : 2^\Theta \longrightarrow [0,1], \qquad G(A) = 1 - F(A^c).$$

Note that $F(\cdot) \leq G(\cdot)$ (by (i) and (ii)). Plausibility functions correspond to upper probabilities, e.g., in Example 2.1,

$$G(A) = \sup_{P \in \mathcal{P}} P(A).$$

Obviously, G is monotone and satisfies

$$G(\emptyset) = 0, \qquad G(\Theta) = 1 \tag{2.2}$$

and

$$G\left(\bigcap_{i=1}^{n} A_i\right) \leq \sum_{\emptyset \neq I \subseteq \{1,2,\ldots,n\}} (-1)^{|I|+1} G\left(\bigcup_I A_i\right). \tag{2.3}$$

Condition (2.3) is referred to as *alternating of infinite order* [1], [11]. The following is an example where the construction of plausibility functions (and hence belief functions) on the real line $\mathbb{R}$ (or more generally on a locally compact, Hausdorff, separable space) is similar to that of probability measures, in which supremum plays the role of integral.

EXAMPLE 2.3. Let $\phi : \mathbb{R} \longrightarrow [0,1]$ be upper semi-continuous. Let G be defined on the class $\mathcal{K}$ of compact sets of $\mathbb{R}$ by:

$$G(K) = \sup_{x \in K} \phi(x), \qquad K \in \mathcal{K}.$$

Then

$$K_n \searrow K \quad \text{in} \quad \mathcal{K} \qquad \text{implies} \qquad G(K_n) \searrow G(K). \tag{2.4}$$

(Where $K_n \searrow K$ means the sequence $\{K_n\}$ is decreasing, i.e., $K_{n+1} \subseteq K_n$ and $K = \cap_n K_n$) □

Indeed, since G is monotone increasing,

$$0 \leq G(K) \leq \inf_n G(K_n).$$

The result is trivial if $\inf_n G(K_n) = 0$. Thus suppose that $\inf_n G(K_n) > 0$. Let $0 < h < \inf_n G(K_n)$ and $A_n = (x : \phi(x) \geq h(x)) \cap K_n$. It follows from the hypothesis that A_n's are compact. By construction, for any n, $A_n \neq \emptyset$, since otherwise, $\forall x \in K_n$, $\phi(x) < h$ ($K_n \neq \emptyset$ as $\inf_n \phi(K_n) > 0$), implying that

$$\sup_{x \in K_n} \phi(x) = G(K_n) \leq h,$$

which contradicts the fact that

$$G(K_n) \geq \inf_{j \geq 1} G(K_j) > h.$$

Thus $A = \bigcap_n A_n \neq \emptyset$.

Now as $A \subseteq K = \bigcap_n K_n$, $G(A) \leq G(K)$. Therefore it suffices to show that $G(A) \geq h$. But this is obvious, since in each A_n, $\phi \geq h$, so that

$$G(A) = \sup_{x \in \bigcap_n A_n} \phi(x) \geq h.$$

The fact that G is alternating of infinite order on $\mathcal{K}$ can be proved directly, but we prefer to give a *probabilistic proof* of this fact in the next section.

The domain of the plausibility function G (or of a belief function F) on any space can be its power set, here $2^{\mathbb{R}}$ (Recall why σ-fields are considered as domains of measures!).

We can extend the domain $\mathcal{K}$ of G to $2^{\mathbb{R}}$ as follows. If A is an open set, define

$$G(A) = \sup\{G(K) : K \in \mathcal{K}, K \subseteq A\},$$

and if $B \subseteq \mathbb{R}$,

$$G(B) = \inf\{G(A), \ A \text{ is open}, \ B \subseteq A\}.$$

It can be shown that the set-function G so defined on $2^{\mathbb{R}}$ is alternating of infinite order [9].

In summary, the plausibility G so constructed on $2^{\mathbb{R}}$ is a $\mathcal{K}$-capacity (with $G(\emptyset) = 0$ and $G(\mathbb{R}) = 1$), alternating of infinite order, where a capacity G is a set-function $G : 2^{\mathbb{R}} \longrightarrow [0,1]$ such that

(a) G is monotone increasing: $A \subseteq B$ implies $G(A) \leq G(B)$.
(b) $A_n \subseteq \mathbb{R}$ and A_n increasing imply $G(\bigcup_n A_n) = \sup_n G(A_n)$.
(c) $K_n \in \mathcal{K}$ and K_n decreasing imply $G(\bigcap_n K_n) = \inf_n G(K_n)$.

Note that Property (b) can be verified for G defined in terms of ϕ above.

More generally, on an abstract space Θ, a plausibility function G is an $\mathcal{F}$-capacity [10], alternating of infinite order, such that $G(\emptyset) = 0$ and $G(\Theta) = 1$, where $\mathcal{F}$ is a collection of subsets of Θ, containing $\emptyset$ and stable under finite unions and intersections.

REMARK 2.2. Unlike measure theory where the σ-additivity property forces one to consider Borel σ-fields as domains for measures, it is possible to define belief functions for all subsets (measurable or not) of the space under consideration. In this spirit, the Choquet functional, namely

$$\int_0^\infty F(\theta : X(\theta) > t)dt$$

can be defined for $X : \Theta \longrightarrow \mathbb{R}^+$, measurable or not: for any t, $(X > t)$, as a subset of Θ, is always in the domain of F (which is 2^Θ); F being monotone increasing, the map $t \to F(X > t)$ is monotone decreasing and hence measurable. Recall that the need to consider non-measurable maps occurs also in some areas of probability and statistics, e.g., in the theory of empirical processes [17]. □

3. Random sets. A random set is a set-valued random element. Specifically, let $(\Omega, \mathcal{A}, P)$ be a probability space, and $(\mathcal{B}, \sigma(\mathcal{B}))$ a measurable space, where $\mathcal{B} \subseteq 2^\Theta$. A mapping S: $\Omega \longrightarrow \mathcal{B}$ which is $\mathcal{A}$-$\sigma(\mathcal{B})$ measurable is called a random set. Its probability law is the probability measure PS^{-1} on $\sigma(\mathcal{B})$. In fact, a random set on Θ is viewed as a probability measure on $(\mathcal{B}, \sigma(\mathcal{B}))$.

For example, for Θ finite and $\mathcal{B} = 2^\Theta$, $\sigma(\mathcal{B})$ is the power set of 2^Θ, and any probability measure (random set S) on Θ is determined by a density function $f : 2^\Theta \longrightarrow [0,1]$, i.e., $f(A) = P(S = A)$ such that $\sum_{A \subseteq \Theta} f(A) = 1$. The "distribution function" F of the random set S is

$$F(A) = P(\{\omega : S(\omega) \subseteq A\}) = \sum_{B \subseteq A} f(B).$$

Thus a belief function on a finite Θ is nothing else but the distribution function of a random set [12]. Indeed, given a set-function F on 2^Θ satisfying (i) and (ii) of the previous section, we see that F is the distribution of a random set S having

$$f(A) = \sum_{B \subseteq A} (-1)^{|A \setminus B|} F(B)$$

as its density (Möbius inversion).

Consider next the case when $\Theta = \mathbb{R}$ (or a locally, compact, Hausdorff, separable space). Take $\mathcal{B} = \mathcal{F}$, the class of all closed sets of $\mathbb{R}$. The σ-field $\sigma(\mathcal{B})$ is taken to be the Borel σ-field $\sigma(\mathcal{F})$ of $\mathcal{F}$, where $\mathcal{F}$ is topologized as usual, i.e., generated by the base

$$\mathcal{F}^K_{A_1,\ldots,A_n} = \mathcal{F}^K \cap \mathcal{F}_{A_1} \cap \cdots \cap \mathcal{F}_{A_n},$$

where $n \geq 0$, K is compact, the Θ_i's are open sets, and

$$\mathcal{F}^K = \{F : F \in \mathcal{F}, F \cap K = \emptyset\}, \quad \mathcal{F}_{A_i} = \{F : F \in \mathcal{F}, F \cap A_i \neq \emptyset\}.$$

See [9]. If $S : \Omega \longrightarrow \mathcal{F}$ is $\mathcal{A}$-$\sigma(\mathcal{F})$ measurable (S is a random closed set), then the counter-part of the distribution function of a real-valued random variable is the set function G, defined on the class of compact sets $\mathcal{K}$ of $\mathbb{R}$ by

$$G(K) = P\left(\{\omega : S(\omega) \cap K \neq \emptyset\}\right), \quad K \in \mathcal{K}.$$

It can be verified that $G(\emptyset) = 0$, $K_n \searrow K \Longrightarrow G(K_n) \searrow G(K)$, and G is alternating of infinite order on $\mathcal{K}$. As in the previous section, G can be extended to $2^{\mathbb{R}}$ to be a $\mathcal{K}$-capacity, alternating of infinite order with $G(\emptyset) = 0$ and $0 \leq G(\cdot) \leq 1$ [9].

As in the case of distribution functions of random variables, an axiomatic definition of the set function G, characterizing a random closed set is possible. Indeed, a *distribution functional* is a set-function G: $\mathcal{K} \longrightarrow [0, 1]$ such that

(i) $G(\emptyset) = 0$.
(ii) $K_n \searrow K$ in $\mathcal{K}$ implies $G(K_n) \searrow G(K)$.
(iii) G is alternating of infinite order on $\mathcal{K}$.

Such a distribution functional characterizes a random closed set, thanks to Choquet's theorem [1], [9], in the sense that there is a (unique) probability measure Q on $\sigma(\mathcal{F})$ such that

$$Q\left(\mathcal{F}_K\right) = G(K), \qquad \forall K \in \mathcal{K}.$$

Thus, a plausibility function is a distribution functional of a random closed set, see also[9], [11].

Regarding Example 2.3 of the previous section, let $\phi : \mathbb{R} \longrightarrow [0, 1]$ be an upper semi-continuous function and $X : (\Omega, \mathcal{A}, P) \longrightarrow [0, 1]$ be a uniformly distributed random variable. Consider the random closed set

$$S(\omega) = \{x \in \mathbb{R} : \phi(x) \geq X(\omega)\}.$$

Its distribution functional is

$$\begin{aligned} G(K) &= P\left(\{\omega : S(\omega) \cap K \neq \emptyset\}\right) \\ &= P\left(\left\{\omega : X(\omega) \leq \sup_{x \in K} \phi(x)\right\}\right) \\ &= \sup_{x \in K} \phi(x). \end{aligned}$$

REMARK 3.1. Suppose that ϕ is arbitrary (not necessarily upper semi-continuous). Then S is a multi-valued mapping with values in $2^{\mathbb{R}}$, and we have

$$\phi(x) = P\left(\{\omega : x \in S(\omega)\}\right) \qquad \forall x \in \mathbb{R},$$

i.e., the *one-point-coverage function* of S [4]. If ϕ is the membership function of a fuzzy subset of $\mathbb{R}$ [20], then this relation provides a probabilistic interpretation of fuzzy sets in terms of random sets. □

Similarly to the maxitive distribution function G above, Zadeh [21] introduced the following concept of *possibility measures.* Let $\phi : U \to [0,1]$ such that $\sup_{u\in U} \phi(u) = 1$. Then the set function $\pi : 2^U \to [0,1]$ defined by $\pi(A) = \sup_{u\in A} \phi(u)$, is called a possibility measure. It is possible to give a connection with random sets. Indeed, if $\mathcal{M} \subseteq 2^U$ denotes the σ-field generated by the semi-algebra (which is also a compact class in $\mathcal{M}$ [9]), $\mathcal{D} = \{M(I, I^c) : I, I^c \in \mathcal{T}\}$, where $\mathcal{T}$ is the collection of all finite subsets of U, and $M(I, I^c) = \{A \in 2^U : I \in A,\ A \cap I^c \neq \emptyset\}$, then there exists a unique probability measure Q on $(2^U, \mathcal{M})$ such that

$$\pi(I) = Q\left(\{A :\ A \in 2^U,\ A \cap I \neq \emptyset\}\right) \qquad \forall I \in \mathcal{T}.$$

As a consequence, there exists a random set $S : (\Omega, \mathcal{A}, P) \longrightarrow (2^U, \mathcal{M})$ such that

$$\pi(A) = \sup\{\pi(I) :\ I \in \mathcal{T},\ I \subseteq A\} \qquad \forall A \in 2^U$$

with $\pi(I) = P\left(\{\omega :\ S(\omega) \cap I \neq \emptyset\}\right)$, for $I \in \mathcal{T}$.

It is interesting to point out here that possibility measures on a topological space, say, $\mathbb{R}^d$, with upper semi-continuous distribution functions are capacities which admit *Radon-Nikodym derivatives* with respect to some canonical capacity. This can be seen as follows.

Let $\Phi : \mathbb{R}^d \longrightarrow [0,1]$ be upper semi-continuous and $\sup_{x\in\mathbb{R}^d} \Phi(x) = 1$. On $\mathcal{B}(\mathbb{R}^d)$, the set-function

$$\mu(A) = \int_A \Phi(x)dx,$$

is a measure with Radon-Nikodym derivative (with respect to Lebesgue measure dx on $\mathbb{R}^d$) $\Phi(x)$, denoted as $d\mu/dx$.

If we replace the integral sign by supremum, then the set-function G on $\mathcal{B}(\mathbb{R}^d)$, defined by

$$G(A) = \sup_{x\in A} \Phi(x),$$

is no longer additive. It is a capacity alternating of infinite order.

Now, observe that

$$G(A) = \int_0^{G(A)} dt = \int_0^{\infty} \nu\{x \in \mathbb{R}^d :\ (1_A\Phi)(x) > t\}dt,$$

where 1_A is the indicator function of A and the set-function ν is defined on $\mathcal{B}(\mathbb{R}^d)$ by

$$\nu(B) = \begin{cases} 1 & \text{if } B \neq \emptyset \\ 0 & \text{if } B = \emptyset. \end{cases}$$

Note that, ν being monotone, the above $G(A)$ is the Choquet integral of the function $1_A\Phi$ with respect to ν. Indeed,

$$\{x : (1_A\Phi)(x) > t\} = A \bigcap \{x : \Phi(x) > t\},$$

which is non-empty if and only if $t \leq \sup_{x\in A} \Phi(x) = G(A)$. Since for each $A \in \mathcal{B}(\mathbb{R}^d)$, $G(A)$ is the Choquet integral of $1_A\Phi$ with respect to the capacity ν, we call Φ the Radon-Nikodym derivative of the capacity G with respect to the capacity ν. Specifically $dG/d\nu = \Phi$ means that

$$G(A) = \int_A \Phi d\nu \qquad \forall A \in \mathcal{B}(\mathbb{R}^d),$$

where this last integral is understood in the sense of Choquet, i.e.,

$$\int_A \Phi d\nu \quad \text{means} \quad \int \nu\{x : (1_A\Phi)(x) > t\}dt,$$

where dt denotes Lebesgue measure on $\mathbb{R}^+$. In other words, G is an indefinite "integral" with respect to ν.

For a reference on measures and integrals in the context of non-additive set-functions, see Denneberg [3]. As far as we know, a Radon-Nikodym theorem for capacities (in fact, for a class of capacities) is given in Graf [6]. We thank I. Molchanov for calling our attention on this reference.

It should be noted that the problem of establishing a Radon-Nikodym theorem for capacity functionals of random sets, or more generally, for arbitrary monotone set-functions, is not only an interesting mathematical problem but is also important for application purposes. For example, such a theorem can provide a simple way to specify models for random sets and foundations for rule-based systems in which uncertainty is non-additive in nature. Obvious open problems are the following:

1. A constructive approach to Radon-Nikodym derivatives of capacity functionals of random sets, similar to the one in constructive measure theory based on derivatives of set-functions (see e.g. Shilov and Gurevich [16]).

2. Extending Radon-Nikodym theorems for capacities to monotone set-functions (such as fuzzy measures).

3. Provide an appropriate counter-part of the concept of conditional expectation in probability theory for the purpose of inference in the face of non-additive uncertainty.

Note that one can generalize Radon-Nikodym derivatives to non-additive set-functions without using the concept of indefinite integral, but only requiring that it coincides with ordinary Radon-Nikodym derivatives of measures, where the set-functions in question become (absolutely continuous) measures. For example, in Huber and Strassen [7], a type of Radon-Nikodym derivative for two 2-alternating capacities μ and ν on a complete, separable, metrizable space is a function f such that

$$\nu(f > t) = \int_{(f>t)} f(x)d\mu(x) \qquad \forall t \in \mathbb{R},$$

where the last integral is taken in the sense of Choquet. In particular, if μ is a probability measure, ν needs not be a measure since the relation

$$\nu(A) = \int_A f(x)d\mu(x)$$

holds only for sets A of the form $A = (f > t)$, $t \in \mathbb{R}$.

4. Reasoning with non–additive set–functions. Belief functions are non-additive set-functions of some special kind. They correspond to distribution functionals of random sets. If we view a belief function as a quantitative representation of a piece of evidence, then the problem of combining evidence can be carried out by using formal associated random sets. For example, two pieces of evidence are represented by two random sets S_1 and S_2. The combined evidence is the random set $S_1 \cap S_2$ from which the combined belief function is derived as the distribution functional of this new random set. This can be done, for example, when pieces of evidence are independent, where by this we mean that the random sets S_1 and S_2 are stochastically independent in the usual sense.

Reasoning or inference with conditional information is a common practice. Recall that probabilistic reasoning and the use of conditional probability measures rest on a firm mathematical foundation and are well justified. The situation for belief functions, and more generally for non-additive set-functions, such as Choquet capacities, monotone set-functions, etc., seems far from satisfactory. Since this is an important problem, we will elaborate on it in some detail.

Let X and Y be two random variables, defined on $(\Omega, \mathcal{A}, P)$. On the product σ-field $\mathcal{B}(\mathbb{R}) \otimes \mathcal{B}(\mathbb{R})$ of $\mathbb{R}^2$, the joint probability measure of (X, Y) is determined by

$$Q(A \times B) = P(X \in A, Y \in B).$$

The marginal probability measures of X and Y are, respectively,

$$Q_X(A) = Q(A \times \mathbb{R})$$

and

$$Q_Y(B) = Q(\mathbb{R} \times B).$$

Suppose that X is a continuous variable, so that $P(X = x) = 0$ for $x \in \mathbb{R}$. Computing probabilities related to Y under an observation $X = x$ requires some mathematical justifications. Specifically, the set-function $B \longrightarrow P(Y \in B|X = x)$ should be viewed as a conditional probability measure of Y given $X = x$. Let us sketch the existence of such conditional probability measures in a general setting. The general problem is the existence of the *conditional expectation* of an integrable random variable Y given X. Since Y is integrable, the set-function on $\mathcal{B}(\mathbb{R})$ defined by

$$M(B) = \int_{(X \in B)} Y(\omega) dP(\omega)$$

is a signed measure, absolutely continuous with respect to Q_X, and hence, by the Radon-Nikodym theorem from the standard measure theory, there exists a $\mathcal{B}$-measurable function f, unique up to a set of Q_X-measure zero, such that

$$\int_{(X \in B)} Y(\omega) dP(\omega) = \int_B f(x) dQ_X(x).$$

When Y is of the form $1_{(Y \in B)}$, we write $Q(B|X) = E\left[1_{(Y \in B)}|X\right]$.

The kernel $K : \mathcal{B}(\mathbb{R}) \times \mathbb{R} \longrightarrow [0,1]$ defined by $K(B, x) = Q(B|x)$ is a probability measure for each fixed x, and is a $\mathcal{B}$-measurable function for each fixed B. We have

$$Q(A \times B) = \int_A K(B, x) dQ_X(x)$$

and

$$Q(\mathbb{R} \times B) = Q_Y(B) = \int_{\mathbb{R}} K(B, x) dQ_X(x).$$

Thus, the conditional probability measure $B \longrightarrow P(Y \in B|X = x)$ is a kernel with the above properties. This can be used as a guideline for defining conditional non-additive set-functions. The σ-additivity of the kernel K, for each x, is weaken to the requirement that it is a monotone set-function. Measures such as Q_X and Q_Y above are replaced by belief functions, or more generally, by monotone set-functions. The Lebesgue integral is replaced by the Choquet integral. Two important problems then arise. The existence of conditional set-functions is related to a Radon-Nikodym theorem for non-additive set-functions, and its analysis involves some form of *integral equations*, where by an integral equation, in the context of non-additive set-functions, we mean an integral equation in which the integral is taken in the Choquet sense. As far as we know, *Choquet-integral equations* in this sense have not been the subject of mathematical research.

REFERENCES

[1] G. CHOQUET, *Theory of capacities*, Ann. Inst. Fourier, Vol. 5, (1953/54), pp. 131–295.

[2] A.P. DEMPSTER, *Upper and lower probabilities induced by a multi-valued mapping*, Ann. Math. Statist., Vol. 38 (1967), pp. 325–339.

[3] D. DENNEBERG, *Non-Additive Measure and Integral*, Kluwer Academic, Dordrecht, 1994.

[4] I.R. GOODMAN, *Fuzzy sets as equivalence classes of random sets*, in Fuzzy Sets and Possibility Theory (R. Yager Ed.), (1982), pp. 327–343.

[5] I.R. GOODMAN, H.T. NGUYEN, AND G.S. ROGERS, *On the scoring approach to admissibility of uncertainty measures in expert systems*, J. Math. Anal. and Appl., Vol. 159 (1991), pp. 550–594.

[6] S. GRAF, *A Radon-Nikodym theorem for capacities*, Journal fuer Reine und Angewandte Mathematik, Vol. 320 (1980), pp. 192–214.

[7] P.J. HUBER AND V. STRASSEN, *Minimax Tests and the Neyman-Pearson lemma for capacities*, Ann. Statist., Vol. 1 (1973), pp. 251–263.

[8] D.V. LINDLEY, *Scoring rules and the inevitability of probability*, Intern. Statist. Rev., Vol. 50 (1982), pp. 1–26.

[9] G. MATHERON, *Random Sets and Integral Geometry*, J. Wiley, New York, 1975.

[10] P.A. MEYER, *Probabilites et Potentiel*, Hermann, Paris, 1966.

[11] I.S. MOLCHANOV, *Limit Theorems for Unions of Random Closed Sets*. Lecture Notes in Mathematics, No. 1561, Springer Verlag, New York, 1993.

[12] H.T. NGUYEN, *On random sets and belief functions*, J. Math. Anal. and Appl., Vol. 65 (1978), pp. 531–542.

[13] A. REVUZ, *Fonctions croissantes et mesures sur les espaces topologiques ordonnes*, Ann. Inst. Fourier, (1956), pp. 187–268.

[14] G. SHAFER, *A Mathematical Theory of Evidence*, Princeton, New Jersey, 1976.

[15] L.S. SHAPLEY, *Cores of convex games*, Intern. J. Game Theory, Vol. 1 (1971), pp. 11–26.

[16] G.E. SHILOV AND B.L. GUREVICH, *Integral, Measure and Derivative: A Unified Approach*, Prentice-Hall, New Jersey, 1996.

[17] A.W. VAN DER VAART AND J.A. WELLER, *Weak Convergence and Empirical Processes*, Springer Verlag, New York, 1996.

[18] L.A. WASSERMAN, *Some Applications of Belief Functions to Statistical Inference*, Ph.D. Thesis, University of Toronto, 1987.

[19] R.R. YAGER, J. KACPRZYK, AND M. FEDRIZZI, *Advances in The Dempster-Shafer Theory of Evidence*, J. Wiley, New York, 1994.

[20] L.A. ZADEH, *Fuzzy sets*, Information and Control, Vol. 8 (1965), pp. 338–353.

[21] L.A. ZADEH, *Fuzzy sets as a basis for a theory of possibility*, Fuzzy Sets and Systems, Vol. 1 (1978), pp. 3–28.

PART III
Theoretical Statistics and Expert Systems

UNCERTAINTY MEASURES, REALIZATIONS AND ENTROPIES*

ULRICH HÖHLE† AND SIEGFRIED WEBER‡

Abstract. This paper presents the axiomatic foundations of uncertainty theories arising in quantum theory and artificial intelligence. Plausibility measures and additive uncertainty measures are investigated. The representation of uncertainty measures by random sets in spaces of events forms a common base for the treatment of an appropriate integration theory as well as for a reasonable decision theory.

Key words. Lattices of Events, Conditional Events, Quantum de Morgan Algebras, Plausibility Measures, Additive Uncertainty Measures, Realizations, Integrals, Entropies.

AMS(MOS) subject classifications. 60D05, 03G10

1. Introduction. After A.N. Kolmogoroff's pioneering work [16] on the mathematical foundations of probability theory had appeared in 1933, in the consecutive years several generalizations of probability theory have been developed for special requirements in quantum or system theory. As examples we quote quantum probability theory [7], plausibility theory [26] and fuzzy set theory [31]. The aim of this paper is to unify and for the first time to give a coherent and comprehensive presentation of these various types of uncertainty theories including new developments related to MV-algebras and Girard-monoids. Stimulated by the fundamental work of J. Łos [17,18] we base our treatment on three basic notions (see also [22]): *Events*, *realizations of events* and *uncertainty measures*. Events constitute the logico-algebraic part, realizations the decision-theoretic part and uncertainty measures the analytic part of a given uncertainty theory. These apparently different pieces are tied together by the fundamental observation that uncertainty measures permit a purely measure-theoretic *representation* by *random sets* in the space of all events. Based on this insight into the axiomatic foundations of uncertainty theories we see that a natural integration and information theory is available for uncertainty measures defined on non necessarily distributive de Morgan algebras. In the case of σ-algebras of subsets of a given set X the integral coincides with Choquet's integral [6] and the entropy of discernibleness with Shannon's entropy. In general the additivity of uncertainty measures does not force the linearity of the integral. Moreover, it is remarkable to see that non trivial, additive uncertainty measures on weakly orthomodular lattices L can be represented by more than one random set in L; i.e. the associated decision theory is *not unique* and consequently fictitious. In particular,

* In memory of Anne Weber

† Bergische Universität, FB Mathematik, D–42097 Wuppertal, Germany
‡ Johannes Gutenberg Universität, FB Mathematik, D–55029 Mainz, Germany

this observation draws attention to the discussion of "hidden variables" in quantum mechanics. If we pass to *plausibility measures*, then this phenomenon does not occur – i.e. the representing random set (resp. regular Borel probability measure) is *uniquely* determined by the given plausibility measure and henceforth offers a non trivial decision theory. Depending on the outcomes of the corresponding uncertainty experiment we are in the position to specify the basic decision whether an event occurs or does not occur. Consequently we can talk about the "stochastic source" of a given uncertainty experiment. Along these lines we develop an information theory which leads to such an important concept as the entropy of discernibleness (resp. indiscernibleness) of events.

Finally we remark that as a by–product of these investigations we obtain a new probability theory on *MV–algebras* [5]. This type of uncertainty theory will have a significant impact on problems involving *measure–free conditioning*. Here we only show that the entropy of discernibleness of unconditional events and of conditional events coincide.

2. Lattices of events. As an appropriate framework for lattices of events we choose the class of *de Morgan algebras* – these are not necessarily distributive, but bounded lattices provided with an order reversing involution. The choice of such a structure is motivated by at least two arguments: First we intend to cover the type of probability theory arising in quantum theory ([29]); therefore we need weakly orthomodular lattices. Secondly, we would like to perform measure–free conditioning of events – an imperative arising quite naturally in artificial intelligence ([8]); this requires the structure of integral, commutative Girard–monoids ([11]). In order to fix notations we start with the following

DEFINITION 2.1. (a) A triple $(L, \leq, ^{\perp})$ is called a *de Morgan algebra* iff $(L, \leq)$ is a lattice with universal bounds $0, 1$, and $^{\perp} : L \longmapsto L$ is an order reversing involution (i.e. $^{\perp}$ is an antitone self–mapping s.t. $(\alpha^{\perp})^{\perp} = \alpha \; \forall \alpha \in L$). In particular, in any de Morgan algebra we can define the concept of *orthogonality* as follows:

$$\alpha \perp \beta \quad \Longleftrightarrow \quad \alpha \leq \beta^{\perp} \text{ and } \beta \leq \alpha^{\perp} \,, \quad \alpha, \beta \in L \,.$$

(b) A quadruple $(L, \leq, \boxplus, ^{\perp})$ is said to be a *quantum de Morgan algebra* iff the following conditions are satisfied:

(QM1) $(L, \leq, ^{\perp})$ is a de Morgan algebra.

(QM2) $(L, \leq, \boxplus)$ is a partially ordered monoid with 0 as unity.

(QM3) $\alpha \boxplus \alpha^{\perp} = 1 \qquad \forall \alpha \in L.$

(c) A de Morgan algebra $(L, \leq, ^{\perp})$ is a *weakly orthomodular lattice* (cf. [3])

iff $(L, \leq, {}^{\perp})$ is subjected to the additional axioms:

$$\text{(OM1)} \quad \alpha \vee \alpha^{\perp} = 1 \qquad \forall \alpha \in L.$$

$$\text{(OM2)} \quad \alpha \leq \beta \,,\, \alpha \perp \gamma \implies \alpha \vee (\beta \wedge \gamma) = (\alpha \vee \gamma) \wedge \beta \,.$$

(d) A triple $(L, \leq, *)$ is called an *integral, commutative Girard–monoid* iff $(L, \leq)$ is a lattice with universal bounds satisfying the following axioms (cf. [11]):

(G1) $(L, *)$ is a commutative monoid such that the upper universal bound 1 acts as unity.

(G2) There exists a further binary operation $\rightarrow : L \times L \longmapsto L$ provided with the subsequent properties:
(1) $\alpha * \beta \leq \gamma \iff \beta \leq \alpha \rightarrow \gamma .$ (Adjunction)
(2) $(\alpha \rightarrow 0) \rightarrow 0 = \alpha .$

(e) An integral, commutative Girard–monoid is called an *MV–algebra* iff $(L, \leq, *)$ satisfies the additional important property

$$\text{(MV)} \quad (\alpha \rightarrow \beta) \rightarrow \beta = \alpha \vee \beta \,.$$

PROPOSITION 2.1. Weakly orthomodular lattices and integral, commutative Girard–monoids are Quantum de Morgan algebras.

PROOF. Let $(L, \leq, {}^{\perp})$ be a weakly orthomodular lattice; then we put $\boxplus = \vee$. Axiom (QM2) is obvious, and (QM3) follows from (OM1). Further, let $(L, \leq, *)$ be an integral, commutative Girard–monoid. Then we can define the following operations:

$$\alpha^{\perp} = \alpha \rightarrow 0 \quad , \quad \alpha \boxplus \beta = (\alpha^{\perp} * \beta^{\perp})^{\perp} \,.$$

It is not difficult to show that $(L, \leq, \boxplus, {}^{\perp})$ is a quantum de Morgan algebra. □

EXAMPLE 2.1. (a) Let H be a Hilbert space. Then the lattice of all *closed* subspaces F of H is a complete, weakly orthomodular lattice. In particular the ortho–complement $F^{\perp}$ of F is given by the orthogonal space of F.
(b) The real unit interval $[0, 1]$ provided with the usual ordering and Łukasiewicz arithmetic conjunction T_m

$$\alpha * \beta = \max(\alpha + \beta - 1, 0) \,, \quad \alpha, \beta \in [0, 1]$$

is a complete MV–algebra. In particular $\alpha \rightarrow 0$ is given by

$$\alpha \rightarrow 0 = 1 - \alpha \qquad \forall \alpha \in [0, 1] \,.$$

(c) Let $\mathbb{B}$ be a Boolean algebra; then $(\mathbb{B}, \leq, \wedge)$ is an MV–algebra. Moreover, every Boolean algebra is also an orthomodular lattice.
(d) Let $L = \{0, a, b, c, d, 1\}$ be a set of six elements. On L we define a structure of an integral, commutative Girard–monoid $(L, \leq, *)$ in the following way:

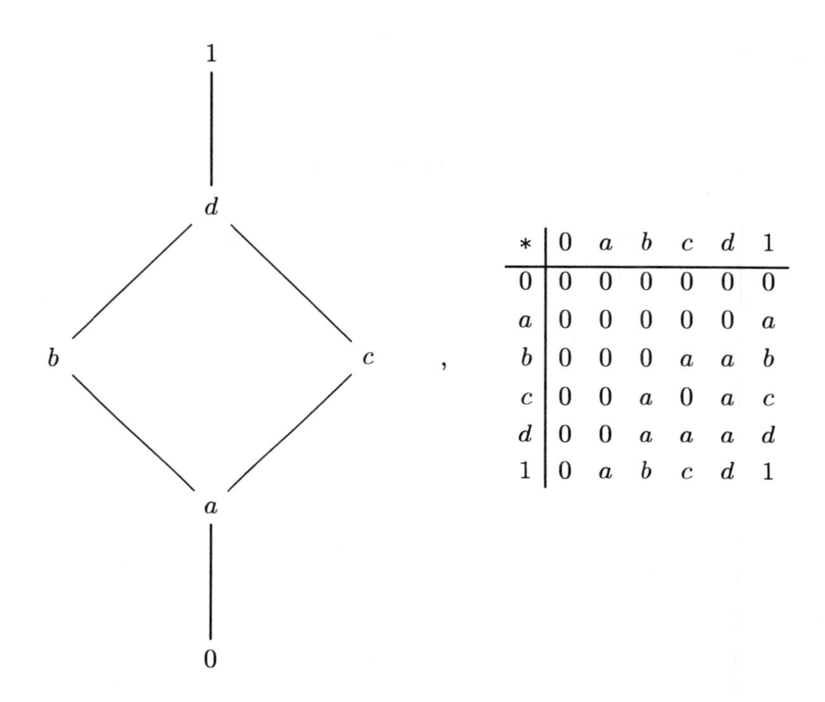

$*$	0	a	b	c	d	1
0	0	0	0	0	0	0
a	0	0	0	0	0	a
b	0	0	0	a	a	b
c	0	0	a	0	a	c
d	0	0	a	a	a	d
1	0	a	b	c	d	1

(e) On the set $L = \{0, a, b, 1\}$ of four elements we can specify the structure of a quantum de Morgan algebra as follows:

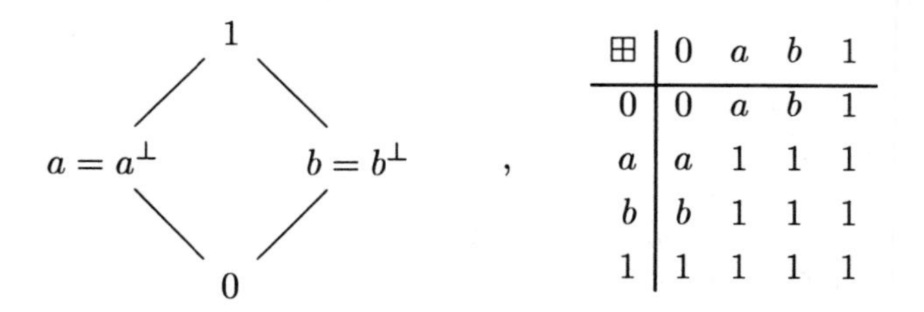

$\boxplus$	0	a	b	1
0	0	a	b	1
a	a	1	1	1
b	b	1	1	1
1	1	1	1	1

Further let $\boxtimes$ be the dual operator corresponding to $\boxplus$ – i.e.

$$\alpha \boxtimes \beta = (\alpha^{\perp} \boxplus \beta^{\perp})^{\perp} .$$

Then $(L, \leq, \boxtimes)$ is not an integral, commutative Girard–monoid. □

HIERARCHY:

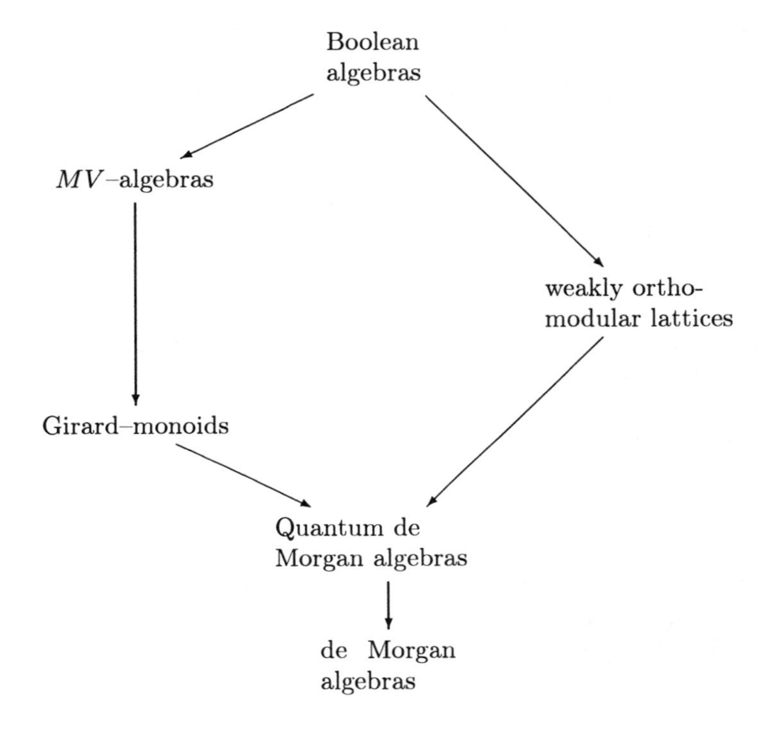

DEFINITION 2.2. (*Conditioning Operator*) Let $(L, \leq, *)$ be an integral, commutative Girard–monoid. A binary operator $| : L \times L \longmapsto L$ is called a *conditioning operator* iff $|$ satisfies the following axioms:

(C1) $\alpha \mid 1 = \alpha \qquad \forall \alpha \in L.$

(C2) $(\beta * (\beta \to \alpha)) \mid \beta = \alpha \mid \beta\,.$

(C3) $(\alpha \mid \beta) \to 0 = (\beta * (\alpha \to 0)) \mid \beta\,.$

(C4) $\alpha_1 \leq \alpha_2 \implies \alpha_1 \mid \beta \leq \alpha_2 \mid \beta \qquad \forall \beta \in L.$

(C5) $\beta_1 \leq \beta_2\;,\; \beta_2 * (\beta_2 \to \alpha) \leq \beta_1 * (\beta_1 \to \alpha)$

$$\implies \alpha \mid \beta_2 \leq \alpha \mid \beta_1\,.$$

LEMMA 2.1. Every conditioning operator fulfills the following property

$$\beta * (\beta \to \alpha) \leq \alpha \mid \beta \leq \beta \to \alpha\,.$$

PROOF. First we observe

$$1 * \Big(1 \to (\beta * (\beta \to \alpha))\Big) = \beta * (\beta \to \alpha) = \beta * \Big(\beta \to (\beta * (\beta \to \alpha))\Big) ;$$

then the inequality $(\beta * (\beta \to \alpha)) \mid 1 \leq (\beta * (\beta \to \alpha)) \mid \beta$ follows from (C5). Now we invoke (C1) and (C2) and obtain:

$$\beta * (\beta \to \alpha) \leq \alpha \mid \beta .$$

Further we observe:

$$1 * \Big(1 \to (\beta * (\alpha \to 0))\Big) = \beta * (\alpha \to 0) = \beta * \Big(\beta \to (\beta * (\alpha \to 0))\Big) ;$$

hence we infer from (C1) and (C5) : $\beta * (\alpha \to 0) \leq (\beta * (\alpha \to 0)) \mid \beta$. Now we embark on (C3) and obtain:

$$\alpha \mid \beta = \Big((\beta * (\alpha \to 0)) \mid \beta \Big) \to 0 \leq (\beta * (\alpha \to 0)) \to 0 = \beta \to \alpha .$$

□

REMARK 2.1. (a) The axiom (C3) implies: $(0 \mid 0) \to 0 = 0 \mid 0$; hence Boolean algebras do *not* admit conditioning operators. But there *exist* conditioning operators on MV–algebras. For example, let L be the real unit interval provided with Łukasiewicz arithmetic conjunction T_m (cf. Example 2.1(b)); then the operator $\mid$ defined by

$$\alpha \mid \beta = \begin{cases} \frac{\min(\alpha,\beta)}{\beta} & , \quad \beta \neq 0 \\ \frac{1}{2} & , \quad \beta = 0 \end{cases} .$$

is a conditioning operator on $([0,1], \leq, T_m)$.

(b) Let $(L, \leq, *)$ be an MV–algebra; then the axiom (MV) implies: $\beta * (\beta \to \alpha) = \alpha \wedge \beta$. Referring to the previous lemma we see that the event α *given* β (i.e. $\alpha \mid \beta$) is between the conjunction $\alpha \wedge \beta$ and the implication $\beta \to \alpha$. This is in accordance with the intuitive understanding of measure–free conditioning. A deeper motivation of the axioms (C1) – (C5) will appear in a forthcoming paper by the authors on Conditioning Operators (cf. [13]).

(c) Every Boolean algebra $(\mathbb{B}, \leq)$ can be embedded into an MV–algebra which admits a conditioning operator. Referring to [8,13] the MV–algebra $(L_c, \preccurlyeq, *)$ of *conditional events* of $\mathbb{B}$ is given by

$$L_c = \{(\alpha, \beta) \in \mathbb{B} \times \mathbb{B} \mid \alpha \leq \beta\} ,$$

$$(\alpha_1, \beta_1) \preccurlyeq (\alpha_2, \beta_2) \iff \alpha_1 \leq \alpha_2 \text{ and } \beta_1 \leq \beta_2 ,$$

$$(\alpha_1, \beta_1) * (\alpha_2, \beta_2) = (\alpha_1 \wedge \alpha_2 , (\alpha_1 \wedge \beta_2) \vee (\beta_1 \wedge \alpha_2)) .$$

It is not difficult to see that $j : \mathbb{B} \longmapsto L_c$ defined by $j(\alpha) = (\alpha, \alpha)$ is an injective MV–algebra–homomorphism and the map $| : L_c \times L_c \longmapsto L_c$ determined by

$$(\alpha_1, \beta_1) \mid (\alpha_2, \beta_2) = (\alpha_1 \wedge \alpha_2 \,, (\alpha_2 \to 0) \vee \beta_1)$$

is a conditioning operator on L_c. Moreover, Copeland's axiom (see also pp. 200–201 in [23])

$$(P \mid R) \mid (Q \mid R) = P \mid (R \wedge Q)$$

is valid in the following sense:

$$\big(j(\alpha) \mid j(\beta)\big) \mid \big(j(\gamma) \mid j(\beta)\big) = j(\alpha) \mid j(\beta \wedge \gamma) \,.$$

As a simple example let us consider the Boolean algebra $\mathbb{B}_2 = \{0, 1\}$ consisting of two elements. Then $(L_c, \preccurlyeq, *)$ is isomorphic to the MV–algebra $\{0, \frac{1}{2}, 1\}$ of three elements. If we view 0 as false and 1 as true, then the *conditional event* $\frac{1}{2} = 0 \mid 0$ can be interpreted as *indeterminate.*
(d) In the construction of (c) we can replace the Boolean algebra by an MV–algebra – i.e. every MV–algebra can be embedded into an integral, commutative Girard–monoid which admits a conditioning operator. The situation is as follows (cf. [13]): Let $(L, \leq, *)$ be an MV–algebra and $(L_c, \preccurlyeq, \circledast)$ be the Girard–monoid of all conditional events of L – i.e.

$$L_c = \{(\alpha, \beta) \in L \times L \mid \alpha \leq \beta\} \,,$$

$$(\alpha_1, \beta_1) \preccurlyeq (\alpha_2, \beta_2) \iff \alpha_1 \leq \alpha_2 \text{ and } \beta_1 \leq \beta_2 \,,$$

$$(\alpha_1, \beta_1) \circledast (\alpha_2, \beta_2) = (\alpha_1 * \alpha_2 \,, (\alpha_1 * \beta_2) \vee (\beta_1 * \alpha_2)) \,.$$

By analogy to (c) the map $j : L \longmapsto L_c$ defined by $j(\alpha) = (\alpha, \alpha)$ is an injective Girard-monoid-homomorphism, and $| : L_c \times L_c \longmapsto L_c$ determined by

$$(\alpha_1, \beta_1) \mid (\alpha_2, \beta_2) = \big(\alpha_2 * \big((\alpha_2 \to \alpha_1) \wedge (\beta_2 \to \beta_1)\big) \,, \alpha_2 \to \beta_1\big)$$

is a conditioning operator on L_c. In the case of $L = \{0, \frac{1}{2}, 1\}$ the Girard–monoid $(L_c, \preccurlyeq, \circledast)$ is already given in Example 2.1(d). □

3. Axioms of uncertainty measures. Let $(L, \leq, {}^\perp)$ be a de Morgan algebra. A map $\mu : L \longmapsto [0, 1]$ is called an *uncertainty measure* on L iff μ satisfies the following axioms

(M1) $\mu(0) = 0 \,, \quad \mu(1) = 1 \,.$ (Boundary Conditions)

$$\text{(M2)} \qquad \alpha \leq \beta \implies \mu(\alpha) \leq \mu(\beta) . \qquad \text{(Isotonicity)}$$

An uncertainty measure μ is called a *plausibility measure* iff μ satisfies the following inequality for any non empty finite subset $\{\alpha_1, \dots, \alpha_n\}$:

$$\text{(PL)} \qquad \mu(\bigwedge_{i=1}^{n} \alpha_i) \leq \sum_{i=1}^{n} (-1)^{i-1} \sum_{1 \leq j_1 < \cdots < j_i \leq n} \mu(\bigvee_{k=1}^{i} \alpha_{j_k}) .$$

If $(L, \leq)$ is a σ–complete lattice (i.e. joins and meets of at most countable subsets of L exist), then an uncertainty measure μ is said to be *σ–smooth* iff μ fulfills the following property

$$\text{(M3)} \qquad \alpha_n \leq \alpha_{n+1} , \quad \sup_{n \in \mathbb{N}} \mu(\alpha_n) = \mu(\bigvee_{n \in \mathbb{N}} \alpha_n) . \qquad (\sigma\text{–Smoothness})$$

PROPOSITION 3.1. Let μ be a *possibility* measure on L (cf. [24]) – i.e. $\mu : L \longmapsto [0,1]$ satisfies the axiom (M1) and the subsequent condition

$$\text{(PO)} \qquad \mu(\alpha \vee \beta) = \max(\mu(\alpha), \mu(\beta)) \qquad \forall \alpha, \beta \in L .$$

Then μ is a plausibility measure.

PROOF.

$$\mu(\bigwedge_{i=1}^{n} \alpha_i) \leq \min_{i=1,\dots,n} \mu(\alpha_i) = \sum_{i=1}^{n} (-1)^{i-1} \sum_{1 \leq j_1 < \cdots < j_i \leq n} \max_{k=1,\dots,i} \mu(\alpha_{j_k}).$$

□

DEFINITION 3.1. (*Additivity*) Let $(L, \leq, \boxplus, ^\perp)$ be a quantum de Morgan algebra. Further let $\boxtimes$ be the dual operation corresponding to $\boxplus$ – i.e. $\alpha \boxtimes \beta = (\alpha^\perp \boxplus \beta^\perp)^\perp$. An uncertainty measure μ on L is said to be *additive* iff μ is provided with the additional property

$$\text{(M4)} \qquad \alpha^\perp \boxtimes (\alpha \boxplus \beta) = \beta , \; \beta^\perp \boxtimes (\alpha \boxplus \beta) = \alpha \implies \mu(\alpha \boxplus \beta) = \mu(\alpha) + \mu(\beta) .$$

The following proposition shows that the axioms of isotonicity and additivity (i.e. (M2) and (M4)) in general are *not* independent from each other.

PROPOSITION 3.2. Let $(L, \leq, \boxplus, ^\perp)$ be a quantum de Morgan algebra satisfying the property

$$\text{(D)} \qquad \alpha \leq \beta \implies \beta \boxtimes (\beta^\perp \boxplus \alpha) = \alpha .$$

Then the axiom (M4) implies (M2).

PROOF. In order to verify (M2) we assume: $\alpha \leq \beta$. Since $^\perp$ is order reversing, the dual inequality $\beta^\perp \leq \alpha^\perp$ holds also. Now we embark on Condition (D) and obtain:

$$\beta^\perp = \alpha^\perp \boxtimes (\alpha \boxplus \beta^\perp) , \quad \alpha = \beta \boxtimes (\beta^\perp \boxplus \alpha) ;$$

hence the relations

$$\beta \boxtimes \alpha^\perp = \alpha^\perp \boxtimes (\alpha \boxplus (\beta \boxtimes \alpha^\perp)) ,$$

$$\alpha = (\beta^\perp \boxplus \alpha) \boxtimes (\alpha^\perp \boxtimes (\alpha \boxplus \beta^\perp))^\perp = (\beta \boxtimes \alpha^\perp)^\perp \boxtimes (\alpha \boxplus (\beta \boxtimes \alpha^\perp))$$

follow; i.e. α and $\beta \boxtimes \alpha^\perp$ satisfy the hypothesis of (M4). Therefore the additivity and the non negativity of μ imply:

$$\mu(\alpha) \leq \mu(\alpha) + \mu(\beta \boxtimes \alpha^\perp) = \mu(\beta) ;$$

hence (M2) is verified. □

PROPOSITION 3.3. Weakly orthomodular lattices and MV–algebras satisfy Condition (D) in Proposition 3.2.

PROOF. If $(L, \leq, ^\perp)$ is a weakly orthomodular lattice, then (D) is an immediate consequence of (OM1) and (OM2). In the case of MV–algebras we infer from Axiom (MV):

$$\beta \boxtimes (\beta^\perp \boxplus \alpha) = \beta * (\beta \to \alpha) = \beta \wedge \alpha ;$$

hence (D) follows immediately. □

MAIN RESULT I. In the case of weakly orthomodular lattices and MV–algebras the *additivity* of uncertainty measures implies already their *isotonicity*.

PROPOSITION 3.4. Let $(L, \leq, *)$ be an MV–algebra. Then every additive uncertainty measure μ on L is a valuation – i.e. μ fulfills the following property

$$\text{(M5)} \qquad \mu(\alpha \vee \beta) = \mu(\alpha) + \mu(\beta) - \mu(\alpha \wedge \beta) .$$

PROOF. Since $(L, \leq, *)$ is an MV–algebra, we derive from Axiom (MV) (see also Proposition 2.1) the following relations:

$$\alpha^\perp \boxtimes (\alpha \boxplus \beta) = (\alpha \to 0) * ((\alpha \to 0) \to \beta) = (\alpha \to 0) \wedge \beta ,$$

$$\beta^\perp \boxtimes (\alpha \boxplus \beta) = (\beta \to 0) * ((\beta \to 0) \to \alpha) = (\beta \to 0) \wedge \alpha ;$$

hence the additivity axiom (M4) is equivalent to

$$\alpha * \beta = 0 \implies \mu(\alpha \boxplus \beta) = \mu(\alpha) + \mu(\beta) .$$

Let us consider $\alpha, \beta \in L$; then we put $\gamma = \beta * (\alpha \to 0)$ and observe: $\gamma = \beta * ((\alpha \wedge \beta) \to 0)$. Now we embark on (MV) and obtain:

$$\begin{aligned} \alpha \boxplus \gamma &= \Big(((\beta * (\alpha \to 0)) \to 0) * (\alpha \to 0)\Big) \to 0 \\ &= (\beta \to \alpha) \to \alpha = \beta \vee \alpha \end{aligned}$$

$$\begin{aligned} (\alpha \wedge \beta) \boxplus \gamma &= \Big(((\beta * ((\alpha \wedge \beta) \to 0)) \to 0) * ((\alpha \wedge \beta) \to 0)\Big) \to 0 \\ &= (\beta \to (\alpha \wedge \beta)) \to (\alpha \wedge \beta) = \beta . \end{aligned}$$

Then the additivity of μ implies:

$$\mu(\alpha \vee \beta) = \mu(\alpha) + \mu(\gamma) = \mu(\alpha) + \mu(\beta) - \mu(\alpha \wedge \beta) .$$

□

PROPOSITION 3.5. Let $(L, \leq, {}^\perp)$ be a *distributive* de Morgan algebra, and μ be an uncertainty measure on L. If μ is a valuation (i.e. μ satisfies (M5)), then μ is a plausibility measure.

PROOF. Since $(L, \leq)$ is distributive, we deduce from (M5) by induction upon n:

$$\mu(\bigwedge_{i=1}^{n} \alpha_i) = \sum_{i=1}^{n} (-1)^{i-1} \sum_{1 \leq j_1 < \cdots < j_i \leq n} \mu(\bigvee_{k=1}^{i} \alpha_{j_k}) .$$

□

EXAMPLE 3.1 (a) Let H be a Hilbert space and $L(H)$ be the lattice of all closed subspaces of H (cf. Example 2.1(a)). With every closed subspace F we associate the corresponding *orthogonal projection* $P_F : H \longmapsto H$. Then every unit vector $x \in H$ (i.e. $\| x \|_2 = 1$) determines a σ–smooth, additive uncertainty measure[1] $\mu : L(H) \longmapsto [0,1]$ by $\mu(F) = \| P_F(x) \|_2^2 = < P_F(x), x >$.
(b) Let $(L, \leq, *)$ be an MV–algebra. A non empty set F is called a *filter* in L iff F satisfies the following conditions

(i) $\alpha \leq \beta$, $\alpha \in F \implies \beta \in F$.
(ii) $\alpha, \beta \in F \implies \alpha * \beta \in F$.
(iii) $0 \notin L$.

[1] In quantum theory (cf. [29]) σ–smooth, additive uncertainty measures are also called probability measures.

According to *Zorn's Lemma* every filter F in L is contained in an appropriate maximal filter U. It is well known (cf. [1]) that the quotient MV–algebra L/U w.r.t. a maximal filter U is a sub–MV–algebra of $([0,1], \leq, T_m)$ (cf. Example 2.1(b)). In order to show that every maximal filter U induces an additive measure on L we fix the following notations: $q_U : L \longmapsto L/U$ denotes the quotient map, and $j_U : L/U \longmapsto [0,1]$ is the embedding. Then the map $\mu : L \longmapsto [0,1]$ determined by $\mu = j_U \circ q_U$ is an additive uncertainty measure on L. In fact, since μ is an MV–algebra–homomorphism, we obtain in the case of $\alpha * \beta = 0$:

$$\mu(\alpha) + \mu(\beta) - 1 \leq 0 ;$$

hence the relation

$$\mu(\alpha \boxplus \beta) = \min(\mu(\alpha) + \mu(\beta), 1) = \mu(\alpha) + \mu(\beta)$$

follows.
(c) Let X be a non empty set; then $L = [0,1]^X$ is a de Morgan algebra w.r.t. the following order reversing involution:

$$f^{\perp}(x) = 1 - f(x) , \quad x \in X, f \in L .$$

Further let $h \in L$ with $\sup_{x \in X} h(x) = 1$. Then h induces a σ–smooth possibility measure μ_h on L by

$$\mu_h(f) = \sup_{x \in X} \min(f(x), h(x)) .$$

According to the terminology proposed by L.A. Zadeh ([32]) the map h is called the possibility distribution of μ_h. Obviously μ_h is not additive.
(d) Let X be a set, $\mathcal{M}$ be a σ–algebra of subsets of X, and let $\zeta(\mathcal{M})$ be the set of all $\mathcal{M}$–measurable functions $f : X \longmapsto [0,1]$. Then $L = \zeta(\mathcal{M})$ is a σ–complete MV–subalgebra of $[0,1]^X$ (cf. (c)). Further let μ be a σ–smooth uncertainty on $\zeta(\mathcal{M})$. We conclude from Theorem 4.1 in [12] that μ is a valuation on $\zeta(\mathcal{M})$ iff there exists an ordinary probability measure P on $\zeta(\mathcal{M})$ and a Markov kernel K such that

$$\mu(f) = \int_X K(x, [0, f(x)[) \, dP(x) \qquad \forall f \in \zeta(\mathcal{M}) .$$

In particular there exist non additive uncertainty measures on $\zeta(\mathcal{M})$ which are still valuations. □

PROPOSITION 3.6. Let $(\mathbb{B}, \leq)$ be a Boolean algebra and $(L_c, \preccurlyeq, *)$ be the MV–algebra of all conditional events of $\mathbb{B}$ (cf. Remark 2.1(c)). Further let μ be a finitely additive probability measure on $\mathbb{B}$. Then there exists a unique additive uncertainty measure $\bar{\mu}$ on L_c making the following diagram commutative

$$\begin{array}{ccc} \mathbb{B} & \xrightarrow{j} & L_c \\ & \searrow^{\mu} & \downarrow{\bar{\mu}} \\ & & [0,1] \end{array}$$

In particular the extension $\bar{\mu}$ is determined by

$$\bar{\mu}((\alpha,\beta)) = \frac{1}{2} \cdot (\mu(\alpha) + \mu(\beta)) .$$

PROOF. It is not difficult to show that $\bar{\mu}$ is an additive uncertainty measure on L_c extending μ. In order to verify the uniqueness of $\bar{\mu}$ we proceed as follows: Because of $(0,1) \rightarrow 0 = (0,1)$ the additivity of $\bar{\mu}$ implies: $\bar{\mu}((0,1)) = \frac{1}{2}$. Applying again the additivity of $\bar{\mu}$ we obtain

$$\bar{\mu}((\alpha,1)) = 1 - \bar{\mu}((0,\alpha \rightarrow 0)) ,$$

$$\bar{\mu}((0,\alpha)) + \bar{\mu}((0,\alpha \rightarrow 0)) = \bar{\mu}((0,1)) = \tfrac{1}{2} .$$

Since $\bar{\mu}$ is also a valuation on $(L_c, \preccurlyeq)$ (cf. Proposition 3.4) and an extension of μ (i.e. $\bar{\mu} \circ j = \mu$), the following relation holds:

$$\begin{aligned} \mu(\alpha) + \frac{1}{2} &= \bar{\mu}((\alpha,\alpha)) + \bar{\mu}((0,1)) = \bar{\mu}((\alpha,1)) + \bar{\mu}((0,\alpha)) \\ &= 1 - \bar{\mu}((0,\alpha \rightarrow 0)) + \bar{\mu}((0,\alpha)) = \frac{1}{2} + 2 \cdot \bar{\mu}((0,\alpha)) ; \end{aligned}$$

i.e. $\bar{\mu}((0,\alpha)) = \frac{1}{2} \cdot \mu(\alpha)$. Now we again use the valuation property and obtain:

$$\begin{aligned} \bar{\mu}((\alpha,\beta)) &= \bar{\mu}((\alpha,\alpha)) + \bar{\mu}((0,\beta)) - \bar{\mu}((0,\alpha)) \\ &= \mu(\alpha) + \frac{1}{2} \cdot (\mu(\beta) - \mu(\alpha)) = \frac{1}{2} \cdot (\mu(\alpha) + \mu(\beta)) . \end{aligned}$$

□

4. Realizations and uncertainty measures. In this section we discuss the decision–theoretic aspect of vague environments. After having specified the "*logic*" of a given environment – i.e. the choice of the lattice of events, we are now facing the problem to find a *mathematical model* which describes the "occurrence" of events. It is a well accepted fact that the *decision*, whether an event occurs, is intrinsically related to observed outcomes of a given uncertainty experiment. Hence the problem reduces to find a mathematical description of outcomes of uncertainty experiments. For the sake of simplicity we first recall the situation in *probability theory.* Let X be a finite sample space of a given *random experiment.* Since we assume classical logic, the lattice of events is given by the (finite) Boolean algebra $\mathcal{P}(X)$ of all subsets A of X. In this context an outcome of the given random experiment is identified with an element $x \in X$. If $x_0 \in X$ is observed, then we say an *event A has occurred* iff $x_0 \in A$. Obviously the event given by the whole sample space X occurs always, while the event determined by the empty set never occurs. Therefore X (resp. $\varnothing$) is called the *sure* (resp. *impossible*) event. Moreover, since $\mathcal{P}(X)$ is finite, we can make the following important observation:

> Every element $x_0 \in X$ (i.e. atom of $\mathcal{P}(X)$) can be identified with a lattice–homomorphism $h_0 : \mathcal{P}(X) \longmapsto \{0,1\}$, and vice versa every lattice–homomorphism $h_0 : \mathcal{P}(X) \longmapsto \{0,1\}$ determines a unique element $x_0 \in X$ by $\{A \in \mathcal{P}(X) \mid x_0 \in A\} = \{A \in \mathcal{P}(X) \mid h_0(A) = 1\}$.

Hence outcomes and lattice–homomorphisms $h : \mathcal{P}(X) \longmapsto \{0,1\}$ are the same things (see also 1.1 and 1.2 in [22]). This motivates the following definition:

DEFINITION 4.1. (*Realizations*) (a) Let $(L, \leq, ^{\perp})$ be a de Morgan algebra. A map $\omega : L \longmapsto \{0,1\}$ is called a *pseudo–realization* iff ω fulfills the conditions:

(R1) $\quad \omega(0) = 0\,, \quad \omega(1) = 1\,.$ (Boundary Conditions)

(R2) $\quad \ell_1 \leq \ell_2 \implies \omega(\ell_1) \leq \omega_2(\ell_2)\,.$ (Isotonicity)

A *realization* is a semilattice–homomorphism $\omega : L \longmapsto \{0,1\}$ – i.e. ω fulfills (R1) and the following axiom

(R3) $\quad \omega(\ell_1 \vee \ell_2) = \max(\omega(\ell_1), \omega(\ell_2)) \qquad \forall\, \ell_1, \ell_2 \in L\,.$

A realization ω is said to be *coherent* iff ω is provided with the additional property

(R4) $\quad \omega(\ell_1 \wedge \ell_2) = \min(\omega(\ell_1), \omega(\ell_2)) \qquad \forall\, \ell_1, \ell_2 \in L\,.$

(b) Let $\omega : L \longmapsto \{0,1\}$ be a pseudo–realization. An event $\ell \in L$ *occurs w.r.t.* ω iff $\omega(\ell) = 1$. An event ℓ is *discernible* from $\ell^{\perp}$ w.r.t. ω iff $\omega(\ell) \neq \omega(\ell^{\perp})$.

REMARK 4.1. (a) Every realization is also a pseudo–realization. Further coherent realizations and lattice–homomorphisms from L to $\{0,1\}$ are the same things. Due to Axiom (R1) the sure (resp. impossible) event always (resp. never) occurs.
(b) Let ω be a realization. If the event $\ell_1 \vee \ell_2$ occurs w.r.t. ω, then at least

ℓ_1 or ℓ_2 must occur.
(c) Let ω be a coherent realization. If the events ℓ_1 and ℓ_2 occur w.r.t. ω, then also the *conjunction* $\ell_1 \wedge \ell_2$ occurs w.r.t. ω. In this sense ω reflects a certain type of *coherency* in the given observation process.
(d) Let $(L, \leq, {}^\perp)$ be a weakly orthomodular lattice and $\omega : L \longmapsto \{0,1\}$ be a realization. If $\ell \in L$ does not occur w.r.t. ω, then ℓ and $\ell^\perp$ are discernible w.r.t. ω (cf. (b)).
(e) Let $L(H)$ be the lattice of all closed subspaces of a given Hilbert space H (cf. Example 2.1(a), Example 3.1(a)). $L(H)$ does *not* admit coherent realizations. □

Now we proceed to describe the special *linkage* between uncertainty measures and realizations. As the reader will see immediately, this linkage is essentially measure theoretical in nature. We prepare the situation as follows:

Let L be the lattice of events; on $\{0,1\}^L$ we consider the product topology τ_p with respect to the discrete topology on $\{0,1\}$. Referring to the Tychonoff–Theorem ([4]) it is easy to see that $(\{0,1\}^L, \tau_p)$ is a totally disconnected, compact, topological space. If we identify subsets of L with their characteristic functions, then every regular Borel probability measure on $\{0,1\}^L$ is called a *random set* in L. Moreover we observe that the subsets $\mathfrak{PR}(L)$ of all pseudo–realizations, $\mathfrak{R}(L)$ of all realizations and $\mathfrak{CR}(L)$ of all coherent realizations are closed w.r.t. τ_p.

In the following considerations we use the relative topology τ_r on $\mathfrak{PR}(L)$ induced by τ_p. Then $(\mathfrak{PR}(L), \tau_r)$ is again a totally disconnected, compact, topological space.

PROPOSITION 4.1. For every uncertainty measure μ on L there exists a regular Borel probability measure ν on the compact space $\mathfrak{PR}(L)$ of all pseudo–realizations satisfying the condition

$$\text{(b1)} \qquad \nu(\{\omega \in \mathfrak{PR}(L) \mid \omega(\ell) = 1\}) = \mu(\ell) \qquad \text{for all} \quad \ell \in L\,.$$

If L is σ–complete and μ σ–smooth, then ν fulfills the further condition

$$\text{(b2)} \qquad \nu(\{\omega \in \mathfrak{PR}(L) \mid \omega(\bigvee_{n\in\mathbb{N}} \ell_n) = 1\,,\ \omega(\ell_n) = 0\ \forall n \in \mathbb{N}\}) = 0$$

$$\text{where } \ell_n \leq \ell_{n+1}\,,\ n \in \mathbb{N}.$$

The proof of Proposition 4.1 is based on two technical Lemmas:

LEMMA 4.1. Let $\{I_j \mid j \in J\}$ be a non empty family of real intervals I_j such that the interior of I_j is non empty for all $j \in J$. Then $\bigcup_{j\in J} I_j$ is a Borel subset of the real line $\mathbb{R}$.

PROOF. Let $\mathcal{P}(\mathbb{R})$ be the power set of $\mathbb{R}$. Further, let $\mathfrak{Z}$ be the set of all subsets $\mathcal{A}$ of $\mathcal{P}(\mathbb{R})$ provided with the following properties:

(i) $\mathcal{A} \neq \varnothing$.

(ii) $A \in \mathcal{A} \implies A$ is an interval.

(iii) interior of $A \neq \varnothing \quad \forall A \in \mathcal{A}$.

(iv) $A_1, A_2 \in \mathcal{A} \implies A_1 \cap A_2 = \varnothing$.

(v) $\bigcup \mathcal{A} \subseteq \bigcup_{j \in J} I_j$.

Obviously $\mathfrak{Z}$ is non empty. We define a partial ordering $\preceq$ on $\mathfrak{Z}$ by

$$\mathcal{A}_1 \preceq \mathcal{A}_2 \iff \forall A_1 \in \mathcal{A}_1 \, \exists A_2 \in \mathcal{A}_2 \text{ s.t. } A_1 \subseteq A_2 .$$

(a) We show that $(\mathfrak{Z}, \preceq)$ is inductively ordered – i.e. any chain in $\mathfrak{Z}$ has an upper bound in $\mathfrak{Z}$ w.r.t. $\preceq$. Therefore let $\{\mathcal{A}_\lambda \mid \lambda \in \Lambda\}$ be a chain in $\mathfrak{Z}$. Then $\mathcal{B} = \bigcup_{\lambda \in \Lambda} \mathcal{A}_\lambda$ is a subset of $\mathcal{P}(\mathbb{R})$ which is partially ordered w.r.t. the set inclusion $\subseteq$. Since $\{\mathcal{A}_\lambda \mid \lambda \in \Lambda\}$ is a chain, we infer from Condition (iv) and the definition of $\preceq$ that the following implication holds

$$(4.1) \quad (A_1, A_2 \in \mathcal{B},\, A_1 \cap A_2 \neq \varnothing) \implies (A_1 \subseteq A_2 \text{ or } A_2 \subseteq A_1) .$$

Further let $\mathfrak{S}$ be the set of all maximal chains $\mathcal{C}$ in $\mathcal{B}$ w.r.t. $\subseteq$; we put $\mathcal{A}_\infty = \{\bigcup \mathcal{C} \mid \mathcal{C} \in \mathfrak{S}\}$. Then it is easy to see that $\mathcal{A}_\infty$ fulfills the conditions (i), (ii), (iii) and (v). In order to verify (iv) we proceed as follows: Let us consider $\mathcal{C}_1, \mathcal{C}_2 \in \mathfrak{S}$ with $(\bigcup \mathcal{C}_1) \cap (\bigcup \mathcal{C}_2) \neq \varnothing$; then there exists $A_1 \in \mathcal{C}_1$ and $A_2 \in \mathcal{C}_2$ s.t. $A_1 \cap A_2 \neq \varnothing$. Referring to (4.1) we assume without loss of generality: $A_1 \subseteq A_2$. Now we again invoke (4.1) and obtain that $\mathcal{C}_1 \cup \{A_2\}$ is a chain in $\mathcal{B}$; hence the maximality of $\mathcal{C}_1$ implies: $A_2 \in \mathcal{C}_1$ – i.e. $A_2 \in \mathcal{C}_1 \cap \mathcal{C}_2$. Using once more (4.1) and the maximality of $\mathcal{C}_1$ and $\mathcal{C}_2$ it is not difficult to verify

$$\{B \in \mathcal{C}_1 \mid A_2 \subseteq B\} = \{B \in \mathcal{C}_2 \mid A_2 \subseteq B\} ;$$

hence we obtain: $\bigcup \mathcal{C}_1 = \bigcup \mathcal{C}_2$; i.e. $\mathcal{A}_\infty$ satisfies (iv). Since every chain can be extended to a maximal chain, the construction of $\mathcal{A}_\infty$ shows that $\mathcal{A}_\infty$ is an upper bound of $\{\mathcal{A}_\lambda \mid \lambda \in \Lambda\}$ w.r.t. $\preceq$.

(b) By virtue of *Zorn's Lemma* $(\mathfrak{Z}, \preceq)$ contains at least a maximal element $\mathcal{A}_0$ w.r.t. $\preceq$. We show that for every $j \in J$ there exists $A \in \mathcal{A}_0$ s.t. $I_j \subseteq A$. Let us assume the contrary – i.e. there exists $j \in J$ such that for all $A \in \mathcal{A}_0$: $I_j \not\subseteq A$. We denote the left (resp. right) boundary point of I_j by r_j (resp. s_j). Then it is not difficult to show that every $A \in \mathcal{A}_0$ provided with the properties

$$A \not\subseteq I_j , \quad A \cap I_j \neq \varnothing$$

contains either r_j or s_j (but not both boundary points); hence the property (iv) implies that

$$\mathcal{D} = \{A \in \mathcal{A}_0 \mid A \not\subseteq I_j , A \cap I_j \neq \varnothing\}$$

has at most cardinality 2. In particular, $A^* = (\cup \mathcal{D}) \cup I_j$ is an interval. Now we are in the position to define a further subset $\mathcal{A}^*$ by

$$\mathcal{A}^* = \{A \in \mathcal{A}_0 \mid A \cap I_j = \varnothing\} \cup \{A^*\} .$$

Obviously $\mathcal{A}^*$ is an element of $\mathfrak{Z}$ and satisfies the inequality $\mathcal{A}_0 \preceq \mathcal{A}^*$. Further, the choice of I_j implies: $\mathcal{A}^* \not\preceq \mathcal{A}_0$ which is a contradiction to the maximality of $\mathcal{A}_0$. Hence the assumption is false – i.e. for every $j \in J$ there exists an interval $A \in \mathcal{A}_0$ s.t. $I_j \subseteq A$. In particular we infer from (v) that $\cup \mathcal{A}_0 = \bigcup_{j \in J} I_j$.

(c) Since the interior of each interval of $\mathcal{A}_0$ is non empty, we choose a rational number $q_A \in A$ for all $A \in \mathcal{A}_0$; hence the countability of $\mathcal{A}_0$ follows from Condition (iv). Therefore we conclude from part (b) that $\bigcup_{j \in J} I_j$ is a Borel subset of $\mathbb{R}$. □

LEMMA 4.2. Let $\{[r_j, s_j[\mid j \in J\}$ be a family of half open, non empty subintervals of $[0,1[$. Then the set

$$\Big(\bigcup_{j \in J} [r_j, s_j[\Big) \cap \complement\Big(\bigcup_{j \in J}]r_j, s_j[\Big)$$

is at most countable.

PROOF. Let B be the set of all real numbers $b \in \bigcup_{j \in J} [r_j, s_j[$ provided with the property $b \notin]r_j, s_j[\ \forall j \in J$. For every $b \in B$ we define an index set $\mathcal{K}_b = \{j \in J \mid b \in [r_j, s_j[\}$ and put $c_b = \sup\{s_j \mid j \in \mathcal{K}_b\}$. Now we choose $b_1, b_2 \in B$ and show:

$$[b_1, c_{b_1}[\cap [b_2, c_{b_2}[= \varnothing \quad \text{whenever} \quad b_1 \neq b_2 . \tag{4.2}$$

For simplicity we consider the case $b_1 < b_2$ and assume that the intersection $[b_1, c_{b_1}[\cap [b_2, c_{b_2}[$ is non empty. Then there exists $j_0 \in \mathcal{K}_{b_1}$ s.t.

$$(r_{j_0} = b_1) < b_2 < s_{j_0} \ (\leq c_{b_1}) \ \text{– i.e. } b_2 \in]r_{j_0}, s_{j_0}[$$

which is a contradiction to $b_2 \in B$. Hence (4.2) is verified. Since every interval $[b, c_b[$ contains a rational number, we conclude from (4.2) that B is at most countable. □

PROOF. (of Proposition 4.1) Let $[0,1[$ be provided with the usual Lebesgue measure λ. Further let μ be an uncertainty measure on L. Because of the axioms (M1) and (M2) we can define a map $\Theta_\mu : [0,1[\longmapsto \mathfrak{PR}(L)$ by

$$[\Theta_\mu(r)](\ell) = \begin{cases} 1 , & r < \mu(\ell) \\ 0 , & r \geq \mu(\ell) \end{cases} , \quad r \in [0,1[.$$

Then we obtain

$$(4.3) \qquad \{r \in [0,1[\mid [\Theta_\mu(r)](\ell) = 1\} = [0, \mu(\ell)[$$

and conclude from Lemma 4.1 that Θ_μ is Borel measurable with respect to the corresponding topologies. Let ν be the image measure of λ under Θ_μ. Then Lemma 4.2 shows that ν is a regular Borel probability measure on $\mathfrak{PR}(L)$. Referring to (4.3) we obtain

$$\nu(\{\omega \in \mathfrak{PR}(L) \mid \omega(\ell) = 1\}) = \lambda([0, \mu(\ell)[) = \mu(\ell) ;$$

i.e. ν satisfies Condition (b1).
Further we assume that L is σ–complete and μ is σ–smooth. Then we obtain for any nondecreasing sequence $(\ell_n)_{n\in\mathbb{N}}$ in L (i.e. $\ell_n \leq \ell_{n+1}$):

$$\mu(\bigvee_{n\in\mathbb{N}} \ell_n) = \sup_{n\in\mathbb{N}} \mu(\ell_n) = \lambda(\bigcup_{n\in\mathbb{N}} \{ r \in [0,1[\mid [\Theta_\mu(r)](\ell_n) = 1 \}) ;$$

hence ν fulfills also Property (b2). □

Let μ be an uncertainty measure on L. It is not difficult to see that regular Borel probability measures ν on the space of all pseudo–realizations satisfying (b1) are *not uniquely* determined by μ and (b1). This fact can be demonstrated by the following simple counterexample. Let $L = \{1, a, a^\perp, b, b^\perp, 0\}$ be the (weakly) orthomodular lattice consisting of six elements:

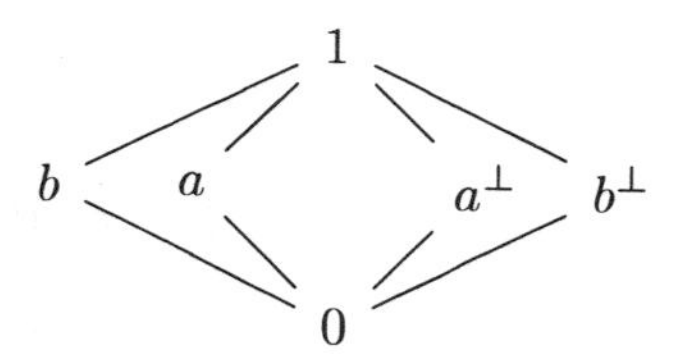

Further let μ be an additive uncertainty measure on L defined as follows:

$$\mu(a) = \mu(a^\perp) = \frac{1}{2}, \quad \mu(b) = \mu(0) = 0 , \ \mu(b^\perp) = \mu(1) = 1 .$$

Now let us consider the subsequent pseudo–realizations on L

$$\omega_1 = \chi_{\{1,a,b\}} , \ \omega_2 = \chi_{\{1,a^\perp,b\}} , \ \omega_3 = \chi_{\{1,a,a^\perp,b\}} , \ \omega_4 = \chi_{\{1,b\}}$$

and the discrete probability measures ν_1 and ν_2 on $\mathfrak{PR}(L)$ determined by

$$\nu_1(\{\omega_1\}) = \nu_1(\{\omega_2\}) = \frac{1}{2}, \quad \nu_2(\{\omega_3\}) = \nu_2(\{\omega_4\}) = \frac{1}{2} .$$

Then ν_1 and ν_2 satisfy Condition (b1) w.r.t. μ.

We can improve the situation, if we restrict our interest to plausibility measures.

THEOREM 4.1. (*Measure Theoretical Representation [10,12]*) For every plausibility measure μ on L there exists a *unique* regular Borel probability measure ν_μ on the compact space $\mathfrak{PR}(L)$ of all pseudo–realizations satisfying the following conditions

(i) ν_μ satisfies (b1) (cf. Proposition 4.1).

(ii) $\nu_\mu(\mathfrak{R}(L)) = 1$ – i.e. the support of ν_μ is contained in the set of all realizations of L.

Moreover, if L is σ–complete and μ is σ–smooth, then ν_μ satisfies also condition (b2) (cf. Proposition 4.1).

PROOF. Let $\mathfrak{A}$ be the set–algebra of all *cylindric* subsets of $\mathfrak{PR}(L)$. In particular, $\mathfrak{A}$ is generated by

$$\mathfrak{E} = \{\{\omega \in \mathfrak{PR}(L) \,|\, \omega(\ell) = 0\} \mid \ell \in L\}.$$

Because of (M1), (M2) and (PL) we can introduce a finitely additive probability measure η_μ on $\mathfrak{A}$ which is determined by $(m, n \in \mathbb{N})$

$$\eta_\mu\Big(\bigcap_{s=1}^{m}\big(\bigcap_{i=1}^{n}\{\omega \in \mathfrak{PR}(L) \mid \omega(\ell_s^*) = 0,\, \omega(\ell_i) = 1\}\big)\Big)$$

$$= \left(\sum_{i=1}^{n}(-1)^{i-1} \sum_{1 \le j_1 < \cdots < j_i \le n} \mu(\ell_0 \vee (\bigvee_{k=1}^{i} \ell_{j_k}))\right) - \mu(\ell_0),$$

where $\ell_0 = \bigvee_{s=1}^{m} \ell_s^*$, $\{\ell_1^*, \ldots, \ell_m^*\} \subseteq L$, $\{\ell_1, \ldots, \ell_n\} \subseteq L$. Since every cylindric set is compact w.r.t. τ_r, we can extend η_μ to a σ–additive Baire probability measure and consequently to a regular Borel probability measure ν_μ on $\mathfrak{PR}(L)$. By definition ν_μ fulfills Condition (b1). Referring again to the construction of η_μ we obtain

$$\eta_\mu(\{\omega \in \mathfrak{PR}(L) \mid \omega(\ell_1) = \omega(\ell_2) = 0,\, \omega(\ell_1 \vee \ell_2) = 1\}) = 0;$$

hence the support of ν_μ is contained in the closed subset $\mathfrak{R}(L)$ of all realizations; i.e. ν_μ fulfills the desired properties. Finally the uniqueness of ν_μ follows from (i), (ii) and the fact that the trace of $\mathfrak{E}$ on $\mathfrak{R}(L)$ – i.e. $\{\{\omega \in \mathfrak{R}(L) \mid \omega(\ell) = 0\} \mid \ell \in L\}$ – is stable w.r.t. finite intersections.

□

The unique Borel probability measure ν_μ constructed in Theorem 4.1 is called the *representing* Borel probability measure of μ.

Further, it is interesting to see that realizations on L *separate points* in L – i.e. for every pair $(\ell_1, \ell_2) \in L \times L$ with $\ell_1 \neq \ell_2$ there exists a realization ω with $\omega(\ell_1) \neq \omega(\ell_2)$; e.g. in the case of $\ell_1 \not\leq \ell_2$ we can use $\omega = \chi_{\{\ell \in L | \ell \not\leq \ell_2\}}$. Hence we deduce from Theorem 4.1:

MAIN RESULT II. Every plausibility measure is the *restriction* of its representing Borel probability measure.

COROLLARY 4.1. Let μ be a plausibility measure on L and ν_μ be the representing Borel probability measure on $\mathfrak{PR}(L)$ (resp. $\mathfrak{R}(L)$). μ is a valuation (i.e. satisfies (M5)) iff the support of ν_μ is contained in the set $\mathfrak{CR}(L)$ of all coherent realizations (i.e. $\nu_\mu(\mathfrak{CR}(L)) = 1$).

PROOF. From the definition of ν_μ we deduce

$$\nu_\mu(\{\omega \in \mathfrak{R}(L) \,|\, \omega(\ell_1) = 1\,,\, \omega(\ell_2) = 1\,,\, \omega(\ell_1 \wedge \ell_2) = 0\,\})$$

$$= \mu(\ell_1) + \mu(\ell_2) - \mu(\ell_1 \vee \ell_2) - \mu(\ell_1 \wedge \ell_2)\,;$$

hence the assertion follows. □

As an immediate consequence from Proposition 3.4, Proposition 3.5 and Corollary 4.1 we obtain:

COROLLARY 4.2. Let $(L, \leq, *)$ be an MV–algebra. Then for every additive uncertainty measure μ on L there exists a unique regular Borel probability measure ν_μ on $\mathfrak{PR}(L)$ satisfying the following conditions

(i) ν_μ fulfills condition (b1) (cf. Proposition 4.1).

(ii) $\nu_\mu(\mathfrak{CR}(L)) = 1$ – i.e. the support of ν_μ is contained in the set of all coherent realizations.

COROLLARY 4.3. Let $(L, \leq, *)$ be an MV–algebra, μ be an additive uncertainty measure and ν_μ be the representing regular Borel probability measure on $\mathfrak{PR}(L)$. Then the continuous map $\varphi^* : \mathfrak{PR}(L) \longmapsto \mathfrak{PR}(L)$ defined by

$$\big[\varphi^*(\omega)\big](\ell) = 1 - \omega(\ell^\perp) \quad \forall \omega \in \mathfrak{PR}(L)$$

is measure preserving w.r.t. ν_μ.

PROOF. Obviously φ^* is an involution leaving the set $\mathfrak{CR}(L)$ invariant. Further let $\varphi^*(\nu_\mu)$ be the image measure of ν_μ under φ^*. If we combine (M4) with (b1), then we obtain:

$$\begin{aligned} \nu_\mu\Big(\varphi^*(\{\omega \mid \omega(\ell) = 1\})\Big) &= \nu_\mu(\{\omega \mid \omega(\ell^\perp) = 0\}) \\ &= 1 - \nu_\mu(\{\omega \mid \omega(\ell^\perp) = 1\}) \\ &= 1 - \mu(\ell^\perp) = \mu(\ell) . \end{aligned}$$

Hence the assertion follows from Corollary 4.2. □

We finish this section with a discussion on realizations in two important special cases:

REMARK 4.2. (*Probabilistic Case*) Let $L = \mathbb{B}$ be a Boolean algebra, μ be a probability measure on $\mathbb{B}$, and let ν_μ be the representing Borel probability measure (cf. Corollary 4.2). Since the support of ν_μ is contained in $\mathfrak{CM}(\mathbb{B})$, all realizations ω coincide ν_μ–almost everywhere with ordinary, characteristic functions of ultrafilters on $\mathbb{B}$. In particular the so–called *Kolmogoroff decision* holds: An event $\ell \in \mathbb{B}$ occurs iff the complement $\ell^\perp$ does not occur. Moreover, if μ is atomless (resp. atomic), then ν_μ–almost everywhere all realizations are determined by free (resp. fixed) ultrafilters. □

REMARK 4.3 (*Possibilistic Case*) Let $\mathcal{P}(X)$ be the ordinary power set of a given, non empty set X and μ_h be a possibility measure on $\mathcal{P}(X)$ – i.e. there exists a map $h : X \mapsto [0,1]$ with $\sup_{x \in X} h(x) = 1$ such that $\mu_h(A) = \sup_{x \in A} h(x)$. In particular μ_h is the restriction of the possibility measure defined on $[0,1]^X$ in Example 3.1(c). Then there exists a Borel measurable map $\Theta_h : [0,1[\longmapsto \mathfrak{R}(\mathcal{P}(X))$ determined by

$$\Theta_h(\alpha) = \omega_\alpha \quad \text{where} \quad \omega_\alpha(A) = \begin{cases} 1, & A \cap \{x \in X \mid h(x) > \alpha\} \neq \varnothing \\ 0, & A \cap \{x \in X \mid h(x) > \alpha\} = \varnothing \end{cases} .$$

It is not difficult to see that the representing Borel probability measure ν_{μ_h} corresponding to μ_h is the image measure of the *Lebesgue measure* under Θ_h. Hence ν_{μ_h}–almost everywhere all realizations of μ_h are of type ω_α. With regard to the Kolmogoroff decision (cf. Remark 4.2) only one part of the bi–implication holds: If the complement of A does not occur, then A occurs. On the other hand, if A occurs, then realizations of possibility measures in general do not contain any information concerning the occurrence of the complement of A. In this sense realizations of possibility measures are *less specific* than those of probability measures. We will return to this point in Section 6. □

We finish this section with a brief discussion showing that basic theorems of the theory of random sets are consequences of Theorem 4.1. Since the proof of Theorem 4.1 is independent from the given order reversing involution $^\perp$ (in particular the natural domain of plausibility measures are bounded lattices), we can make the following observations:

REMARK 4.4. (*Random Sets*) Let $(L, \leq)$ be a bounded lattice such that the sublattice $L \setminus \{1\}$ is a freely generated join–semilattice (cf. [15]). Further let X be the set of free generators of $L \setminus \{1\}$. Since *realizations* preserve finite joins, the set $\mathfrak{R}(L)$ of all realizations on L is homeomorphic to the power set $\mathcal{P}(X)\, (\simeq \{0,1\}^X)$ of X. Hence we conclude from Theorem 4.1 that *random sets* in X (i.e. regular Borel probability measures on $\mathcal{P}(X)$) and *plausibility measures* on L are the *same* things. □

REMARK 4.5. (*Random Closed Sets [19]*) Let (X, τ) be a locally compact, topological space and $\mathfrak{K}(X)$ be the lattice of all τ–compact subsets of X. Then we put $\mathfrak{L} = \mathfrak{K}(X) \cup \{X\}$ and observe that $\mathfrak{L}$ is a bounded lattice w.r.t. the set inclusion $\subseteq$. A realization ω on $\mathfrak{L}$ is called *smooth* iff for every downward directed sub–family $\mathcal{F}$ of $\mathfrak{L}$ the following implication holds

$$\omega(K) = 1 \quad \forall K \in \mathcal{F} \quad \Longrightarrow \quad \omega(\cap \mathcal{F}) = 1 \, .$$

The set of all smooth realizations ω on $\mathfrak{L}$ is denoted by $\mathfrak{R}_s(\mathfrak{L})$. A realization $\omega \in \mathfrak{R}(\mathfrak{L})$ is *non trivial* (w.r.t. τ) iff there exists $K \in \mathfrak{K}(X) \setminus \{\varnothing\}$ s.t. $\omega(K) = 1$. If X is *not compact*, then $\mathfrak{R}(\mathfrak{L})$ contains a unique trivial realization ω_0 where ω_0 is given by $\omega_0(X) = 1$, $\omega_0(K) = 0 \;\forall K \in \mathfrak{K}(X)$. Moreover it is not difficult to show that *non empty τ–closed* subsets of X and *smooth, non trivial realizations* are the same things. In particular, if ω is a (smooth) non trivial realization, then[2]

$$F_\omega = X \cap \complement\{x \in X \mid \exists V \in \mathfrak{K}(X) : x \in V^o, \omega(V) = 0\}$$

is non empty and τ–closed; and vice versa, if F is a non empty τ–closed subset of X, then $\omega_F : \mathfrak{L} \longmapsto \{0,1\}$ determined by

$$\omega_F(K) = \begin{cases} 1 & : \quad K \cap F \neq \varnothing \\ 0 & : \quad K \cap F = \varnothing \end{cases} , \qquad K \in \mathfrak{L}$$

is a smooth, non trivial realization. Obviously the smoothness of ω implies: $\omega = \omega_{F_\omega}$. Further let us consider the following self–mapping $\Xi : \mathfrak{R}(\mathfrak{L}) \longmapsto \mathfrak{R}(\mathfrak{L})$ defined by

$$[\Xi(\omega)](K) = \omega_{F_\omega}(K) = \begin{cases} 1 & : \quad K \cap F_\omega \neq \varnothing \\ 0 & : \quad K \cap F_\omega = \varnothing \end{cases} , \qquad \omega \neq \omega_0 \, ,$$

$\Xi(\omega_0) = \omega_o$. Then we obtain for $K \in \mathfrak{L} \setminus \{\varnothing\}$:

$$\text{(4.4)} \quad \{\omega \in \mathfrak{R}(\mathfrak{L}) \mid [\Xi(\omega)](K) = 0\} = \bigcup_{\{L \in \mathfrak{K}(X) : K \subseteq L^o\}} \{\omega \in \mathfrak{R}(\mathfrak{L}) \mid \omega(L) = 0\} .$$

Obviously Ξ is Baire–Borel–measurable. Further let μ be a plausibility measure on $\mathfrak{L}$, ν_μ its representing Borel probability measure on $\mathfrak{R}(\mathfrak{L})$, and let $\bar{\eta}_\mu$ be the restriction of ν_μ to the σ–algebra of all Baire subsets of $\mathfrak{R}(\mathfrak{L})$ (w.r.t. the relative topology induced by τ_r). Referring to (4.4) we obtain

[2] V^o denotes the interior of V.

that Ξ is measure preserving (i.e. $\Xi(\nu_\mu) = \bar{\eta}_\mu$) iff the given plausibility measure is *upper semicontinuous* in the following sense:

$$\mu(K) = \inf\{\mu(L) \mid K \subseteq L^o\} \qquad \forall K \in \mathfrak{L} \setminus \{\varnothing\} . \tag{4.5}$$

Now we additionally assume that (X, τ) is metrizable and countable at infinity. Further let $\bar{\eta}_\mu^*$ be the outer measure corresponding to $\bar{\eta}_\mu$. Then $\bar{\eta}_\mu^*(\mathfrak{R}_s(\mathfrak{L})) = 1$ iff μ fulfills (4.5). Hence representing Borel probability measures of *upper semicontinuous plausibility measures* μ on $\mathfrak{L}$ satisfying the condition

$$\sup\{\mu(K) \mid K \in \mathfrak{K}(X)\} = 1 \tag{4.6}$$

and non empty *random closed sets* in X (w.r.t. τ) are the *same* things. In this sense Choquet's theorem (cf. 2.2 in [19]) is a consequence of Theorem 4.1. □

5. Integration with respect to uncertainty measures.

DEFINITION 5.1. (*L–Valued Real Numbers*) Let $(L, \leq, ^\perp)$ be a σ–complete de Morgan algebra and $\mathbb{Q}$ be the set of all rational numbers. A map $F : \mathbb{Q} \longmapsto L$ is called an *L–valued real number* iff F satisfies the following conditions:

(D1) $$\bigvee_{n \in \mathbb{N}} F(-n) = 1 , \qquad \bigwedge_{n \in \mathbb{N}} F(n) = 0 .$$

(D2) $$\bigvee_{r < r'} F(r') = F(r) \quad \forall r \in \mathbb{Q} . \qquad \text{(Right–continuity)}$$

An L–valued real number F is said to be *non negative* iff $\bigwedge_{n \in \mathbb{N}} F(-\frac{1}{n}) = 1$.

REMARK 5.1. (a) Right-continuous, non increasing probability distribution functions and $[0,1]$–valued real numbers are the same things. Further, if the underlying lattice L is given by the lattice of all closed subspaces of a given Hilbert space (cf. Example 2.1(a)), then L–valued real numbers are precisely *resolutions of the identity* (cf. [30]).

(b) Let X be a non empty set and $\mathcal{M}$ be a σ–algebra of subsets of X. Then $\mathcal{M}$–valued real numbers can be identified with $\mathcal{M}$–measurable, real valued functions with domain X. In particular, if $\varphi : X \longmapsto \mathbb{R}$ is an $\mathcal{M}$–measurable function, then $F_\varphi : \mathbb{Q} \longmapsto \mathcal{M}$ defined by

$$F_\varphi(r) = \varphi^{-1}(]r, +\infty[) , \quad r \in \mathbb{Q}$$

is an $\mathcal{M}$–valued, real number; and vice versa, if $F : \mathbb{Q} \longmapsto \mathcal{M}$ is an $\mathcal{M}$–valued, real number, then the map $\varphi_F : X \longmapsto \mathbb{R}$ defined by

$$\varphi_F(x) = \inf\{r \in \mathbb{Q} \mid x \notin F(r)\} , \quad x \in X$$

is an $\mathcal{M}$–measurable function, and the following relation holds:

$$F(r) = \bigcup_{n \in \mathbb{N}} F(r + \frac{1}{n}) = \{x \in X \mid r < \varphi_F(x)\} .$$

(c) Every element $\ell \in L$ determines an L–valued, real number F_ℓ by

$$F_\ell(r) = \begin{cases} 0 & : \quad 1 \leq r < +\infty \\ \ell & : \quad 0 \leq r < 1 \\ 1 & : \quad -\infty < r < 0 \end{cases} .$$

(d) Since real numbers can be considered as Dedekind cuts in $\mathbb{Q}$, every (ordinary) real number $\alpha \in \mathbb{R}$ can be identified with an L–valued, real number H_α as follows:

$$H_\alpha(r) = \begin{cases} 1 & : \quad r < \alpha \\ 0 & : \quad \alpha \leq r \end{cases} , \quad r \in \mathbb{Q} .$$

□

On the set $\mathcal{D}(\mathbb{R}, L)$ of all L–valued real numbers we define a partial ordering $\preccurlyeq$ by

$$F_1 \preccurlyeq F_2 \iff F_1(r) \leq F_2(r) \qquad \forall r \in \mathbb{Q} .$$

LEMMA 5.1. (a) $(\mathcal{D}(\mathbb{R}, L), \preccurlyeq)$ is a conditionally σ–complete lattice – i.e. the supremum (resp. infimum) exists for every at most countable subset having an upper (resp. lower) bound w.r.t. $\preccurlyeq$.
(b) The map $j : L \longmapsto \mathcal{D}(\mathbb{R}, L)$ defined by $j(\ell) = F_\ell$ (cf. Remark 5.1(c)) is an order preserving embedding – i.e. $j(\ell_1) \preccurlyeq j(\ell_2) \iff \ell_1 \leq \ell_2$.
(c) If the binary meet operation $\wedge$ is distributive over countable joins, then the infimum and supremum of $\{F_1, F_2\}$ ($\subseteq \mathcal{D}(\mathbb{R}, L)$) exist and are given by

$$(F_1 \wedge F_2)(r) = \bigvee_{r < r'} (F_1(r') \wedge F_2(r')) , \quad (F_1 \vee F_2)(r) = F_1(r) \vee F_2(r) .$$

PROOF. For any countable subset $\mathfrak{S} = \{F_n \mid n \in \mathbb{N}\}$ we define maps $\bigvee_{n \in \mathbb{N}} F_n : \mathbb{Q} \longmapsto L$, $\bigwedge_{n \in \mathbb{N}} F_n : \mathbb{Q} \longmapsto L$ by

$$(\bigvee_{n \in \mathbb{N}} F_n)(r) = \bigvee_{n \in \mathbb{N}} F_n(r) , \quad (\bigwedge_{n \in \mathbb{N}} F_n)(r) = \bigvee_{r < r'} (\bigwedge_{n \in \mathbb{N}} F_n(r')) .$$

If $\mathfrak{S}$ has an upper (resp. lower) bound, then $\bigvee_{n \in \mathbb{N}} F_n$ (resp. $\bigwedge_{n \in \mathbb{N}} F_n$) are L–valued real numbers; hence $(\mathcal{D}(\mathbb{R}, L), \preccurlyeq)$ is conditionally σ–complete. The assertion (b) is trivial. In order to verify (c) we proceed as follows: Since $(L, \leq, {}^\perp)$ is a de Morgan algebra, the distributivity of $\wedge$ over countable joins implies the distributivity of the binary join operation $\vee$ over countable meets. Hence for L–valued numbers F_1 and F_2 the following relation is valid:

$$\begin{aligned} \bigvee_{n\in\mathbb{N}} (F_1(-n) \wedge F_2(-n)) &= (\bigvee_{n\in\mathbb{N}} F_1(-n)) \wedge (\bigvee_{n\in\mathbb{N}} F_2(-n)) \\ \bigwedge_{n\in\mathbb{N}} (F_1(n) \vee F_2(n)) &= (\bigwedge_{n\in\mathbb{N}} F_1(n)) \vee (\bigwedge_{n\in\mathbb{N}} F_2(n)) \end{aligned} ;$$

i.e. the supremum and infimum of $\{F_1, F_2\}$ exist. □

With every L–valued real number we can associate a τ_r–continuous map $F^{(q)} : \mathfrak{PR}(L) \longmapsto [-\infty, +\infty]$ by

$$F^{(q)}(\omega) = \inf\{r \in \mathbb{Q} \mid \omega(F(r)) = 0\} , \quad \omega \in \mathfrak{PR}(L) .$$

In particular, the following relations hold:

$$\{\omega \mid F^{(q)}(\omega) > r\} = \bigcup_{n\in\mathbb{N}} \{\omega \mid \omega(F(r + \frac{1}{n})) = 1\} . \tag{5.1}$$

$$\{\omega \mid F^{(q)}(\omega) < r\} = \bigcup_{n\in\mathbb{N}} \{\omega \mid \omega(F(r - \frac{1}{n})) = 0\} . \tag{5.2}$$

$F^{(q)}$ is called the *quasi–inverse function* of F (cf. [28,27,9]). An L–valued real number F is non negative iff its quasi–inverse functions is non negative.

LEMMA 5.2. Let $(L, \leq, ^\perp)$ be a σ–complete de Morgan algebra, and μ be an uncertainty measure on L. Further let ν_i be a regular Borel probability measure on $\mathfrak{PR}(L)$ satisfying Condition (b1) w.r.t. μ (cf. Proposition 4.1) $(i = 1, 2)$. Then for every non negative, L–valued, real number the following relation holds

$$\int_{\mathfrak{PR}(L)} F^{(q)} \, d\nu_1 = \int_{\mathfrak{PR}(L)} F^{(q)} \, d\nu_2 .$$

PROOF. Let $F^{(q)}(\nu_i)$ be the image probability measure on $\bar{\mathbb{R}}^+ = [0, +\infty]$ of ν_i under $F^{(q)}$. Further we infer from (5.1) and (b1) :

$$\nu_1(\{\omega \mid F^{(q)}(\omega) > r\}) = \sup_{n\in\mathbb{N}} F(r + \frac{1}{n}) = \nu_2(\{\omega \mid F^{(q)}(\omega) > r\}) ;$$

hence the image measures $F^{(q)}(\nu_1)$ and $F^{(q)}(\nu_2)$ coincide and are denoted by η. Then we obtain:

$$\int_{\mathfrak{P}\mathfrak{R}(L)} F(q)\, d\nu_1 = \int_{\bar{\mathbb{R}}^+} x\, d\eta = \int_{\mathfrak{P}\mathfrak{R}(L)} F^{(q)}\, d\nu_2 .$$

□

By virtue of the previous lemma we define the *integral* of a non negative, L–valued real number F w.r.t. an uncertainty measure μ on L by

$$\int_L F\, d\mu = \int_{\mathfrak{P}\mathfrak{R}(L)} F^{(q)}\, d\nu ,$$

where ν satisfies Condition (b1) w.r.t. μ.

Since the Lebesgue measure plays a significant role in the representation of uncertainty measures by regular Borel probability measures (cf. Proof of Proposition 4.1), we ask the following question: Does there exist a special relationship between the integral w.r.t. uncertainty measures and certain types of integrals w.r.t. the Lebesgue measure? A *complete* answer will be given infra in Proposition 5.1. First we need a technical lemma.

Let $\mathfrak{P}\mathfrak{R}(L)_\infty$ be the set of all σ–smooth pseudo–realizations on L – i.e.

$$\mathfrak{P}\mathfrak{R}(L)_\infty = \{\omega \in \mathfrak{P}\mathfrak{R}(L) \mid \sup_{n\in\mathbb{N}} \omega(\ell_n) = \omega(\bigvee_{n\in\mathbb{N}} \ell_n) \text{ where } \ell_n \leq \ell_{n+1}\} .$$

Further let $\mathfrak{B}(\mathfrak{P}\mathfrak{R}(L))$ be the σ–algebra of all Borel subsets of $\mathfrak{P}\mathfrak{R}(L)$ and $\mathfrak{B}_\infty$ be the trace of $\mathfrak{B}(\mathfrak{P}\mathfrak{R}(L))$ on $\mathfrak{P}\mathfrak{R}(L)_\infty$ – i.e.

$$\mathcal{M} \in \mathfrak{B}_\infty \iff \exists\, \mathcal{B} \in \mathfrak{B}(\mathfrak{P}\mathfrak{R}(L)) \text{ s.t. } \mathcal{M} = \mathfrak{P}\mathfrak{R}(L)_\infty \cap \mathcal{B} .$$

LEMMA 5.3. Let $\mathfrak{B}([0,+\infty[)$ be the σ–algebra of all Borel subsets of $[0,+\infty[$ and F be a non negative, L–valued, real number. Then the set

$$\mathcal{M} = \{(\omega,t) \in \mathfrak{P}\mathfrak{R}(L)_\infty \times [0,+\infty[\mid \omega(\bigvee_{t<r} F(r)) = 1\}$$

is measurable w.r.t. the product–σ–algebra $\mathfrak{B}_\infty \otimes \mathfrak{B}([0,+\infty[)$.

PROOF. (a) Let F be a non negative, L–valued, real number with finite range; then there exists $n \in \mathbb{N}$ and finite subsets

$$\{\ell_i \mid i = 0,1,\ldots,n\} \subseteq L , \{r_i \mid i = 0,1,\ldots,n\} \subseteq \mathbb{Q}$$

such that

$$\begin{array}{l} 1 = \ell_0 > \ell_1 > \ell_2 > \cdots > \ell_{n-1} > \ell_n = 0 \\ 0 = r_0 \leq r_1 < r_2 < \cdots < r_{n-1} < r_n \end{array} ,$$

$$F(r) = \begin{cases} 1 & : \; r < r_1 \\ \ell_{i-1} & : \; r_{i-1} \leq r < r_i \;,\; i = 2,3,\ldots,n \\ 0 & : \; r_n \leq r \end{cases} .$$

Because of Axiom (R1) the system

$$\Big\{\{\omega \in \mathfrak{PR}(L) \mid \omega(\ell_{i-1}) = 1 \;,\; \omega(\ell_i) = 0\} \mid i = 1,2,\ldots,n\Big\}$$

forms a measurable partition of $\mathfrak{PR}(L)$. Moreover we obtain

$$\mathcal{M} = \bigcup_{i=1}^{n} \{\omega \in \mathfrak{PR}(L)_\infty \mid \omega(\ell_{i-1}) = 1 \;,\; \omega(\ell_i) = 0\} \times [0, r_i[\;;$$

hence $\mathcal{M}$ is measurable w.r.t. $\mathfrak{B}_\infty \otimes \mathfrak{B}([0,+\infty[)$.
(b) Let F be an arbitrary, non negative, L–valued, real number. We fix $n \in \mathbb{N}$ and put $r_{i,k} = \frac{i}{2^n} + k$ for $i = 0,1,\ldots,2^n$ and $k = 0,1,\ldots,n$. Now for each $n \in \mathbb{N}$ we can define a further non negative, L–valued, real number G_n by

$$G_n(r) = \begin{cases} 1 & : \; r < 0 \\ F(r_{i-1,k}) & : \; r_{i-1,k} \leq r < r_{i,k}, \; i = 1,\ldots,2^n \;,\; k = 0,\ldots,n. \\ 0 & : \; n+1 \leq r \end{cases}$$

Due to the right continuity of F (cf. (D2)) we obtain $\bigvee_{n \in \mathbb{N}} G_n(r) = F(r)$ for all $r \in \mathbb{Q}$. Now we apply the σ–smoothness of pseudo-realizations and obtain:

$$\mathcal{M} = \bigcup_{n \in \mathbb{N}} \{(\omega,t) \in \mathfrak{PR}(L)_\infty \times [0,+\infty[\mid \omega(\bigvee_{t<r} G_n(r)) = 1\} \;;$$

hence the assertion follows from part (a). □

PROPOSITION 5.1. Let μ be a σ–smooth uncertainty measure on L and $F : \mathbb{Q} \longmapsto L$ be a non negative, L–valued, real number. Further let λ be the ordinary Lebesgue measure on $[0,+\infty[$. Then the following relation is valid

$$\int_L F \, d\mu = \int_0^{+\infty} \mu(\bigvee_{t<r} F(r)) \, d\lambda(t) \, .$$

PROOF. Let $\Theta_\mu : [0,1[\longmapsto \mathfrak{PR}(L)$ be the Borel measurable map constructed in the proof of Proposition 4.1. Further let λ be the Lebesgue measure on $[0,1[$; then $\nu = \Theta_\mu(\lambda)$ satisfies Condition (b1) w.r.t. μ (cf. Proof of Proposition 4.1). Since μ is σ–smooth, the range of Θ_μ is contained in $\mathfrak{PR}(L)_\infty$; i.e. ν can be extended to a measure on $\mathfrak{PR}(L)_\infty$ w.r.t.

$\mathfrak{B}_\infty$. Further, it is not difficult to show that for any (σ–smooth) pseudo–realization ω on L the following relation holds:

$$F^{(q)}(\omega) = \sup\{r \in \mathbb{Q} \mid \omega(F(r)) = 1\} = \lambda(\{t \in [0,+\infty[\mid \omega(\bigvee_{t<r} F(r)) = 1\}) .$$

Now we apply Fubini's Theorem to Lemma 5.3 and obtain:

$$\begin{aligned}\int_L F\,d\mu &= \int_{\mathfrak{PR}(L)} F^{(q)}\,d\nu \\ &= \int_{\mathfrak{PR}(L)_\infty} \lambda(\{t \in [0,+\infty[\mid \omega(\bigvee_{t<r} F(r)) = 1)\,d\nu(\omega) \\ &= \nu \otimes \lambda(\{(\omega,t) \in \mathfrak{PR}(L)_\infty \times [0,+\infty[\mid \omega(\bigvee_{t<r} F(r)) = 1\}) \\ &= \int_0^{+\infty} \nu(\{\omega \in \mathfrak{PR}(L)_\infty \mid \omega(\bigvee_{t<r} F(r)) = 1\})\,d\lambda(t) \\ &= \int_0^{+\infty} \mu(\bigvee_{t<r} F(r))\,d\lambda(t) .\end{aligned}$$

□

COMMENT. If μ is not a σ–smooth uncertainty measure, then a modification of the proof of the previous proposition leads to the following relation:

$$\int_L F\,d\mu = \int_0^{+\infty} \sup_{t<r} \mu \circ F(r)\,d\lambda(t) ,$$

where F denotes a non negative, L–valued, real number.

REMARK 5.2. (*Hilbert Spaces, Measurable Subsets*) (a) Let $L(H)$ be the weakly orthomodular lattice of all closed subspaces of a given Hilbert space (cf. Example 2.1(a)). Further let x be a unit vector of H and μ_x be the σ–smooth, additive uncertainty measure on $L(H)$ induced by x according to Example 3.1(a). If F is an $L(H)$–valued, non negative real number, then a positive, not necessarily bounded, linear operator $T : H \longmapsto H$ is determined by

$$(5.3) \qquad <\int_0^{+\infty} z\,dFx\,, x> = <T(x), x> = \int_{L(H)} F\,d\mu_x\,, \qquad \| x \|_2 = 1 .$$

In this context F is the (unique) resolution of identity of T, and the relation (5.3) is called the *spectral representation* of T.
(b) Let X be a non empty set, $\mathcal{M}$ be a σ–algebra of subsets of X, and let μ be a σ–smooth uncertainty measure on $\mathcal{M}$. Further let F be a non negative, $\mathcal{M}$–valued real number, and φ_F be the measurable function corresponding to F (cf. Remark 5.1(b)). Then Proposition 5.1 shows that $\int_{\mathcal{M}} F\, d\mu$ coincides with the Choquet integral of φ_F w.r.t. μ (cf. [6,25,14,21]). Further let $j : X \longmapsto \mathfrak{PR}(\mathcal{M})$ be the natural embedding determined by

$$[j(x)](A) = \begin{cases} 1 & : \quad x \in A \\ 0 & : \quad x \notin A \end{cases} .$$

Obviously the following diagram is commutative

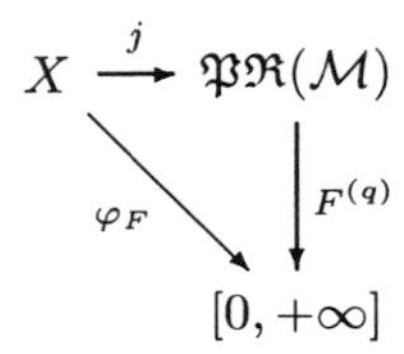

Therefore, if we extend φ_F to the quasi–inverse mapping $F^{(q)}$ of F, then the *Choquet integral* of φ_F w.r.t. μ can be viewed as the ordinary *Lebesgue integral* of $F^{(q)}$ w.r.t. an ordinary regular Borel probability measure ν on $\mathfrak{PR}(\mathcal{M})$ which represents μ in the sense of Proposition 4.1. Similar, but not the same ideas appeared also under the name "representation of fuzzy measures" in [20,21]. □

PROPOSITION 5.2. Let μ be an uncertainty measure on L. (a) The relation $\int_L F_\ell\, d\mu = \mu(\ell)$ holds for all $\ell \in L$. (b) μ is σ–smooth iff for every order bounded, non decreasing sequence $(F_n)_{n\in\mathbb{N}}$ of non negative, L–valued real numbers F_n the following relation holds:

$$\sup_{n\in\mathbb{N}} \int_L F_n\, d\mu = \int_L \Big(\bigvee_{n\in\mathbb{N}} F_n \Big)\, d\mu . \qquad \text{(B. Levi Property)}$$

PROOF. The assertion (a) follows from $F_\ell^{(q)}(\omega) = \chi_A(\omega)$ where $A = \{\omega \in \mathfrak{PR}(L) \mid \omega(\ell) = 1\}$. Further we observe:

$$\bigvee_{n\in\mathbb{N}} F_{\ell_n} = F_{\bigvee\limits_{n\in\mathbb{N}} \ell_n} .$$

Since the integral w.r.t. μ is an extension of μ to the set of all non negative, L–valued real numbers, the B. Levi property always implies the σ–smoothness of μ. On the other hand, if μ is σ–smooth and ν is a regular Borel probability measure on $\mathfrak{PR}(L)$ satisfying (b1) and (b2) w.r.t. μ, then we obtain

$$\sup_{n \in \mathbb{N}} F_n^{(q)}(\omega) = \Big(\bigvee_{n \in \mathbb{N}} F_n\Big)^{(q)}(\omega) \qquad \nu - \text{almost everywhere ;}$$

hence the B. Levi property follows. □

PROPOSITION 5.3. Let $(L, \leq, {}^{\perp})$ be a de Morgan algebra such that the binary meet operator $\wedge$ is distributive over countable joins. Further let μ be an uncertainty measure on L. Then the following assertions are equivalent:

(i) μ is a plausibility measure.

(ii) For every non empty, finite subset $\{F_1, \dots, F_n\}$ of non negative L–valued real numbers the subsequent relation is valid:

$$\int_L \Big(\bigwedge_{i=1}^{n} F_i\Big)\, d\mu \leq \sum_{i=1}^{n} (-1)^{i-1} \sum_{1 \leq j_1 < \dots < j_i \leq n} \int_L \Big(\bigvee_{k=1}^{i} F_{j_k}\Big)\, d\mu \, .$$

PROOF. It is not difficult to verify the following relations:

$$\begin{aligned} \Big(\bigvee_{k=1}^{i} F_{j_k}\Big)^{(q)}(\omega) &= \max_{k=1,\dots,i} F_{j_k}^{(q)}(\omega)\,, && \forall\, \omega \in \mathfrak{R}(L) \\ \Big(\bigwedge_{i=1}^{n} F_i\Big)^{(q)}(\omega) &\leq \min_{i=1,\dots,n} F_i^{(q)}(\omega)\,, && \forall\, \omega \in \mathfrak{PR}(L) \end{aligned}\, .$$

Hence the implication (i) $\Longrightarrow$ (ii) follows from Theorem 4.1. Since the integration w.r.t. μ is an extension of μ the implication (ii) $\Longrightarrow$ (i) is obvious. □

Let $(L, \leq, {}^{\perp})$ be a σ-complete de Morgan algebra such that the binary meet operator $\wedge$ is distributive over at most countable joins. Since ${}^{\perp}$ is an order reversing involution, the binary join operator $\vee$ is distributive over countable meets. Therefore we are in the position to introduce a semigroup operation $\oplus$ on the set $\mathcal{D}^{+}(\mathbb{R}, L)$ of all non negative, L–valued real numbers as follows:

$$(F_1 \oplus F_2)(r) = \bigvee_{r = r_1 + r_2} (F_1(r_1) \wedge F_2(r_2)) \qquad \forall\, r \in \mathbb{Q}\, .$$

LEMMA 5.4. Let $(L, \leq, {}^{\perp})$ be a σ–complete de Morgan algebra. Further we assume that the binary meet operator $\wedge$ is distributive over at most countable joins. If F_1, F_2 are non negative, L–valued real numbers, then the following relation holds:

$$(F_1 \oplus F_2)^{(q)}(\omega) = F_1^{(q)}(\omega) + F_2^{(q)}(\omega) \qquad \forall\, \omega \in \mathfrak{CR}(L)\, .$$

PROOF. It is sufficient to show that

$$\mathcal{G} := \{\omega \in \mathfrak{CM}(L) \mid F_1^{(q)}(\omega) + F_2^{(q)}(\omega) \neq (F_1 \oplus F_2)^{(q)}(\omega)\}$$

is an open subset of the first category of $\mathfrak{CM}(L)$. The openness of $\mathcal{G}$ follows from the continuity of quasi–inverse functions and the continuity of the usual addition on $[0, +\infty]$. Further we put for each $r \in \mathbb{Q}$

$$\mathcal{K}_r = \bigcap_{r_1+r_2=r} \{\omega \in \mathfrak{CM}(L) | \min(\omega(F_1(r_1)), \omega(F_2(r_2))) = 0, \omega(F_1 \oplus F_2)(r) = 1\}.$$

The distributivity of $\wedge$ over at most countable joins implies that $\mathcal{K}_r$ is a closed, nowhere dense subset of $\mathfrak{CM}(L)$. Now we observe:

$$\{\omega \in \mathfrak{CM}(L) \mid F_1^{(q)}(\omega) + F_2^{(q)}(\omega) < r\} \subseteq \bigcap_{r_1+r_2=r} \{\omega \in \mathfrak{CM}(L) \mid \omega(F_1(r_1)) = 0 \text{ or } \omega(F_2(r_2)) = 0\},$$

$$\{\omega \in \mathfrak{CM}(L) \mid r < F_1^{(q)}(\omega) + F_2^{(q)}(\omega)\} \subseteq \{\omega \in \mathfrak{CM}(L) \mid r < (F_1 \oplus F_2)^{(q)}(\omega)\};$$

hence we obtain: $\mathcal{G} \subseteq \bigcup_{r\in\mathbb{Q}} \mathcal{K}_r$; i.e. $\mathcal{G}$ is a subset of the first category of $\mathfrak{CM}(L)$. □

PROPOSITION 5.4. (*Additivity of the Integral*) Let $(L, \leq, ^{\perp})$ be a σ–complete de Morgan algebra such that the binary meet operator $\wedge$ is distributive over at most countable joins. Further let μ be an uncertainty measure satisfying $(M5)$ (i.e. μ is a valuation on L). Then for any pair (F_1, F_2) of non negative, L–valued real numbers the following relation holds:

$$\int_L (F_1 \oplus F_2) d\mu = \Big(\int_L F_1 \, d\mu\Big) + \Big(\int_L F_2 \, d\mu\Big).$$

PROOF. The assertion follows immediately from Proposition 3.5, Corollary 4.1 and Lemma 5.4. □

COROLLARY 5.1. Let $(L, \leq, *)$ be a σ–complete MV–algebra.[3] Then the integral w.r.t. an additive uncertainty measure on L is additive.

PROOF. The assertion follows immediately from Proposition 3.4 and Proposition 5.4. □

REMARK 5.3. (*Multiplication*) Let us consider non negative, L–valued, real numbers F and G provided with the following property

$$\bigwedge_{n\in\mathbb{N}} (F(n) \vee G(n)) = 0. \tag{5.4}$$

[3] In particular $(L, \leq, *)$ is a semisimple MV–algebra (cf. [2]).

Then the *product* $F \odot G$ is defined by

$$(F \odot G)(r) = \begin{cases} 1 & : \quad r < 0 \\ \bigvee_{r_1 \cdot r_2 > r} (F(r_1) \wedge G(r_2)) & : \quad 0 \leq r \end{cases} .$$

If we identify non negative, real numbers t with non negative, L–valued, real numbers H_t in the sense of Remark 5.1(d), then it is easy to see that H_t and G satisfy (5.4). In particular, we obtain for $0 < t$

$$(t \odot G)(r) := (H_t \odot G)(r) = G(\frac{r}{t}) \qquad \forall r \in \mathbb{Q} .$$

Further, it is not difficult to show

$$(t \odot G)^{(q)}(\omega) = t \cdot G^{(q)}(\omega) \qquad \forall \omega \in \mathfrak{PM}(L) ;$$

hence the relation $\int_L t \odot G \, d\mu = t \cdot \int_L G \, d\mu$ follows. □

We finish this section with the discussion of *densities* of uncertainty measures.

REMARK 5.4. (*Densities*) According to Remark 5.1(c) we can identify every event $\ell \in L$ with a non negative, L–valued, real number F_ℓ. Then the pair (G, F_ℓ) satisfies (5.4) for any non negative, L–valued, real number G. Hence we can introduce the concept of densities of an uncertainty measure μ_1 w.r.t. a given uncertainty μ_2 as follows: A non negative, L–valued, real number G is called a *density* of μ_1 w.r.t. μ_2 if and only if the relation

$$\mu_1(\ell) = \int_L F_\ell \odot G \, d\mu_2$$

holds for all $\ell \in L$. If the binary meet operation is distributive over at most countable joins, then $(F_\ell \odot G)(r) = \ell \wedge G(r)$ holds for all $r \in \mathbb{Q}$. Referring to Proposition 5.1 we obtain that G is a density of μ_1 w.r.t. μ_2 iff the following relation is valid

$$\mu_1(\ell) = \int_0^{+\infty} \mu_2(\ell \wedge (\bigvee_{t<r} G(r))) \, d\lambda(t) \qquad \forall \ell \in L ,$$

where λ denotes the Lebesgue measure. □

6. Information theory based on plausibility measures. Let μ be a plausibility measure on L. Then we can introduce for each realization $\omega : L \longmapsto \{0, 1\}$ the *maximal information of discernibleness*, resp. *maximal* *information of indiscernibleness*, at least in two ways:

$$\begin{aligned}
e_\mu^{(1)}(\omega) &= \sup_{\ell \in L}\{-(ln(\mu(\ell) - \mu(\ell \wedge \ell^\perp))) \cdot \omega(\ell) \cdot (1 - \omega(\ell^\perp))\} \\
e_\mu^{(2)}(\omega) &= \sup_{\ell \in L}\{-(ln(\mu(\ell \vee \ell^\perp) - \mu(\ell))) \cdot (1 - \omega(\ell)) \cdot \omega(\ell^\perp)\} \\
s_\mu^{(1)}(\omega) &= \sup_{\ell \in L}\{-(ln(\mu(\ell) + \mu(\ell^\perp) - \mu(\ell \vee \ell^\perp))) \cdot \omega(\ell) \cdot \omega(\ell^\perp)\} \\
s_\mu^{(2)}(\omega) &= \sup_{\ell \in L}\{-(ln(1 - \mu(\ell \vee \ell^\perp))) \cdot (1 - \omega(\ell)) \cdot (1 - \omega(\ell^\perp))\}
\end{aligned}$$

We observe that the maps $e_\mu^{(i)} : \mathfrak{R}(L) \longmapsto [0, +\infty]$ and $s_\mu^{(i)} : \mathfrak{R}(L) \longmapsto [0, +\infty]$ $(i = 1, 2)$ are lower semicontinuous w.r.t. the topology induced by τ_r on $\mathfrak{R}(L)$; hence $e_\mu^{(i)}$, $s_\mu^{(i)}$ are Borel measurable $(i = 1, 2)$. Since *entropies* are mean values of *information functions*, we can use the μ representing Borel probability measure[4] ν_μ on $\mathfrak{R}(L)$ (cf. Theorem 4.1) and define the entropies of discernibleness (resp. indiscernibleness) as follows:

$$\begin{aligned}
E_\mu^{(1)} &= \int_{\mathfrak{R}(L)} e_\mu^{(1)}\, d\nu_\mu \quad , \quad & E_\mu^{(2)} &= \int_{\mathfrak{R}(L)} e_\mu^{(2)}\, d\nu_\mu \\
S_\mu^{(1)} &= \int_{\mathfrak{R}(L)} s_\mu^{(1)}\, d\nu_\mu \quad , \quad & S_\mu^{(2)} &= \int_{\mathfrak{R}(L)} s_\mu^{(2)}\, d\nu_\mu
\end{aligned}$$

In accordance with the previous interpretation $E_\mu^{(1)}$ (resp. $E_\mu^{(2)}$) is called the *entropy of discernibleness of the first* (resp. *second*) *kind* determined by the plausibility measure μ. $S_\mu^{(1)}$ (resp. $S_\mu^{(2)}$) is said to be the *entropy of indiscernibleness of the first* (resp. *second*) *kind* determined by μ.

PROPOSITION 6.1. (*Shannon's Entropy*) Let X be a non empty set, μ be an atomic probability measure on the power set of X, and let $\mathfrak{A}$ be the set of atoms of μ. Then the entropies of discernibleness and indiscernibleness determined by μ are given by

$$E_\mu^{(1)} = E_\mu^{(2)} = \sum_{x_n \in \mathfrak{A}} (-ln(\mu(\{x_n\}))) \cdot \mu(\{x_n\}) \; , \quad S_\mu^{(1)} = S_\mu^{(2)} = 0 \, .$$

PROOF. Let $\mathcal{P}(X)$ be the power set of X. We define a map $\Theta : X \longmapsto \mathfrak{R}(\mathcal{P}(X))$ by

$$\Theta(x) = \omega_x \, , \quad \omega_x(A) = \begin{cases} 1 & : \; x \in A \\ 0 & : \; x \notin A \end{cases} \, , \qquad x \in X \, .$$

Then Θ is trivially Borel measurable and the image measure $\Theta(\mu)$ of μ

[4] Because of $\nu_\mu(\mathfrak{R}(L)) = 1$ we view ν_μ as a measure on $\mathfrak{R}(L)$.

under Θ coincides with the Borel probability measure representing μ (cf. Remark 4.2). Now we observe

$$s_{\mu}^{(1)} \circ \Theta = s_{\mu}^{(2)} \circ \Theta = 0 \,, \quad e_{\mu}^{(1)} \circ \Theta(x) = e_{\mu}^{(2)} \circ \Theta(x) = -ln(\{x\}) \,, \; x \in X;$$

hence the assertion follows. □

PROPOSITION 6.2. (*Possibility Measures*) Let X be an ordinary, non empty set, $h : X \mapsto [0,1]$ be a map, and let μ_h be the possibility measure on the ordinary power set $\mathcal{P}(X)$ of X induced by h (cf. Remark 4.3). Then the entropies of discernibleness and indiscernibleness are given by

$$\begin{aligned} E_{\mu_h}^{(1)} &= 0 \quad , \quad & E_{\mu_h}^{(2)} &= \int_0^1 (-\,ln(1 - \sup\{h(h^{-1}([0,\alpha]))\})) \; d\alpha \\ S_{\mu_h}^{(2)} &= 0 \quad , \quad & S_{\mu_h}^{(1)} &= \int_0^1 (-\,ln(\inf\{h(h^{-1}(]\alpha,1[))\})) \; d\alpha \end{aligned}$$

PROOF. Let us consider the measurable map $\Theta_h : [0,1[\longmapsto \mathfrak{R}(\mathcal{P}(X))$ defined in Remark 4.3. Referring to the equivalence

$$\omega_\alpha(A) = 0 \iff A \subset \{x \in X \mid h(x) \leq \alpha\} \iff \mu_h(A) \leq \alpha \quad (A \in \mathcal{P}(X))$$

we easily establish the following relations

$$\begin{aligned} e_{\mu_h}^{(1)} \circ \Theta_h(\alpha) &= 0 \qquad \forall \alpha \in [0,1[\\ e_{\mu_h}^{(2)} \circ \Theta_h(\alpha) &= -\,ln(1 - \sup\{h(h^{-1}([0,\alpha]))\}) \\ s_{\mu_h}^{(1)} \circ \Theta_h(\alpha) &= -\,ln(\inf\{h(h^{-1}(]\alpha,1[))\}) \end{aligned}$$

By definition we have $s_{\mu_h}^{(2)} = 0$ (cf. Remark 4.3). Since ν_{μ_h} is the image measure of the Lebesgue measure under Θ_h, the assertion follows. □

EXAMPLE 6.1. Let X be the half open interval $]0,1]$, $\mathcal{P}(X)$ be the power set of X, and for each $n \in \mathbb{N}$ let $h_n : X \longmapsto [0,1]$ be a map defined by

$$h_n(x) = \frac{i}{n} \quad \text{whenever} \quad x \in \left]\frac{i-1}{n}, \frac{i}{n}\right] , \quad i = 1,2,\ldots,n \,.$$

Further let μ_{h_n} be the possibility measure on $\mathcal{P}(X)$ induced by h_n (cf. Remark 4.3). Then the entropies w.r.t. μ_{h_n} (cf. Proposition 6.2) are given by

$$E_{\mu_{h_n}}^{(2)} = S_{\mu_{h_n}}^{(1)} = \ln\Big(\Big(\frac{n^n}{n!}\Big)^{\frac{1}{n}}\Big) \,.$$

In particular: $\lim_{n\to\infty} E_{\mu_{h_n}}^{(2)} = \lim_{n\to\infty} S_{\mu_{h_n}}^{(1)} = 1 \,.$ □

It is remarkable to see that in probabilistic environments both types of entropies of discernibleness are the same and coincide with Shannon's entropy. If we pass to possibilistic (or more general to plausibilistic) environments, then new, powerful types of entropies (e.g. entropies of indiscernibleness) arise.

We finish this section with an information–theoretic discussion of additive uncertainty measures on MV–algebras.

PROPOSITION 6.3. Let μ be an additive uncertainty measure on an MV–algebra $(L, \leq, *)$. Then the following relations hold:

(i) $E_\mu^{(1)} = E_\mu^{(2)}, \quad S_\mu^{(1)} = S_\mu^{(2)}$.

(ii) If the order reversing involution has a (unique) fixed point, then $\frac{1}{2} \cdot \ln(2) \leq S_\mu^{(i)} \quad (i = 1, 2)$.

PROOF. We infer from the additivity of μ :

$$\mu(\ell) - \mu(\ell \wedge \ell^\perp) = \mu(\ell * \ell), \quad \mu(\ell \vee \ell^\perp) - \mu(\ell) = \mu(\ell^\perp * \ell^\perp),$$

$$\mu(\ell) + \mu(\ell^\perp) - \mu(\ell \vee \ell^\perp) = \mu(\ell \wedge \ell^\perp) = 1 - \mu(\ell \vee \ell^\perp);$$

hence we obtain for all $\omega \in \mathfrak{CR}(L)$:

$$\begin{aligned} e_\mu^{(1)}(\omega) &= e_\mu^{(2)}(\omega), \\ s_\mu^{(2)}(\omega) &= s_\mu^{(1)} \circ \varphi^*(\omega), \quad \text{where} \quad \varphi^*(\omega) = 1 - \omega(\ell^\perp). \end{aligned}$$

Since φ^* is measure preserving (cf. Corollary 4.3), the relation (i) follows immediately. Further let $\frac{1}{2}$ be the unique element of L with $\frac{1}{2} = \frac{1}{2}^\perp$. Then we observe

$$\begin{aligned} (\ell \wedge \ell^\perp) \to \frac{1}{2} &= (\ell \to \frac{1}{2}) \vee (\ell^\perp \to \frac{1}{2}) \\ &= (\ell \to \frac{1}{2}) \vee (\frac{1}{2} \to \ell) = 1; \end{aligned}$$

hence $\ell \wedge \ell^\perp \leq \frac{1}{2}$ for all $\ell \in L$. If we denote by $\mathcal{A}$ the set of all coherent realizations ω with $\omega(\frac{1}{2}) = 1$, then we obtain:

$$\ln(2) \cdot \chi_{\mathcal{A}}(\omega) \leq \sup_{\ell \in L}(-\ln(\mu(\ell \wedge \ell^\perp)) \cdot \omega(\ell) \cdot \omega(\ell^\perp)) \quad \forall \omega \in \mathfrak{CR}(L);$$

hence the relation (ii) follows. □

PROPOSITION 6.4. Let $(\mathbb{B}, \leq)$ be a Boolean algebra and $(L_c, \preccurlyeq, *)$ be the MV–algebra of all conditional events of $\mathbb{B}$ (cf. Remark 2.1(c)). Further let μ be a probability measure on $\mathbb{B}$ and $\bar{\mu}$ be its unique extension to L_c (cf. Proposition 3.6). Then the entropies of discernibleness determined by μ and $\bar{\mu}$ coincide, and the entropies of indiscernibleness are given by $\frac{1}{2} \cdot (\ln(2) + E_{\mu}^{(1)})$.

PROOF. Let $\nu_{\bar{\mu}}$ be the representing Borel probability measure of $\bar{\mu}$, and $\varrho : \mathfrak{CR}(L_c) \longmapsto \mathfrak{CR}(\mathbb{B})$ be the restriction map. Then ϱ is continuous and the image measure $\varrho(\nu_{\bar{\mu}})$ coincide with the representing Borel probability measure of μ. If ω is a coherent realization on L_c, then

$$\omega((\alpha,\beta)) = 1 \text{ and } \omega((\alpha,\beta)^{\perp}) = 0 \iff \omega((\alpha,\alpha)) = 1 \,;$$

hence the relation $e_{\mu}^{(1)} \circ \varrho(\omega) = e_{\bar{\mu}}^{(1)}(\omega)$ follows from $(\alpha,\beta) * (\alpha,\beta) = (\alpha,\alpha)$ for all $\omega \in \mathfrak{CR}(L_c)$. Therefore we obtain:

$$E_{\mu}^{(1)} = \int_{\mathfrak{CR}(\mathbb{B})} e_{\mu}^{(1)} \, d\varrho(\nu_{\bar{\mu}}) = \int_{\mathfrak{CR}(L_c)} e_{\mu}^{(1)} \circ \varrho \, d\nu_{\bar{\mu}}$$

$$= \int_{\mathfrak{CR}(L_c)} e_{\bar{\mu}}^{(1)} \, d\nu_{\bar{\mu}} = E_{\bar{\mu}}^{(1)} \,.$$

In order to verify $S_{\bar{\mu}}^{(1)}$ we proceed as follows: Let $\mathcal{M}$ be the set of all coherent realizations ω on L_c provided with $\omega((0,1)) = 1$. Then we obtain immediately

$$\mathcal{M} \cap \{\omega \in \mathfrak{CR}(L_c) \mid \omega((\alpha,\beta)) = 1\} = \{\omega \in \mathfrak{CR}(L_c) \mid \omega((0,\beta)) = 1\}.$$

Further, we define a map $\Psi : \mathfrak{CR}(\mathbb{B}) \longmapsto \mathfrak{CR}(L_c)$ by

$$[\Psi(\omega)]((\alpha,\beta)) = \omega(\beta) \qquad \forall \omega \in \mathfrak{CR}(\mathbb{B}) \,.$$

Then Ψ is continuous w.r.t. the corresponding topologies, and the range of Ψ coincides with $\mathcal{M}$. Obviously the conditional probability measure $\nu_{\bar{\mu}}(\omega \mid \mathcal{M})$ is the image measure $\Psi(\nu_{\mu})$ of ν_{μ} under Ψ. Because of $s_{\bar{\mu}}^{(1)} \circ \Psi = \ln(2) + e_{\mu}^{(1)}$ we obtain

$$S_{\bar{\mu}}^{(1)} = \int_{\mathfrak{CR}(L_c)} \chi_{\mathcal{M}} \cdot s_{\bar{\mu}}^{(1)} \, d\nu_{\bar{\mu}} = \frac{1}{2} \cdot \int_{\mathfrak{CR}(L_c)} s_{\bar{\mu}}^{(1)} \, d\nu_{\bar{\mu}}(\omega \mid \mathcal{M})$$

$$= \frac{1}{2} \cdot \int_{\mathfrak{CR}(\mathbb{B})} s_{\bar{\mu}}^{(1)} \circ \Psi \, d\nu_{\mu} = \frac{1}{2} \cdot \left(\ln(2) + E_{\mu}^{(1)}\right) .$$

□

MAIN RESULT III. In the case of Boolean algebras measure–free conditioning (i.e. the addition of conditional events in the sense of Remark 2.1(c)) does not *change* the *entropies* of discernibleness.

7. Concluding remarks. Uncertainty measures form a common basis for plausibility theory on de Morgan algebras, probability theory on weakly orthomodular lattices and a new type of uncertainty theory on MV–algebras. In contrast to the case of weakly orthomodular lattices the *decision theory* associated with *additive uncertainty measures* on *MV–algebras* is *not fictitious* and has various interesting applications. Among others we here advocate applications of this decision theory to problems of *measure–free conditioning* (cf. [8]).

REFERENCES

[1] L.P. Belluce, *Semi–simple algebras of infinite valued logic and bold fuzzy set theory*, Can. J. Math., 38 (1986), pp. 1356–1379.

[2] L.P. Belluce, *Semi–simple and complete MV–algebras*, Algebra Universalis, 29 (1992), pp. 1–9.

[3] G. Birkhoff, *Lattice Theory*, Amer. Math. Soc. Colloquium Publications, 3rd Edition (Amer. Math. Soc., RI, 1973).

[4] N. Bourbaki, Eléments de Mathématiques, Livre III, *Topologie Générale* (Hermann, Paris, 1962).

[5] C.C. Chang, *Algebraic analysis of many valued logics*, Trans. Amer. Math. Soc., 88 (1958), pp. 467–490.

[6] G. Choquet, *Theory of capacities*, Ann. Inst. Fourier, 5 (1953), pp. 131–295.

[7] St.P. Gudder, *Stochastic Methods in Quantum Mechanics* (North–Holland, New York, 1979).

[8] I.R. Goodman, H.T. Nguyen and E.A. Walker, *Conditional Inference and Logic for Intelligent Systems – A Theory of Measure–Free Conditioning* (North–Holland, Amsterdam, New York, 1991).

[9] U. Höhle, *Representation theorems for L–fuzzy quantities*, Fuzzy Sets and Systems, 5 (1981), pp. 83–107.

[10] U. Höhle, *A remark on entropies with respect to plausibility measures*. In: Cybernetics and System Research (ed. R. Trappl), pp. 735–738 (North–Holland, Amsterdam, 1982).

[11] U. Höhle, *Commutative, residuated ℓ–monoids*. In: U. Höhle and E.P. Klement, Eds., Nonclassical Logics and Their Applications to Fuzzy Subsets, pp. 53–106 (Kluwer Academic Publishers, Dordrecht, Boston, 1995).

[12] U. Höhle and E.P. Klement, *Plausibility measures – A general framework for possibility and fuzzy probability measures*. In: Aspects of Vagueness (eds. H.J. Skala et al.), pp. 31–50 (Reidel, Dordrecht, 1984).

[13] U. Höhle and S. Weber, *On conditioning operators*. In: The Mathematics of Fuzzy Sets, Volume II, Handbook of Fuzzy Sets Methodology, (eds. D. Dubois and H. Prade), To Appear, (Kluwer Academic Publishers, Dordrecht, Boston, 1998).

[14] J.E. Honeycutt, *On an abstract Stieltjes measure*, Ann. Inst. Fourier, 21 (1971), pp. 143–154.

[15] P.T. Johnstone, *Stone Spaces* (Cambridge University Press, Cambridge, London, 1982).

[16] A.N. Kolmogoroff, *Grundbegriffe der Wahrscheinlichkeitsrechnung*, Ergebnisse der Mathematik und ihrer Grenzgebiete Bd. 2 (Springer–Verlag, Berlin, 1933).

[17] J. Łos, *On the axiomatic treatment of probability*, Coll. Math., 3 (1955), pp. 125–137.

[18] J. Łos, *O ciatach zdarzeń i ich definicji w aksjomatycznej teorii prawdopodobieństwa*, Studia Logica, 9 (1960), pp. 95–132 (English translation: *Fields of events and their definitions in the axiomatic treatment of probability theory.*

In: Selected Translations of Mathematical Statistics and Probability, Vol. 7, pp. 17–39 (Inst of Math. Stat., Amer. Math. Soc., Providence, RI, 1968)).

[19] G. MATHERON, *Random Sets and Integral Geometry* (Wiley, New York, 1975).

[20] T. MUROFUSHI, *Fuzzy measures and Choquet's integral.* In: Proc. 9th Symposium on Applied Functional Analysis 1986 (Ed. H. Umegaki), pp. 58–69.

[21] T. MUROFUSHI AND M. SUGENO, *Fuzzy t–conorm integral with respect to fuzzy measures: Generalization of Sugeno integral and Choquet integral*, Fuzzy Sets and Systems, 42 (1991), pp. 57–71.

[22] J. NEVEU, *Bases Mathématiques du Calcul des Probabilités* (Masson, Paris, 1964).

[23] J. PFANZAGL, *Theory of Measurement*, 2nd ed. (Physica–Verlag, Wűrzburg, Wien 1971).

[24] H. PRADE, *Nomenclature of fuzzy measures*, Proceedings of the 1st International Seminar on Fuzzy Set Theory, Linz (Austria) September 1979 (ed. E.P. Klement), pp. 8–25.

[25] A. REVUZ, *Fonctions croissantes et mesures sur les espace topologiques ordonnés*, Ann. Inst. Fourier, 6 (1955), pp. 187–269.

[26] G. SHAFER, *A Mathematical Theory of Evidence* (Princeton University Press, Princeton, NJ, 1976).

[27] H. SHERWOOD AND M.D. TAYLOR, *Some PM structures on the set of distribution functions*, Rev. Roumaine Math. Pures Appl., 19 (1974), pp. 1251–1260.

[28] A. SKLAR, *Fonctions de répartition à n dimensions et leurs marges*, Publ. Inst. Statist. Univ. Paris VIII (1959), pp. 229–231.

[29] V.S. VARADARAJAN, *Geometry of Quantum Theory*, 2nd edition (Springer–Verlag, New York, Berlin, 1985).

[30] K. YOSIDA, *Functional Analysis*, 2nd edition (Springer–Verlag, New York, Berlin, 1966).

[31] L.A. ZADEH, *Fuzzy sets*, Information and Control, 8 (1965), pp. 338–353.

[32] L.A. ZADEH, *Fuzzy sets as a basis for a possibility theory*, Fuzzy Sets and Systems, 1 (1978), pp. 3–28.

RANDOM SETS IN DECISION–MAKING

HUNG T. NGUYEN AND NHU T. NGUYEN*

Abstract. As its title indicates, this contribution aims at presenting some typical situations in decision-making in which random sets appear naturally and seem to play a useful role.

Key words. Capacity, Capacity Functional, Choquet Integral, Distribution Functional, Maximum Entropy, Random Sets, Set-Function.

AMS(MOS) subject classifications. 60D05, 62C99, 28E05

1. Introduction. Within probability theory, random sets are simply a special class of random elements, namely set-valued random elements. In the context of Euclidean spaces, random sets are generalizations of random variables or random vectors. Unlike other generalizations, such as probability theory on Banach spaces, the theory of random sets does not seem to attract probabilists, let alone statisticians. There exist several books devoted to random sets, e.g., Harding and Kendall [12] , Matheron [14], Stoyan et al. [23], Hall [9], and Molchanov [16].

From time to time, research articles on random sets appear in the Annals of Probability and in the Annals of Statistics. The references in Hall [9] cover almost completely the literature on random sets and their applications. Recently, besides stochastic geometry, image analysis and mathematical morphology, interest in random sets has appeared in Engineering fields such as expert systems and data fusion. As such, it seems reasonable to focus more systematically on both the theory and applications of random sets. It is our hope that this workshop will trigger further research on random sets and their applications especially in developing models for statistical applications.

2. Generalities on random sets. A random set is a set-valued random element. Specifically, a random set S is a map from some space Ω to some collection Θ of subsets of some set $\mathcal{U}$, such that S is $\mathcal{A}$-$\mathcal{B}$ measurable, where $\mathcal{A}$ and $\mathcal{B}$ are σ-fields on Ω and Θ, respectively. For a probability measure P on $(\Omega, \mathcal{A})$, the probability law of S is the induced probability measure $P_S = PS^{-1}$ on $(\Theta, \mathcal{B})$. In practice, a probability measure on $(\Theta, \mathcal{B})$ is viewed as a random set on $\mathcal{U}$. When $\mathcal{U}$ is the Euclidean space $\mathbb{R}^d$, random sets are generalizations of random vectors. As such, it is expected that the usual method of introducing probability measures on $\mathbb{R}^d$ via distribution functions can be carried over to the set-valued case. In fact, as we will see, this is possible, since Θ, as a collection of subsets, shares basic topological and algebraic properties with $\mathbb{R}^d$, namely Θ is a lattice with

* Department of Mathematical Sciences, New Mexico State University, Las Cruces, New Mexico 88003-8001.

some suitable topology. These topological and algebraic properties can be used to define the concept of distribution functionals, generalizing distribution functions of random vectors, and to establish the counter-part of Lebesgue-Stieljes theory. To see this, let us proceed in a tutorial manner!

When $\Theta = \mathbb{R}$ (Θ consists of singletons of $\mathbb{R}$), the distribution function of S is the real-valued function F defined on $\mathbb{R}$ by

$$F(x) = P\left(\{\omega : S(\omega) \leq x\}\right), \qquad \forall x \in \mathbb{R}.$$

By construction, F satisfies the following conditions:

(i) $\lim_{x \to -\infty} F(x) = F(-\infty) = 0$, $\lim_{x \to \infty} F(x) = F(\infty) = 1$.

(ii) F is right continuous on $\mathbb{R}$, i.e., for any $x \in \mathbb{R}$, if $y \searrow x$ then $F(y) \searrow F(x)$.

(iii) F is monotone non-decreasing: $x \leq y$ implies $F(x) \leq F(y)$.

The classical result is this (for a proof, see, e.g. Durrett [3]). Let

$$C(x) = \{y : y \in \mathbb{R}, y \leq x\} \qquad (\text{here } = (-\infty, x])$$

and $\mathcal{B}(\mathbb{R})$ be the σ-field generated by the $C(x)$'s, $x \in \mathbb{R}$ (the so called Borel σ-field of $\mathbb{R}$). If a given function F, defined on $\mathbb{R}$, satisfies the above conditions (i), (ii), and (iii), then F uniquely determines a probability measure μ on $\mathcal{B}(\mathbb{R})$ such that

$$F(x) = \mu\left(C(x)\right) \qquad \forall x \in \mathbb{R}.$$

Thus to specify the probability law (measure) of a random variable, or to propose a model for a random quantity, it suffices to specify a distribution function.

For the case when $\Theta = \mathbb{R}^d$, $d > 1$ (singletons of $\mathbb{R}^d$), we note that $\mathbb{R}^d$ is a topological space, and a partially ordered set, where, by abuse of notation, the order relation on $\mathbb{R}^d$ is written as

$$x \leq y \qquad \text{iff} \qquad x_i \leq y_i, \quad i = 1, 2, \ldots, d,$$

$x = (x_1, x_2, \ldots, x_d)$ and $y = (y_1, y_2, \ldots, y_d)$. In fact, $\mathbb{R}^d$ is a semi-lattice (inf), where

$$\inf(x, y) = x \wedge y = (x_1 \wedge y_1, \ldots, x_d \wedge y_d).$$

By analogy with the case when $d = 1$, consider $F : \mathbb{R}^d \longrightarrow [0, 1]$ defined by

$$F(x_1, \ldots, x_d) = P\left(S_1 \leq x_1, \ldots, S_d \leq x_d\right),$$

where $S = (S_1, \ldots, S_d) : \Omega \longrightarrow \mathbb{R}^d$ is a d-dimensional random vector. Obviously, by construction, F satisfies the counter-parts of (i) and (ii), namely (i′) $\lim_{x_j \to -\infty} F(x_1, \ldots, x_d) = 0$ for each j,

$$\lim_{x_1 \to \infty, \ldots, x_d \to \infty} F(x_1, \ldots, x_d) = 1.$$

(ii′) F is right continuous on $\mathbb{R}^d$, i.e. for any $x \in \mathbb{R}^d$, if $y \in \mathbb{R}^d$ and $y \searrow x$ (i.e. $y_i \searrow x_i$, $i = 1, 2, \ldots, d$), then $F(y) \searrow F(x)$.

As for the counter-part of (iii), first note that F is monotone non-decreasing, i.e., if $x \leq y$ (on $\mathbb{R}^d$), then $F(x) \leq F(y)$, by construction of F in terms of P and S. Now, also by construction of F, we see that for $x = (x_1, \ldots, x_d) \leq y = (y_1, \ldots, y_d)$,

$$\begin{aligned} P(x < S \leq y) &= F(y) - \sum_{i=1}^{d} F_i \\ &\quad + \sum_{i<j} F_{ij} \pm \cdots \pm \cdots + (-1)^d F(x) \geq 0, \end{aligned}$$

where $F_{ij\cdots k}$ denotes the value $F(z_1, z_2, \ldots, z_d)$ with $z_i = x_i, \ldots, z_k = x_k$ and the others $z_t = y_t$.

Note that $P(x < S \leq y)$ can be written in a more compact form as follows (see Durrett [3]). Let the set of vertices of $(x_1, y_1] \times \cdots \times (x_d, y_d]$ be

$$V = \{x_1, y_1\} \times \cdots \times \{x_d, y_d\},$$

and for $v \in V$, let

$$sgn\,(v) = (-1)^{\alpha(v)},$$

where $\alpha(v)$ is the number of the x_i's in v, then

$$P(x < S \leq y) = \sum_{v \in V} sgn\,(v) F(v).$$

Of course, being a probability value, $P(x < S \leq y) \geq 0$ for any $x \leq y$.

For a probability measure μ on $(\mathbb{R}^d, \mathcal{B}(\mathbb{R}^d))$ to be uniquely determined through a given function $F : \mathbb{R}^d \to [0, 1]$, by

$$\mu(C(x)) = F(x), \qquad \forall x \in \mathbb{R}^d,$$

where $C(x) = \{y \in \mathbb{R}^d : y \leq x\}$, we not only need that $F(x)$ satisfies (i′), (ii′) and be monotone non-decreasing, but also the additional condition (iii′) For any $x \le y$ in $\mathbb{R}^d$,

$$\sum_{v \in V} sgn\,(v) F(v) \ge 0.$$

Again for a proof, see, e.g., Durrett [3]. Here is an example where F satisfies (i′), (ii′) and is monotone non-decreasing, but fails to satisfy (iii′). Consider

$$F(x,y) = \begin{cases} 1 & \text{if } x \ge 0,\ y \ge 0, \text{ and } x + y \ge 1 \\ 0 & \text{otherwise,} \end{cases}$$

then for $(1/4, 1/2) \le (1,1)$, we have

$$F(1,1) - F\left(\frac{1}{4}, 1\right) - F\left(1, \frac{1}{2}\right) + F\left(\frac{1}{4}, \frac{1}{2}\right) < 0.$$

We can group the monotone non-decreasing property and (iii′) into a single condition:

(iv) For $a, a_1, \ldots, a_n$ in $\mathbb{R}^d$, $n \ge 0$,

$$F(a) \quad \ge \sum_{\emptyset \ne I \subseteq \{1,2,\ldots,n\}} (-1)^{|I|+1} F\left(a \wedge_I a_i\right) \quad \forall a, a_1, \ldots, a_n \in \mathbb{R}^d,$$

where $|I|$ denotes the cardinality of I.

The condition (iv) is stronger than monotone non-decreasing and is referred to as *monotone of infinite order* or *totally monotone* (Choquet [1] and Revuz [19]). Note that for $d = 1$, the real line $\mathbb{R}$ is a chain, so that total monotonicity is the same as monotonicity. For $d > 1$, monotone functions might not be totally monotone. For example, let $d = 2$ and

$$F(x,y) = \begin{cases} 1 & \text{for } x \ge 0,\ y \ge 0, \text{ and } \max(x,y) \ge 1 \\ 0 & \text{otherwise.} \end{cases}$$

Let $a = (x,y)$, $a_1 = (x_1, y_1)$, $a_2 = (x_2, y_2)$ with

$$0 \le x_1 < 1 < x < x_2, \qquad 0 \le y_2 < 1 < y < y_1,$$

then

$$F(a) = F(a \wedge a_1) = F(a \wedge a_2) = 1, \qquad F(a \wedge a_1 \wedge a_2) = F(x_1, y_2) = 0,$$

and hence F is not totally monotone.

Now consider the case when Θ is a collection of subsets of some set U. Θ is partially ordered set-function. Suppose that Θ is stable under finite intersections, so that it is a semi-lattice, where $\inf(A,B) = A \cap B$. To carry conditions on a distribution function to this case, we need to define an appropriate topology for Θ (to formalize the concept of right continuity and to define an associated Borel σ-field $\mathcal{B}$ on Θ, so that $C(A) = \{B \in \Theta : B \subseteq A\} \in \mathcal{B}$ for all $A \in \Theta$). The condition of monotone of infinite order (iv) is expressed in terms of the lattice structure of Θ. Various situations of this setting can be found in Revuz [19]. When Θ is the power of U, or Θ is a Borel σ-field, say, of $\mathbb{R}^d$, the dual of a set-function F, defined on Θ, monotone of infinite order, is the set-function G, defined on Θ, by

$$G(A) = 1 - F(A^c), \qquad A^c \in \Theta,$$

where A^c denotes the set-complement of A. Such a set-function G is *alternating of infinite order.*

In the well-developed theory of *random closed sets* on locally compact, Hausdorff, separable spaces, or on Polish spaces (see, e.g. Matheron [14], Molchanov [16], Wasserman [24], and Stoyan [23]), the set-functions used to introduce probability measures (laws) are normalized *Choquet capacities*, alternating of infinite order (called capacity functionals). Specifically, let $\Theta = \mathcal{F}$, the class of closed sets of, say, $\mathbb{R}^d$. $\mathcal{F}$ is topologized by the so-called hit-and-miss topology. The capacity functional of a random closed set S is the set-function G, defined in the class $\mathcal{K}$ of all compact sets of $\mathbb{R}^d$, by

$$G(K) = P(S \cap K \neq \emptyset), \qquad K \in \mathcal{K}.$$

The probability law of a random closed set S is completely determined by a set-function G, defined on $\mathcal{K}$, satisfying:

(a) $G(\emptyset) = 0$, $0 \leq G(\cdot) \leq 1$.
(b) If $K_n \searrow K$ in $\mathcal{K}$, then $G(K_n) \searrow G(K)$.
(c) (Alternating of infinite order) For all $K, K_1, \ldots, K_n$ in $\mathcal{K}$, n integer, $\Delta_n(K, K_1, \ldots, K_n) \geq 0$, where

$$\Delta_1(K, K_1) = G(K \cup K_1) - G(K), \quad \ldots,$$

$$\begin{aligned} \Delta_n(K, K_1, \ldots, K_n) &= \Delta_{n-1}(K, K_1, \ldots, K_{n-1}) \\ &\quad - \Delta_{n-1}(K \cup K_n; K_1, \ldots, K_{n-1}). \end{aligned}$$

While the theory of random sets remains entirely within the abstract theory of probability, there exist certain special features which deserve special studies. In order to suggest models for random sets, we might be forced to consider distribution functionals which are set-functions rather than point-functions. Specifying multivariate distribution functions is not a trivial task. Fortunately, the unconstrained form of Frechet's problem was solved, see Schweizer and Sklar [22]. As for the case of random sets, the main ingredient is non-additive set-functions. For example, what are the counter-parts of usual concepts associated with distribution functions of random vectors? An integral with respect to a non-additive set-function can be taken to be the so-called Choquet integral; when Θ is finite, the "density" of a distribution functional F of a random set is its Möbius inverse, but what happens if Θ is infinite? See, however, Graf [7] for a Radon-Nikodym theorem for capacities. Another problem arises: essentially we are dealing with set-valued functions. An integration theory of set-valued function is needed. Obviously set-valued analysis is needed.

3. On distributions of random sets. In this section we discuss some basic issues concerning distributions of random sets as set-functions. Also, we discuss the result of Graf [7] on the existence of Radon-Nikodym derivatives of set-functions.

A well-known example of capacity functional of a random closed set is the following.

Let $f : \mathbb{R}^d \to [0,1]$ be a upper semi-continuous function. Then

$$\mu(A) = \sup_A f(x), \qquad A \in \mathcal{K},$$

is the capacity functional of the random set $S : \Omega \to \mathcal{F}$, where $S(\omega) = \{x \in \mathbb{R}^d : f(x) \geq \alpha(\omega)\}$ and $\alpha : (\Omega, \mathcal{A}, P) \longrightarrow [0,1]$, being a uniformly distributed random variable. Indeed,

$$P(S \cap A \neq \emptyset) = \mu(A), \qquad \forall A \in \mathcal{K}.$$

It then follows from Choquet's theorem (e.g. Matheron [14]) that among other properties, $\mu(\cdot)$ is alternating of infinite order. This set-function μ is very special since $\mu(\cup_I A_i) = \sup_I \mu(A_i)$ for any index set I. A more general class of set-functions appearing, e.g., in the theory of extremal stochastic processes (see, e.g. Molchanov [16]), consists of set-functions μ defined on a σ-field $\mathcal{U}$ of some space $\mathcal{X}$, $\mu : \mathcal{U} \to [0, \infty)$, $\mu(\emptyset) = 0$ and *maxitive*, i.e.,

$$\mu(A \cup B) = \max\{\mu(A), \mu(B)\}, \qquad \forall A, B \in \mathcal{U}.$$

Note that such set-functions are monotone increasing, and the above maxitivity property holds for any finite collections $A_1, \ldots, A_n$, $n \geq 2$. This class of set-functions is quite large. A general construction of maxitive set-functions goes as follows. If μ is maxitive (we always assume that $\mu(\emptyset) = 0$), then for each $t \geq 0$, $\mathcal{J}_t = \{A \in \mathcal{U} : \mu(A) \leq t\}$ is an ideal in $\mathcal{U}$. Moreover, the family $\{\mathcal{J}_t, t \geq 0\}$ is increasing: $s < t$ implies that $\mathcal{J}_s \subseteq \mathcal{J}_t$. And we have

$$\mu(A) = \inf\{t \geq 0 : A \in \mathcal{J}_t\}.$$

If $\{\eta_t, t \geq 0\}$ is an increasing family of ideals in $\mathcal{U}$, then the set-function μ defined by

$$\mu(A) = \inf\{t \geq 0 : A \in \eta_t\}$$

is maxitive. Similarly to the situation for outer measures, it may happen that $\{\eta_t, t \geq 0\}$ is different than the canonical $\{\mathcal{J}_t, t \geq 0\}$. They do coincide if and only if $\{\eta_t, t \geq 0\}$ is "right-continuous", in the sense that,

$$\eta_t = \cap_{s>t}\eta_s, \qquad \forall t \geq 0.$$

Here are several examples. Let η be an ideal in $\mathcal{U}$ and $f : X \to [0, \infty)$ measurable. Let

$$\eta_t = \{A \in \mathcal{U} : A \cap (f > t) \in \eta\}, \qquad t \geq 0.$$

Then

$$\mu(A) = \inf\{t \geq 0 : A \in \eta_t\}$$

is maxitive. If η is the σ-ideal of sets of P-probability zero, where P is a probability measure on $(\mathcal{X}, \mathcal{U})$, and f is measurable and bounded, then

$$\mu(A) = ||f1_A||_\infty = \inf\{t \geq 0 : P(A \cap (f > t)) = 0\}$$

is maxitive.

In analysis, the Hausdorff dimension on a metric space has a similar representation, and hence maxitive. Let $(\mathcal{X}, d)$ be a metric space, and $d(A)$ be the diameter of $A \subset \mathcal{X}$. For $\alpha \geq 0$, the outer measure μ_α is defined as

$$\mu_\alpha(A) = \lim_{\varepsilon \to 0}\left[\inf \sum_n d^\alpha(B_n)\right],$$

where the infimum is taken over all countable coverings of A by closed balls B_n such that $d(B_n) < \varepsilon$. If $\eta_\alpha = \{A : \mu_\alpha(A) = 0\}$, then $\{\eta_\alpha, \alpha \geq 0\}$ is an increasing family of σ-ideals in the power set of $\mathcal{X}$, the Hausdorff dimension is defined by

$$D(A) = \inf\{\alpha \geq 0 : \mu_\alpha(A) = 0\} = \inf\{\alpha \geq 0 : A \in \eta_\alpha\}.$$

It turns out that maxitive set-functions are alternating of infinite order.

Let μ be a maxitive set-function on $(\mathcal{X}, \mathcal{U})$. Since μ is monotone increasing, it suffices to verify that, $\forall n \geq 2$, $\forall A_1, \ldots, A_n \in \mathcal{U}$, we have

$$\mu\left(\bigcap_{i=1}^n A_i\right) \leq \sum_{\emptyset \neq I \subseteq \{1,\ldots,n} (-1)^{|I|+1}\mu\left(\bigcup_I A_i\right).$$

Without loss of generality, we assume that

$$0 \leq m_n = \mu(A_n) \leq m_{n-1} = \mu(A_{n-1}) \leq \cdots \leq m_1 = \mu(A_1).$$

We have

$$\sum_{\emptyset \neq I \subseteq \{1,\ldots,n\}} (-1)^{|I|+1}\mu\left(\bigcup_I A_i\right) = \sum_{k=1}^n (-1)^{k+1} \sum_{I \in I(k)} \mu\left(\bigcup_I A_i\right),$$

where $I(k) = \{I \subseteq \{1,\dots,n\} : |I| = k\}$. Let $I_i(k) = \{I \in I(k) : m(I) = i\}$, where $m(I) = \min\{i : i \in I\}$. Observe that, since μ is maxitive, for $I \in I_i(k)$,

$$\mu\left(\bigcup_{j\in I} A_j\right) = m_i, \quad i = 1,\dots,n-k+1,$$

and there are $\binom{n-i}{k-1}$ elements in $I_i(k)$. Thus,

$$\begin{aligned}
\sum_{\emptyset\neq I\subseteq\{1,\dots,n\}} (-1)^{|I|+1}\mu\left(\bigcup_I A_i\right) &= \sum_{k=1}^{n}(-1)^{k+1}\sum_{i=1}^{n-k+1}\binom{n-i+1}{k-1} m_i \\
&= \sum_{i=1}^{n}\left[\sum_{k=1}^{n-i+1}\binom{n-i+1}{k-1}(-1)^{k+1}\right] m_i \\
&= \sum_{i=1}^{n}\left[\sum_{k=0}^{n-i}\binom{n-i}{k}(-1)^{k}\right] m_i \\
&= m_n = \mu(A_n) \geq \mu\left(\bigcap_{i=1}^{n} A_i\right),
\end{aligned}$$

by observing that

$$\sum_{k=0}^{n-i}\binom{n-i}{k}(-1)^k = (1-1)^{n-i} = 0 \quad \text{for } i = 1,\dots n-1.$$

Consider again $f : \mathbb{R}^d \to [0,1]$ upper semi-continuous and the set function $\mu : \mathcal{B}(\mathbb{R}^d) \to [0,1]$,

$$\mu(A) = \sup_A f(x).$$

Also, let $\nu_0 : \mathcal{B}(\mathbb{R}^d) \to [0,1]$,

$$\nu_0(A) = \begin{cases} 1 & \text{if } A \neq \emptyset \\ 0 & \text{if } A = \emptyset. \end{cases}$$

Then both μ and ν_0 are "capacities" (in fact, subadditive pre-capacities) considered in Graf [7], i.e., satisfying the following:

i) $\mu(\emptyset) = 0$,
ii) $\mu(A \cup B) \leq \mu(A) + \mu(B)$,
iii) $A \subseteq B$ implies that $\mu(A) \leq \mu(B)$,
(iv) if $A_n \nearrow A$ then $\mu(A_n) \nearrow \mu(A)$.

Observe that

$$\mu(A) = \int_A f d\nu_0, \qquad \forall A \in \mathcal{B}(\mathbb{R}^d),$$

where the integral is taken in Choquet's sense, i.e.

$$\begin{aligned}\int_A f d\nu_0 &= \int_{\mathbb{R}^d} (f 1_A) d\nu_0 \\ &= \int_0^\infty \nu_0 \left(A \cap (f \geq t)\right) dt + \int_{-\infty}^0 [\nu_0(A \cap (f \geq t)) - \nu_0(A)] dt.\end{aligned}$$

In this case, f is referred to as the *Radon-Nikodym derivative* of μ with respect to ν_0, and is denoted as $f = d\mu/d\nu_0$.

If μ and ν are two "capacities" (on a measurable space $(\mathcal{X}, \mathcal{U})$) in the above sense, and $\mu(A) = \int_A f d\nu$, then obviously, $\nu(A) = 0$ implies that $\mu(A) = 0$. However, unlike the situation for measures, this *absolute continuity* (denoted as $\mu << \nu$) among capacities is not sufficient for the existence of a Radon-Nikodym derivative. In fact, the main result in Graf [7] is this. A necessary and sufficient condition for the capacity μ to admit a Radon-Nikodym derivative with respect to a capacity ν is that $\mu << \nu$ *and* (μ, ν) has the following "strong decomposition property" (SDP): for every $t \in [0, \infty)$, there exists $A(t) \in \mathcal{U}$ such that

(i) $t[\nu(A) - \nu(B)] \leq \mu(A) - \mu(B)$ for $A, B \in \mathcal{U}$ and $B \subset A \subset A(t)$,
(ii) $t[\nu(A) - \nu(A \cap A(t))] \geq \mu(A) - \mu(A \cap A(t))$ for any $A \in \mathcal{U}$.

Note that the SDP is automatically satisfied for measures in view of the Hahn decomposition (of a signed measure in terms of a partition of the space $\mathcal{X}$). Recall that if μ and ν are measures, then for any $t \geq 0$, $t\nu - \mu$ is a signed measure, and hence, by the Hahn decomposition, there exists $A(t) \in \mathcal{U}$ such that

$$(t\nu - \mu)|\mathcal{U}_{A(t)} \leq 0 \quad \text{and} \quad (t\nu - \mu)|\mathcal{U}_{A^c(t)} \geq 0,$$

where A^c denotes the set complement of A in $\mathcal{X}$, and the notation $\nu|\mathcal{U}_A \leq \mu|\mathcal{U}_A$ means $\nu(B) \leq \mu(B)$ for any $B \in \mathcal{U}_A$, where $\mathcal{U}_A = \{B \in \mathcal{U} : B \subseteq A\}$. Observe that the family $\{A(t), t \geq 0\}$ has the following Hahn decomposition property (HDP):

$$t\nu|\mathcal{U}_{A(t)} \leq \mu|\mathcal{U}_{A(t)} \quad \text{and} \quad t\nu|\mathcal{U}_{A^c(t)} \geq \mu|\mathcal{U}_{A^c(t)}.$$

The HDP is called the weak decomposition property (WDP) in Graf [7].

The following example shows that the HDP may fail for capacities. Let $(\mathcal{X}, \mathcal{U}) = (\mathbb{R}, \mathcal{B}(\mathbb{R}))$. For $n \geq 1$, let $G_n = \{1/k : k \text{ integer}, k \geq n\}$, and $G = G_1 \cup \{0\}$. Define $\mu, \nu : \mathcal{B}(\mathbb{R}) \longrightarrow [0, 1]$ by

$$\mu(A) = \begin{cases} \min\left\{1, \frac{1}{2}\sum\{x : x \in A \cap G\}\right\}, & \text{if } A \cap G_1 \neq \emptyset \\ 0 & \text{if } A \cap G_1 = \emptyset, \end{cases}$$

$$\nu(A) = \begin{cases} \min\{x : x \in A \cap G\}, & \text{if } A \cap G_1 \neq \emptyset \\ 0 & \text{if } A \cap G_1 = \emptyset. \end{cases}$$

Then $\mu << \nu$ and $\nu << \mu$, but (μ, ν) does not have the HDP. This can be seen as follows. Suppose that (μ, ν) has the HDP. Then for $t = 1$, we have $\nu(A) \leq \mu(A)$ for any $A \in \mathcal{U}_{A(1)}$ and $\nu(A) \geq \mu(A)$ for any $A \in \mathcal{U}_{A^c(1)}$. Now $G_n \subset A^c(1)$ for any $n \geq 2$. Indeed, if $G_n \cap A(1) \neq \emptyset$ for some $n \geq 2$, then for $1/k \in G_n \cap A(1)$, we have $\nu\left(\{\frac{1}{k}\}\right) \leq \mu\left(\{\frac{1}{k}\}\right)$. But by construction of μ and ν, we have $\nu\left(\{\frac{1}{k}\}\right) = \frac{1}{k}$ and $\mu\left(\{\frac{1}{k}\}\right) = \frac{1}{2k}$, a contradiction. Thus $\nu(G_n) \geq \mu(G_n)$. But, by construction, $\nu(G_n) = \frac{1}{n}$ and $\mu(G_n) = 1$, a contradiction.

Here is an example showing that absolute continuity and HDP are not sufficient for the existence of a Radon-Nikodym derivative. In other words, the SDP is stronger than the WDP (or HDP).

With the notation of the previous example, let ν as before, and

$$\mu(A) = \begin{cases} 0 & \text{if } A \cap G_1 = \emptyset \\ 1 & \text{if } A \cap G_1 \neq \emptyset. \end{cases}$$

It can be checked that $\mu << \nu$ and $\nu << \mu$. Moreover, (μ, ν) has the HDP. Indeed, the

$$A(t) = \left\{\frac{1}{n} : n \geq t\right\}, \quad t \geq 0, \quad A(0) = \mathbb{R},$$

is a Hahn decomposition for (μ, ν), since for $t > 0$ and $A \subset A(t)$, we have

$$t\nu(A) = t \sup_{x \in A \cap G} \{x\} \leq t \sup_{n \geq t} \left\{\frac{1}{n}\right\} \leq 1 = \mu(A),$$

and for $A \subset A^c(t)$,

$$\begin{aligned} t\nu(A) &= t \sup_{x \in A \cap G} \{x\} = \frac{m}{t} \quad \text{for some } m < t \\ &\geq 1 = \mu(A). \end{aligned}$$

The above is trivial for $t = 0$.

However, μ does not admit a Radon-Nikodym derivative with respect to ν. Indeed, assume the contrary, and let $f = d\mu/d\nu$, i.e. $f : \mathbb{R} \to [0, \infty)$, measurable, such that

$$\mu(A) = \int_0^\infty \nu(A \cap (f \geq t))dt, \qquad \forall A \in \mathcal{B}(\mathbb{R}).$$

Now for $A = \{\frac{1}{n}\}$,

$$\begin{aligned} \mu(A) &= \int_0^\infty \nu\left(\{\frac{1}{n}\} \cap (f \geq t)\right) dt = \int_0^{f(1/n)} \nu\left(\{\frac{1}{n}\}\right) dt \\ &= \int_0^{f(1/n)} \frac{1}{n} dt = \frac{1}{n} f\left(\frac{1}{n}\right). \end{aligned}$$

But, by construction, $\mu\left(\{\frac{1}{n}\}\right) = 1$. Thus

$$f\left(\frac{1}{n}\right) = n \qquad \forall n \geq 1.$$

But then

$$\begin{aligned}
\mu(G) &= \int_0^\infty \nu\left(G \cap (f \geq t)\right) dt \\
&= \sum_{n=1}^\infty \int_{n-1}^n \nu\left(G \cap (f \geq t)\right) dt \\
&\geq \sum_{n=1}^\infty \int_{n-1}^n \nu\left(G \cap (f \geq n)\right) dt \\
&= \sum_{n=1}^\infty \int_{n-1}^n \nu\left(\left\{\frac{1}{n} : f\left(\frac{1}{k}\right) \geq n\right\}\right) dt \\
&= \sum_{n=1}^\infty \int_{n-1}^n \mu\left(\left\{\frac{1}{k} : k \geq n\right\}\right) dt \\
&= \sum_{n=1}^\infty \sup\left\{\frac{1}{k} : k \geq n\right\} = \sum_{n=1}^\infty \frac{1}{n} = \infty.
\end{aligned}$$

On the other hand, by construction, $\mu(G) = 1$, a contradiction.

The capacities in Graf's sense are σ-subadditive, i.e.

$$\mu\left(\cup_n A_n\right) \leq \sum_n \mu(A_n) \qquad \forall A_n \in \mathcal{U}.$$

However, a *sub-measure*, i.e. a set-function $\mu : \mathcal{U} \to [0, \infty)$, such that $\mu(\emptyset) = 0$, μ is monotone increasing and σ-subadditive, need not be a capacity. For example, let $(\mathcal{X}, \mathcal{U}) = (\mathbb{R}^d, \mathcal{B}(\mathbb{R}^d))$ and

$$\mu(A) = \begin{cases} 0 & \text{if } A = \emptyset \\ 1 & \text{if } \emptyset \neq A \neq \mathbb{R}^d \\ 2 & \text{if } A = \mathbb{R}^d \end{cases}$$

A submeasure μ is an outer measure when $\mathcal{U} = 2^{\mathcal{X}}$. For submeasures, the following is a necessary condition for the existence of Radon-Nikodym derivatives.

We say that a pair (μ, ν) of submeasures has the *curve level property* (CLP) if there exists a decreasing family $\{A(t) : t \geq 0\} \subset \mathcal{U}$ such that

(i) $\mu\left(\cap_{n \geq 1} A(n)\right) = \nu\left(\cap_{n \geq 1} A(n)\right) = 0$ and
(ii) for $s < t$ and $A \in \mathcal{U}$,

$$\begin{aligned}
s[\nu(A \cap A(s)) - \nu(A \cap A(t))] &\leq \mu(A \cap A(s)) - \mu(A \cap A(t)) \\
&\leq t[\nu(A \cap A(s)) - \nu(A \cap A(t))].
\end{aligned}$$

Indeed, let $f = d\mu/d\nu$. Then (μ,ν) has the CLP with $A(t) = (f \geq t)$. Indeed, since $f(x) < \infty$ for all $x \in \mathcal{X}$, we have that $\cap_{n\geq 1} A(n) = \emptyset$, and hence (i) is satisfied. Next,

$$\begin{aligned}
&\mu(A \cap A(s) - \mu(A \cap A(t)) \\
&= \int_0^\infty [\nu(A \cap A(s) \cap (f \geq \alpha)) - \nu(A \cap A(t) \cap (f \geq \alpha))] d\alpha \\
&\geq \int_0^s [\nu(A \cap A(s) \cap (f \geq \alpha)) - \nu(A \cap A(t) \cap (f \geq \alpha))] d\alpha \\
&= \int_0^s [\nu(A \cap A(s)) - \nu(A \cap A(t))] d\alpha \\
&= s[\nu(A \cap A(s)) - \nu(A \cap A(t))].
\end{aligned}$$

Also

$$\begin{aligned}
&\mu(A \cap A(s) - \mu(A \cap A(t)) \\
&= \int_0^\infty [\nu(A \cap A(s) \cap (f \geq \alpha)) - \nu(A \cap A(t) \cap (f \geq \alpha))] d\alpha \\
&= \int_0^t [\nu(A \cap A(s) \cap (f \geq \alpha)) - \nu(A \cap A(t))] d\alpha \\
&\leq \int_0^t [\nu(A \cap A(s)) - \nu(A \cap A(t))] d\alpha \\
&= t[\nu(A \cap A(s)) - \nu(A \cap A(t))].
\end{aligned}$$

Note that the CLP for (μ,ν) implies the absolute continuity of μ with respect to ν. It turns out that the curve level property above is also sufficient for μ to have a Radon-Nikodym derivative with respect to ν. The proof of this will be published elsewhere.

4. Random sets in confidence region estimation. The problem of computations of various numerical characteristics of random sets appears in various contexts of statistical inference, such as in the theory of coverage processes (see Hall [9]), and in the optimal choice of confidence region estimation procedures (see Robbins [20]). In many situations, one is interested in the expected measure of random sets. Specifically, let S be a random set with values in $\mathcal{C} \subseteq \mathcal{B}(\mathbb{R}^d)$. Let $\sigma(\mathcal{C})$ be a σ-field on $\mathcal{C}$. One is interested in the expected value $E(\mu(S))$, where μ denotes the Lebesgue measure on $\mathbb{R}^d$. Of course, for this to make sense, $\mu(S)$ has to be random variable. This is indeed the case, for example, when S depends upon a finite number of random variables. In general, the distribution of $\mu(S)$ seems difficult to obtain. One then proceeds directly to the computation of $E(\mu(S))$ as follows. The *one-point coverage function* of S is defined as $\pi : \mathbb{R}^d \longrightarrow [0,1]$,

$$\pi(x_1,\ldots,x_d) = P(\{\omega : (x_1,\ldots,x_d) \in S(\omega)\}).$$

The main result in Robbins [20] is this. If the map $g : \mathbb{R}^d \times \mathcal{C} \longrightarrow \{0,1\}$, defined by $g(x, A) = 1_A(x)$, is $\mathcal{B}(\mathbb{R}^d) \otimes \sigma(\mathcal{C})$–measurable, then

$$E(\mu(S)) = \int_{\mathbb{R}^d} \pi(x) d\mu(x).$$

By considering many-point coverage functions, i.e.,

$$\pi_n(u_1, \ldots, u_n) = P(\{\omega : \{u_1, \ldots, u_n\} \subseteq S(\omega)\}),$$

for $u_i \in \mathbb{R}^d$, $i = 1, 2, \ldots, n$, higher order moments of $\mu(S)$ are obtained in a similar way, under suitable measurability conditions. Specifically, for $n \geq 1$,

$$E[\mu(S)]^n = \int_{\mathbb{R}^{nd}} \pi(u_1, \ldots, u_n) \otimes_{i=1}^n \mu_i(du_1, \ldots, du_n),$$

where $\otimes_{i=1}^n \mu_i$ denotes the product measure $\mu \otimes \cdots \otimes \mu$ on $\mathbb{R}^{nd}$.

When $n = 1$, the measurability of $g(x, A)$ implies that $\pi(x)$ is measurable and $\mu(S)$ is a random variable. Viewing $\left(g(x, \cdot), x \in \mathbb{R}^d\right)$ as a 0-1 stochastic process on the measurable space $(\mathcal{C}, \sigma(\mathcal{C}))$, we see that the measurability condition on g is simply that this stochastic process is a measurable process. If S is a compact-valued random set on $\mathbb{R}^d$, then the process g is measurable, and hence $\mu(S)$ is a random variable. For a general condition on the measurability of the process g, see Debreu [2].

Obviously, being unaware of Robbins' result, Matheron [14] established the same measurability condition and deduced Robbins' formula, but the purpose was to obtain a tractable criterion for almost sure continuity for random closed sets. According to Matheron, any random closed set S is measurable in the following sense: the map $g : \mathbb{R}^d \times \mathcal{F} \longrightarrow \{0,1\}$ is $\mathcal{B}(\mathbb{R}^d) \otimes \mathcal{B}(\mathcal{F})$–measurable. Thus, if μ denotes a positive measure (σ-finite) on $\left(\mathbb{R}^d, \mathcal{B}(\mathbb{R}^d)\right)$ (or more generally, on a locally compact, Hausdorff, separable space), then the map

$$F \in \mathcal{F} \longrightarrow \mu(F) = \int_{\mathbb{R}^d} g(x, F) d\mu(x)$$

is a positive random variable whose expected value is $\int_{\mathbb{R}^d} P(x \in S) d\mu(x)$.

Note that for $d = 1$ and $S(\omega) = [0, X(\omega)]$, where X is a non-negative random variable, Robbins' formula becomes

$$E(X) = \int_0^\infty P(X > t) dt.$$

Also, note that in sample surveys, e.g., Hajek [8], a sampling design on a finite population $\mathcal{U} = \{u_1, u_2, \ldots, u_N\}$ is a probability measure Q on its power set. The inclusion probabilities (first and second order) are precisely the values of the coverage functions, indeed

$$\alpha(u) = \sum_{A \ni u} Q(A) = P(u \in S),$$

$$\beta(u,v) = \sum_{A \ni u,v} Q(A) = P\left(\{u,v\} \subseteq S\right),$$

where S is a random set on $2^{\mathcal{U}}$ with probability law Q. These inclusion probabilities are related to the concept of measure of spread for sampling probabilities (defined as the entropy of the random set S).

5. Decision–making with imprecise probabilities. In its simplest form, a decision problem consists of choosing an action among a collection of relevant actions $\mathbb{A}$ in such a way that "utility" is maximized. Specifically, let Θ denote the set of possible "states of nature" where the unknown, true state is denoted by θ_0. A specific utility function is a map $u : \mathbb{A} \times \Theta \longrightarrow \mathbb{R}$, where $u(a, \theta)$ is interpreted as the "payoff" when action a is taken and nature presents θ. In a Bayesian framework, the knowledge about Θ is represented as a probability measure P_0 on Θ. The expected value $E_{P_0}(u(a,\cdot))$ can be used to make a choice as to which a to take. The optimal action is the one that maximizes $E_{P_0}(u(a,\cdot))$ over $a \in \mathbb{A}$. In many situations, the probability measure P_0 is only known partially, say, $P_0 \in \mathcal{P}$, a specified class of probability measures on Θ, so that $F \leq P_0 \leq G$, where $F = \inf_{\mathcal{P}} P$ and $G = \sup_{\mathcal{P}} P$. There are situations in which the lower envelop F turns out to be a set-function, totally monotone, i.e., a distribution functional of some random set S. Decision-making in this imprecise probabilistic knowledge can be carried out by using one of the following approaches.

(a) Choquet integral. From a minimax viewpoint, we choose the action that maximizes

$$\inf \left\{E_p\left(u(a,\cdot)\right) \,:\, p \in \mathcal{P}\right\}.$$

Consider the case when Θ is finite, say, $\Theta = \{\theta_1, \theta_2, \ldots, \theta_n\}$. Suppose that F is monotone of infinite order. For each a, rename the elements of Θ if necessary so that

$$u(a,\theta_1) \leq u(a,\theta_2) \leq \cdots \leq u(a,\theta_n).$$

Let $A_i = \{\theta_i, \theta_{i+1}, \ldots, \theta_n\}$ and

$$h(\theta_i) = F(A_i) - F\left(A_i \setminus \{\theta_i\}\right).$$

Then h is a probability density on Θ such that

$$P_h(A) = \sum_{B \subseteq A_i} m(B) - \sum_{B \subseteq A_i \setminus \{\theta_i\}} m(B) = \sum_{\theta_i \in B \subseteq A_i} m(B),$$

where m is the Möbius inverse of F, i.e., $m : 2^\Theta \longrightarrow [0,1]$,

$$m(A) = \sum_{B \subseteq A} (-1)^{|A \setminus B|} F(B).$$

Now, for each $t \in \mathbb{R}$ and a density g such that $P_g \in \mathcal{P}$, we have

$$P_h\left(u(a,\cdot) > t\right) \leq P_g\left(u(a,\cdot) > t\right)$$

since $(u(a,\cdot) > t)$ is of the form $\{\theta_i, \theta_{i+1}, \dots, \theta_n\}$. Thus

$$E\left[P_h\left(u(a,\cdot) > t\right)\right] \leq E\left[P_g\left(u(a,\cdot) > t\right)\right]$$

for all densities g such that $P_g \in \mathcal{P}$. But, by construction of h

$$\begin{aligned} &E[P_h(u(a,\cdot) > t)] \\ &= \sum_{i=1}^{n} u(a,\theta_i) \left[F(\{\theta_i, \dots, \theta_n\}) - F(\{\theta_{i+1}, \dots, \theta_n\})\right] \\ &= \int_0^\infty F\left(u(a,\cdot) > t\right) dt + \int_{-\infty}^0 \left[F\left(u(a,\cdot) > t\right) - 1\right] dt, \end{aligned}$$

which is nothing else than the *Choquet integral* of the function $u(a,\cdot)$ with respect to the monotone set-function F, denoted as $E_F\left(U(a,\cdot)\right)$.

We have

$$E_F\left(U(a,\cdot)\right) = \inf\{E_P\left(U(a,\cdot)\right) : P \in \mathcal{P}\}.$$

For example, let $\Theta = \{\theta_1, \theta_2, \theta_3, \theta_4\}$ and

$$F(A) = \sum_{B \subseteq A} m(B), \qquad A \subseteq \Theta,$$

where

$$m\left(\{\theta_1\}\right) = 0.4, \qquad m(\{\theta_2\}) = f(\{\theta_3\}) = 0.2,$$

$$m(\{\theta_4\}) = 0.1, \qquad m\left(\{\theta_1, \theta_2, \theta_3, \theta_4\}\right) = 0.1,$$

and $m(A) = 0$ for any other subset A of Θ; then, we have

$$h(\theta_1) = 0.5, \quad h(\theta_2) = h(\theta_3) = 0.2, \quad h(\theta_4) = 0.1.$$

If

$$u(a,\theta_1) = 1, \quad u(a,\theta_2) = 5, \quad u(a,\theta_3) = 10, \quad u(a,\theta_4) = 20,$$

then

$$E_F[u(a,\cdot)] = \sum_{i=1}^{4} u(a,\theta_i)h(\theta_i) = 5.5.$$

When Θ is infinite and F is monotone of infinite order, we still have

$$E_F\left(U(a,\cdot)\right) = \inf\{E_P\left(U(a,\cdot)\right) : P \in \mathcal{P}\},$$

although, unlike the infinite case, the inf might not be attained (see Wasserman [24]).

From the above, we see that the concept of Choquet integral can be used as a tool for decision-making based on expected utility in situations where imprecise probabilistic knowledge can be modeled as a distribution functional of some random set. For example, consider another situation with imprecise information. Let X be a random variable, defined on $(\Omega, \mathcal{A}, P)$, and $g : \mathbb{R} \to \mathbb{R}^+$ be a measurable function. We have

$$E\left(g(X)\right) = \int_\Omega g(X(\omega))dP(\omega) = \int_{\mathbb{R}} g(x)dP_X(x),$$

where $P_X = PX^{-1}$.

Suppose that to each random experiment ω, we can only assert that the outcome $X(\omega)$ is some interval, say $[a,b]$. This situation can be described by a multi-valued mapping $\Gamma : \Omega \longrightarrow \mathcal{F}(\mathbb{R})$, the class of closed subsets of $\mathbb{R}$. Moreover, for each $\omega \in \Omega$, $X(\omega) \in \Gamma(\omega)$. As in Wasserman [24], the computation, or approximation of $E[g(X)]$ is carried out as follows.

Since

$$g\left(X(\omega)\right) \in g\left(\Gamma(\omega)\right) = \{g(x) : x \in \Gamma(\omega)\},$$

we have

$$\begin{aligned} g_*(\omega) &= \inf\{g(x) : x \in \Gamma(\omega)\} \leq g\left(X(\omega)\right) \\ &\leq \sup\{g(x) : x \in \Gamma(\omega)\} = g^*(\omega). \end{aligned}$$

Thus, $g(x)$ is such that $g_* \leq g(X) \leq g*$ and hence formally,

$$E(g_*) \leq E\left(g(X)\right) \leq E(g^*).$$

For this to make sense, we first need to see whether g_* and g^* are measurable functions.

Let $\mathcal{B}(\mathbb{R})$ be the Borel σ-field on $\mathbb{R}$. The multi-valued mapping Γ is said to be strongly measurable if

$$B_* = \{\omega : \Gamma(\omega) \subseteq B\} \in \mathcal{A}, \quad B^* = \{\omega : \Gamma(\omega) \cap B \neq \emptyset\} \in \mathcal{A},$$

for all $B \in \mathcal{B}(\mathbb{R})$. If Γ is strongly measurable, then the measurability of g implies that of g_* and g^*, and conversely. Indeed, suppose that Γ is strongly measurable and g is measurable. For $\alpha \in \mathbb{R}$,

$$\Gamma(\omega) \subseteq \{x : g(x) \geq \alpha\} = g^{-1}([\alpha, \infty)),$$

and hence $\omega \in \left[g^{-1}([\alpha,\infty))\right]_*$. If $\omega \in \left[g^{-1}([\alpha,\infty))\right]_*$, then $\Gamma(\omega) \subseteq g^{-1}([\alpha,\infty))$, that is, for all $x \in \Gamma(\omega)$, $g(x) \geq \alpha$ implying that $\inf_{x\in\Gamma(\omega)} g(x) \geq \alpha$, and hence $g_*(\omega) \in [\alpha,\infty)$ or $\omega \in g_*^{-1}([\alpha,\infty))$. Therefore

$$g_*^{-1}([\alpha,\infty)) = \left[g^{-1}([\alpha,\infty))\right]_*.$$

Since g is measurable, $g^{-1}([\alpha,\infty)) \in \mathcal{B}(\mathbb{R})$, and since Γ is strongly measurable, $\left[g^{-1}([\alpha,\infty))\right]_* \in \mathcal{A}$. The measurability of g^* follows similarly. For the converse, let $A \in \mathcal{B}(\mathbb{R})$. Then $1_A = f$ is measurable and

$$f_*(\omega) = \begin{cases} 1 & \text{if } \Gamma(\omega) \subseteq A \\ 0 & \text{otherwise} \end{cases}$$

Thus, $f^{-1}(\{1\}) = A_*$, and by hypothesis, $A_* \in \mathcal{A}$. Similarly $A^* \in \mathcal{A}$.

If we let $F_* : \mathcal{B}(\mathbb{R}) \longrightarrow [0,1]$ be defined by

$$F_*(B) = P(\{\omega : \Gamma(\omega) \subseteq B\}) = P(B_*),$$

then

$$\begin{aligned} E_*(g(X)) &= \int_\Omega g_*(\omega)dP(\omega) = \int_0^\infty P(\{\omega : g_*(\omega) > t\})\,dt \\ &= \int_0^\infty P\left(\{g_*^{-1}(t,\infty)\}\right)dt = \int_0^\infty P\left[g^{-1}(t,\infty)\right]_* dt \\ &= \int_0^\infty P\left(\{\omega : \Gamma(\omega) \subseteq g^{-1}(t,\infty)\}\right)dt \\ &= \int_0^\infty F_*\left(g^{-1}(t,\infty)\right)dt = \int_0^\infty F_*(g > t)dt. \end{aligned}$$

Note that Γ is assumed to be strongly measurable, so that F_* is well-defined on $\mathcal{B}(\mathbb{R})$. Similarly, letting

$$F^*(B) = P(\{\omega : \Gamma(\omega) \cap B \neq \emptyset\}) = P(B^*),$$

we have

$$E^*(g(X)) = \int_0^\infty F^*(g > t)dt.$$

In the above situation, the set-functions F^* and F_* are known (say, Γ is observable but X is not). Although F^* and F_* are not probability measures, they can be used for approximate inference procedures. Here Choquet integrals with respect to monotone set-functions represent some practical quantities of interest.

(b) Expectation of a function of a random set. Lebesgue measure of a closed random set on $\mathbb{R}^d$ is an example of a function of a random set in the infinite case. Its expectation can be computed from knowledge of the one-point coverage function of the random set. For the finite case, say $\Theta = \{\theta_1, \theta_2, \ldots, \theta_n\}$, a random set $S : (\Omega, \mathcal{A}, P) \longrightarrow 2^\Theta$ is characterized by $F : 2^\Theta \longrightarrow [0,1]$, where

$$F(A) = P(\omega : S(\omega) \subseteq A) = \sum_{B \subseteq A} P(\omega : S(\omega) = B) = \sum_{B \subseteq A} m(B).$$

By Möbius inverse, we have

$$m(A) = \sum_{B \subseteq A} (-1)^{|A \setminus B|} F(B).$$

It can be shown that $m : 2^\Theta \longrightarrow \mathbb{R}$ is a probability "density," i.e., $m : 2^\Theta \longrightarrow [0,1]$ and $\sum_{A \subseteq \Theta} m(A) = 1$ if and only if $F(\emptyset) = 0$, $F(\Theta) = 1$, and F is monotone of infinite order.

Let $\phi : 2^\Theta \longrightarrow \mathbb{R}$. Then $\phi(S)$ is a random variable whose expectation is

$$E(\phi(S)) = \sum_{A \subseteq \Theta} \phi(A) m(A).$$

Consider, as above, $F = \inf_{\mathcal{P}} P$ and suppose F is monotone of infinite order, and $\mathcal{P} = \{P \in \mathbb{P} : F \leq P\}$, where $\mathbb{P}$ denotes the class of all probability measures on Θ. For example, let $0 < \varepsilon < 1$, $P_0 \in \mathbb{P}$, and consider

$$\mathcal{P} = \{\varepsilon P + (1-\varepsilon) P_0 : P \in \mathbb{P}\}.$$

Then F is monotone of infinite order. Indeed, $\mathcal{A}$ is a subset of 2^Θ and $\bigcup_{A \in \mathcal{P}} A \neq \emptyset$. Then, there is a $P \in \mathbb{P}$ such that $P\left(\bigcup_{A \in \mathcal{A}} A\right) = 0$, so that

$$\begin{aligned} F\left(\bigcup_{A \in \mathcal{A}} A\right) &= (1-\varepsilon) P_0 \left(\bigcup_{A \in \mathcal{A}} A\right) \\ &\geq \sum_{\emptyset \neq \mathcal{T} \subseteq \mathcal{A}} (-1)^{|\mathcal{T}|+1} P_0 \left(\bigcap_{\mathcal{T}} A\right) \\ &\geq \sum_{\emptyset \neq \mathcal{T} \subseteq \mathcal{A}} (-1)^{|\mathcal{T}|+1} F \left(\bigcap_{\mathcal{T}} A\right). \end{aligned}$$

If $\cup_{\mathcal{A}} A = \Theta$, then $F(\cup_{\mathcal{A}} A) = 1$ and

$$\sum_{\emptyset \neq \mathcal{T} \subseteq \mathcal{A}} (-1)^{|\mathcal{T}|+1} F\left(\bigcap_{\mathcal{T}} A\right) \leq 1 - \varepsilon$$

unless each $A = \Theta$, in which case

$$\sum_{\emptyset \neq \mathcal{T} \subseteq \mathcal{A}} (-1)^{|\mathcal{T}|+1} F\left(\bigcap_{\mathcal{T}} A\right) = 1.$$

In any case, F is monotone of infinite order. Next, if $F \leq P$, then

$$\begin{aligned} f(\theta) &= \frac{1}{\varepsilon}\left[P(\{\theta\}) - F(\{\theta\})\right] \\ &= \frac{1}{\varepsilon}\left[P(\{\theta\}) - (1-\varepsilon)P_0(\{\theta\})\right] \geq 0 \end{aligned}$$

and $\sum_{\theta \in \Theta} f(\theta) = 1$. Thus

$$P = (1-\varepsilon)P_0 + \varepsilon Q \quad \text{with} \quad Q(A) = \sum_{\theta \in A} f(\theta).$$

Hence $\mathcal{P} = \{P \in \mathbb{P} : F \leq P\}$.

Let $\mathcal{D}$ be the class of density functions on Θ such that $f \in \mathcal{D}$ if and only if $P_f \in \mathcal{P}$, where $P_f(A) = \sum_{\theta \in A} f(A)$. In the above decision procedure, based on the Choquet integral, we observe that

$$E_F(u(a,\cdot)) = E_{P_f}(u(a,\cdot)) \quad \text{for some } f \in \mathcal{D}$$

(depending not only on F but also on $u(a,\cdot)$). In other words, the Choquet integral approach leads to the selection of a density in $\mathcal{D}$ that defines the expected utility.

Now, observe that, to each $f \in \mathcal{D}$, one can find many set-functions $\phi : 2^{\Theta} \longrightarrow \mathbb{R}$ such that

$$E_{P_f}(u(a,\cdot)) = E(\phi(S)),$$

where S is the random set with density m. Indeed, define ϕ arbitrarily on every element of 2^{Θ}, except for some A for which $m(A) \neq 0$, and set

$$\phi(A) = \frac{1}{m(A)}\left[E_{P_f}(u(a,\cdot)) - \sum_{B \neq A} \phi(B)m(B)\right].$$

The point is this. Selecting ϕ and considering $E(\phi(S)) = \sum_A \phi(A)m(A)$ as expected utility is a general procedure. In practice, the choice of ϕ is guided by additional subjective information. For example, for $\rho \in [0,1]$, consider

$$\phi_{\rho,a}(A) = \rho \max\{u(a,\theta) : \theta \in A\} + (1-\rho)\min\{u(a,\theta) : \theta \in A\}$$

and use

$$E_m(\phi_{\rho,a}) = \sum_{A \subseteq \Theta} \phi_{\rho,a}(A)m(A)$$

for decision-making. Note that $E_m(\phi_{\rho,a}) = E_{P_g}(u(a,\cdot))$ where $g \in \mathcal{D}$ is given by

$$g(\theta) = \rho \sum_{A \in \mathcal{A}} m(A) + (1-\rho) \sum_{B \in \mathcal{B}} m(B),$$

where

$$\mathcal{A} = \{A : \theta_i \in A \subseteq \{\theta_1, \ldots, \theta_i\}\}, \quad \mathcal{B} = \{B : \theta_i \in B \subseteq \{\theta_i, \ldots, \theta_n\}\},$$

and recall that, for a given action a, the θ_i's are ordered so that

$$u(a,\theta_1) \le u(a,\theta_2) \le \cdots \le u(a,\theta_n).$$

(c) Maximum entropy principle. Consider the situation as in (b). There are other ways to select an element of $\mathcal{D}$ to form the ordinary expectation of $u(a,\cdot)$, for example by using the well-known maximum entropy principle in statistical inference. Recall that, usually, the constraints in an entropy maximization problem are given in the form of a known expectation and higher order moments. For example, with $\Theta = \{\theta_1, \theta_2, \ldots, \theta_n\}$, the canonical density on Θ which maximizes the entropy

$$H(f) = -\sum_{i=1}^{n} f_i \log f_i$$

subject to

$$f_i \ge 0, \quad \sum_{i=1}^{n} f_i = 1, \quad \text{and} \quad \sum_{i=1}^{n} \theta_i f_i = \alpha,$$

is given by

$$f_j = e^{-\beta\theta_j}/\Phi(\beta), \qquad j = 1, 2, \ldots, n,$$

where $\Phi(\beta) = \sum_{i=1}^{n} e^{-\beta\theta_i}$ and β is the unique solution of

$$d \log \Phi(\beta)/d\beta = -\alpha.$$

Now, our optimization problem is this:

$$\text{Maximize} \quad H(f) \quad \text{subject to} \quad f \in \mathcal{D}.$$

Of course, the principle of maximum entropy is sound for any kind of constrains! The problem is with the solution of the maximization problem! Note that, for $F = \inf_{\mathcal{P}} P$, $\mathcal{D} \neq \emptyset$, since the distribution functional F of a random set is convex, i.e.,

$$F(A \cup B) \geq F(A) + F(B) - F(A \cap B)$$

and

$$\mathcal{P} = \{P \in \mathbb{P} : F \leq P\} \neq \emptyset$$

(see Shapley [21]).

Here is an example. Let m be a density on 2^{Θ} with $m(\{\theta_i\}) = \alpha_i$, $i = 1, 2, \ldots, n$ and $m(\Theta) = \varepsilon = 1 - \sum_{i=1}^{n} \alpha_i$. If we write $\varepsilon = \sum_{i=1}^{n} \varepsilon_i$ with $\varepsilon_i \geq 0$, then $f \in \mathcal{D}$ if and only if it is of the form

$$f(\theta_i) = \alpha_i + \varepsilon_i, \qquad i = 1, 2, \ldots, n.$$

So the problem is to find the ε_i's so that f maximizes $H(g)$ over $g \in \mathcal{D}$. Specially, the problem is this. Determine the ε_i's which maximize

$$H(\varepsilon_1, \ldots, \varepsilon_n) = -\sum_{i=1}^{n} (\alpha_i + \varepsilon_i) \log(\alpha_i + \varepsilon_i).$$

The following observations show that nonlinear programming techniques are not needed. For details, see Nguyen and Walker [17]. There exists exactly one element $f \in \mathcal{D}$ having the largest entropy. That density is given by the following algorithm. First, rename the θ_i's so that $\alpha_1 \leq \alpha_2 \leq \cdots \leq \alpha_n$. Then

$$f(\theta_i) = \alpha_i + \varepsilon_i$$

with

$$\varepsilon_i \geq 0, \qquad \sum_{i=1}^{k} \varepsilon_i = m(\Theta),$$

and

$$\alpha_1 + \varepsilon_1 = \cdots = \alpha_k + \varepsilon_k \leq \alpha_{k+1} \leq \cdots \leq \alpha_n.$$

The construction of f is as follows. Setting

$$\delta_i = \alpha_k - \alpha_i, \qquad i = 1, 2, \ldots, k,$$

where k is the maximum index such that

$$\sum_{i=1}^{k} \delta_i \leq m(\Theta);$$

letting

$$\varepsilon_i = \delta_i + \frac{1}{k}\left(\varepsilon - \sum_{i=1}^{k} \delta_i\right), \qquad i = 1, 2, \ldots, k$$

and $\varepsilon_i = 0$ for $i > k$.

The general case is given in Meyerowitz et al. [15]. It calculates the maximum entropy density f directly from the distribution functional F as follows. Inductively define a decreasing sequence of subsets Θ_i of Θ, and numbers γ_i, as follows, quitting when Θ_i is empty:

(i) $\Theta_0 = \Theta$.
(ii) $\gamma_i = \max\{[F(A \cup \Theta_i^c) - F(\Theta_i^c)]/|A| : \emptyset \neq A \subseteq \Theta_i\}$.
(iii) A_i is the largest subset of Θ_i such that

$$F(A_i \cup \Theta_i^c) - F(\Theta_i^c) = \gamma_i |A_i|$$

(note that there is a unique such A_i).
(iv) $\Theta_{i+1} = \Theta_i \setminus A_i$.
(v) Set $f(\theta) = \gamma_i$ for $\theta \in A_i$.

REMARK 5.1. The history of the above entropy maximization problem is interesting. Starting in Nguyen and Walker [17] as a procedure for decision-making with belief functions, where only the solutions of some special cases are given, the general algorithm is presented in Meyerowitz et al. [15] (in fact, they presented two algorithms). One of their algorithms (the above one) was immediately generalized to the case of convex capacities in Jaffray [10]. Very recently, it was observed in Jaffray [11] that the general algorithm for convex capacities is the same as the one given in Dutta and Ray [4] in the context of game and welfare theory! □

6. Random sets in uncertainty modeling. Fuzziness is a type of uncertainty encountered in modeling linguistic information (in expert systems and control engineering). A formal connection with random sets was pointed out in Goodman [5]. If $f : U \longrightarrow [0,1]$ is the membership function of a fuzzy subset of U, then there exists a random set $S : (\Omega, \mathcal{A}, P) \longrightarrow 2^U$ such that

$$f(u) = P(\{\omega : u \in S(\omega)\}), \qquad \forall u \in U,$$

i.e., $f(\cdot)$ is the one-point coverage function of S. Indeed, it suffices to consider a random variable X, defined on some probability space $(\Omega, \mathcal{A}, P)$, uniformly distributed on the unit interval $[0,1]$, and define

$$S(\omega) = \{u \in U : f(u) \geq X(\omega)\}.$$

This probabilistic interpretation of fuzzy sets does not mean that fuzziness is captured by randomness! However, among other things, it suggests a very realistic way for obtaining membership functions for fuzzy sets. First of all, the specification of a membership function to a fuzzy concept can be done in many different ways. For example, by a statistical survey, or by experts (thus, subjectively). The subjectivity of experts in defining f can be understood through

$$f(u) = \mu\left(\{u \in S\}\right),$$

where S is a multi-valued mapping, say, $S : \Omega \longrightarrow 2^U$, and μ is a monotone set-function on U, not necessarily a probability measure. This is Orlowski's model [18]. In a given decision-making problem, the multi-valued mapping S is easy to specify, and the subjectivity of an expert is captured by the set-function μ. Thus, as in game theory, we are led to consider very general set-functions in uncertainty modeling and decisions. For recent works on integral representation of set-functions, see e.g., Gilboa and Schmeidler [6] and Marinacci [13].

REFERENCES

[1] G. Choquet, *Theory of capacities*, Ann. Inst. Fourier, 5 (1953/54), pp. 131–295.

[2] G. Debreu, *Integration of correspondences*, Proc. 5th Berkeley Symp. Math. Statist. Prob., 2 (1967), pp. 351.

[3] R. Durrett, *Probability: Theory and Examples*, (Second Edition), Duxbury Press, Belmont, 1996.

[4] B. Dutta and D. Ray, *A concept of egalitarianism under participation constraints*, Econometrica, 57 (1989), pp. 615–635.

[5] I.R. Goodman, *Fuzzy sets as equivalence classes of random sets*, in Fuzzy Sets and Possibility Theory (R. Yager Ed.), (1982), pp. 327–343.

[6] I. Gilboa and D. Schmeidler, *Canonical representation of set-functions*, Math. Oper. Res., 20 (1995), pp. 197–212.

[7] S. Graf, *A Radon-Nikodym theorem for capacities*, Journal fuer Reine und Angewandte Mathematik, 320 (1980), pp. 192–214.

[8] J. Hajek, *Sampling from a Finite Population*, Marcel Dekker, New York, 1981.

[9] P. Hall, *Introduction to the Theory of Coverage Processes*, J. Wiley, New York, 1988.

[10] J.Y. Jaffray, *On the maximum-entropy probability which is consistent with a convex capacity*, Intern. J. Uncertainty, Fuzziness and Knowledge-Based Systems, 3 (1995), pp. 27–33.

[11] J.Y. Jaffray (1996), *A complement to an maximum entropy probability which is consistent with a convex capacity*, Preprint.

[12] E.F. Harding and D.G. Kendall (Eds.), *Stochastic Geometry*, J. Wiley, New York, 1973.

[13] M. Marinacci (1996), *Decomposition and representation of coalitional games*, Math. Oper. Res., 21 (1996), pp. 1000–1015.

[14] G. Matheron, *Random Sets and Integral Geometry*, John Wiley, New York, 1975.

[15] A. Meyerowitz, F. Richman and E.A. Walker, *Calculating maximum entropy probability densities for belief functions*, Intern. J. Uncertainty, Fuzziness and Knowledge-Based Systems, 2 (1994), pp. 377–390.

[16] I.S. Molchanov, *Limit Theorems for Unions of Random Closet Sets*, Lecture Notes in Math. No. 1561, Springer-Verlag, Berlin, 1993.

[17] H.T. NGUYEN AND E.A. WALKER, *On decision-making using belief functions*, In Advances in the Dempster-Shafer Theory of Evidence (R. Yager et al. Eds.) J. Wiley, New York 1994, pp. 312–330.

[18] S.A. ORLOWSKI, *Calculus of Decomposable Properties, Fuzzy Sets and Decisions*, Allerton Press, New York, 1994.

[19] A. REVUZ, *Fonctions croissantes et mesures sur les espaces topologiques ordonnes*, Ann. Inst. Fourier VI (1956), pp. 187–268.

[20] H.E. ROBBINS, *On the measure of a random set*, Ann. Math. Statist. 15 (1944), pp. 70–74.

[21] L.S. SHAPLEY, *Cores of convex games*, Intern. J. Game Theory, 1 (1971), pp. 11–26.

[22] B. SCHWEIZER AND A. SKLAR, *Probabilistic Metric Spaces*, North-Holland, Amsterdam, 1983.

[23] D. STOYAN, W.S. KENDALL AND J. MECKE, *Stochastic Geometry and Its Applications*, (Second Edition), J. Wiley, New York, 1995.

[24] L.A. WASSERMAN, *Some Applications of Belief Functions to Statistical Inference*, Ph.D. Thesis, University of Toronto, 1987.

RANDOM SETS UNIFY, EXPLAIN, AND AID KNOWN UNCERTAINTY METHODS IN EXPERT SYSTEMS

VLADIK KREINOVICH*

Abstract. Numerous formalisms have been proposed for representing and processing uncertainty in expert systems. Several of these formalisms are somewhat *ad hoc*, in the sense that some of their formulas seem to have been chosen rather arbitrarily.

In this paper, we show that random sets provide a natural general framework for describing uncertainty, a framework in which many existing formalisms appear as particular cases. This interpretation of known formalisms (e.g., of fuzzy logic) in terms of random sets enables us to justify many "ad hoc" formulas. In some cases, when several alternative formulas have been proposed, random sets help to choose the best ones (in some reasonable sense).

One of the main objectives of expert systems is not only to *describe* the current state of the world, but also to provide us with reasonable *actions*. The simplest case is when we have the *exact* objective function. In this case, random sets can help in choosing the proper method of "fuzzy optimization."

As a test case, we describe the problem of choosing the best tests in technical diagnostics. For this problem, feasible algorithms are possible.

In many real-life situations, instead of an *exact* objective function, we have several participants with *different* objective functions, and we must somehow reconcile their (often conflicting) interests. Sometimes, standard approaches of game theory are not working. We show that in such situations, random sets present a working alternative. This is one of the cases when particular cases of random sets (such as fuzzy sets) are not sufficient, and general random sets are needed.

Key words. Random Sets, Fuzzy Logic, Expert Systems, Maximum Entropy, Fuzzy Optimization, Technical Diagnostics, Cooperative Games, von Neumann-Morgenstern Solution.

AMS(MOS) subject classifications. 60D05, 03B52, 04A72, 90C70, 90D12, 94C12, 94A17, 68T35

1. Main problems: Too many different formalisms describe uncertainty, and some of these formalisms are not well justified. Many different formalisms have been proposed for representing and processing uncertainty in expert systems (see, e.g., [30]); some of these formalisms are not well justified. There are two problems here:

1.1. First problem: Justification. First of all, we are not sure whether all these formalisms are truly adequate for describing uncertainty. The fact that these formalisms have survived means that they have been successful in application and thus, that they do represent (exactly or approximately) some features of expert reasoning. However, several of the

---

* Department of Computer Science, University of Texas at El Paso, El Paso, TX 79968. This work was partially supported by NASA grant No. NAG 9-757. The author is greatly thankful to O.N. Bondareva, P. Fishburn, O.M. Kosheleva, H.T. Nguyen, P. Suppes, and L.A. Zadeh for stimulating discussions, and to J. Goutsias and to the anonymous referees for valuable suggestions.

formalisms are somewhat *ad hoc* in the sense that some of their formulas (such as combination formulas in Dempster-Shafer formalism or in fuzzy logic) seem to have been chosen rather arbitrarily. It may, in principle, turn out that these formulas are only approximations to other formulas that describe human reasoning much better. It is therefore desirable to either *explain* the original formulas (from some reasonable first principles), or to *find* other formulas (close to the original ones) that can be thus explained.

1.2. Second problem: Combination. In many cases, different pieces of knowledge are represented in different formalisms. To process all knowledge, it is therefore desirable to combine (*unify*) these formalisms.

1.3. What we are planning to do. In this paper, we will show that both problems can be solved (to a large extent) in the framework of random sets: Namely, random sets provide a natural general framework for describing uncertainty, a framework of which many existing formalisms appear as particular cases. This interpretation of known formalisms (like fuzzy logic) in terms of random sets enables us to justify many "ad hoc" formulas.

In this paper, due to size limitations, we will mainly talk about the first (*justification*) problem. As a result of our analysis, we will conclude that most existing formalisms can be justified within the same framework: of *random sets*.

The existence of such a unified framework opens possibilities for solving the second (*combination*) problem. In this paper, we will show how this general random set framework can be helpful in solving two *specific* classes of problems: *decision making* (on the example of technical diagnostics) and *conflict resolution* (on the example of cooperative games). In contrast to these (well-defined and algorithmically analyzed) *specific* solutions, a *general* random-set-based description of the possible combined formalisms has only been developed *in principle* and requires further research before it can be practically used.

2. Random sets: Solution to the main problems.

2.1. Uncertainty naturally leads to random sets.

2.1.1. In rare cases of complete knowledge, we know the exact state of the system. In rare cases, we know the *exact* state of a system, i.e., in more mathematical terms, we know exactly which element s of the set of all states A describes the current state of the system.

In the majority of the situations, however, we only have a *partial* knowledge about the system, i.e., we have *uncertainty*.

2.1.2. If all statements about the system are precise, then such incomplete knowledge describes a set of possible states. Usually, only a *part* of the knowledge is formulated in precise terms. In some situations, however, *all* the knowledge is formulated in precise terms. In such situations, the knowledge consists of one or several *precise* statements $E_1(s), \dots, E_n(s)$ about the (unknown) state s.

For example, if we are describing the temperature s in El Paso at this very moment, a possible knowledge base may consist of a single statement $E_1(s)$, according to which this temperature s is in the nineties, meaning that it is between 90 and 100 Fahrenheit.

We can describe such a knowledge by the set S of all states $s \in A$ for which all these statements $E_1(s), \dots, E_n(s)$ are true; e.g., in the above example of a single temperature statement, S is the set of all the weather states s in which the temperature is exactly in between 90 and 100.

2.1.3. Vague statements naturally lead to random sets of possible states. Ideally, it would be great if all our knowledge was precise. In reality, an important part of knowledge consists of expert statements, and experts often cannot express their knowledge in precise terms. Instead, they make statements like "it will be pretty hot tomorrow," or, even worse, "it will most probably be pretty hot tomorrow." Such *vague* statements definitely carry some information. The question is: how can we formalize such vague statements? If a statement $P(s)$ about the state s is precise, then, for every state s, we can definitely tell whether this state does or does not satisfy this property $P(s)$. As a result, a precise statement characterizes a subset S of the set of all states A. In contrast to precise statements, for vague statements $P(s)$, we are often not sure whether a given state does or does not satisfy this "vague" property: e.g., intuitively, there is no strict border between "pretty hot" and "not pretty hot": there are some intermediate values about which we are not sure whether they are "pretty hot" or not. In other words, a vague statement means that an expert is not sure which set S of possible values is described by this statement. For example, the majority of the people understand the term "pretty hot" pretty well; so, we can ask different people to describe exactly what this "pretty hot" means. The problem is that different people will interpret the same statement by different subsets $S \subseteq A$ (not to mention that some people will have trouble choosing any formalization at all).

Summarizing, a vague statement is best represented not by a *single* subset S of possible states, but by a *class* $\mathcal{S}$ of possible sets $S \subseteq A$. Some sets from this class are more probable, some are less probable. It is natural to describe this "probability" by a number from the interval $[0, 1]$. For example, we can ask several people to interpret a vague statement like "pretty hot" in precise terms, and for each set S (e.g., for the set $[80, 95]$), take the fraction of those people who interpret "pretty hot" as belonging to this particular set S, as the probability $p(S)$ of this set (so that, e.g., if 10% of the people interpret "pretty hot" as "belonging to $[80, 95]$," we take $p([80, 95]) = 0.1$). Another possibility is to ask a single "interpreter" to provide us with the "subjective" probabilities $p(S)$ that describe how probable each set $S \in \mathcal{S}$ is. In both cases, we get a probability measure on a class of sets, i.e., a *random set.*

COMMENT. Our approach seems to indicate that random sets are a *reasonable* general description of uncertainty. However, this does not mean that random sets are the *only* possible general description of uncertainty. Another reasonable possibilities could include *interval-valued* probability measures, in which instead of a single probability of an event, we get its lower and upper probabilities; *Bayesian*-type approach, in which in addition to the *interval* of possible values of probability, we have (second-order) *probabilities* of different probability values; etc.

In this paper, we selected random sets mainly for one *pragmatic* reason: because the theory of random sets is, currently, the most developed and thus, reduction to random sets is, currently, most useful. It may happen, however, that further progress will show that interval-valued or second order probabilities provide an even better general description of uncertainty.

2.2. If random sets are so natural, why do we need other formalisms for uncertainty. Since random sets are such a natural formalism for describing uncertainty, why not use them? The main reason why generic random sets are rarely used (and other formalisms are used more frequently) becomes clear if we analyze how much computer memory and computer time we need to process general random sets.

Indeed, for a system with n possible states, there exist 2^n possible sets $S \subseteq A$. Hence, to describe a generic random set, we need $2^n - 1$ real numbers $p(S) \geq 0$ corresponding to different sets S (we need $2^n - 1$ and not 2^n because we have a relation $\sum p(S) = 1$).[1] For sets A of moderate and realistic size (e.g., for $n \approx 300$), this number $2^n - 1$ exceeds the number of particles in the known Universe; thus, it is impossible to store all this information. For the same reason, it is even less possible to process it.

Thus, for practical use in expert systems, we must use *partial* information about these probabilities. Let us show that three formalisms for uncertainty — statistics, fuzzy, and Dempster-Shafer — can indeed be interpreted as such partial information. (This fact has been known for some time, but unfortunately many researchers in uncertainty are still unaware of it; so, without going into technical detail, we will briefly explain how these formalisms can be interpreted in the random set framework.)

2.3. Three main uncertainty formalisms as particular cases of random set framework.

2.3.1. Standard statistical approach.

Description. In the standard statistical approach, we describe uncertainty by assigning a probability $P(s)$ to each *event* s. (The sum of these

[1] It is quite possible that some experts believe that knowledge is inconsistent, and thus, $p(\emptyset) > 0$. If we assume that all experts believe that knowledge is consistent, then we get $p(\emptyset) = 0$ and thus, only $2^n - 2$ real numbers are needed to describe a random set.

probabilities must add up to 1).

Interpretation in terms of random sets. This description is clearly a particular case of a random set, in which only one-element sets have non-zero probability $p(\{s\}) = P(s)$, and all other sets have probability 0. For this description, we need $n - 1$ numbers instead of $2^n - 1$; thus, this formalism is quite feasible.

Limitations. The statistical description is indeed a *particular* case not only in the mathematical sense, but also from the viewpoint of common sense reasoning: Indeed, e.g., if only two states are possible, and we know nothing about the probability of each of them, then it is natural to assign equal probability to both; thus, $P(s_1) = P(s_2)$ and hence, $P(s_1) = P(s_2) = 0.5$. Thence, within the standard statistical approach, we are unable to distinguish between the situation in which we know nothing about the probabilities, and the situations like tossing a coin, in which we are absolutely sure that the probability of each state is exactly 0.5.

2.3.2. Fuzzy formalism.

Description. Statistical approach has shown that if we store only one real number per state, we get a computationally feasible formalism. However, the way this is done in the standard statistical approach is too restrictive. It is therefore desirable to store some partial information about the set without restricting the underlying probability measure $p(S)$ defined on the class of sets S.

A natural way to do this is to assign, to each such probability measure, the values

$$\mu(s) = P(s \in S) = P(\{S \mid s \in S\}) = \sum_{S \ni s} p(S).$$

These values belong to the interval $[0, 1]$ and do not necessarily add up to 1.

In fuzzy formalism (see, e.g., [10,24]), numbers $\mu(s) \in [0,1]$, $s \in A$ (that do not necessarily add up to 1), form a *membership function* of a *fuzzy set.*

Interpretation in terms of random sets. Hung T. Nguyen has shown [22] that every membership function can be thus interpreted, and that, moreover, standard (initially *ad hoc*) operations with fuzzy numbers can be interpreted in this manner.

For example, the standard fuzzy interpretation of the vague property "pretty hot" is as follows: we take, say, $A = [60, 130]$, and assign to every value $s \in A$, a number $\mu(s)$ that describes to what extent experts believe that s is pretty hot.

The random set interpretation is that with different probabilities, "pretty hot" can mean belonging to different subsets of the set $[60, 130]$, and, for example, $\mu(80)$ is the total probability that 80 is pretty hot.

Limitations. In this interpretation, a fuzzy set contains only a *partial* information about uncertainty; this information contains the probabilities $\mu(s)$ that different states s are possible, but it does not contain, e.g., probabilities that a pair (s, s') is possible. It may happen that the possibility of a state s (i.e., the fact that $s \in S$) renders some other states impossible. E.g., intuitively, if two values $s < s'$ correspond to "pretty hot," then any temperature in between s and s' must also correspond to "pretty hot"; thus, the probability measure $p(S)$ should be only located on convex sets S. However, "fine" information of this type cannot be captured by a fuzzy description.

2.3.3. Dempster-Shafer formalism.

Description and interpretation in terms of random sets. This formalism (see, e.g., [32]) is the closest to random sets: each expert's opinion is actually represented by the values ("masses") $m(S) \geq 0$ assigned to sets $S \subseteq A$ that add up to 1 and thus form a random set. The only difference from the general random set framework is that instead of representing the entire knowledge as a random set, this formalism represents each expert's opinion as a random set, with somewhat *ad hoc* combination rules.

Limitations. Since this formalism is so close to the general random set framework, it suffers (although somewhat less) from the same computational complexity problems as the general formalism; for details and for methods to overcome this complexity, see. e.g., [15].

2.4. Random set interpretation explains seemingly ad hoc operations from different uncertainty formalisms. Operations from standard statistics are usually pretty well justified. The Dempster-Shafer formalism is also mainly developed along statistical lines and uses *mostly* well-justified operations.[2] Since both formalisms are explicitly formulated in terms of probabilities, these justifications, of course, are quite in line with the random set interpretation.

The only formalism in which operations are mainly *ad hoc* and need justification is fuzzy formalism. Let us show that its basic operations can be naturally justified by the random set interpretation.

The basic operations of fuzzy formalism are operations of *fuzzy logic* that describe *combinations* of different parts of knowledge. For example, if we know the membership functions $\mu_1(s)$ and $\mu_2(s)$ that describe two "vague" statements $E_1(s)$ and $E_2(s)$, what membership function $\mu(s)$ corresponds to their conjunction $E_1(s) \& E_2(s)$? In other words, how to interpret "and" and "or" in fuzzy logic?

2.4.1. "And"– and "or"–operations of fuzzy logic: Extremal approach. Let us first show how the simplest "and" and "or" operations

[2] It should be mentioned that Dempster's formalism is not *completely* justified along statistical/probabilistic lines because it performs some *ad hoc* conditioning.

$f_{\&}(a,b) = \min(a,b)$ and $f_{\vee}(a,b) = \max(a,b)$ can be explained (in this description, we follow [22] and [2]):

DEFINITION 2.1. Assume that a (crisp) set A is given. This set will be called a *Universum*.

- By a *random set*, we mean a probability measure P on a class 2^A of all subsets S of A.
- By a *fuzzy set* C, we mean a map (*membership function*) $\mu : A \to [0,1]$.
- We say that a random set P *represents* a fuzzy set C with a membership function $\mu(s)$, if for every $s \in A$, $P(s \in S) = \mu(s)$.

If we have *two* membership functions, this means that we actually have *two* unknown sets. To describe this uncertainty in probabilistic terms, we therefore need a probability measure on a set $2^A \times 2^A$ of all *pairs* of sets.

DEFINITION 2.2. By a *random pair of sets*, we mean a probability measure P on a class $2^A \times 2^A$ of all pairs (S_1, S_2) of subsets of A. We say that a random pair P *represents* a pair of fuzzy sets (C_1, C_2) with membership functions $\mu_1(s)$ and $\mu_2(s)$ if for every $s \in A$: $P(s \in S_1) = \mu_1(s)$ and $P(s \in S_2) = \mu_2(s)$.

We are interested in the membership functions $\mu_{C_1 \cap C_2}(s)$ and $\mu_{C_1 \cup C_2}(s)$. It is natural to interpret these numbers as $P(s \in S_1 \cap S_2)$ and $P(s \in S_1 \cup S_2)$. The problem is that these numbers are *not* uniquely defined by μ_1 and μ_2. So, instead of a single value, we get a whole class of possible values. However, this class has a very natural bound:

PROPOSITION 2.1. Let C_1 and C_2 be fuzzy sets with membership functions $\mu_1(s)$ and $\mu_2(s)$. Then, the following is true:

- For every random pair P that represents (C_1, C_2), and for every $s \in A$, $P(s \in S_1 \cap S_2) \leq \min(\mu_1(s), \mu_2(s))$.
- There exists a random pair P that represents (C_1, C_2) and for which for every $s \in A$, $P(s \in S_1 \cap S_2) = \min(\mu_1(s), \mu_2(s))$.

So, min is an *upper bound* of possible values of probability. This min is exactly the operation originally proposed by Zadeh to describe "and" in fuzzy logic. Similarly, for union, max turns out to be the *lower bound* [2].

COMMENT. One of the main original reasons for choosing the operations $\min(a,b)$ and $\max(a,b)$ as analogues of "and" and "or" is that these are the only "and" and "or" operations that satisfy the intuitively clear requirement that for every statement E, both $E\&E$ and $E \vee E$ have the same degree of belief as E itself. However, this reason does not mean that min and max are the *only* reasonable "and" and "or" operations: in the following, we will see that other requirements (also seemingly reasonable) lead to different pairs of "and" and "or" operations.

2.4.2. "And"– and "or"–operations of fuzzy logic: Maximum entropy approach. In the previous section, we considered *all possible* probability distributions consistent with the given membership function. As a result, for each desired quantity (such as $P(s \in S_1 \cap S_2)$), we get an *interval* of possible values.

In some situations, this interval may be too wide (close to $[0,1]$), which makes the results of this "extremal" approach rather meaningless. It is therefore necessary to consider only *some* probability measures, ideally, the measures that are (in some reasonable sense) "most probable."

The (reasonably known) formalization of this idea leads to the so-called *maximum entropy (MaxEnt) method* in which we choose the distribution for which the entropy $-\sum p(S) \cdot \log_2(p(S))$ is the largest possible (this method was originally proposed in [9]; for detailed derivation of this method, see, e.g., [4,19,7]).

In this approach, it is natural to take the value $f_\&(\mu_1(s), \mu_2(s))$ as the membership function $\mu_{C_1 \cap C_2}(s)$ corresponding to the intersection $C_1 \cap C_2$, where for each a and b, the value $f_\&(a,b)$ is defined as follows: As a Universum A, we take a 2-element set $A = \{E_1, E_2\}$; for each set $S \subseteq A$, we define $p(S)$ as a probability that all statements $E_i \in S$ are true and all statements $E_i \notin S$ are false: e.g., $p(\emptyset) = P(\neg E_1 \& \neg E_2)$ and $p(\{E_1\}) = P(E_1 \& \neg E_2)$, where $\neg E$ denotes the false of statement E. We consider only distributions for which $P(E_1) = a$ for $P(E_2) = b$. In terms of $p(S)$, this means that $p(\{E_1\}) + p(\{E_1, E_2\}) = a$ and $p(\{E_2\}) + p(\{E_1, E_2\}) = b$. Among all probability distributions $p(S)$ with this property, we choose the one p_{ME} with the maximal entropy. The value $P_{ME}(E_1 \& E_2)$ corresponding to this distribution p_{ME} is taken as the desired value of $f_\&(a,b)$. Similarly, we can define $f_\vee(a,b)$ as $P_{ME}(E_1 \vee E_2)$ for the MaxEnt distribution $p_{ME}(S)$. We will call the resulting operations $f_\&(a,b)$ and $f_\vee(a,b)$ *MaxEnt operations.*

To find $f_\&(a,b)$ and $f_\vee(a,b)$, we thus have to solve a conditional nonlinear optimization problem with four unknowns $p(S)$ (for $S = \emptyset$, $\{E_1\}$, $\{E_2\}$, and $\{E_1, E_2\}$). This problem has the following explicit solution (see, e.g., [18]):

PROPOSITION 2.2. For & and $\vee$, the MaxEnt operations are $f_\&(a,b) = a \cdot b$ and $f_\vee(a,b) = a + b - a \cdot b$.

These operations have also been originally proposed by Zadeh. It is worth mentioning that both the pair of operations $(\min(a,b), \max(a,b))$ coming from the extremal approach and this pair of MaxEnt operations are in good accordance with the general group-theoretic approach [14,16,17] for describing operations $f_\&$ and $f_\vee$ which can be optimal w.r.t. different criteria.

In [18], a general description of MaxEnt operations is given. For example, for implication $A \to B$, the corresponding MaxEnt operation is $f_\to(a,b) = 1 - a + a \cdot b$, which coincides with the result of a step-by-step application of MaxEnt operations $f_{\&}$ and $f_{\vee}$ to the formula $B \vee \neg A$ (which is a representation of $A \rightarrow B$ in terms of &, $\vee$, and $\neg$).

For equivalence $A \equiv B$, the corresponding MaxEnt operation is $f_{\equiv}(a,b) = 1 - a - b + 2a \cdot b$. Unlike $f_{\rightarrow}$, the resulting expression *cannot* be obtained by a step-by-step application of $f_{\&}$, $f_{\vee}$, and $f_{\neg}$ to any propositional formula.

Similar formulas can be obtained for logical operations with three, four, etc., arguments. In particular, the MaxEnt analogues of $f_{\&}(a, b, \ldots, c)$ and $f_{\vee}(a, b, \ldots, c)$ turn out to be equal to the results of consequent applications of the corresponding *binary* operations, i.e., to $a \cdot b \cdot \cdots \cdot c$ and to $f_{\vee}(\cdots (f_{\vee}(a, b), \cdots), c)$.

2.4.3. Extremal and MaxEnt approaches are not always applicable. At first glance, MaxEnt seems to be a reasonable approach to choosing "and" and "or" operations. It has indeed lead to successful expert systems (see, e.g., Pearl [26]). Unfortunately, the resulting operations are not always the ones we need.

For example, if we are interested in *fuzzy control* applications (see, e.g., [23,10,24]), then it is natural to choose "and" and "or" operations for which the resulting *control* is the most stable or the most smooth. Let us describe this example in some detail.

In some control problems (e.g., in *tracking* a spaceship), we are interested in the most *stable* control, i.e., in the control that would, after the initial deviation, return us within the prescribed distance of the original trajectory in the shortest possible time. We can take this time as the measure of stability.

In some other problems, e.g., in *docking* a spaceship to the Space Station, we are not that worried about the time, but we want the controlled trajectory to be as smooth as possible, i.e., we want the derivative dx/dt to be as small as possible. In such problems, we can take, e.g., the "squared average" of the derivative $\int (dx/dt)^2\,dt$ as the numerical characteristic of smoothness that we want to minimize.

The simplest possible controlled system is a 1D system, i.e., a system characterized by a single parameter x. For 1D systems, fuzzy control rules take the form $A_1(x) \rightarrow B_1(u), \ldots, A_n(x) \rightarrow B_n(u)$, where $A_i(x)$ and $B_i(u)$ are fuzzy predicates characterized by their membership functions $\mu_{Ai}(x)$ and $\mu_{Bi}(u)$. For these rules, the standard fuzzy control methodology prescribes the use of a control value $\bar{u} = (\int u \cdot \mu(u)\,du)/(\int \mu(u)\,du)$, where $\mu(u) = f_{\vee}(r_1, \ldots, r_n)$ and $r_i = f_{\&}(\mu_{Ai}(x), \mu_{Bi}(u))$.

The resulting control $\bar{u}(x)$, and therefore, the smoothness and stability of the resulting trajectory $x(t)$, depend on the choice of "and" and "or" operations. It turns out that thus defined control is the most *stable* if we use $f_{\&}(a,b) = \min(a,b)$ and $f_{\vee}(a,b) = \min(a+b, 1)$, while the *smoothest* control is attained for $f_{\&}(a,b) = ab$ and $f_{\vee}(a,b) = \max(a,b)$ (for proofs, see, e.g., [17,31]). None of these pairs can be obtained by using extremal or

MaxEnt approaches (some of these operations actually correspond to the *minimal* entropy [28,29]).

3. Random sets help in choosing appropriate actions. One of the main objectives of expert systems is not only to *describe* the current state of the world, but also to provide us with reasonable *actions*. Let us show that in choosing an appropriate action, random sets are also very helpful.

3.1. Random sets and fuzzy optimization. The simplest case of choosing an action is when we have the *exact* objective function. Let us describe such a situation in some detail. We have a set A of all actions that are (in principle) possible (i.e., about which we have no reasons to believe that these actions are impossible). For each action $s \in A$, we know exactly the resulting gain $f(s)$. We also know that our knowledge is only partial; in particular, as we learn more about the system, some actions from the set A which we now consider possible may turn out to be actually impossible. In other words, not all actions from the set A may be actually possible.

If we know exactly which actions are possible and which actions are not, then we know the set C of possible actions, and the problem of choosing the best action becomes a conditional optimization problem: to find $s \in A$ for which $f(s) \to \max$ under the condition that $s \in C$. In many real-life situations, we only vaguely know which actions are possible and which are not, so the set C of possible actions is a "vague" set. If we formalize C as a fuzzy set (with a membership function $\mu(s)$), we get a problem of optimizing $f(s)$ under the fuzzy condition $s \in C$. This informal problem is called *fuzzy optimization.* Since we are not sure what the conditions are, the answer will also be vague. In other words, the answer that we are looking for is not a single state s, but a *fuzzy set* $\mu_D(s)$ (that becomes a one-element crisp set when C is a crisp set with a unique maximum of $f(s)$).

There are several heuristic methods of defining what it means for a fuzzy set $\mu_D(s)$ to be the solution of the fuzzy optimization problem. Most of these methods are not well justified. To get a well justified method, let us use the random set interpretation of fuzzy sets. In this interpretation, $\mu(s)$ is interpreted as the probability that $s \in S$ for a random set S, and $\mu_D(s)$ is the probability that a conditional maximum of the function f on a set S is attained at s, i.e., the probability

$$P(s \in S \,\&\, f(s) = \max_{s' \in S} f(s')).$$

PROPOSITION 3.1. [2] Let C be a fuzzy set with a membership function $\mu(s)$ and let $f : A \to \mathbb{R}$ be a function. Then, the following is true:

- For every random set $p(S)$ that represents the fuzzy set C, and for every $s \in A$,

$$P(s \in S \,\&\, f(s) = \max_{s' \in S} f(s')) \leq \mu_{DP}(s),$$

where

$$\mu_{DP}(s) = \min(\mu(s), 1 \;-\; \sup_{s' :\, f(s') > f(s)} \mu(s')).$$

- For every $s \in A$, there exists a random set $p(S)$ that represents C and for which

$$P(s \in S \,\&\, f(s) = \max_{s' \in S} f(s')) \leq \mu_{DP}(s).$$

Just like Proposition 2.1 justified the choice of min and max as $\&-$ and $\vee-$operations, this result justifies the use of $\mu_{DP}(x)$ as a membership function that describes the result of fuzzy optimization.

3.2. Random sets are helpful in intelligent control. A slightly more complicated case is when the objective function is *not* exactly known. An important class of such situations is given by *intelligent control* situations [23]. Traditional expert systems produce a list of possible actions, with "degree of certainty" attached to each. For example, if at any given moment of time, control is characterized by a single real-valued variable, the output of an expert system consists of a number $\mu(u)$ assigned to every possible value u of control; in other words, this output is a membership function. In order to use this output in automatic control, we must select a single control value u; this selection of a single value from a fuzzy set is called a *defuzzification*.

The most widely used defuzzification method (called "centroid") chooses the value $\bar{u} = (\int u \cdot \mu(u)\, du)/(\int \mu(u)\, du)$. Centroid defuzzification is a very successful method, but it is usually only heuristically justified. It turns out that the random-set interpretation of fuzzy sets (described above), together with the MaxEnt approach, leads exactly to centroid defuzzification; for precise definitions and proofs, see [18].

4. Case study: Technical diagnostics.

4.1. Real–life problem: Which tests should we choose. A typical problem of technical diagnostics is as follows: the system does not work, and we need to know which components malfunction. Since a system can have lots of components, and it is very difficult (or even impossible) to check them one by one, usually, some tests are undertaken to narrow down the set of possibly malfunctioning components.

Each of the tests brings us some additional information, but costs money. The question is: within a given budget C, how can we choose the set of test that will either solve our problem or at least, that would bring us, on average, the maximum information.

This problem and methods of solving it was partially described in [12,6,27].

To select the tests, we must estimate the information gain of each test. To be able to do that, we must first describe what we mean by *information.*

Intuitively, an information means a *decrease in uncertainty.* Thus, in order to define the notion of information, we will first formalize the notion of *uncertainty.*

4.2. To estimate the information gain of each test, we must estimate the amount of uncertainty in our knowledge. The amount of uncertainty in our knowledge is usually defined as the average number of binary ("yes"-"no") questions that we need to ask to gain the complete knowledge of the system, i.e., in our case, to determine the set F of faulty components. The only thing we know about this set is that it is non-empty, because otherwise, the system would not be faulty and we would not have to do any testing in the first place.

In the ideal situation, when for each non-empty subset F of the set of all components $\{1,\ldots,n\}$, we know the probability $p(F)$ that components from F and only these components are faulty, then we can determine this uncertainty as the *entropy* $-\sum p(F)\log_2(p(F))$. In reality, we only have a *partial* information about these probabilities, i.e., we only know a *set* $\mathcal{P}$ of possible distributions.

There exist many different probability distributions that are consistent with our information, and these distributions have different entropy. Therefore, it may happen that in say, 5 questions (in average) we will get the answer, or it may happen that we will only get the complete picture in 10 questions. The only way to guarantee that after a certain (average) amount of questions, we will get the complete information, is to take the *maximum* of possible entropies.

In mathematical terms, we thus define the *uncertainty* U of a situation characterized by the set $\mathcal{P}$ of probability measures as its *worst-case* entropy, i.e., as the *maximum* possible value of entropy for all distributions $\{p(F)\}\in\mathcal{P}$.

In particular, in technical diagnostics, the number of components n is usually large. As a result, the number 2^n is so astronomical that it is impossible to know 2^n-1 numbers corresponding to all possible non-empty sets F. What we usually know instead is the probability p_i of each component i to be faulty: $p_i=\sum\{p(F)\,|\,i\in F\}$. Since it is quite possible that two or more component are faulty at the same time, the sum of these n probabilities is usually greater than 1.

EXAMPLE 4.1. Let us consider a simple example, in which a system consists of two components and each of them can be faulty with probability 20%; we will also assume that the failures of these two components are independent events.

In this example, we get:

- a non-faulty situation with probability $80\% \cdot 80\% = 0.64$;
- both components faulty with probability $20\% \cdot 20\% = 0.04$;
- the first component only faulty with probability $20\% \cdot 80\% = 0.16$; and
- the second component only faulty also with probability 0.16.

Totally, the probability of a fault is $1 - 0.64 = 0.36$. Hence:

- the probability $p(\{1,2\})$ that both components are faulty can be computed as $p(\{1,2\}) = 0.04/0.36 = 1/9$;
- the probability that the first component only is faulty is $p(\{1\}) = 0.16/0.36 = 4/9$, and
- the probability that the second component only is faulty is $p(\{2\}) = 4/9$.

Among all faulty cases, the first component fails in $p_1 = 1/9{+}4/9 = 5/9$ cases, and the second component fails also in $1/9 + 4/9 = 5/9$ cases. Here, $p_1 + p_2 = 10/9 > 1$.

This inequality is in perfect accordance with the probability theory, because when we add p_1 and p_2, we thus add some events *twice*: namely, those events in which both components fail. □

We arrive at the following problem:

DEFINITION 4.1. By an *uncertainty estimation problem for diagnostics*, we mean the following problem: *We know* the probabilities $p_1, \ldots, p_n$, and *we want to find* the maximum U of the expression $-\sum p(F)\log(p(F))$ (where F runs over all possible non-empty subsets $F \subseteq \{1, \ldots, n\}$) under the conditions that $\sum p(F) = 1$ and $\sum\{p(F) \mid i \in F\} = p_i$ for all $i = 1, \ldots, n$.

The following algorithm solves this problem:

PROPOSITION 4.1. The following algorithm solves the uncertainty estimation problem for diagnostics:

1. Find a real number α from the equation

$$\prod_{i=1}^{n}(1 + \alpha(1 - p_i)) = (1 + \alpha)^{n-1}$$

   by using a bisection method; start with an interval $[0, 1/\prod(1-p_i)]$.
2. Compute the value

$$U = -\sum_{i=1}^{n} p_i \cdot \log\left(\frac{p_i}{1 + \alpha(1 - p_i)}\right) - \log_2(\alpha) \cdot \left(\sum_{i=1}^{n} p_i - 1\right).$$

(For a proof, see [12,6,27].)

4.3. The problem of test selection: To select the tests, we must estimate the information gain of each test. Now that we know how to estimate the *uncertainty* of each situation, let us describe how to estimate the *information gain* of each test t.

After each test, we usually decrease the uncertainty. For every possible outcome r of each test t, we usually know the probability $p(r)$ of this outcome, and we know the (conditional) probabilities of different components to be faulty under the condition that the test t resulted in r. In this case, for each outcome r, we can estimate the *conditional* information gain of the test t as the *difference* $I_t(r) = U - U(r)$ between the original uncertainty U and the uncertainty $U(r)$ after the test. It is, therefore, natural to define the information gain I_t of the test t as the *average* value of this conditional gain: $I_t = \sum_r p(r) \cdot I_t(r)$.

Since we know how to solve the problem of estimating uncertainty (and thus, how to estimate U and $U(r)$ for all possible outcomes r), we can, therefore, easily estimate the information gain I_t of each test t.

4.4. How to select a sequence of tests. Usually, a single test is not sufficient for diagnostics, so, we need to select not a *single* test, but a *sequence* of tests. Tests are usually independent, so, the total amount of information that we get from a set S of tests is equal to the sum of the amounts of informations that we gain from each one of them. Hence, if we denote the total number of tests by T, the cost of the tth test by c_t, and the amount of information gained by using the tth test by I_t, we can reformulate the problem of choosing the optimal set of tests as follows: among all sets $S \subseteq \{1, \ldots, T\}$ for which

$$\sum_{t \in S} c_t \leq C,$$

find a set for which

$$\sum_{t \in S} I_t \to \max.$$

As soon as we know the values c_t and I_t that correspond to each test, this problem becomes a *knapsack problem* well known in computer science. Although in general, this problem formulated *exactly* is computationally intractable (NP-hard) [5], there exist many efficient heuristic algorithms for solving this problem *approximately* [5,3].

From the practical viewpoint, finding an "almost optimal" solution is OK. This algorithm was actually applied to real-life technical diagnostics in manufacturing (textile coloring) [12,6] and in civil engineering (hoisting crane) [27].

5. Random sets help in describing conflicting interests. In Section 3, we considered the situations in which we either know the exact objective function $f(s)$ or at least in which we know *intuitively* which action is best. In many real-life situations, instead of a precise idea of which actions are best, we have several participants with different objective functions, and we must somehow reconcile their (often conflicting) interests.

We will show that sometimes, standard approaches of game theory (mathematical discipline specifically designed to solve conflict situations) are not working, and that in such situations, random sets present a working alternative.

5.1. Traditional approach to conflict resolution. Game theory was invented by von Neumann and Morgenstern [21] (see also [25,8]) to assist in conflict resolution, i.e., to help several participants (*players*), with different goals, to come to an agreement.

5.1.1. To resolve conflicts, we must adopt some standards of behavior. A normal way to resolve a conflict situation with many players is that first, several players find a compromise between their goals, so they form a *coalition*; these coalitions merge, split, etc., until we get a coalition that is sufficiently strong to impose its decision on the others (this is, e.g., the way a multi-party parliament usually works).

The main problem with this coalition formation is that sometimes it goes on and on and never seems to stop: indeed, when a powerful coalition is formed, outsiders can often ruin it by promising more to some of its minor players; thus, a new coalition is formed, etc. This long process frequently happens in multi-party parliaments.

How to stop this seemingly endless process? In real economic life, not all outputs and not all coalitions that are mathematically possible are considered: there exist legal restrictions (like anti-trust law) and ethical restrictions (like "good business practice") that represent the ideas of social justice, social acceptance, etc. Von Neumann and Morgenstern called these restrictions the "standard of behavior". So, in real-life conflict situations, we look not for an arbitrary outcome, but only for the outcome that belongs to some *a priori* fixed set S of "socially acceptable" outcomes, i.e., outcomes that are in accordance with the existing "standard of behavior". This set S is called a *solution*, or *NM-solution*.

For this standard of behavior to work, we must require two things:

- First, as soon as we have achieved a "socially acceptable" outcome (i.e., outcome x from the set S), no new coalition can force a change in this decision (as long as we stay inside the social norm, i.e., inside the state S).
- Second, if some coalition proposes an outcome that is not socially acceptable, then there must always exist a coalition powerful enough to enforce a return to the social norm.

In this framework, conflict resolution consists of two stages:

- first, the society selects a "standard of behavior" (i.e., a NM solution S);
- second, the players negotiate a compromise solution within the selected set S.

Let us describe Neumann-Morgenstern's formalization of the idea of "standard of behavior."

5.1.2. The notion of a "standard of behavior" is traditionally formalized as a NM solution.

General case. Let us denote the total number of players by n. For simplicity, we will identify players with their ordinal numbers, and thus, identify the set of all players with a set $N = \{1, \ldots, n\}$. In these terms, a *coalition* is simply a subset $C \subseteq N$ of the set N of all players.

Let us denote the set of all possible outcomes by X.

To formalize the notion of an NM-solution, we need to describe the enforcement abilities of different coalitions. The negotiating power of each coalition C can be described by its ability, given an outcome x, to change it to some other outcome y. We will denote this ability by $x <_C y$. In these terms, the above requirements on a "standard of behavior" S can be formalized as follows:

DEFINITION 5.1. By a *conflict situation*, we mean a triple $(N, X, \{<_C\}_{C \subseteq N})$, where:

N is a finite set whose elements are called *players* or *participants*;
X is a set whose elements are called *outcomes*;
$<_C$ for every *coalition* C (i.e., for every subset $C \subseteq N$), is a binary relation on the set X.

A set $S \subseteq X$ is called a *NM-solution* if it satisfies the following two conditions:

1. If $x, y \in C$, then for every coalition C, $x \not<_C y$.
2. If $x \notin S$, then there exists a coalition C and an outcome $y \in S$ for which $x <_C y$.

COMMENT. One can easily see that the definition of an NM-solution depends only on the *union*

$$< = \bigcup_C <_C$$

of binary relations $<_C$ that correspond to different coalitions. Thus, we can reformulate this definition as follows:

DEFINITION 5.2. Let $(N, X, \{<_C\})$ be a conflict situation. We say that an outcome y *dominates* an outcome x (and denote it by $x < y$) if there exists a coalition C for which $x <_C y$.

DEFINITION 5.3. A subset $S \subseteq X$ of the set of all outcomes is called a *NM-solution* if it satisfies the following two conditions:

1. if x and y are elements of S, then y cannot dominate x;
2. if x doesn't belong to S, then there exists an outcome y belonging to S that dominates x ($x < y$).

Important particular case: Cooperative games. The most thoroughly analyzed conflict situations are so-called *cooperative games*, in which cooperation is, in principle, profitable to all players. An outcome is usually described by the gains $x_1 \geq 0, \ldots, x_n \geq 0$ (called *utilities*) of all the players. In these terms, each outcome $x \in X$ is an $n-$dimensional vector $(x_1, \ldots, x_n)$ called a *payoff vector*. The total amount of gains of all the players is bounded by the maximal amount of money that the players can gain by cooperating; this amount is usually denoted by $v(N)$. In these terms, the set of all possible outcomes is the set of all vectors x_i for which $\sum x_i \leq v(N)$.

For cooperative games, the "enforcing" binary relation $<_C$ is usually described as follows: For every coalition C, we can determine the largest possible amount of money $v(C)$ that this coalition can gain in the hypothetical situation when all its members work together and all the others work against them. The function v that assigns the value $v(C)$ to each coalition C is called a *characteristic function* of the game.

The values $v(C)$ that correspond to different coalitions C must satisfy the following natural requirement: If two disjoint coalitions C and C' join forces, they can gain at least the same amount of money as when they acted separately. Thus, $v(C \cup C') \geq v(C) + v(C')$.

In terms of a characteristic function, a coalition can force the transition from x to y if the following two conditions hold:

- first, when C *can* gain for itself this new amount of money, i.e., when the total amount of money gained by the coalition C in the outcome y does not exceed $v(C)$;
- second, when all players from the coalition C gain by going from x to y ($x_i < y_i$ for all $i \in C$), and are thus interested in this transition.

Let us describe such conflict situations formally:

DEFINITION 5.4. Let n be a positive integer and $N = \{1, \ldots, n\}$. By a *cooperative game*, we mean a function $v : 2^N \to [0, \infty)$ for which $v(C \cup C') \geq v(C) + v(C')$ for disjoint C and C'. For each cooperative game, we can define the conflict situation $(N, X, \{<_C\})$ as follows:

- X is the set of all $n-$dimensional vectors $x = (x_1, \ldots, x_n)$ for which $x_i \geq 0$ and $\sum x_i = v(N)$.
- $x <_C y$ if $\sum \{y_i \,|\, i \in C\} \leq v(C)$ and $x_i < y_i$ for all $i \in C$.

5.2. Sometimes, this traditional approach to conflict resolution does not work. Von Neumann and Morgenstern have shown that NM-solutions exist for many reasonable cooperative games, and have conjectured that such a solution exists for every cooperative game. It turned out, however, that there exist games without NM-solutions (see [20,25]).

At first glance, it seems that for such conflict situations, no "standard of behavior" is possible, and thus, endless coalition re-formation is inevitable. We will show, however, that in such situations, not the original idea of "standard of behavior" is inconsistent, but only its deterministic formalization, and that in a more realistic *random-set* formalization, a "standard of behavior" always exists.

5.3. A more realistic formalization of the "standard of behavior" requires random sets. Von Neumann-Morgenstern's formalization of the notion of the "standard of behavior" (described above) was based on the simplifying assumption that this notion is deterministic, i.e., that about every possible outcome $x \in X$, either everyone agrees that x is socially acceptable, or everyone agrees that x is not socially acceptable. In reality, there are many "gray zone" situations, in which different lawyers and experts have different opinions. Thus, the actual societal "standard of behavior" is best described not by a single set S, but by a *class* $\mathcal{S}$ of sets that express the views of different experts.

Some opinions (and sets S) are more frequent, some are rarer. Thus, to adequately describe the actual standard of behavior, we must know not only this class $\mathcal{S}$, but also the *frequencies* (*probabilities*) $p(S)$ of different sets S from this class. In other words, a more adequate formalization of the "standard of behavior" is not a *set* $S \subseteq X$, but a *random set* $p(S)$.

Since S is a *random* set, we *cannot* anymore demand that the resulting outcome x is socially acceptable for all the experts; what we *can* demand is that this outcome should be socially acceptable for the *majority* of them, or, better yet, for an *overwhelming majority*, i.e., that $P(\{S \mid x \in S\}) > 1 - \varepsilon$ for some (small) $\varepsilon > 0$.

Similarly, we can reformulate the definition of a NM-solution. The first condition of the original definition was that if y dominates x, then it is impossible that both outcomes x and y are socially acceptable. A natural "random" analogue of this requirement is as follows: if y dominates x, then the overwhelming majority of experts believe that x and y cannot be both socially acceptable, i.e., $P(\{S \mid x \in S \,\&\, y \in S\}) < \varepsilon$.

The second condition was that if x is not socially acceptable, then we can enforce socially acceptable y, i.e., if $x \notin S$, then $\exists y (y \in S \,\&\, x < y)$. We would like to formulate a "random" analogue of this notion as requiring a similar property for "almost all" elements of S's complement, but unfortunately, the set S can be non-measurable [25] and the probability measure on its complement can be difficult to define. To overcome this difficulty, let us reformulate the second condition in terms of *all* x, not only $x \notin S$. This can be easily done: the second condition means that for every x there exists a $y \in S$ that is either equal to x or dominates x. This reformulated condition can be easily modified for the case when S is a random set: for every $x \in X$, according to the overwhelming majority of experts, either x is already socially acceptable, or there exists another outcome y that is socially acceptable and that dominates x:

$$P(\{S \mid \exists y \in S(x < y \vee x = y\}) > 1 - \varepsilon.$$

Thus, we arrive at the definition described in the following section.

5.4. Random sets help in conflict resolution.

5.4.1. Definitions and the main result.

DEFINITION 5.5. Let $(N, X, \{<_C\})$ be a conflict situation, and let $\varepsilon \in (0,1)$ be a real number. A random set $p(S)$, $S \subseteq X$, is called a *(random) $\varepsilon-$solution* if it satisfies the following two conditions:

1. if $x < y$ then $P(\{S \mid x \in S \,\&\, y \in S\}) < \varepsilon$;
2. for every $x \in X$, $P(\{S \mid \exists y \in S(x < y \vee x = y\}) > 1 - \varepsilon$.

COMMENT. If p is a *degenerate* random set, i.e., $p(S_0) = 1$ for some set $S_0 \subseteq X$, then p is an $\varepsilon-$solution if and only if S_0 is a NM-solution.

PROOF. If $p(S)$ is degenerate, then all the probabilities are either 0 or 1; so the inequality $P(\{S \mid x \in S \,\&\, y \in S\}) < \varepsilon$ means that it is simply impossible that x and y both belong to S, and the fact that $P(\{S \mid \exists y \in S(x < y \vee x = y\}) > 1 - \varepsilon$ means that such a y really exists. □

PROPOSITION 5.1. [13] For every cooperative game and for every $\varepsilon > 0$, there exists an $\varepsilon-$solution.

Before we prove this result about cooperative games, let us show that a similar result is true not only for cooperative games, but also for arbitrary conflict situations that satisfy some natural requirements.

5.4.2. This result holds not only for cooperative games, but for arbitrary natural conflict situations. To describe these "naturality" requirements, let us recall the definition of a *core* as the set of all non-dominated outcomes. It is easy to see that every outcome from the core belongs to every NM-solution.

DEFINITION 5.6. We say that a conflict situation is *natural* if for every outcome x that doesn't belong to the core, there exists infinitely many different outcomes y_n that dominate x.

MOTIVATION. If x doesn't belong to the core, this means that some coalition C can force the change from x to some other outcome y. We can then take an arbitrary probability $p \in [0,1]$ and then, with probability p

undertake this "forcing" and with probability $1-p$ leave the outcome as is. The resulting outcomes (it is natural to denote them by $p * y + (1-p) * x$) are different for different values of p, and they all dominate x, because C can always force x into each of them. So, in natural situations, we really have infinitely many $y_n > x$.

PROPOSITION 5.2. Conflict situations that correspond to cooperative games are natural.

PROOF. The proof follows the informal argument given as a motivation of the naturality notion: if $x \in X$ and x is not in the core, this means that $x <_C y$ for some y and for some coalition C. But one can easily check that if $x <_C y$, then $x <_C p \cdot y + (1-p) \cdot x$ for all $p \in (0, 1]$. Therefore, $x < p \cdot y + (1-p) \cdot x$ for all such p. □

PROPOSITION 5.3. For every natural conflict situation and for every $\varepsilon > 0$, there exists an ε–solution.

PROOF. Due to the definition of an ε–solution, every outcome from the core belongs to the solution with probability 1. As for other outcomes, some of them may belong to the solution, some of them may not. Let's consider the simplest possible random set with this property: this random set contains all the points from the core with probability 1, and as for all the other points, the probability that it contains each of them is one and the same (say α), and the events corresponding to different points belonging or not to this random set are independent. Formally: $P(\{S \mid x \in S\}) = \alpha$, $P(\{S \mid x \notin S\}) = 1 - \alpha$, and

$$P(\{S \mid x_1 \in S, \ldots, x_k \in S, y_1 \notin S, \ldots, y_m \notin S\}) = \alpha^k \cdot (1-\alpha)^m.$$

All the values $\chi_S(x)$ of the (random) *characteristic function* of the random set S for x outside the core are independent, and this function thus describes a *white noise*.

We want to prove that for an appropriate value of α, this random set is an ε–solution. Let's prove the first condition first.

If $x < y$, then x cannot belong to the core, so either y belongs to the core, or both do not. If y belongs to the core, then y belongs to S with probability 1, so $P(\{S \mid x \in S \,\&\, y \in S\}) = P(\{S \mid x \in S\}) = \alpha$. Therefore, for $\alpha < \varepsilon$, we have the desired property.

If neither x, nor y belong to the core, then $P(\{S \mid x \in S \,\&\, y \in S\}) = \alpha^2$. If $\alpha < \varepsilon$, then automatically $\alpha^2 < \alpha < \varepsilon$, and thus, the first condition is satisfied.

Let's now check the second condition. If x belongs to the core then $x \in S$ with probability 1, so, we can take $y = x$. If x does not belong to the core then, according to the definition of a natural conflict situation there exist infinitely many different outcomes y_i such that $y_i > x$. If at least one of these outcomes y_i belongs to the core, then $y = y_i$ belongs to S and dominates x with probability 1. To complete the proof, it is thus sufficient to consider the case when all outcomes y_i are from outside the core.

In this case, if y_i belongs to S for some i, then for this S, we get an element $y = y_i \in S$ that dominates x. Hence, for every integer m, the desired probability P that such a dominating element $y \in S$ exists is greater than or equal to the probability P_m that one of the elements $y_1, \ldots, y_m$ belongs to S. But $P_m = 1 - P(\{S \,|\, y_1 \notin S, \ldots, y_m \notin S\}) = 1 - (1-\alpha)^m$, so $P \geq 1 - (1-\alpha)^m$ for all m. By choosing a sufficiently large m, we can conclude that $P > 1 - \varepsilon$.

So, for an arbitrary $\alpha < \varepsilon$, the "white noise" random set defined above is an ε−solution. □

5.4.3. Discussion: Introduction of random sets is in complete accordance with the history of game theory. To our viewpoint, the main idea of this section is in complete accordance with the von Neumann-Morgenstern approach to game theory. Indeed, where did they start? With zero-sum games, where if we limit ourselves to deterministic strategies, then not every game has a stable solution. Then, they noticed that in real life, when people do not know what to do, they sometimes flip coins and choose an action at random. So, they showed that if we add such "randomized" strategies, then *every zero-sum game has a stable solution.*

Similarly, when it turned out that in some games, no set of outcomes can be called socially acceptable, we noticed that in reality, whether an outcome is socially acceptable or not is sometimes not known. If we allow such "randomized" standards of behavior, then we also arrive at the conclusion that *every cooperative game has a solution.*

5.5. Fuzzy solutions are not sufficient for conflict resolution. Is it reasonable to use such a complicated formalism as generic random sets? Maybe, it is sufficient to use other, simpler formalisms for expressing experts' uncertainty like fuzzy logic. In this section, we will show that fuzzy solutions are not sufficient, and general random sets are indeed necessary.

HISTORICAL COMMENT. A definition of a fuzzy NM-solution was first proposed in [1]; the result that fuzzy solutions are not sufficient was first announced in [11].

5.5.1. Fuzzy NM-Solution: Motivations. We want to describe the "vague" "standard of behavior" S, i.e., the standard of behavior in which for some outcomes $x \in X$, we are not sure whether this outcome is acceptable or not. In fuzzy logic formalism, our degree of belief in a statement E about which we may be unsure is described by a number $t(E)$ from the interval $[0, 1]$: 0 corresponds to "absolutely false," 1 corresponds to "absolutely true," and intermediate values describe uncertainty. Therefore, within this formalism, a "vague" set S can be described by assigning, for every $x \in X$, a number from the interval $[0, 1]$ that describe our degree of belief that this particular outcome x is acceptable. This number is usually denoted by $\mu(x)$, and the resulting function $\mu : X \to [0,1]$ describes a *fuzzy set*.

We would like to find a fuzzy set S for which the two properties describing NM solution are, to a large extent, true. These two properties are:

1. if $x < y$, then x and y cannot both be elements of S, i.e., $\neg(x \in S \,\&\, y \in S)$;
2. for every $x \in X$, we have $\exists y(y \in S \,\&\, (x < y \vee x = y))$.

In order to interpret these conditions for the case when S is a fuzzy set, and thus, when the statements $x \in S$ and $y \in S$ can have degree of belief different from 0 or 1, we must define logical operations for intermediate truth values (i.e., for values $\in (0,1)$).

Quantifiers $\forall x E(x)$ and $\exists x E(x)$ mean, correspondingly, infinite "and" $E(x_1) \,\&\, E(x_2) \ldots$ and infinite "or" $E(x_1) \vee E(x_2) \vee \ldots$. For most "and" and "or" operations, e.g.,, for $f_\&(a,b) = a \cdot b$ or $f_\vee(a,b) = a + b - a \cdot b$, an infinite repetition of "and" leads to a meaningless value 0, and a meaningless repetition of "or" leads to a meaningless value 1. It can be shown (see, e.g., [2]) that under certain reasonable conditions, the only "and" and "or" operations that do not always lead to these meaningless values after an infinite iteration are $\min(a,b)$ and $\max(a,b)$. Thus, as a degree of belief $t(E\&F)$ in the conjunction $E\&F$, we will take $t(E\&F) = \min(t(E), t(F))$ and, similarly, we will take $t(E \vee F) = \max(t(E), t(F))$. For negation, we will take the standard operation $t(\neg E) = 1 - t(E)$.

Thus, we can define

$$t(\forall x E(x)) = \min(t(E(x_1)), t(E(x_2)), \ldots) = \inf_x t(E(x))$$

and, similarly,

$$t(\exists x E(x)) = \max(t(E(x_1)), t(E(x_2)), \ldots) = \sup_x t(E(x)).$$

Thus, to get the degree of belief of a complex logical statement that uses propositional connectives and quantifiers, we must replace & by min, $\vee$ by max, $\neg$ by $t \to 1 - t$, $\forall$ by inf, and $\exists$ by sup.

As a result, for both conditions that define a NM-solution, we will get a degree of belief t that describes to what extent this condition is satisfied. We can, in principle, require that the desired fuzzy set S satisfies these conditions with degree of belief at least 1, or at least 0.99, or at least 0.9, or a least t_0 for some "cut-off" value t_0. The choice of this "cut-off" t_0 is reasonably arbitrary; the only thing that we want to guarantee is that our belief that S is a solution should exceed our degree of belief that S is not a NM-solution. In other words, we would like to guarantee that $t > 1 - t$; this inequality is equivalent to $t > 1/2$. Thus, we can express the "cut-off" degree of belief as $t_0 = 1 - \varepsilon$ for some $\varepsilon \in [0, 1/2)$.

So, we arrive at the following definition:

5.5.2. Fuzzy NM–solution: Definition and the main result.

DEFINITION 5.7. Let $(N, X, \{<_C\})$ be a conflict situation, and let $\varepsilon \in [0, 1/2)$ be a real number. A fuzzy set $S \subseteq X$ with a membership function $\mu : X \to [0, 1]$ is called a *fuzzy ε–solution* iff the following two conditions hold:

1. if $x < y$, then $1 - \min(\mu(x), \mu(y)) \geq 1 - \varepsilon$;
2. for every $x \in X$, $\sup\{\mu(y) \mid x < y \text{ or } y = x\} \geq 1 - \varepsilon$.

The following result shows that this notion does not help when there is no NM-solution:

PROPOSITION 5.4. A conflict situation has a fuzzy ε–solution iff it has a (crisp) NM-solution.

PROOF. It is easy to check that if S is a NM-solution, then its characteristic function is a fuzzy ε–solution.

Vice versa, let a fuzzy set S with a membership function $\mu(x)$ be a fuzzy ε–solution. Let us show that the set $S_0 = \{x \mid \mu(x) > \varepsilon\}$ is then an NM-solution. Indeed, from 1., it follows that if $x < y$, then both x and y cannot belong to S_0: otherwise both values $\mu(x)$ and $\mu(y)$ would be $> \varepsilon$, and thus, their minimum would also be $> \varepsilon$, and $1-$ this minimum would be $< 1 - \varepsilon$.

From the second condition of Definition 5.7, it follows that no matter what small $\delta > 0$ we take, for every x, there exists a y for which $\mu(y) < 1 - \varepsilon - \delta$, and either $x < y$ or $y = x$.

But $\varepsilon < 1/2$, so $1 - \varepsilon - \delta > \varepsilon$ for sufficiently small δ; so, if we take y_0 that corresponds to such δ, we will get a $y \in S_0$ for which $x < y$ or $y = x$.

So, the set S_0 satisfies both conditions of the NM-solution and thus, it is a NM-solution. □

REFERENCES

[1] O.N. BONDAREVA AND O.M. KOSHELEVA, *Axiomatics of core and von Neumann-Morgenstern solution and the fuzzy choice*, Proc. USSR National conference on optimization and its applications, Dushanbe, 1986, pp. 40–41 (in Russian).

[2] B. BOUCHON-MEUNIER, V. KREINOVICH, A. LOKSHIN, AND H.T. NGUYEN, *On the formulation of optimization under elastic constraints (with control in mind)*, Fuzzy Sets and Systems, vol. 81 (1996), pp. 5–29.

[3] TH.H. CORMEN, CH.L. LEISERSON, AND R.L. RIVEST, *Introduction to algorithms*, MIT Press, Cambridge, MA, 1990.

[4] G.J. ERICKSON AND C.R. SMITH (eds.), *Maximum-entropy and Bayesian methods in science and engineering*, Kluwer Acad. Publishers, 1988.

[5] M.R. GAREY AND D.S. JOHNSON, *Computers and intractability: A guide to the theory of NP-completeness*, W.F. Freeman, San Francisco, 1979.

[6] R.I. FREIDZON*et al.*, *Hard problems: Formalizing creative intelligent activity (new directions)*, Proceedings of the Conference on Semiotic aspects of Formalizing Intelligent Activity, Borzhomi–88, Moscow, 1988, pp. 407–408 (in Russian).

[7] K. HANSON AND R. SILVER, Eds., *Maximum Entropy and Bayesian Methods, Santa Fe, New Mexico, 1995*, Kluwer Academic Publishers, Dordrecht, Boston, 1996.

[8] J.C. HARSHANYI, *An equilibrium-point interpretation of the von Neumann-Morgenstern solution and a proposed alternative definition*, In: *John von Neumann and modern economics*, Claredon Press, Oxford, 1989, pp. 162–190.

[9] E.T. JAYNES, *Information theory and statistical mechanics*, Phys. Rev. 1957, vol. 106, pp. 620–630.

[10] G.J. KLIR AND BO YUAN, *Fuzzy Sets and Fuzzy Logic*, Prentice Hall, NJ, 1995.

[11] O.M. KOSHELEVA AND V.YA. KREINOVICH, *Computational complexity of game-theoretic problems*, Technical report, Informatika center, Leningrad, 1989 (in Russian).

[12] V.YA. KREINOVICH, *Entropy estimates in case of a priori uncertainty as an approach to solving hard problems*, Proceedings of the IX National Conference on Mathematical Logic, Mathematical Institute, Leningrad, 1988, p. 80 (in Russian).

[13] O.M. KOSHELEVA AND V.YA. KREINOVICH, *What to do if there exist no von Neumann-Morgenstern solutions*, University of Texas at El Paso, Department of Computer Science, Technical Report No. UTEP-CS-90-3, 1990.

[14] V. KREINOVICH, *Group-theoretic approach to intractable problems*, In: Lecture Notes in Computer Science, Springer-Verlag, Berlin, 1990, vol. 417, pp. 112–121.

[15] V. KREINOVICH *et al.*, *Monte-Carlo methods make Dempster-Shafer formalism feasible*, in [32], pp. 175–191.

[16] V. KREINOVICH AND S. KUMAR, *Optimal choice of &- and $\vee$-operations for expert values*, Proceedings of the 3rd University of New Brunswick Artificial Intelligence Workshop, Fredericton, N.B., Canada, 1990, pp. 169–178.

[17] V. KREINOVICH *et al.*, *What non-linearity to choose? Mathematical foundations of fuzzy control*, Proceedings of the 1992 International Conference on Fuzzy Systems and Intelligent Control, Louisville, KY, 1992, pp. 349–412.

[18] V. KREINOVICH, H.T. NGUYEN, AND E.A. WALKER, *Maximum entropy (MaxEnt) method in expert systems and intelligent control: New possibilities and limitations*, In: [7].

[19] V. KREINOVICH, *Maximum entropy and interval computations*, Reliable Computing, vol. 2 (1996), pp. 63–79.

[20] W.F. LUCAS, *The proof that a game may not have a solution*, Trans. Amer. Math. Soc., 1969, vol. 136, pp. 219–229.

[21] J. VON NEUMANN AND O. MORGENSTERN, *Theory of games and economic behavior*, Princeton University Press, Princeton, NJ, 1944.

[22] H.T. NGUYEN, *Some mathematical tools for linguistic probabilities*, Fuzzy Sets and Systems, vol. 2 (1979), pp. 53–65.

[23] H.T. NGUYEN *et al.*, *Theoretical aspects of fuzzy control*, J. Wiley, N.Y., 1995.

[24] H. T. NGUYEN AND E. A. WALKER, *A First Course in Fuzzy Logic*, CRC Press, Boca Raton, Florida, 1996.

[25] G. OWEN, *Game theory*, Academic Press, N.Y., 1982.

[26] J. PEARL, *Probabilistic Reasoning in Intelligent Systems*, Morgan Kaufmann, San Mateo, CA, 1988.

[27] D. RAJENDRAN, *Application of discrete optimization techniques to the diagnostics of industrial systems*, University of Texas at El Paso, Department of Mechanical and Industrial Engineering, Master Thesis, 1991.

[28] A. RAMER AND V. KREINOVICH, *Maximum entropy approach to fuzzy control*, Proceedings of the Second International Workshop on Industrial Applications of Fuzzy Control and Intelligent Systems, College Station, December 2–4, 1992, pp. 113–117.

[29] A. RAMER AND V. KREINOVICH, *Maximum entropy approach to fuzzy control*, Information Sciences, vol. 81 (1994), pp. 235–260.

[30] G. SHAFER AND J. PEARL (eds.), *Readings in Uncertain Reasoning*, Morgan Kauf-

mann, San Mateo, CA, 1990.

[31] M.H. SMITH AND V. KREINOVICH, *Optimal strategy of switching reasoning methods in fuzzy control*, in [23], pp. 117–146.

[32] R.R. YAGER, J. KACPRZYK, AND M. PEDRIZZI (Eds.), *Advances in the Dempster-Shafer Theory of Evidence*, Wiley, N.Y., 1994.

LAWS OF LARGE NUMBERS FOR RANDOM SETS

ROBERT L. TAYLOR* AND HIROSHI INOUE†

Abstract. The probabilistic study of geometrical objects has motivated the formulation of a general theory of random sets. Central to the general theory of random sets are questions concerning the convergence for averages of random sets which are known as laws of large numbers. General laws of large numbers for random sets are examined in this paper with emphasis on useful characterizations for possible applications.

Key words. Random Sets, Laws of Large Numbers, Tightness, Moment Conditions.

AMS(MOS) subject classifications. 60D05, 60F15

1. Introduction. The idea of random sets was probably in existence for some time. Robbins (1944, 1945) appears to have been the first to provide a mathematical formulation for random sets, and his early works investigated the relationships between random sets and geometric probabilities. Later, Kendall (1974) and Matheron (1975) provided a comprehensive mathematical theory of random sets which was greatly influenced by the geometric probability prospective.

Numerous motivating examples and possible applications for modeling random objects are provided in Chapter 9 of Cressie (1991). Many of these examples and applications lead to the need for statistical inference for random sets, and in particular, estimation of the expected value of a random set. The natural estimator for the expected value is the sample average of random sets, and the convergence of the sample average is termed the law of large numbers. With respect to laws of large numbers, the pioneering works of Artstein and Vitale (1975), Cressie (1978), Hess (1979, 1985), Artstein and Hart (1981), Puri and Ralescu (1983, 1985), Gine, Hahn and Zinn (1983), Taylor and Inoue (1985a, 1985b) and Artstein and Hansen (1985) are cited among many others. Several laws of large numbers for random sets will be presented in this paper. Characterizations of necessary conditions and references to examples and applications will be provided to illustrate the practicality and applicability of these results.

2. Definitions and properties. In this section definitions and properties for random sets are given. The range of definitions and properties is primarily focused on the material necessary for the laws of large numbers for random sets. However, it is important to first distinguish between fuzzy sets and random sets.

A fuzzy set is a subset whose boundaries may not be identifiable with certainty. More technically, a *fuzzy set* in a space $\mathcal{X}$ is defined as

$$\{(x, u(x)) : x \in \mathcal{X}\} \tag{2.1}$$

* Department of Statistics, University of Georgia, Athens, GA 30602, U.S.A.

† School of Management, Science University of Tokyo, Kuki, Saitama 346, Japan.

where $0 \leq u(x) \leq 1$ for each $x \in \mathcal{X}$. The function u: $\mathcal{X} \to [0,1]$ is referred to as the *membership function.* A function u: $\mathcal{X} \to \{0,1\}$ determines membership in a set with certainty whereas, other functions allow for (more or) less certain membership. For example, in image processing with black, white and varying shades of gray images,

(2.2) $\{x : b \leq u(x) \leq 1\}$ may designate the interior of an object

(2.3) $\{x : 0 \leq u(x) \leq a\}$ may designate areas which are probably not part of the object

and

(2.4) $\{x : a < u(x) < b\}$ may designate possibly (weighted by $u(x)$) part of the object

where $u(x)$ corresponds to darkness at a particular point of the image.

It is convenient to define

$$\mathcal{F}(\mathcal{X}) \equiv \text{ the set of functions: } \mathcal{X} \to [0,1] \tag{2.5}$$

and to require that

$$\{x \in \mathcal{X} : u(x) \geq a\} \text{ to be compact for each } a > 0 \tag{2.6}$$

and

$$\{x \in \mathcal{X} : \quad u(x) = 1\} \neq \emptyset. \tag{2.7}$$

Note that (2.6) calls for topological considerations which are easily addressed in this paper by restricting $\mathcal{X}$ to be a separable Banach space (or $\mathbb{R}^m$, the m-dimensional Euclidean space). Fuzzy sets are often denoted as $\tilde{A}_i \equiv \{(x, u_i(x)) : x \in \mathcal{X}\}$, $i \geq 1$, and fuzzy set operations can be defined as

$$\tilde{A}_1 \bigcup \tilde{A}_2 \equiv \{(x, \max\{u_1(x), u_2(x)\}) : x \in \mathcal{X}\} \tag{2.8}$$

$$\tilde{A}_1 \bigcap \tilde{A}_2 \equiv \{(x, \min\{u_1(x), u_2(x)\}) : x \in \mathcal{X}\} \tag{2.9}$$

and

$$\tilde{A}_1^c \equiv \{(x, 1 - u_1(x)) : x \in \mathcal{X}\} \ .$$

In contrast to fuzzy sets whose boundaries may not be identified with certainty, random sets involve sets whose boundaries can be identified with certainty but whose selections are (randomly) determined by a probability distribution. Several excellent motivating examples of random sets are provided in Chapter 9 of Cressie (1991). One such example is primate skulls where the general (or average) skull characteristics are of interest. In this example, $\mathcal{X} = \mathbb{R}^3$, and each skull can be precisely measured (that is, membership in a 3-dimensional set can be determined with certainty for each skull). The skulls are assumed to be observed random sets from a distribution of sets, that is, $X_i(w) \subset \mathbb{R}^3, 1 \leq i \leq n$, where $X_i : \Omega \to \mathbb{R}^3$ for some probability space $(\Omega, \mathcal{A}, \mathcal{P})$. Then the convergence of the sample average of the primate skulls to an average skull for the population becomes the major consideration. To adequately address this situation, several basic definitions and properties will be needed to provide a framework for random sets. These definitions and properties will be listed briefly, and additional details with respect to this development can be obtained from the references.

Let $\mathcal{X}$ be a real, separable Banach space, let $\mathcal{K}(\mathcal{X})$ denote all non-empty compact subsets of $\mathcal{X}$ and let $\mathcal{C}(\mathcal{X})$ denote all non-empty convex, compact subsets of $\mathcal{X}$. For $A_1, A_2 \in \mathcal{K}(\mathcal{X})$ (or $\in \mathcal{C}(\mathcal{X})$) and $\lambda \in \mathbb{R}$, Minkowski addition and scalar multiplication are given by

$$A_1 + A_2 = \{a_1 + a_2 : a_1 \in A_1, a_2 \in A_2\} \tag{2.10}$$

and

$$\lambda A_1 = \{\lambda a_1 : a_1 \in A_1\}. \tag{2.11}$$

Neither $\mathcal{K}(\mathcal{X})$ nor $\mathcal{C}(\mathcal{X})$ are linear spaces even when $\mathcal{X} = \mathbb{R}$. For example, let $\lambda_1 = 1, \lambda_2 = -1$ and $A = [0, 1]$, and observe that

$$\begin{aligned} \{0\} = 0A = (\lambda_1 + \lambda_2)A &\neq \lambda_1 A + \lambda_2 A_2 \\ &= [0,1] + [-1,0] \\ &= [-1,1]. \end{aligned}$$

The Hausdorff distance between two sets $A_1, A_2 \in \mathcal{K}(\mathcal{X})$ can be defined as

$$\begin{aligned} d(A_1, A_2) &= \inf\{r > 0 : A_1 \subset A_2 + rU \text{ and } A_2 \subset A_1 + rU\} \\ &= \max\{\sup_{a \in A_1} \inf_{b \in A_2} \|a - b\|, \sup_{b \in A_2} \inf_{a \in A_1} \|a - b\|\} \end{aligned} \tag{2.12}$$

where $U = \{x : x \in \mathcal{X} \text{ and } \|x\| \leq 1\}$. With the Hausdorff metric, $\mathcal{K}(\mathcal{X})$ (and $\mathcal{C}(\mathcal{X})$) is a complete, separable metric space. The open subsets of $\mathcal{K}(\mathcal{X})$ generate a collection of Borel subsets, $\mathcal{B}(\mathcal{K}(\mathcal{X}))$ ($\equiv \sigma$-field generated by the open subsets of the d metric topology on subsets of $\mathcal{K}(\mathcal{X})$), and random sets can be defined as Borel measurable functions from a probability space to $\mathcal{K}(\mathcal{X})$.

DEFINITION 2.1. Let $(\Omega, \mathcal{A}, P)$ be a probability space. A *random set* X is a function: $\Omega \to \mathcal{K}(\mathcal{X})$ such that $X^{-1}(B) \in \mathcal{A}$ for each $B \in \mathcal{B}(\mathcal{K}(\mathcal{X}))$.

Definition 2.1 has the equivalent formulation that a random set is the limit of simple random sets, that is,

$$\lim_{n\to\infty} d\Big(X(\omega), \sum_{i=1}^{m_n} k_{ni} I_{E_{ni}}(\omega)\Big) = 0$$

where $k_{n1}, ..., k_{nm_n} \in \mathcal{K}(\mathcal{X})$ and $E_{n1}, ..., E_{nm_n}$ are disjoint sets in Ω for each n. Most distributional properties (independence, identical distributions, etc.) of random sets have obvious formulations and will not be repeated here. However, the expectation of a random set has a seemingly strange convexity requirement. For a set A let coA denote the convex hull of A. The mapping of a set to its convex hull is a continuous operation of $\mathcal{K}(\mathcal{X})$ to $\mathcal{C}(\mathcal{X})$ (cf: Matheron (1975), p. 21). Moreover, let $A = \{0, 1\} \in \mathcal{K}(\mathbb{R})$. Then by (2.10)

$$\frac{A + A + \cdots + A}{n} = \left\{\frac{i}{n} : 1 \le i \le n\right\} \to [0,1] = coA \tag{2.13}$$

in $\mathcal{K}(\mathcal{X})$ with the Hausdorff metric d. Thus, for the sample average to converge to the expected value in the laws of large numbers, convexity may be needed for the expectation. The Shapley–Forkman Theorem (cf: Artstein and Vitale (1975)) is stated next and better illustrates the convexity property associated with the Minkowski addition. Let $||A|| = \sup\{||a|| : a \in A\}$ where $A \in \mathcal{K}(\mathcal{X})$.

PROPOSITION 2.1. Let $A_i \in \mathcal{K}(\mathbb{R}^m), 1 \le i \le n$, such that $\sup_i ||A_i|| \le M$. Then

$$d\Big(\sum_{i=1}^{n} A_i, co\sum_{i=1}^{n} A_i\Big) \le \sqrt{m}M. \tag{2.14}$$

(Note the lack of dependence on n.)

Artstein and Hansen (1985) provided the following lemma for Banach spaces which dictates that the laws of large numbers in $\mathcal{K}(\mathcal{X})$ must involve convergence to a convex, compact subset of $\mathcal{X}$.

LEMMA 2.1. Let $A_1, A_2, ...$ be a sequence of compact subsets in a Banach space and let A_0 be compact and convex. If

$$\frac{1}{n}(coA_1 + \cdots + coA_n) \text{ converges to } A_0$$

as $n \to \infty$, then

$$\frac{1}{n}(A_1 + \cdots + A_n) \text{ converges to } A_0$$

as $n \to \infty$.

Several embedding theorems exist for embedding the convex, compact subsets of $\mathcal{X}$ into Banach spaces. The Rådström Embedding Theorem (cf: Rådström (1952)) provides a normed linear space U and a linear isometry j such that $j : C(\mathcal{X}) \to C_b$. Hörmander (1954) provided more specifically that $j : C(\mathcal{X}) \to C_b(\text{ball } \mathcal{X}^*)$ where $C_b(\text{ball } \mathcal{X}^*)$ denotes the space of all bounded, continuous functions on the unit ball of $\mathcal{X}^*(= \{x^* \in \mathcal{X}^* : ||x^*|| \leq 1\})$ with the sup norm topology. Giné, Hahn and Zinn (1983) observed that $C(\mathcal{X})$ could be embedded into $C_b(\text{ball } \mathcal{X}^*, \mathcal{W}^*)$, the Banach space of functions defined on the unit ball of $\mathcal{X}^*$ and continuous in the weak topology.

The mapping of $\mathcal{K}(\mathcal{X})$ into $\mathcal{C}(\mathcal{X})$ and the isometric identification of $\mathcal{C}(\mathcal{X})$ as a subspace of a separable Banach space allow many of probabilistic results for Banach spaces to be used for obtaining results for random sets. In particular, the expected value may be defined by the Pettis integral (or the Bochner integral when the first absolute moment exists).

DEFINITION 2.2. A random set X in $\mathcal{K}(\mathcal{X})$ is said to have (Pettis) expected value $E(coX) \in \mathcal{C}(\mathcal{X})$ if and only if

$$f(jE(coX)) = E(f(j(coX))) \tag{2.15}$$

for all $f \in U^*$ where j is the Rådström embedding function of $\mathcal{C}(\mathcal{X})$ into U.

When $E||coX|| = E||X|| < \infty$, then the Pettis and Bochner integrals exist, and $E(coX)$ may be calculated by $\lim_{n\to\infty} E(coX_n)$ where

$$X_n = \sum_{j=1}^{m_n} k_{nj} I_{E_{nj}} \quad (\text{and} \quad EcoX_n = \sum_{j=1}^{m_n} cok_{nj} P(E_{nj}))$$

is a sequence of simple functions such that $E||coX - coX_n|| \to 0$ as $n \to \infty$ (Debreu (1966)).

While useful in establishing properties of random sets and proving theorems, the embedding techniques can transform consideration from a finite dimensional space (like $\mathbb{R}^m$) to an infinite-dimensional Banach space. To show that $\mathcal{C}(\mathbb{R}^m)$ (and hence $\mathcal{K}(\mathbb{R}^m)$) can not be examined in the finite-dimensional setting, for $A \in \mathcal{C}(\mathbb{R}^m)$, let

$$h_A(x) = \max_{u \in A}(x \cdot u) = \max_{u \in A}(\sum_{i=1}^{m} x_i u_i) \tag{2.16}$$

where $x \in S^{m-1} = \{y \in \mathbb{R}^m : ||y|| = 1\}$. The function $h_A(x)$ is a continuous function on S^{m-1} and is called the support function of the set A. There is a one-to-one, norm preserving identification of h_A and $A \in \mathcal{C}(\mathbb{R}^m)$. Hence, $\mathcal{C}(\mathbb{R}^m)$ is isometric to the closed cone of support functions. A troublesome aspect of this identification (and well as Hörmander's identification) is the lack of Banach space geometric conditions which Hoffmann–Jørgensen and Pisier (1976) showed was required in obtaining Banach space versions of many of the laws of large numbers for random variables. However, the finite-dimensional properties of $\mathbb{R}^m$ can be used as will be shown in the next section.

3. Laws of large numbers for random sets. Several laws of large numbers for random sets in $\mathcal{K}(\mathcal{X})$ are examined in this section. An attempt will be made to cite places in the literature where detailed proofs are presented.

The most common form for the law of large numbers is when the random sets are identically distributed. Artstein and Vitale (1975), Puri and Ralescu (1983, 1985) and Taylor and Inoue (1985a, 1985b) provide more details on the proofs and development of Theorems 3.1 and 3.2.

THEOREM 3.1. Let $\{X_n\}$ be a sequence of i.i.d. random sets in $\mathcal{K}(\mathcal{X})$. If $E||X_1|| < \infty$, then

$$\frac{1}{n}\sum_{i=1}^{n} X_i \to E(coX_1) \text{ w. prob. one.} \tag{3.1}$$

While different proofs have originated historically, Mourier's (1953) SLLN for i.i.d. random elements on Banach spaces (and one of the embedding results) insures that

$$\frac{1}{n}\sum_{i=1}^{n} coX_i \to E(coX_1) \text{ w. prob. one,}$$

and Lemma 2.1 removes the convexity restriction.

Similarly, Adler, Rosalsky and Taylor's (1991) WLLN provides the next result.

THEOREM 3.2. Let $\{X_n\}$ be a sequence of i.i.d. random sets in $\mathcal{K}(\mathcal{X})$. Let $\{a_n\}$ and $\{b_n\}$ be positive constants such that

$$\frac{b_n}{a_n}\uparrow, \ \frac{b_n}{na_n} \to \infty, \ \sum_{i=1}^{n} a_i = O(na_n) \tag{3.2}$$

and

$$\sum_{i=1}^{n} \frac{b_j}{i^2 a_i} = O\left(\frac{b_n}{\sum_{i=1}^{n} a_i}\right).$$

If $nP[||X_1|| > \frac{b_n}{a_n}] \to 0$, then

$$d\left(\frac{1}{b_n}\sum_{i=1}^{n} a_i X_i, \frac{1}{b_n}\sum_{i=1}^{n} a_i E(coX_1 I_{[||X_1|| \le \frac{b_n}{a_n}]})\right) \to 0 \text{ in prob.} \tag{3.3}$$

Other conditions on a_n and b_n may be given (cf: Alder, Rosalsky and Taylor (1991)). Next, let $\{\{x\} : x \in \mathcal{X}\} \equiv S(\mathcal{X}) \subset \mathcal{K}(\mathcal{X})$ be the subset of singleton sets. Observe that $d(\{x\},\{y\}) = ||x-y||$ for all $x, y \in \mathcal{X}$ and that $S(\mathcal{X})$ is isometric to $\mathcal{X}$. Thus, Theorems 3.1 and 3.2 are in their sharpest forms since the ranges of $\{X_n\}$ could be restricted to $S(\mathcal{X})$ and Kolmogorov's SLLN and Feller's WLLN would apply. Moreover, Example 4.1.1 of Taylor (1978) shows that the identical distribution condition in Theorems 3.1 and 3.2 can not be easily dropped for infinite-dimensional spaces without geometric conditions on $\mathcal{X}$ or other distributional conditions.

Tightness and moment conditions can be used to provide other laws of large numbers. A family of random sets, $\{X_\alpha, \alpha \in \Lambda\}$, is said to be *tight* if for each $\epsilon > 0$ there exists a compact subset of $\mathcal{K}(\mathcal{X})$, D_ϵ, such that $P[X_\alpha \notin D_\epsilon] < \epsilon$ for all $\alpha \in \Lambda$. If $\{X_\alpha\}$ are i.i.d. random sets, then $\{X_\alpha\}$ is tight, but the converse is not true. Daffer and Taylor (1982) added a moment condition to tightness in obtaining strong laws of large numbers in Banach spaces. Specifically, they defined the family of random sets $\{X_\alpha, \alpha \in \Lambda\}$ to be *compactly uniformly integrable* (CUI) if for each $\epsilon > 0$ there exists a compact subset of $\mathcal{K}(\mathcal{X}), \mathcal{D}_\epsilon$, such that

$$E||X_\alpha I_{[X_\alpha \notin D_\epsilon]}|| < \epsilon \quad \text{for all } \alpha \in \Lambda.$$

Compact uniform integrability implies tightness, but not conversely. Compact uniform integrability also implies that $\{||X_\alpha||\}$ are uniformly integrable and is equivalent to uniform integrability for real-valued random variables. The basic Banach space foundation for Theorem 3.3 is from Daffer and Taylor (1982).

THEOREM 3.3 (TAYLOR AND INOUE, 1985A). Let $\{X_n\}$ be a sequence of independent random sets in $\mathcal{K}(\mathcal{X})$ which are compactly, uniformly integrable. If

$$\sum_{n=1}^{\infty} \frac{1}{n^p} E||X_n||^p < \infty \quad \text{for some } 1 \le p \le 2, \tag{3.4}$$

then

$$d\Big(\frac{1}{n}\sum_{i=1}^{n} X_i, \frac{1}{n}\sum_{i=1}^{n} E(coX_i)\Big) \to 0 \quad \text{w. prob. one.}$$

If $\sup_\alpha E||X_\alpha||^p < \infty$ for some $p > 1$, then (3.4) holds. Next, it will be shown that the uniformly bounded p^{th} ($p > 1$) moments implies CUI for random sets in $\mathcal{K}(\mathbb{R}^m)$. Lemma 3.1 is a major result in establishing this fact. The finite-dimensional property of $\mathbb{R}^m$ provides the result easily, but the following proof is instructive on the metric topology of $\mathcal{K}(\mathbb{R}^m)$.

LEMMA 3.1. Let $c > 0$ be fixed. Then

$$B_c = \{k \in \mathcal{K}(\mathbb{R}^m) : ||k|| \le c\}$$

is compact in $\mathcal{K}(\mathbb{R}^m)$ with the Hausdorff distance metric d.

PROOF. For each positive integer $t \geq 1$, there exists $x_{t1}, ..., x_{tn_t} \in \mathbb{R}^m$ such that for each $x \in \mathbb{R}^m$, with $||x|| \leq c$, then $||x - x_{tj}|| < \frac{1}{2t}$ for some $1 \leq j \leq n_t$. Define $k_{tj} = \{y \in \mathbb{R}^m : ||y - x_{tj}|| \leq \frac{1}{2t}\}$, $1 \leq j \leq n_t$. Let $B_{tl}, l = 1, ..., 2^{n_t}$ denote the subsets of $\{t1, ..., tn_t\}$, and define $A_{tl} = \bigcup_{tj \in B_{tl}} k_{tj}$. Next, it will be shown that $B_c = \{k \in \mathcal{K}(\mathbb{R}^m) : ||k|| \leq c\}$ is totally bounded (and hence d-compact) by showing that $A_{tl}, 1 \leq l \leq 2^{n_t}$, is a finite ϵ-net for B_c. Let $\epsilon > 0$ be given and choose t such that $\frac{1}{t} < \epsilon$. Let $k \in B_c$. Since k is compact and $k \subset \bigcup_{y \in k}\{x : ||y - x|| < \frac{1}{2t}\}$, there exists $y_1, ..., y_p \in k$ such that $k \subset \bigcup_{j=1}^{p}\{x : ||y_j - x|| < \frac{1}{2t}\}$. Moreover, there exists $B_{tl_0} = \{ti_1, ..., ti_p\}$ such that $||y_j - x_{ti_j}|| < \frac{1}{2t}$, $1 \leq j \leq p$. For $z \in k$, $||z - y_j|| < \frac{1}{2t}$ for some j implies that $||z - x_{ti_j}|| < \frac{1}{t}$ and

$$k \subset A_{tl_0} + \frac{1}{t}U, \tag{3.5}$$

where $U = \{x \in \mathbb{R}^m : ||x|| \leq 1\}$. For $x \in A_{tl_0}, ||x - x_{ti_j}|| < \frac{1}{2t}$ for some j, $1 \leq j \leq p$. Hence, $||x - y_j|| < \frac{1}{t}$ and

$$A_{tl_0} \subset k + \frac{1}{t}U. \tag{3.6}$$

From (3.5) and (3.6) it follows that $d(k, A_{tl_0}) \leq \frac{1}{t} < \epsilon$. Thus, $\{A_{t1}, ..., A_{t2^n}\}$ is a finite ϵ-net for $B_c = \{k \in \mathcal{K}(\mathbb{R}^m) : ||k|| \leq c\}$, and therefore, B_c is compact. □

If $\sup_\alpha E||X_\alpha||^p < \infty$ for some $p > 0$, then a straight-forward application of Markov's inequality yields

$$P[||X_\alpha|| > c] \leq \frac{E||X_\alpha||^p}{c^p}$$

and tightness by Lemma 3.1 for $\mathcal{K}(\mathbb{R}^m)$. If $\sup_\alpha E||X_\alpha||^p < \infty$, $p > 1$, then Hölder inequality yields

$$E||X_\alpha I_{[||X_\alpha||>c]}|| \leq (E||X_\alpha||^p)^{1/p}(P[||X_\alpha|| > c])^{\frac{p-1}{p}}$$

and uniform compact integrability for $\mathcal{K}(\mathbb{R}^m)$. Moreover, UCI (or tightness and $p > 1$ moments) is somewhat necessary as shown by Example 5.2.3 of Taylor (1978).

Many applications require convergent results for weighted sums of the form

$$\sum_{i=1}^{n} a_{ni}X_i.$$

For the next results, it will be assumed that $\{a_{ni}; 1 \leq i \leq n, n \geq 1\}$ is an array of non-negative constants such that $\sum_{i=1}^{n} a_{ni} \leq 1$ for all n. Recall that j denotes the isometry of $C(\mathcal{X})$ into a separable normed linear space U.

THEOREM 3.4 (TAYLOR AND INOUE, 1985B). Let $\{X_n\}$ be a sequence of CUI random sets in $\mathcal{K}(\mathcal{X})$ and let $\max_{1\leq i\leq n} a_{ni} \to 0$ as $n \to \infty$. For each $f \in U^*$

$$\left|\sum_{i=1}^{n} a_{nk}(f(j(coX_i) - Ej(coX_i))\right| \to 0 \text{ in prob.} \tag{3.7}$$

if and only if

$$d\left(\sum_{i=1}^{n} a_{ni}X_i, \sum_{i=1}^{n} a_{ni}EcoX_i\right) \to 0 \text{ in prob.} \tag{3.8}$$

If $\{X_n\}$ are independent CUI random sets in $\mathcal{K}(\mathcal{X})$, then (3.7) is obtained. For convergence with probability one in (3.8), a stronger moment condition and a more restrictive condition on the weights are needed.

THEOREM 3.5 (TAYLOR AND INOUE, 1985B). Let $\{X_n\}$ be a sequence of independent, CUI random sets in $\mathcal{K}(\mathcal{X})$. Let X be a r.v. such that $P[||X_n|| > t] \leq P[X > t]$ for all n and $t > 0$. If $EX^{1+1/\gamma}$ for some $\gamma > 0$ and $\max_{1\leq i\leq n} a_{ni} = \theta(n^{-\gamma})$, then

$$d\left(\sum_{i=1}^{n} a_{ni}X_i, \sum_{i=1}^{n} a_{ni}E(coX_i)\right) \to 0 \text{ w. prob. one.}$$

The conditions in Theorem 3.4 and Theorem 3.5 are sharp for r.v.'s, (cf: Section 3.4 of Taylor (1978)), and hence, must be sharp for random sets (via the previous argument for $S(\mathcal{X})$).

Central limit theorems are available also, especially for $\mathcal{K}(\mathbb{R}^m)$ where the special properties of $\mathcal{X} = \mathbb{R}^m$ may be used to simplify verification of involved hypotheses. Gine, Hahn and Zinn (1983) and Weil (1982) obtained the standard i.i.d. form given by Theorem 3.6.

THEOREM 3.6. Let $X, X_1, X_2, ...$ be i.i.d. random sets in $\mathcal{K}(\mathbb{R}^m)$. If $E||X||^2 < \infty$, then there exists a Gaussian process Z on $C_b(\text{ball } \mathcal{X}^*)$ with the covariance of coX such that

$$\sqrt{n}d\left(\frac{X_1 + \cdots + X_n}{n}, E(coX)\right) \to ||Z|| \text{ in distribution.}$$

With similar conditions to Theorem 3.6, a law of the iterated logarithm is also available in Gine, Hahn and Zinn (1983) along with more general central limit theorems. Also, Molchanov, Omey and Kozarovitzky (1995) obtained a renewal theorem for i.i.d. random sets in $\mathcal{C}(\mathbb{R}^m)$.

4. Arrays of random fuzzy sets. There are many recent results for laws of large of number for random fuzzy variables and random fuzzy sets, see for example, Stein and Talati (1981), Kruse (1982), Miyakoshi and Shimbo (1984), Puri and Ralescu (1986), Klement, Puri and Ralescu (1986), Inoue (1990), Inoue (1991), and Inoue and Taylor (1995). A strong law of large numbers for random fuzzy sets will be presented in this section. The framework for this result will have three important differences from the previous results which have been cited in this paper. First, the random sets are allowed to also be fuzzy. Secondly, the more general structure is used rather than the sequential structure. Thirdly, the assumption of independence is replaced by exchangeability.

Recall in Section 2 (cf: (2.5), (2.6) and (2.7)) that a fuzzy set in $\mathcal{X}$ is characterized as a function $u : \mathcal{X} \to [0,1]$ such that

$$\{x \in \mathcal{X} : \quad u(x) \geq a\} \quad \text{is compact for each } a > 0$$

and

$$\{x \in \mathcal{X} : \quad u(x) = 1\} \neq \emptyset$$

and that $\mathcal{F}(\mathcal{X})$ denotes the set of all such fuzzy sets (all such functions $u : \mathcal{X} \to [0,1]$). A linear structure in $\mathcal{F}(\mathcal{X})$ can be defined by the following operations:

$$(u+v)(x) = \sup_{y+z=x} \min[(u(y), v(z)] \tag{4.1}$$

$$(\lambda u)(x) = \begin{cases} u(\lambda^{-1}x) & \text{if } \lambda \neq 0 \\ I_{\{0\}}(x) & \text{if } \lambda = 0 \end{cases} \tag{4.2}$$

where $u, v \in \mathcal{F}(\mathcal{X})$ and $\lambda \in R$.

For each fuzzy set $u \in \mathcal{F}(\mathcal{X})$, its *a-level set* is defined by $La(u) = \{x \in \mathcal{X} : \ u(x) \geq a\}$, $a \in (0,1]$. A metric d_r (with $1 \leq r \leq 2$) can be defined on $\mathcal{F}(\mathcal{X})$ by

$$d_r(u,v) = \left[\int_0^1 d_H^r(La(u), La(v)) da \right]^{1/r}. \tag{4.3}$$

Klement, Puri and Ralescu (1986) showed that $\mathcal{F}(\mathcal{X})$ with the metric d_r is complete and separable.

Let $(\Omega, \mathcal{A}, \mathcal{P})$ be a probability space and let $X : \Omega \to \mathcal{F}(\mathcal{X})$ such that for each $a \in (0,1]$ and each $\omega \in \Omega$ $Xa(\omega) = \{x \in \mathcal{X}\text{: } X(\omega)(x) \geq a\} \in \mathcal{K}(\mathcal{X})$. A *random fuzzy set* has the property that $Xa(\omega)$ (as a function of ω) is a random set in $\mathcal{K}(\mathcal{X})$ for each $a \in (0,1]$. Similarly, coX can be defined as a function $\Omega \to \mathcal{F}(\mathcal{X})$ such that $LacoX = co\{x \in \mathcal{X} : \ X(\omega)(x) \geq a\}$ for each a, and Klement, Puri and Ralescu (1986) showed that $La(EcoX) = ELa(coX)$ for each $a \in (0,1]$. Finally, when $X_n, n \geq 1$, and X are random fuzzy sets such that

$$\sup_{a>0} d_H\Big(La(X_n), La(X)\Big) \le Y \qquad P \text{ a.s.} \tag{4.4}$$

and $E|Y|^r < \infty$, then convergence of the random sets $La(X_n)$ to LaX for each $a \in (0,1]$, yields $d_r(X_n, X) \to 0$ by the dominated convergence theorem. Consequently, laws of large numbers for random fuzzy sets follows from LLN's for random sets under appropriate integrability conditions. Hence, many LLN's are easily available for random fuzzy sets. This section will close with some results on exchangeable random fuzzy sets.

A sequence of random fuzzy sets $\{X_n\}$ is said to be *exchangeable* if for each $n \ge 1$ and each permutation π of $\{1, 2, ..., n\}$

$$P[X_1a \in B_1, ..., X_na \in B_n] = P[X_{\pi 1}a \in B_1, ..., X_{\pi n}a \in B_n] \tag{4.5}$$

for each $a > 0$ and for all Borel subsets $B_1, ..., B_n$ from $\mathcal{K}(\mathcal{X})$. Clearly, exchangeability implies identical distributions. Moreover, i.i.d. fuzzy random sets are exchangeable, but the converse need not be true.

An infinite sequence $\{X_n\}$ is exchangeable if and only if it is a mixture of sequences of i.i.d. random fuzzy sets (or equivalently, $\{X_n\}$ is conditional i.i.d. with respect to some σ-field). A useful calculational version of this characterization is available in the following extension of the results in Section 4.2 of Taylor, Daffer and Patterson (1985). Let M be collection of probability measures on the Borel subsets of a complete, separable metric space and let M be equipped with the σ-field generated by the topology of weak convergence of the probability measures. If $\{X_n\}$ is an infinite sequence of exchangeable random elements, then there exists a probability measure μ on the Borel subsets of M such that (imprecisely stated)

$$P(B) = \int_M P_V(B) d\mu(P_V) \tag{4.6}$$

for any $B \in \sigma\{X_n : n \ge 1\}$ and $P_V(B)$ is calculated as if the random elements $\{X_n\}$ are i.i.d. with respect to P_V. The following result follows from Theorem 6.2.5 of Taylor, Daffer and Patterson (1986).

THEOREM 4.1. Let $\{X_{ni}\}$ be random fuzzy sets in $\mathcal{F}(\mathcal{X})$ such that for some compact subset D of $\mathcal{K}(\mathcal{X})$

$$P[X_{ni}a \in \mathrm{D}] = 1 \ \text{ for all } n \text{ and } i \text{ and for all } a > 0. \tag{4.7}$$

Let $\{X_{ni} : i \ge 1\}$ be exchangeable for each n and let $\{a_{ni}\}$ be nonnegative constants such that $\sum_{i=1}^{\infty} a_{ni} \le 1$ for all i and $\sum_{n=1}^{\infty} \exp(-\alpha/A_n) < \infty$ for each $\alpha > 0$ where $A_n = \sum_{i=1}^{\infty} a_{ni}^2$. If for each $n > 0$

$$\sum_{n=1}^{\infty} \mu_n(\{P_V : d(E_V(coX_{n1}a, E(coX_{n1}a)) > \eta\}) < \infty \tag{4.8}$$

for each $a > 0$, then

$$d_r\left(\sum_{i=1}^{\infty} a_{ni}X_{ni}, \sum_{i=1}^{\infty} a_{ni}E(coX_{n1})\right) \to 0 \text{ completely}$$

for each r, $1 \le r \le 2$.

If random fuzzy sets in $\mathbb{R}^m = \mathcal{X}$ are being considered, and it is reasonable to believe that they are contained in a bounded region, then Lemma 3.1 provides for the existence of the set D for Condition (4.7). Taylor and Hu (1987) have results which relate the conditional means in (4.8) precisely to the laws of large numbers for exchangeable random variables. For a sequence $\{X_n\}$ of exchangeable random variables, they showed that

$$\frac{1}{n}\sum_{i=1}^{n} X_i \to c \text{ w. prob. one} \tag{4.9}$$

if and only if

$$\mu(\{P_v : E_v X_1 = c\}) = 1 \tag{4.10}$$

where $E_v|X_1| < \infty$ for almost all P_v. A genetic example will better illustrate Condition (4.8).

Let $\{Y_i\}$ be a sequence of i.i.d. random variables and let $\{Z_n\}$ be a sequence of random variables such that $\{Y_i\}$ and $\{Z_n\}$ are independent of each other. Then $X_{ni} = Y_i + Z_n$, for $i \ge 1$ and $n \ge 1$ defines an array of random variables which are row-wise exchangeable (for example, condition on Z_n for each row). Moreover, $E_V X_{ni} = EY_1 + Z_n$. For

$$\frac{1}{n}\sum_{i=1}^{n} X_{ni} = \frac{1}{n}\sum_{i=1}^{n} Y_i + Z_n \text{ to converge to } EY_1$$

it is necessary for Z_n to converge to 0. Condition (4.8) requires this convergence to be fast enough to achieve complete convergence. Specifically for this example,

$$\sum_{n=1}^{\infty} P\Big[|Z_n| > \eta\Big] < \infty$$

for each η would yield

$$\sum_{n=1}^{\infty} P\left[\left|\frac{1}{n}\sum_{i=1}^{n} X_{ni} - EY_1\right| > \varepsilon\right] < \infty$$

for each $\varepsilon > 0$.

REFERENCES

[1] A. Adler, A. Rosalsky and R.L. Taylor, *A weak law for normed weighted sums of random elements in Rademacher type p Banach spaces*, J. Mult. Anal, 37 (1991), 259–268.

[2] Z. Artstein and R. Vitale, *A strong law of large numbers for random compact sets*, Ann. Probab., 3 (1975), 879–882.

[3] Z. Artstein and S. Hart, *Law of large numbers for random sets and allocation processes*, Mathematics of Operations Research, 6 (1981), 485–492.

[4] Z. Artstein and J.C. Hansen, *Convexification in limit laws of random sets in Banach spaces*, Ann. Probab., 13 (1985), 307–309.

[5] N. Cressie, *A strong limit theorem for random sets*, Suppl. Adv. Appl. Probab., 10 (1978), 36–46.

[6] N. Cressie, *Statistics for spatial data*, Wiley, New York (1991).

[7] P.Z. Daffer and R.L. Taylor, *Tightness and strong laws of large numbers in Banach spaces*, Bull. of Inst. Math., Academia Sinica, 10 (1982), 251–263.

[8] G. Debreu, *Integration of correspondences*, Proc. Fifth Berkeley Symp. Math. Statist. Prob., 2 (1966), Univ. of California Press, 351–372.

[9] E. Gine, M.G. Hahn and J. Zinn, *Limit theorems for random sets: An application of probability in Banach space results*, In Probability in Banach Spaces IV (A. Beck and K. Jacobs, eds.), Lecture Notes in Mathematics, 990, Springer, New York (1983), 112–135.

[10] C. Hess, *Theoreme ergodique et loi forte des grands nombres pour des ensembles aleatoires*, Comptes Redus de l'Academie des Sciences, 288 (1979), 519–522.

[11] C. Hess, *Loi forte des grands nombres pour des ensembles aleatoires non bornes a valeurs dans un espace de Banach separable*, Comptes Rendus de l'Academie des Sciences, Paris, Serie I (1985), 177–180.

[12] J. Hoffmann–Jørgensen and G. Pisier, *The law of large numbers and the central limit theorem in Banach spaces*, Ann. Probab., 4 (1976), 587–599.

[13] L. Hörmander, *Sur la fonction d'appui des ensembles convexes dans un espace localement convexe*, Ark. Mat., 3 (1954), 181–186.

[14] H. Inoue, *A limit law for row-wise exchangeable fuzzy random variables*, Sino-Japan Joint Meeting on Fuzzy Sets and Systems (1990).

[15] H. Inoue, *A strong law of large numbers for fuzzy random sets*, Fuzzy Sets and Systems, 41 (1991), 285–291.

[16] H. Inoue and R.L. Taylor, *A SLLN for arrays of row-wise exchangeable fuzzy random sets*, Stoch. Anal. and Appl., 13 (1995), 461–470.

[17] D.G. Kendall, *Foundations of a theory of random sets*, In Stochastic Geometry (ed. E.F. Harding and D.G. Kendall), Wiley (1974), 322–376.

[18] E.P. Klement, M.L. Puri and D. Ralescu, *Limit theorems for fuzzy random variables*, Proc. R. Soc. Lond., A407 (1986), 171–182.

[19] R. Kruse, *The strong law of large numbers for fuzzy random variables*, Info. Sci., 21 (1982), 233–241.

[20] G. Matheron, *Random Sets and Integral Geometry*, Wiley (1975).

[21] M. Miyakoshi and M. Shimbo, *A strong law of large numbers for fuzzy random variables*, Fuzzy Sets and Syst., 12 (1984), 133–142.

[22] I.S. Molchanov, E. Omey and E. Kozarovitzky, *An elementary renewal theorem for random compact convex sets*, Adv. Appl. Probab., 27 (1995), 931–942.

[23] E. Mourier, *Eléments aleatories dan un espace de Banach*, Ann. Inst. Henri Poincare, 13 (1953), 159–244.

[24] M.L. Puri and D.A. Ralescu, *A strong law of large numbers for Banach space-valued random sets*, Ann. Probab., 11 (1983), 222–224.

[25] M.L. Puri and D.A. Ralescu, *Limit theorems for random compact sets in Banach spaces*, Math. Proc. Cambridge Phil. Soc., 97 (1985), 151–158.

[26] M.L. Puri and D.A. Ralescu, *Fuzzy random variables*, J. Math. Anal. Apl., 114 (1986), 409–422.

[27] H. RÅDSTRÖM, *An embedding theorem for spaces of convex sets*, Proc. Amer. Math. Soc., 3 (1952), 165–169.

[28] H.E. ROBBINS, *On the measure of a random set*, Ann. Math. Statist., 14 (1944), 70–74.

[29] H.E. ROBBINS, *On the measure of a random set II*, Ann. Math. Statist., 15 (1945), 342–347.

[30] W.E. STEIN AND K. TALATI, *Convex fuzzy random variables*, Fuzzy Sets and Syst., 6 (1981), 271–283.

[31] R.L. TAYLOR, *Stochastic Convergence of Weighted Sums of Random Elements in Linear Spaces*, Lecture Notes in Mathematics, V672 (1978), Springer–Verlag, Berlin–New York.

[32] R.L. TAYLOR, P.Z. DAFFER AND R.F. PATTERSON, *Limit theorems for sums of exchangeable variables*, Rowman & Allanheld, Totowa N. J. (1985).

[33] R.L. TAYLOR AND H. INOUE, *A strong law of large numbers for random sets in Banach spaces*, Bull. Inst. Math., Academia Sinica, 13 (1985a), 403–409.

[34] R.L. TAYLOR AND H. INOUE, *Convergence of weighted sums of random sets*, Stoch. Anal. and Appl., 3 (1985b), 379–396.

[35] R.L. TAYLOR AND T.-C. HU, *On laws of large numbers for exchangeable random variables,* Stoch. Anal. and Appl., 5 (1987), 323–334.

[36] W. WEIL, *An application of the central limit theorem for Banach space-valued random variables to the theory of random sets*, Z. Wharsch. v. Geb., 60 (1982), 203–208.

GEOMETRIC STRUCTURE OF LOWER PROBABILITIES

PAUL K. BLACK*

Abstract. Lower probability theory has gained interest in recent years as a basis for representing uncertainty. Much of the work in this area has focused on lower probabilities termed belief functions. Belief functions correspond precisely to random set representations of uncertainty, although more general classes of lower probabilities are available, such as lower envelopes and 2–monotone capacities. Characterization through definitions of monotone capacities provides a mechanism by which different classes of lower probabilities can be distinguished. Such characterizations have previously been performed through mathematical representations that offer little intuition about differences between lower probability classes. An alternative characterization is offered that uses geometric structures to provide a direct visual representation of the distinctions between different classes of lower probabilities, or monotone capacities. Results are presented in terms of the shape of lower probabilities that provide a characterization of monotone capacities that, for example, distinguishes belief functions from other classes of lower probabilities. Results are also presented in terms of the size, relative size, and location-scale transformation of lower probabilities that provide further insights into their general structure.

Key words. Barycentric Coordinate System, Belief Functions, Coherent Lower Probabilities, Lower Envelopes, Möbius Transformations, Monotone Capacities, 2–Monotone Lower Probabilities, Undominated Lower Probabilities.

AMS(MOS) subject classifications. 04A72, 05A99, 20M99, 60D05, 60A05

1. Introduction. Much of the recent work on issues that relate to the theory of lower probability (e.g., Smith [17], Dempster [8], Shafer [15], Levi [11], Wolfenson and Fine [24], Kyburg [10], Seidenfeld, Schervish, and Kadane [13], Wasserman and Kadane [23], Walley [21], Black [1,3]) has foundations in Choquet's 1954 [7] treatise on capacities. Lower probability measures correspond to simple monotone capacities, which can be defined from more basic properties of lower probabilities: Let F be a finite algebra for a finite space Θ. A lower probability P_* is a set function on (Θ, F) that, together with a dual set function (the upper probability function P^*), satisfies the following basic properties:

(i) Duality of the lower and upper probability functions (A^c is used to represent the complement of A with respect to Θ, etc.):

$$P_*(A) = 1 - P^*(A^c) \qquad \forall A \subseteq \Theta\,.$$

(ii) Non-negativity:

$$P_*(A) \geq 0 \qquad \forall A \subseteq \Theta\,.$$

(iii) Extreme Values:

$$P_*(\emptyset) = 0 \quad \text{and} \quad P_*(\Theta) = 1\,.$$

* Neptune and Company, Inc., 1505 15th Street, Los Alamos, New Mexico 87544.

(iv.a) Super-additivity of P_* (for disjoint events):

$$P_*(A) + P_*(B) \leq P_*(A \cup B) \qquad \forall A,\, B \subseteq \Theta \quad \text{s.t.} \quad A \cap B = \emptyset.$$

(iv.b) Sub-additivity of P^* (for disjoint events):

$$P^*(A) + P^*(B) \geq P^*(A \cup B) \qquad \forall A,\, B \subseteq \Theta \quad \text{s.t.} \quad A \cap B = \emptyset.$$

(v) Basic Monotone Order:

$$\text{if } A \subset B \text{ then } P_*(A) \leq P_*(B)\,.$$

Basic monotone order is given as a property, although it is a consequence of the other basic properties. It should also be recognized that the upper probability function dominates the lower probability function (i.e., $P^*(A) \geq P_*(A) \quad \forall A \subseteq \Theta$). Further constraints can be imposed on lower probability measures that permit distinctions among subclasses. Some of the more important subclasses include: Undominated lower probabilities (Papamarcou and Fine [12]); lower envelopes or coherent lower probabilities (Walley and Fine [22]); monotone capacities of order at least 2; and, monotone capacities of order infinity or belief functions (Dempster [8], Shafer [15]). Further discussion of lower probability subclasses is provided in Walley [21, 22] and Black [1].

Lower probability functions yield a decomposition via Möbius inversion (e.g., Shafer [12], Chateauneuf and Jaffray [6], Jaffray [9]) that provides a representation in terms of values attached to each subset of the underlying space. There is a direct correspondence between a lower probability function and its decomposition. Shafer [15] demonstrated that the decomposition is non-negative everywhere for belief functions. By comparison with Choquet [7], belief functions are monotone capacities of order infinity. It is this class of lower probabilities that corresponds to random sets. Other classes, which contain the class of belief functions, include lower probability functions that do not yield a non-negative decomposition for all subsets. This suggests that lower probability theory requires a more broad calculus than that offered by any theory of belief functions or random sets. Different, but overlapping, classes of lower probability can be characterized through monotone order properties and a number of other mathematical mechanisms [21, 22]. However, these formalisms do not appear to offer intuitions or insights into the differences between classes. The purpose of this paper is to provide an alternative characterization of lower probability through the use of geometric structures. This approach yields a more appealing characterization of differences between classes of lower probabilities, particularly for small dimensional problems that can be visualized directly.

2. Monotone capacities and lower probabilities. There is a direct link between lower probability functions and monotone capacities. Of particular importance for random set theory are monotone capacities of infinite order, often termed belief functions. A mathematical form of this property is given by Inequality 2.1. Consider a space $\Theta = \{A_1, \ldots, A_n\}$. Then a capacity, or lower probability, defined over Θ is of infinite order if the following inequality holds for all collections of subsets of Θ (where $|I|$ represents the cardinality of a set I):

$$P_* \left(A_1 \cup \cdots \cup A_k\right) \geq \sum_{I \subseteq \{1,\ldots,k\}} (-1)^{|I|+1} P_* \left(\bigcap_{i \in I} A_i \right). \tag{2.1}$$

By convention, the order of a belief function is termed infinite. However, it should be noted that infinite order corresponds to order n for a finite space Θ of cardinality n. Given the definition above, terms such as "monotone capacity of maximal order," or "complete monotone capacity," may be more descriptive. If there exists at least one collection of subsets of Θ for which Inequality 2.1 is not satisfied, then the capacity, or lower probability, is termed of finite order. In general, a finite order capacity may be termed of order $k < n$ if the smallest collection of subsets for which Inequality 2.1 is not satisfied is of cardinality $k + 1$.

Several interesting subclasses of lower probability have been identified. Dempster [8] identified three different classes of convex sets of probability distributions: The class of all convex sets of probability distributions; the class of lower probability functions known as lower envelopes or coherent lower probabilities; and, the class of belief functions. The second class is more easily characterized as one of a closed convex set of probability measures that are formed solely from inequalities on probabilities of events. The former class consists of more general convex sets of probability distributions from which the second class can be obtained through infimum and supremum operations on the general convex set. It has been suggested that lower envelopes, which can be provided a definition in terms of coherence (Papamarcou and Fine [12], Walley [21, 22]), correspond to the minimum requirement for lower probabilities. However, Papamarcou and Fine [12] have indicated that there may be a role for undominated lower probabilities. Given the above discussion, the following hierarchy of classes of lower probability measures is suggested:

C_L Lower probabilities;
C_E Lower envelopes or coherent lower probabilities;
C_2 Monotone capacities of order at least 2; and,
C_B Belief functions, or infinite order monotone capacities.

Note that the lower ordered subclasses are contained in the listed predecessors. Many other subclasses could be defined, but these subclasses capture important differences that lend themselves to geometric examination. For example, belief functions correspond to random sets, the class C_2 includes a minimal monotone order requirement in the sense of Inequality 2.1, coherent lower probabilities are fully dominated, and the class C_L includes lower probabilities that are not dominated by any probability measure, as well as partially dominated lower probabilities.

As indicated in the introduction, a decomposition exists for lower probabilities. This decomposition can be framed in terms of a linear combination of lower probability values as follows (consider $\Theta = \{A_1, \ldots, A_n\}$ as before, and any subsets A, B of Θ):

$$m(A) = \sum_{B \subseteq A} (-1)^{|A \setminus B|} P_*(B) . \tag{2.2}$$

In belief function terminology the function m is often termed the basic probability assignment (Shafer [15]). This term may be reasonable for belief functions because the decomposition is non-negative everywhere for this subclass of lower probabilities. This explains the direct correspondence between belief functions and random sets. The term m–function is used herein to better accommodate decomposition of other lower probabilities. Equation 2.2 is invertible, so that lower probabilities can be obtained directly from a specified m–function.

$$P_*(A) = \sum_{B \subseteq A} m(B) . \tag{2.3}$$

Other useful functions include the upper probability function and the commonality function:

$$P^*(A) = \sum_{B \cap A \neq \emptyset} m(B) , \tag{2.4}$$

$$Q(A) = \sum_{A \subseteq B} m(B) . \tag{2.5}$$

There is a large computational burden involved with using Equations 2.2 through 2.5 and related transformations. This burden can be relieved to some extent by taking advantage of matrix factorizations that involve 2×2 base matrices with entries of 0, 1, and −1 only and Krönecker products (Thoma [19], Black and Martin [5], Smets [16], and Black [1]). Further relief can be obtained through approximations to belief functions (Black [1], Black and Eddy [4], Black and Martin [5]). The formulas presented thus far indicate how classes of lower probabilities can be differentiated. However, they offer no intuition beyond the mathematics. A geometric approach is now taken that provides an alternative characterization of lower probabilities that yields insights into the structures of the different classes identified. The main focus is the distinction between infinite and finite monotone capacities. These are labeled respectively C_F (for finite monotonicity) and C_B (for infinite monotonicity or belief functions). From the class definitions offered earlier it is clear that $C_F + C_B = C_L$, which may also be labeled C_{F+B}.

In the remaining discussion the terms Bel and Pl are used to represent respectively the lower and upper probability functions for all monotone capacities. This simplification is used both for ease of recognition of lower and upper probability functions and to avoid complications in notation relating to distinctions between finite and infinite order monotone capacities.

3. Shape of lower probabilities. A set of probability distributions for a given finite space $\Theta = \{A_1, \ldots, A_n\}$ of cardinality n occupies a subspace of the full space of possible probability distributions for Θ. The full probability space can be given a geometric interpretation in terms of regular simplexes and the barycentric coordinate system. The full space is characterized by a regular simplex of order n with vertices that correspond to a probability of one for each of the n possible outcomes, and by sub-simplexes that connect the vertices. The vertices of the full simplex are connected by n sub-simplexes of order $n-1$ that correspond to the n cardinal $n-1$ subsets of Θ. More generally, the number of sub-simplexes of order k, that connect the set of vertices of Θ, is defined by the binomial expansion.

Probability distributions are represented by a point in the simplex for which the values assigned to each element of Θ sum to one; i.e., $\{Pr(A_1), \ldots, Pr(A_n)\}$. For example, equal probabilities for each event correspond to the center of gravity, or the barycenter, of the coordinate system: $\{1/n, \ldots, 1/n\}$. Any subspace of the full probability space represents a subset of probability distributions. The focus here is on sets that can be formed solely by lower probabilities on subsets.

Although it is not possible to provide geometric illustrations of lower probabilities that are undominated, it is possible to provide illustrations of belief functions and lower probability functions in C_B and C_F. Even in these cases, however, it is difficult to picture the convex sets for frames of cardinality greater than three. Figures 1a and 1b present two examples: The first represents a convex set in C_B; and, the second a convex set in C_F.

The examples presented in Figures 1a and 1b were contrived to demonstrate that convex sets that are infinitesimally close may have different monotone capacity order properties. In particular, the first is a monotone capacity of order infinity (i.e., a belief function), whereas the second is not (it is a monotone capacity of order two). It appears that, in a problem of cardinality three at least, a convex set of probability distributions can be given a belief function representation if the shape of the convex set is like the shape of the encompassing simplex.

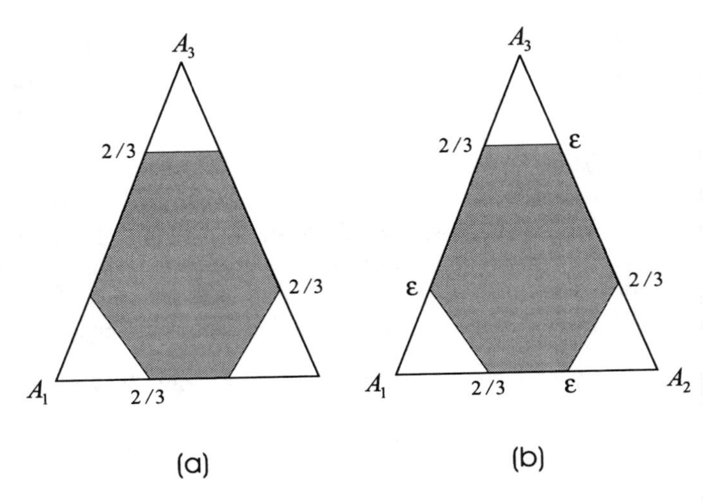

FIG. 3.1. *(a): Example of a convex set in* C_B. *(b): Example of a convex set in* C_F.

To make further progress with the geometry of lower probabilities it will prove convenient to introduce some notation and terminology. A reduction of the frame is required that admits m–function values without requiring normalization. Because many convenient terms have been taken in probability and belief function theory, including reduction, this form of reduction will be referred to as a *contraction* of the frame. A contraction is not unlike a conditioning of the frame (that is, some subset of the frame is deemed impossible), but there is no requirement that a proper m–function results, and normalization is not performed. The contraction operation affects the m–function as follows:

DEFINITION 3.1 (CONTRACTION OF BASIC PROBABILITY ASSIGNMENT). Suppose an m–function m_Θ is defined over $\Theta = \{A_1, \ldots, A_n\}$. Consider a subset B of Θ that consists of a collection of elements A_i. A *contraction* with respect to B of m_Θ, denoted $m^c_{\Theta \setminus B}$, is defined as follows:

$$m^c_{\Theta \setminus B}(A) = m_\Theta(A), \qquad \forall A \subseteq B . \tag{3.1}$$

The effect of performing a contraction is that the m–function assignment for all subsets that include B are removed from consideration, and the remaining subsets maintain their initial values. Consequently, the frame or space has been contracted, although a proper belief function does not, in general, result from this operation. Equation 3.1 provides the foundation for defining contractions of the belief, plausibility, and commonality functions. The forms are provided below in terms of m_Θ, but they also follow from the usual group of belief function transformations applied to $m^c_{\Theta \setminus B}$. Both the m–function and the belief function take the same values in m_Θ

and $m^c_{\Theta \setminus B}$ on subsets that are admitted by the latter function. There are differences, however, in the plausibility and commonality functions that are of interest:

$$Pl^c_{\Theta \setminus B}(A) = \sum_{\substack{A \cap D \neq \emptyset \\ D \subseteq \Theta \setminus B}} m_\Theta(D) , \tag{3.2}$$

$$Q^c_{\Theta \setminus B}(A) = \sum_{\substack{A \subseteq D \\ D \subseteq \Theta \setminus B}} m_\Theta(D) . \tag{3.3}$$

Note that the contracted functions are not restricted to any subclass of lower probability measures; they are applicable to the full lower probability class C_L. A few other results follow from the definitions of the contracted functions that are useful in the developments that follow:

RESULT 3.1. Let $\Theta = \{A_1, \ldots, A_n\}$, and consider a subset $B = \{A_1, \ldots, A_k\}$, where $k < n$, then:

$$Bel\,(A_1, \ldots, A_k) = \sum_{i=1}^{k} Pl^c_{A_1,\ldots,A_k}(A_i) . \tag{3.4}$$

PROOF. By expansion of both sides in terms of m–function decomposition. □

Note that Result 3.1 can be applied to any subset B of Θ by simple reordering of the elements of Θ. The next result provides a relationship between the contracted plausibility and commonality functions.

RESULT 3.2. Let $\Theta = \{A_1, \ldots, A_n\}$, and $B \subset \Theta$ such that $A_j, A_k \notin B$, then:

$$Pl^c_{\Theta \setminus B}(A_j) - Pl^c_{\Theta \setminus B \setminus A_k}(A_j) = Q^c_{\Theta \setminus B}(A_j, A_k) . \tag{3.5}$$

PROOF. By expansion of the left hand side in terms of m-function decomposition and comparison with decomposition of right hand side. □

In general, the convex set of probability distributions formed by lower and upper probability bounds on all subsets of Θ, when embedded in a simplex as described earlier, contains $n!$ vertices. These vertices correspond to the extreme probability distributions in the convex set. The extreme probability distributions can be derived from the m–function and the lower probability function specifications by considering the limiting probabilities on the n events taken in each of their $n!$ possible orderings.

For example, taking the events in their lexicographic order, the lowest probability that event A_1 can attain is $m(A_1)$, and, given that probability assignment, the lowest probability that event A_2 can attain is $m(A_2) + m(A_1, A_2)$ because $Bel\,(A_1, A_2) = m(A_1) + m(A_2) + m(A_1, A_2)$, etc.

The following result provides a general formula for determining the probability distributions at the vertices.

RESULT 3.3. Let $\Theta = \{A_1, \ldots, A_n\}$. Consider a set of probability distributions on C_{B+F}. The extreme probability distributions over Θ are given by the following formulation for each of the $n!$ possible orderings $\{(1), (2), \ldots, (n)\}$ of Θ:

$$\{Pl^c_{A_{(1)}}(A_{(1)}), \ldots, Pl^c_{\Theta \setminus A_{(n)}}(A_{(n-1)}), Pl_\Theta(A_{(n)})\}\ . \tag{3.6}$$

PROOF. Without loss of generality, assume the lexicographic ordering of Θ. Denote $m(A_i)$ with $a_i, m(A_i, A_i)$ with $a_{i,j}$, etc.

1. The lowest probability of A_1 is $a_1 = Pl^c_{A_1}(A_1)$.
2. Given a_1, the lowest possible probability of A_2 is $a_2 + a_{1,2} = Pl^c_{A_1,A_2}(A_2)$.
3. Suppose the lowest possible probability of A_k $(k < n)$ is $Pl^c_{A_1,\ldots,A_k}(A_k)$, then the lowest possible probability of A_{k+1} is

$$Bel\ (A_{k+1}) - \sum_{i=1}^{k} Pl^c_{A_1,\ldots,A_i}(A_i) = Pl^c_{A_1,\ldots,A_k,A_{k+1}}(A_{k+1}).$$

4. And, the lowest possible probability of A_n is $Pl^c_\Theta(A_n) = Pl_\Theta(A_n)$.

□

Distances between the extreme probability coordinates can be specified in terms of the n possible directions associated with the probability space. For example, in the direction corresponding to A_1, there is a path from each of the $(n-1)!$ vertices that are associated with the upper probability of A_1 to the corresponding opposite base vertices that are associated with the lower probability of A_1. Each of these paths connects n vertices, in which case each path consists of $(n-1)$ edges. (It should be noted that this is the general case. Any of these edges may have zero length depending on the specific convex set. For example, the vacuous belief function which corresponds to the full probability space has n vertices (as opposed to $n!$), and only one edge that connects a vertex to its opposite base.) There are n events in total, and each edge appears in the system in two directions, in which case there are $n!\,(n-1)/2$ edges in the convex system. Other paths are possible, but they are not unidirectional in the sense indicated, and hence connect more vertices. The following result defines the lengths of the edges in each direction:

RESULT 3.4. Given the conditions of Result 3.3, the lengths of the edges that connect the vertices of Θ are given by $Q^c_{\Theta\setminus B}(A_j, A_k)$, for all $B \subset \Theta$ and $|B| < n-1$, and for all distinct pairs A_j, A_k such that $A_j, A_k \notin B$.

PROOF. Without loss of generality, assume lexicographic order of Θ: Each edge has common probability on all but 2 adjoining events, i.e., their coordinate systems are defined by the points $\{1, \ldots, j, k, \ldots, n\}$ and $\{1, \ldots, k, j, \ldots, n\}$ for all $\{j, k\}$ pairs $\{j, k\} \in \{\{1, 2\}, \ldots, \{n-1, n\}\}$.

The length between two directly connected vertices is the probability difference on the 2 events. These probability differences are (using Result 3.3) $Pl^c_{\Theta\setminus B}(A_j) - Pl^c_{\Theta\setminus B\setminus A_k}(A_j)$, where $B = \{A_1, \ldots, A_{j-1}\}$ for $j = 1$ to $n-1$.

From Result 3.2, this is the same as $Pl^c_{\Theta\setminus B}(A_k) - Pl^c_{\Theta\setminus B\setminus A_j}(A_k)$, where $B = \{A_1, \ldots, A_{k-1}\}$ for $k = 1$ to $n-1$, which, from Result 3.3, is equal to $Q^c_{\Theta\setminus B}(A_j, A_k)$. □

The lengths of the edges can be characterized by particular contracted commonality functions. It was indicated earlier that there are $n!\,(n-1)/2$ edges in the convex system. Another way to realize this same result is that each of the $n!$ vertices is connected directly to $(n-1)$ other vertices; the connection is bi-directional in which case the total number of edges is, as stated, $n!\,(n-1)/2$.

The next result shows that there are the same number of edges in each subscript category of Q. Some notation is needed first: Let $N(i)$ represent the number of edges characterized by $Q^c_{\Theta\setminus B}(A_j, A_k)$, where B is of cardinality $i \in \{0, n-2\}$.

RESULT 3.5. Given the conditions of Results 3.3 and 3.4, the number of edges in the convex system is $n!\,(n-1)/2$, and $N(0) = N(1) = \cdots = N(n-2)$, in which case $N(i) = n!/2$ for each $i \in \{0, n-2\}$.

PROOF. The result is a consequence of the lengths of edges in terms of the contracted commonality function, and symmetry in the arguments. □

The convex system contains edges that have the same lengths. The reduction is realized because, for example, several vertices associated with the upper probability of an event A_j connect with the vertices that correspond to the upper probability of a different event A_k. Let $L(i)$ represent the number of different lengths characterized by $Q^c_{\Theta\setminus B}(A_j, A_k)$, where B is of cardinality i, $i \in \{0, n-2\}$.

RESULT 3.6. Given the conditions of Results 3.3 and 3.4, the number of different lengths in the convex system is given by (for $i \in \{0, n-2\}$):

$$L(i) = \binom{n-i}{2}\binom{n}{n-i}. \tag{3.7}$$

PROOF. The number of different lengths is given by the number of possible values taken by $Q^c_{\Theta \setminus B}$. Recall that B is of cardinality i. The result is now a matter of combinatorics; the number of possible ways of choosing (A_j, A_k) from $\Theta \setminus B$, and the number of ways of choosing B of cardinality i from Θ. □

Result 3.7 provides the relationship between the number of edges and the number of different lengths for each cardinality of $B, i \in \{0, n-2\}$.

RESULT 3.7. Given the conditions of Results 3.5 and 3.6, the relationship between the number of lengths and the number of edges in the convex system is given by (for $i \in \{0, n-2\}$):

$$\frac{N(i)}{L(i)} = i!\,(n-i-2)!\,. \tag{3.8}$$

PROOF. By expansion of both terms. □

The results presented so far will prove useful in the development of the relationship between the geometry of a monotone capacity and its monotone order. The results will be presented in terms of the summation of the lengths of edges corresponding to specific cardinalities of subsets of the underlying frame. Consequently, it helps first to define a function S, similarly to N and L, to be the sum of the lengths associated with subsets B of cardinality i. Let

$$S(i) = \sum_{j,k} Q^c_{\Theta \setminus B}(A_j, A_k)\,, \tag{3.9}$$

where $B \subset \Theta$ is of cardinality i, $i \in \{0, n-2\}$, $A_j, A_k \notin B$. The summations $S(i)$ correspond to the distinct lengths for each contraction level, and not to all edges in the system.

The results and formulas presented so far deal with the geometry of the full monotone capacity specified over some frame $\Theta = \{A_1, \ldots, A_n\}$. The results can also be applied to substructures of Θ. It is helpful once again to define some intermediary functions, this time in terms of subsets of Θ. First, consider $\Theta = \{A_1, \ldots, A_n\}$, and $K \subset \Theta$, where K is of cardinality $k < n$; consider the substructure corresponding to the subset K of Θ, and consider the number of edges that connect elements in K, the number of lengths in K, and the sum of the lengths of the edges in K. The following formulas, which are analogous to the formulas presented in Results 3.5 through 3.7, apply to each such substructure (where $B \subset K \subset \Theta$ is of cardinality i, $i \in \{0, k-2\}, A_j, A_k \notin B$):

$$N_K(i) = \frac{k!}{2}\,, \tag{i}$$

$$L_K(i) = \binom{k-i}{2}\binom{k}{k-i}\,, \tag{ii}$$

(*iii*)
$$S_K(i) = \sum_{j,k} Q^c_{K \setminus B}(A_j, A_k) .$$

Proof of these formulations follows trivially by substitution of k for n in Results 3.5 through 3.7. The following example is offered for clarification of the underlying geometry of the substructures. Consider a frame $\Theta = \{A_1, \ldots, A_4\}$. Following the geometric presentation, Θ has 36 edges, or 24 distinct lengths. Without loss of generality, consider a subset K of Θ of cardinality three. The corresponding substructure has six edges, or six distinct lengths. There are four possible substructures of Θ with cardinality three. Each of the four substructures has six edges, but 12 further edges are created by their conjoining (these edges correspond to lower and upper probabilities on the doubleton events). Consequently, there are 36 edges in the full space of cardinality four (four multiplied by six, plus 12). The lengths of the edges in the substructures are formed by considering the contracted commonality function applied to the substructures of Θ. This is clear because substructures must contain edges that exist in the full structure. Consequently, the lengths of their edges cannot be different and the contracted commonality function applied to the subset K completes the formulation. Equivalently, the contracted commonality function is applied to $\Theta \setminus K^c$.

Of further interest is the particular hyperplane that is associated with a substructure. Each subset of Θ is associated with a lower and upper probability hyperplane. In this geometric representation, the edges of the substructure under consideration correspond to the lower probability hyperplane. This can be understood most easily by realization that certain elements of the frame are omitted from a substructure and that the lengths of each edge of the substructure, therefore, cannot contain information related to the omitted elements. The upper probability hyperplane still contains such information, whereas the lower probability hyperplane does not. The effect is similar to conditioning without normalization, but the form of conditioning is through the lower probability hyperplane of the conditioning event and this corresponds to neither Bayes nor Dempster conditioning (Black [1,2]).

The following Theorem relates the geometry of a monotone capacity, in terms of the distinct lengths contained in the convex system, to the monotone order of the capacity. A Corollary to this Theorem is then provided that provides the same basic result in terms of all of the edges of the convex system, and not just the distinct lengths.

THEOREM 3.1. Given the conditions of Results 3.3 and 3.4, if a monotone capacity is specified over $\Theta = \{A_1, \ldots, A_n\}$ such that $m(D) \geq 0$ for all $D \subset K \subseteq \Theta$, and where K is of cardinality $k \leq n$, then $m(K) \geq 0$ if and only if the following is true:

$$\sum_{i=0}^{k-2}(-1)^i\, S_K(i) \geq 0\,. \tag{3.10}$$

PROOF. The proof hinges on expansion of $S_k(i)$.

$$S_K(i) = \sum_{j,l} Q^c_{K\setminus B}(A_j, A_l) = \sum_{j,l}\{\sum_{\substack{A_j, A_l \subseteq D\\ D\subseteq K\setminus B}} m(D)\}$$

$$= L_K(i) \sum_{c=2}^{k-i} \frac{\binom{k-2-i}{c-2}}{\binom{k}{c}} \sum_{|D|=c} m(D)\,.$$

Therefore,

$$\sum_{i=0}^{k-2}(-1)^i\, S_K(i) = \sum_{i=0}^{k-2}(-1)^i L_K(i) \sum_{c=2}^{k-i} \frac{\binom{k-2-i}{c-2}}{\binom{k}{c}} \sum_{|D|=c} m(D)$$

$$= \sum_{c=2}^{k}\left\{\sum_{i=0}^{k-c}(-1)^i \binom{k-c}{i}\right\}\binom{c}{2}\sum_{|D|=c} m(D) = \binom{k}{2}\, m(K)\,.$$

□

COROLLARY 3.1. Given the conditions of Results 3.3 and 3.4, if a monotone capacity is specified over $\Theta = \{A_1, \ldots, A_n\}$ such that $m(D) \geq 0$ for all $D \subset K \subseteq \Theta$, and where K is of cardinality $k \leq n$, then $m(K) \geq 0$ if and only if the following is true:

$$\sum_{i=0}^{k-2}(-1)^i \binom{k-2}{i} \frac{N_K(i)}{L_K(i)}\, S_K(i) \geq 0\,. \tag{3.11}$$

PROOF. By expansion of inserted terms, which are a function of k only. □

As stated, Corollary 3.1 deals with all of the edges, whereas the first result, Theorem 3.1, deals only with the distinct lengths. A simple extension of the results presented demonstrates that order of a monotone capacity can be bounded directly through the geometry of the capacity:

COROLLARY 3.2. Given the conditions of Theorem 3.1, if

$$\sum_{i=0}^{k-2}(-1)^i\, S_K(i) < 0\,,$$

for some subset K of Θ of cardinality k, then m_Θ is of order less than k.

PROOF. By Theorem 3.1, if the forms hold for a particular subset K of Θ of cardinality k, then $m_\Theta(K)$ is less than zero, hence the monotone capacity cannot be of order k. □

COROLLARY 3.3. Given the conditions of Theorem 3.1, if

$$\sum_{i=0}^{k-2}(-1)^i \binom{k-2}{i} \frac{N_K(i)}{L_K(i)} S_K(i) < 0 ,$$

for some subset K of Θ of cardinality k, then m_Θ is of order less than k.

PROOF. Immediate from Corollaries 3.1 and 3.2. □

Implicit in the proofs of these corollaries is that belief function substructures also correspond to belief functions. The operation of substructuring corresponds to a type of conditioning operation that maintains monotone capacity status, although the order of the substructure monotone capacities might be different than the order of the full capacity. The following example illustrates some of these concepts:

> Consider a frame $\Theta = \{A_1, \ldots, A_4\}$ with basic probability assignment of zero to all the singleton events, zero to all the doubleton events, 1/2 to all the tripleton events and, hence, -1 to Θ. Consequently, the basic probability assignment is symmetric in its arguments, the belief and plausibility of each singleton event are zero and 1/2, and the belief and plausibility of each doubleton event are 0 and 1. This monotone capacity is of order two. However, all substructures of cardinality three yield monotone order infinity representations.

Corollary 3.2, together with the example above, shows that the order of a monotone capacity can be bounded by failure of Inequality 3.1, but that the order cannot be obtained exactly from the geometry. The example above provides a four cardinal case that is monotone order two, but the failure of Inequality 3.1 occurs only with the full four cardinal frame and not with any substructure of the frame. However, the results presented are sufficient to distinguish between belief functions and other monotone capacities, as shown in Corollary 3.4:

COROLLARY 3.4. Given the conditions of Results 3.3 and 3.4, if a monotone capacity is specified over $\Theta = \{A_1, \ldots, A_n\}$, then it is a belief function if and only if the following is true for all subsets K of Θ of cardinality

k, $k \in \{3, \ldots, n\}$:

$$\sum_{i=0}^{k-2} (-1)^i \; S_K(i) \geq 0 \, ,$$

or, equivalently, if and only if,

$$\sum_{i=0}^{k-2} (-1)^i \binom{k-2}{i} \frac{N_K(i)}{L_K(i)} \; S_K(i) \geq 0 \, .$$

PROOF. The result is a direct consequence of Theorem 3.1, and Corollaries 3.1 and 3.2. □

Corollary 3.4 simply states, in terms of the lengths of edges of the full geometric structure and all of the substructures of cardinality at least three, that the m–function values must all be non-negative.

4. Size of lower probabilities. The previous sections have provided discussion of geometric results related to the lengths of edges of convex sets that are monotone capacities. The results are further extended in this section in terms of the relative size of a convex set in C_{B+F}. Size of a convex set in C_{B+F} is defined simply as the sum of lengths of edges. The following definition uses the notation introduced earlier:

DEFINITION 4.1 (SIZE OF A MONOTONE CAPACITY). The size Z_Θ of a monotone capacity on Θ of cardinality n is defined as follows:

$$Z_\Theta = \sum_{i=0}^{n-2} \frac{N(i)}{L(i)} \; S(i) \, . \tag{4.1}$$

Some edges may coincide, and these edges are counted multiple times in the summation. Other edges may have zero length. For example, consider the vacuous functions of cardinality 3 and 4, denoted V_3 and V_4. The size of V_3 is 3; the size of V_4 is 12. V_4 has six apparent edges of length one. It also has edges of length zero, which correspond to lower probabilities on the doubleton subsets. All edges are shared by two faces. Equation 4.1 has a useful expansion:

RESULT 4.1.

$$Z_\Theta = \binom{n-1}{2} \sum_{c=2}^{n} c \Big\{ \sum_{|D|=c} m(D) \Big\} \, . \tag{4.2}$$

PROOF. Recall that:

$$S(i) = L(i) \sum_{c=2}^{n-i} \frac{\binom{n-2-i}{c-2}}{\binom{n}{c}} \sum_{|D|=c} m(D) .$$

After some manipulation, it can be shown that

$$\sum_{i=0}^{n-2} \frac{N(i)}{L(i)} S(i) = \binom{n-1}{2} \sum_{c=2}^{n} c \sum_{|D|=c} m_\Theta(D) .$$

□

Result 4.1 provides a formulation for the size of any lower probability measure. In particular, the size $Z_{0,n}$ of V_n can be recovered from Equation 4.2 as follows:

RESULT 4.2. The size $Z_{0,n}$ of a vacuous belief function of cardinality n, may be given as follows:

$$Z_{0,n} = n \binom{n-1}{2} . \tag{4.3}$$

PROOF. V_n is defined by $m(\Theta) = 1, m(B) = 0, B \subset \Theta$. Therefore, the form in Equation 4.2 reduces to the form in Equation 4.3. □

The relative size can now be defined in terms of the size of a convex set relative to the size of V_n of the same cardinality. Clearly, the relative size of V_n is one.

DEFINITION 4.2 (RELATIVE SIZE OF A MONOTONE CAPACITY). The relative size $R_{\Theta,n}$ of a monotone capacity on Θ of cardinality n is defined as:

$$R_{\Theta,n} = \frac{Z_\Theta}{Z_{0,n}} = \sum_{i=0}^{n-2} \frac{N(i)}{L(i)} S(i) \Big/ n \binom{n-1}{2} . \tag{4.4}$$

Some further results are now possible in terms of the relative size.

RESULT 4.3. The relative size $R_{\Theta,n}$ of a monotone capacity may be given as follows:

$$R_{\Theta,n} = \frac{1}{n} \sum_{c=2}^{n} c \sum_{|D|=c} m_\Theta(D) . \tag{4.5}$$

PROOF. By substitution of Equation 4.2 in Equation 4.4 followed by cancellation of like terms. □

RESULT 4.4. The relative size $R_{\Theta,n}$ of a monotone capacity is less than or equal to 1.

PROOF. By expansion of Equation 4.5. □

THEOREM 4.1. If a monotone capacity is specified over $\Theta = \{A_1, \ldots, A_n\}$ such that $m(D) \geq 0$ for all proper subsets $D \subset \Theta$ and if its relative size $R_{\Theta,n}$ is greater than or equal to $(n-1)/n$, then it is of infinite order.

PROOF. Under the conditions of the Theorem, a monotone capacity is of infinite order if and only if $m(\Theta) \geq 0$. Now,

$$\begin{aligned} R_{\Theta,n} &= \frac{1}{n}\sum_{c=2}^{n} c \sum_{|D|=c} m_{\Theta}(D) \\ &= \frac{1}{n}\left\{2\sum_{i,j} a_{i,j} + 3\sum_{i,j,k} a_{i,j,k} + \cdots + nm(\Theta)\right\}. \end{aligned}$$

Therefore, the statement of the Theorem specifies that,

$$\left\{2\sum_{i,j} a_{i,j} + 3\sum_{i,j,k} a_{i,j,k} + \cdots + nm(\Theta)\right\} \geq n-1\,,$$

or,

$$m(\Theta) \geq (n-1) - \left\{2\sum_{i,j} a_{i,j} + 3\sum_{i,j,k} a_{i,j,k} + \cdots + (n-1)\sum_{|D|=n-1} m(D)\right\}.$$

The right most term is less than or equal to $(n-1)$ under the conditions of the Theorem. Therefore, $m(\Theta)$ is greater than or equal to zero. □

The results presented thus far rely on differences of sums of lengths of edges, and provide geometric interpretations on the bounds of belief function status. The following result relates lengths of edges to simple differences between sums of belief function and plausibility function values.

RESULT 4.5. The following relationship exists between the size Z_{Θ} of a convex set and belief and plausibility function values on the singleton events of a frame $\Theta = \{A_1, \ldots, A_n\}$:

$$Z_{\Theta} = \binom{n-1}{2}\left\{\sum_{D\in\Theta} Pl(D) - \sum_{D\in\Theta} Bel(D)\right\}. \tag{4.6}$$

PROOF. The right hand side of Equation 4.6 may be written as:

$$\binom{n-1}{2}\left\{2\sum_{i,j} a_{i,j} + 3\sum_{i,j,k} a_{i,j,k} + \cdots + nm(\Theta)\right\}.$$

This is also an expansion of Z_Θ. □

Consequently, size can be obtained from the difference between the plausibility and belief function values on the singleton events of Θ. All results presented based on sums of lengths of edges for size can be framed in terms of differences in lower and upper probabilities.

5. Location–scale transformations of monotone capacities. Results have been provided in the previous sections that give some insight into the characterization of belief functions versus their finite monotone counterparts. In particular, in the previous section results were offered that indicate that sufficiently large monotone capacities are belief functions. However, belief function status has more to do with shape than size, as indicated in Section 3. The final piece of the puzzle concerns simple transformations of monotone capacities that brings the notions of shape and size together. These transformations are termed location-scale transformations for reasons that will become obvious:

DEFINITION 5.1 (LOCATION–SCALE TRANSFORMATION OF A MONOTONE CAPACITY). A location-scale transformation of a monotone capacity is defined as follows in terms of a transformation applied to the m–function. Denote the new function Sm, let A represent singleton events in Θ, and let D represent other subsets of Θ:

$$Sm(A) = 0\,,$$
$$Sm(D) = m(D)\Big/1 - \sum_{B\in\Theta} m(B)\,.$$

This formulation explains why the transformation is considered a location and scale transformation. The m–function values on the singletons are location shifted to zero, whereas the values for the remaining subsets are scale shifted to redefine a proper m–function. The transformation has the effect of enlarging the convex set while maintaining the relative differences between belief and plausibility function values. The convex set is effectively enlarged such that lower probability boundaries of V_n for the singleton events are reached exactly by the transformed, or enlarged, convex set. Consequently, the transformed convex set is the same shape as the original, but is maximally enlarged to fit as closely as possible to V_n. Of importance is the effect of this transformation on the belief and plausibility function values assigned to the singleton events. This is because Equation 4.6 includes singleton events only, and this Equation forms a basis for the size relationships developed in the previous section. Consequently, the following results are provided for the singleton events (denote the transformed belief and plausibility functions $SBel$ and SPl, respectively):

$$SBel(A) = 0\,, \tag{5.1}$$

$$SPl(A) = \frac{Pl(A) - Bel(A)}{1 - \sum_{B\in\Theta} m(B)}\,. \tag{5.2}$$

This formulation also explains why the transformation is considered a location and scale transformation. The belief function values on the singletons are location shifted to zero, and the plausibility function values are location and scale shifted. Equation 4.6 can be expressed using the transformed functions as follows:

$$SZ_\Theta = \binom{n-1}{2}\left\{\sum_{A\in\Theta} SPl(A) - \sum_{A\in\Theta} SBel(A)\right\}. \tag{5.3}$$

Expansion of the right hand side of Equation 5.3, together with results from the previous section relating to relative size, yields the following result:

RESULT 5.1. The relative size $SR_{\Theta,n}$ of a location-scale transformed monotone capacity may be written as follows:

$$SR_{\Theta,n} = \frac{1}{n}\left\{\frac{2\sum_{i,j} a_{i,j} + 3\sum_{i,j,k} a_{i,j,k} + \cdots + nm(\Theta)}{\sum_{i,j} a_{i,j} + \sum_{i,j,k} a_{i,j,k} + \cdots + m(\Theta)}\right\}. \tag{5.4}$$

PROOF. The right hand side of Equation 4.5 may be written:

$$SZ_\Theta = \binom{n-1}{2}\left\{\sum_{A\in\Theta} SPl(A)\right\} = \binom{n-1}{2}\left\{\sum_{A\in\Theta} \frac{Pl(A) - Bel(A)}{1 - \sum_{B\in\Theta} m(B)}\right\}.$$

Note that the relative size of a monotone capacity is different than the size by a factor of $n(n-1)(n-2)/2$. Consequently, the above form may be written in terms of the relative size:

$$SR_{\Theta,n} = \frac{1}{n}\left\{\sum_{A\in\Theta} \frac{Pl(A) - Bel(A)}{1 - \sum_{B\in\Theta} m(B)}\right\}.$$

Expansion of the terms in the right hand side provides the result. □

Result 5.1 can be used to provide a generalization of Theorem 4.1. The generalization admits more belief functions under the specified conditions.

THEOREM 5.1. If a monotone capacity is specified over $\Theta = \{A_1, \ldots, A_n\}$ such that $m(D) \geq 0$ for all $D \subset \Theta$, and if its relative location-scale transformed size $SR_{\Theta,n}$ is greater than or equal to $(n-1)/n$ then it is a belief function.

PROOF. The result follows from application of Theorem 4.1 to the location-scale transformed belief function. □

Theorem 5.1 provides a sufficient condition for infinite order status. A necessary condition does not exist as shown by the following result.

RESULT 5.2. Let $\Theta = \{A_1, \ldots, A_n\}$. If a monotone capacity specified over Θ is of infinite order, then its relative location-scale transformed size $SR_{\Theta,n}$ is greater than or equal to $2/n$.

PROOF. It is required to show that

$$SR_{\Theta,n} = \frac{1}{n}\left\{\frac{2\sum_{i,j} a_{i,j} + 3\sum_{i,j,k} a_{i,j,k} + \cdots + nm(\Theta)}{\sum_{i,j} a_{i,j} + \sum_{i,j,k} a_{i,j,k} + \cdots + m(\Theta)}\right\} \geq \frac{2}{n},$$

or

$$\left\{2\sum_{i,j} a_{i,j} + \cdots + nm(\Theta)\right\} \geq 2\left\{\sum_{i,j} a_{i,j} + \sum_{i,j,k} a_{i,j,k} + \cdots + m(\Theta)\right\}$$

which provides the result for an infinite order capacity. □

Theorem 5.1 and Result 5.2 show that there is a range of relative (location-scale transformed) sizes from $2/n$ to $(n-1)/n$ for which both infinite and finite monotone capacities are possible.

The next section provides some examples that highlight some of the results indicated. For 3–cardinal problems a lower probability can be constructed as a six sided figure embedded in an equilateral triangle. For 4–cardinal problems a 3–dimensional figure with, in general, 24 vertices, 36 edges, and 14 faces can be constructed, although with some difficulty. The following examples indicate that some intuitions and insights are also possible in greater dimensions.

6. Examples. Application of the results of the previous sections indicates that a general 3–cardinal monotone capacity can be characterized by a convex figure that has six vertices and six edges. Stronger results are possible for 3–cardinal problems that allow a complete distinction between belief functions and other lower probabilities. For a 3–cardinal problem the geometry is such that the monotone capacity is a belief function if the sum of the lengths of edges that correspond to the lower probability lines for the singleton events is greater than the sum of the lengths of the edges that correspond to their upper probability lines. In this sense, a cardinal three monotone capacity is of infinite order if the shape of the convex set is more like, or similar to, the shape of the full probability space V_3 than is its inverted form. Also, any 3–cardinal problem that has size greater than 2/3 must be a belief function. All 3–cardinal lower probabilities that are not belief functions are monotone of order 2. There are no distinctions at this level between lower envelopes and 2–monotone capacities, and all lower probabilities are fully dominated.

Observations on 4–cardinal problems become more interesting. The same general requirement for a belief function applies; that is, similarity with the encompassing space. However, the similarity is in terms of the linear combination of contraction levels measured through the contracted commonality function. A 4–cardinal problem has 24 vertices and 36 edges, with 24 distinct edge-lengths. It is comprised of four 3–cardinal substructures, which is perhaps best seen through the vacuous lower probability function. There are three contraction levels. In effect, the linear combination of interest is one that considers the lowest and highest contraction levels against the middle contraction level. Similarly to the 3–cardinal case, if a lower probability of interest is a belief function, then its physical inversion is not a belief function in the same space. Although the 4–cardinal structure should be symmetric in its geometric characterization, its 3–cardinal substructures are not.

Suppes [18] provides an example of a 4–cardinal lower probability that is not 2–monotone. The example was brought about by considering the probabilistic outcomes of independent tosses of two coins the first of which is "known" to be fair, and the second of which is "known" to be biased in favor of heads. It is not difficult to show that this example has six distinct vertices, 12 distinct edges and eight distinct faces in the corresponding regular simplex. Because of the built in symmetry of this example, the edges are all of the same length (1/4), and the faces are all triangular. The structure looks like two same size pyramids placed back to back. This lower probability is completely dominated, and hence is fully coherent, but it is not 2–monotone. The size of the convex set presented is 3, and the relative size is 1/4.

The next example is of a lower probability that is not a lower envelope, previously presented in Walley [20, 21], Walley and Fine [22], and Papamarcou and Fine [12]. This example occurs in seven dimensions. Consider a frame $\Theta = \{1,2,3,4,5,6,7\}$, subsets $\mathbf{X} = \{\{134\},\{245\},\{356\},\{467\},\{571\},\{612\},\{723\}\}$, and define the following lower probability function:

$$Bel(A) = \begin{cases} 0, & \text{if} \quad (\forall B \in \mathbf{X})\ B \cap A^c \neq \emptyset \\ 1/2, & \text{if} \quad (\exists B \in \mathbf{X})\ B \subset A \quad \text{and} \quad A \neq \emptyset \\ 1, & \text{if} \quad A = \Theta \end{cases} .$$

The seven subsets in $\mathbf{X}$ are symmetric in a certain sense. Each element is contained in three subsets, and exactly one element overlaps between any pair of subsets. Singleton elements and doubleton subsets take lower probability values of zero. The seven tripleton subsets of $\mathbf{X}$ take lower probability values of 1/2, whereas the remaining tripleton subsets take lower probability values of zero. All quadrupleton subsets that contain a tripleton from $\mathbf{X}$ also take lower probability values of 1/2, the remaining seven quadrupleton subsets taking lower probability values of zero. All higher dimensional subsets take lower probability values of 1/2.

Like the previous examples in this section, this is a lower probability measure. That is, it is monotone, and it satisfies super-additivity. Given that it is a lower probability measure, it can be given context in terms of its size and relative size as presented in Sections 4 and 5. The size of this lower probability is 52.5, and the relative size is 1/2.

It is not difficult to show that this example has 21 distinct vertices (it also appears to have 63 distinct edges and 44 distinct faces in the corresponding regular simplex, the latter from Euler's theorem). Because of the built in symmetry of this example, the edges are all of the same length (1/2), and the faces are all regular 5 dimensional simplexes.

This lower probability system has size and structure, yet it is undominated (this can be realized most easily, perhaps, by recognizing the symmetry in the system and by considering equal probability of 1/7 on each element). That is, it has size, yet it contains no probability distributions. It should be realized that the probability simplex is a subset of all possible functions for the defined frame. That is, the six dimensional simplex, in this example, is a subspace of the seven dimensional representation of the possibilities for the frame. What happens with this example is that the convex set is outside of the probability simplex. It has size because it contains measures, but none of these measures are probability distributions. Consequently, it is undominated by a probability distribution. This helps provide an understanding of the meaning of undominated lower probabilities. Coherent lower probabilities are, therefore, convex sets of probability distributions that are fully contained within the probability simplex. This notion conforms with the idea that a coherent lower probability must be formable using *infimum* and *supremum* operations on an auxiliary probability set. That is, all distributions must be contained within the probability simplex.

Papamarcou and Fine (1986) provide more general forms of this type of example, in which they show that a similar example cannot be found in four dimensions or fewer, but can be found in seven dimensions or greater. The cases of five and six dimensions are considered unresolved. It is interesting then to consider a similar example for a frame of cardinality six. The set $\mathbf{X}$ can be constructed similarly (but with six subsets). The relative size of this convex set is also 1/2, and the number of vertices is 15. Using a symmetry argument, it is clear that one probability distribution falls within this convex set of size 1/2, that is, the probability distribution that assigns equal probability of 1/6 to each element. This does not resolve the unresolved problem because it may be possible to construct other types of examples. It however illuminates the point that a convex set that has size may contain a single probability distribution. Much like the previous example in seven dimensions this causes a problem. However, in this case the convex system is not completely undominated. It is also not coherent.

That is, a further distinction can be made that falls between the class of undominated lower probability measures and coherent lower probability measures. This class can be thought of in terms of partial coherence, where some distributions in the lower probability are contained in the probability simplex, but others are not. For the example given, only one distribution falls within the probability simplex.

7. Summary. Previous characterizations of monotone capacities and lower probabilities have relied on mathematical formulas that offer little in the way of insight into the nature of lower probability. Various classes of lower probabilities can be defined, some of which can be distinguished through these mathematical formalisms. These formalisms present difficulties both from the point of view of intuition and insight, and from the point of view of computer algorithms for checking for lower probability subclasses. The results presented, based on geometric structures, offer a more intuitive and insightful characterization of monotone capacities, or lower probabilities. They also offer some computational challenges for systems based on lower probabilities as a possible means by which infinite monotone order, for example, can be verified geometrically as opposed to through mathematical representations alone.

This initial set of results for characterizing monotone capacities opens many areas for further research. Some computational challenges have been noted. More complete characterizations in terms of faces, or higher dimensional hyperplanes, may also be possible. The main result focuses on distinctions between belief functions and other lower probabilities. Greater refinements in the results may facilitate better identification of finite order. Also, relative size can be used as a measure of partial ignorance that provides a type of information measure that is very different than more traditional probability-based information measures. If such a measure is to be used, however, some adjustment needs to be made for lower probabilities that are only partially dominated. Also, the geometric structures can be used to characterize different types of conditioning, such as Bayes' conditioning or Dempster's conditioning (see, for example, Seidenfeld and Wasserman [14], Black [1,2]. Further investigation can only lead to a stronger understanding of the structure of lower probabilities.

REFERENCES

[1] P.K. Black, *An Examination of Belief Functions and Other Monotone Capacities*, Ph.D. Dissertation, Department of Statistics, Carnegie Mellon University, 1996.

[2] P.K. Black, *Methods for 'conditioning' on sets of joint probability distributions induced by upper and lower marginals*, Presented at the 149th Meeting of the American Statistical Association, 1988.

[3] P.K. Black, *Is Shafer general Bayes ?*, Proceedings of the 3rd Workshop on Uncertainty in Artificial Intelligence, pp. 2–9, Seattle, Washington, July 1987.

[4] P.K. BLACK AND W.F. EDDY, *The implementation of belief functions in a rule–based system*, Technical Report 371, Department of Statistics, Carnegie Mellon University, 1986.

[5] P.K. BLACK AND A.W. MARTIN, *Shipboard evidential reasoning algorithms*, Technical Report 90–11, Decision Science Consortium, Inc., Fairfax, Virginia, 1990.

[6] A. CHATEAUNEUF AND J.Y. JAFFRAY, *Some characterizations of lower probabilities and other monotone capacities*, Unpublished Manuscript, Groupe de Mathematique Economiques, Universite Paris I, 12, Place du Pantheon, Paris, France, 1986.

[7] G. CHOQUET, *Theory of capacities*, Annales de l'Institut Fourier, V (1954), pp. 131–295.

[8] A.P. DEMPSTER, *Upper and lower probabilities induced by a multivalued mapping*, The Annals of Mathematical Statistics, 38 (1967), pp. 325–339.

[9] J.Y. JAFFRAY, *Linear utility theory for belief functions*, Unpublished Manuscript, Universite P. et M. Curie (Paris 6), 4 Place Jussieu, 75005 Paris, France, 1989.

[10] H.E. KYBURG, *Bayesian and non–Bayesian evidential updating*, Artificial Intelligence, 31 (1987), pp. 279–294.

[11] I. LEVI, *The Enterprise of Knowledge*, MIT Press, 1980.

[12] A. PAPAMARCOU AND T.L. FINE, *A note on undominated lower probabilities*, Annals of Probability, 14 (1986), pp. 710–723.

[13] T. SEIDENFELD, M.J. SCHERVISH, AND J.B. KADANE, *Decisions without ordering*, Technical Report 391, Department of Statistics, Carnegie Mellon University, Pittsburgh, PA 15213, 1987.

[14] T. SEIDENFELD AND L.A. WASSERMAN, *Dilation for sets of probabilities*, Annals of Statistics, 21 (1993), pp. 1139–1154.

[15] G. SHAFER, *A Mathematical Theory of Evidence*, Princeton University Press, 1976.

[16] P. SMETS, *Constructing the pignistic probability function in the context of uncertainty*, Uncertainty in Artificial Intelligence V, M. Henrion, R.D. Shachter, L.N. Kanal, and J.F. Lemmer (Eds.), North–Holland, Amsterdam (1990), pp. 29–40.

[17] C.A.B. SMITH, *Consistency in statistical inference and decision* (with Discussion), Journal of the Royal Statistical Society, Series B, 23 (1961), pp. 1–25.

[18] P. SUPPES, *The measurement of belief*, Journal of the Royal Statistical Society, Series B, 36 (1974), pp. 160–191.

[19] H.M. THOMA, *Factorization of Belief Functions*, Doctoral Dissertation, Department of Statistics, Harvard University, 1989.

[20] P. WALLEY, *Coherent lower (and upper) probabilities*, Tech. Report No. 22, Department of Statistics, University of Warwick, Coventry, U.K., 1981.

[21] P. WALLEY, *Statistical Reasoning with Imprecise Probabilities*, Chapman–Hall, 1991.

[22] P. WALLEY AND T.L. FINE, *Towards a frequentist theory of upper and lower probabilities*, Annals of Statistics, 10 (1982), pp. 741–761.

[23] L.A. WASSERMAN AND J.B. KADANE, *Bayes' theorem for Choquet capacities*, Annals of Statistics, 18 (1990), pp. 1328–1339.

[24] M. WOLFENSON AND T.L. FINE, *Bayes–like decision making with upper and lower probabilities*, Journal of the American Statistical Association, 77 (1982), pp. 80–88.

SOME STATIC AND DYNAMIC ASPECTS OF ROBUST BAYESIAN THEORY

TEDDY SEIDENFELD*

Abstract. In this presentation, I discuss two features of robust Bayesian theory that arise, naturally, when considering more than one decision maker:

(1) On a question of statics – what opportunities are there for Bayesians to engage in cooperative decision making while conforming the group to a (mild) Pareto principle and expected utility theory?

(2) On a question of dynamics – what happens, particularly in the short run, when a collection of Bayesian opinions are updated using Bayes' rule, where each opinion is conditioned on shared data?

In connection with the first problem – to allow for a Pareto efficient cooperative group – I argue for a relaxation of the "ordering" postulate in expected utility theory. I outline a theory of partially ordered preferences, developed in collaboration with Jay Kadane and Mark Schervish (1995), that relies on sets of probability/utility pairs rather than a single such pair for making robust Bayesian decisions.

In connection with the second problem – looking at the dynamics of sets of probabilities under Bayesian updating – Tim Herron, Larry Wasserman, and I report on an anomalous phenomenon. We call it "dilation," where conditioning on new shared evidence is certain to enlarge the range of opinions about an event of common interest. Dilation stands in contrast to well-known results about the asymptotic merging of Bayesian opinions. It leads to other puzzles too, e.g., is it always desirable within robust Bayesian decision making to collect "cost-free" data prior to making a terminal decision?

The use of a set of probabilities to represent opinion, rather than the use of a single probability to do so, arises in the theory of random sets, e.g., when the random objects are events from a finite powerset. Then, as is well known, the set of probabilities is just that determined by the lower probability of a belief function. Dilation applies to belief functions, too, as I illustrate.

Key words. Axioms of Group Decision Theory, Belief Functions, Dilation for Sets of Probabilitites, Upper and Lower Probabilities.

AMS(MOS) subject classifications. 62A15, 62C10, 62F35

1. An impossibility for Pareto efficient Bayesian cooperative decision–making and a representation for partially ordered preferences using sets of probability/utility pairs. In this section, I outline a decision theory for partially ordered preferences, developed in collaboration with J.B. Kadane and M.J. Schervish (1995). We modify the foundational approach of Ramsey (1931), deFinetti (1937), Savage (1954), and Anscombe–Aumann (1963) by giving axioms for a theory of preference in which not all acts need be comparable. The goal is a representation of preference in terms of a set $\mathcal{S}$ of probability/utility pairs where one act

* Departments of Philosophy and Statistics, Carnegie Mellon University, Pittsburgh, PA 15213.

is preferred to another if and only if it carries greater expected utility for each probability/utility pair in $\mathcal{S}$.

Our theory is formulated so that partially ordered preferences may arise from a Pareto rule applied in cooperative group settings, as illustrated below. Then these are *indeterminate* group preferences: under the Pareto rule, the group has no preferences beyond what the relation captures. Also, our theory applies when the partial order stems from an incomplete Bayesian elicitation of an agent's preferences. Then the partial order reflects imprecise preferences: the agent has additional preferences which may be elicited. In the latter case, our theory helps to identify what these extra preferences might be. (I borrow the distinction between indeterminate and imprecise probability from I. Levi (1985).)

The well known theories mentioned in the second sentence, above, are formulated in terms of a binary preference relation, defined over pairs of acts. They lead to an expected utility representation for preference in terms of a *single* subjective probability/utility pair, (P, U). In these theories, act A_1 is not preferred to act A_2, $A_1 \precsim A_2$, just in case the P–expected U–utility of act A_1 is not greater than that for act A_2, denoted as $E_{P,U}[A_1] \leq E_{P,U}[A_2]$. However, these theories require that each pair of acts be comparable under the preference relation, $\precsim$.

By contrast, we establish that preferences which satisfy our axioms can be represented by a *set* of cardinal utilities. Moreover, in the presence of an axiom relating to state-independent utility, these partially ordered preferences are represented by a *set* of probability/utility pairs. (The utilities are almost state-independent, in a sense which we make precise). Our goal is to focus on preference alone and to extract whatever probability and/or utility information is contained in the preference relation when that is merely a partial order. This is in sharp contrast with the contemporary approach to Bayesian robustness in Statistics (so well illustrated in the excellent work of J.O. Berger, 1985) that starts with a class of "priors" or "likelihoods," and a single loss-function, in order to derive preferences from these probability/utility assumptions. We work in the reverse direction: from preference to probability and utility.

I begin §1.1 with a challenge for strict (normative) Bayesian theory. It is captured in an impossibility result about cooperative, Pareto efficient group decisions under the Bayesian paradigm. The result provides one motivation for our alternative approach, outlined in Section §1.3. Our theory of partially ordered preferences is based upon Anscombe and Aumann's theory of subjective expected utility, which I summarize in §1.2.

1.1. Impossibility of Pareto efficient, cooperative Bayesian compromises. Consider Dick and Jane, two Bayesian agents whose preferences for acts $\precsim_k$ $(k = 1, 2)$ are coherent. Specifically, assume that their preferences satisfy the Anscombe–Aumann theory, described in detail below in §1.2. According to the Representation Theorem for the Anscombe–

Aumann theory of preference, Dick's (Jane's) preferences for acts is summarized in terms of a single probability/utility pair $\langle P_1, U_1 \rangle$ (and $\langle P_2, U_2 \rangle$ for Jane's preferences) according to simple subjective expected utility. That is, as mentioned above, the Anscombe–Aumann (subjective expected utility) representation theorem asserts that there exists a real-valued (state-independent) cardinal utility U defined over "outcomes" of acts and a personal probability P defined over "states," so that

$$A_1 \precsim_k A_2 \quad \text{iff} \quad E_{P,U}[A_1] \leq E_{P,U}[A_2].$$

Suppose that Dick and Jane want to form a cooperative partnership. For clarity, assume that prior to entering the partnership they confer thoroughly. They share and discuss their opinions and values first, so that whatever they might learn from each other is already reflected in the coherent preferences each brings to the partnership.

How do their individual preferences constrain their collective preference? As a necessary condition for a consensus grounded on shared preferences, we adopt a *Pareto* rule applied to the common, strict preferences that Dick and Jane hold. This forms a (group) partial order $\prec_G$ over acts according to the following Pareto rule for consensus of strict preferences:

$$\text{If } A_1 \prec_k A_2 \ (k = 1, 2), \quad \text{then } A_1 \prec_G A_2\,.$$

REMARK 1.1. It is evident that $\prec_G$ is irreflexive and transitive, making it a strict partial order. □

Which Anscombe–Aumann preferences agree with the partial order $\prec_G$? That is, which probability/utility pairs $\langle P_G, U_G \rangle$ assign greater expected utility to act A_2 than to act A_1 whenever $H_1 \prec_G H_2$? The answer is reported by the following theorem.

THEOREM 1.1. Let Dick and Jane agree on the order of two rewards or "sure" outcomes. (That is, assume $\mathbf{r}_* \prec_k \mathbf{r}^*$, $k = 1, 2$, for two constant reward acts.) If $P_1 \neq P_2$ and $U_1 \neq U_2$, then the only Anscombe–Aumann coherent extensions of $\prec_G$ are $\precsim_1$ and $\precsim_2$.

The theorem asserts that, apart from $\precsim_1$ and $\precsim_2$, there are no Anscombe–Aumann preferences that extend the partial order $\prec_G$ – there are no Pareto efficient compromises between Dick and Jane that satisfy the Anscombe–Aumann theory of coherent preferences.

How can expected utility theory be modified to permit Pareto efficient compromises that accord with the modified (normative) theory? That is, how can expected utility theory be changed so that the same normative standards apply to individuals as to Pareto efficient (cooperative) groups? In §1.3, below, I offer one solution to this problem that involves sets of probabilities and utilities rather than a single probability/utility pair. A special case of this theory turns out to be a class of sets of probabilities (i.e., those which dominate a belief function) familiar from the theory of random sets. I discuss some anomalous dynamics for these sets of probabilities in Part 2 of this paper.

1.2. The Anscombe–Aumann theory of "horse lotteries". In this subsection, I sketch the Anscombe–Aumann theory, which forms the background for our relaxation of expected utility. An Anscombe–Aumann act, denoted by "H," also called a "horse lottery," is a function from states to (simple) probability distributions over a set of rewards. That is, outcomes are simple lotteries over a set of rewards. A lottery L is nothing more than a probability p on a collection of prizes or rewards. (To say the lottery is *simple* means that p assigns probability 1 to some *finite* set of rewards.) The table below schematizes horse lottery acts as rows, in which columns represent the (n-many) states of uncertainty. The outcome of choosing H_i when state s_j obtains is the outcome – lottery L_{ij}.

TABLE 1
Horse lottery acts as functions from states to (lottery) outcomes.

	s_1	s_2	$\cdots$	s_j	$\cdots$	s_n
H_1	L_{11}	L_{12}	$\cdots$	L_{1j}	$\cdots$	L_{1n}
H_2	L_{21}	L_{22}	$\cdots$	L_{2j}	$\cdots$	L_{2n}
$\vdots$	$\vdots$	$\vdots$	$\ddots$	$\vdots$	$\ddots$	$\vdots$
H_m	L_{m1}	L_{m2}	$\cdots$	L_{mj}	$\cdots$	L_{mn}

Next, define the convex combination of two acts, denoted with "$\oplus$" between acts,

$$xH_1 \oplus (1-x)H_2 = H_3 \qquad (0 \leq x \leq 1)$$

by the convolution of their two probability distributions,

$$p_{3j}(\bullet) = xp_{1j} + (1-x)p_{2j} \qquad (j = 1, \ldots, n).$$

Operation $\oplus$ may be interpreted as the use of (value-neutral) compound chances to create new acts. For example, H_3 can be the act obtained by flipping a coin loaded for "heads up" with chance x. If the coin lands "heads" receive H_1, while if it lands "tails" receive H_2.

The Anscombe–Aumann theory is given by five axioms regulating preferences $\precsim$ over acts:

AXIOM 1. Preference is a *weak order*. That is, $\precsim$ is a binary relation that is reflexive, transitive, and each pair of acts is comparable.

In light of the transitivity of preference, denote by $\approx$ the induced (transitive) indifference relation, defined by $H_1 \approx H_2$ if and only if $H_1 \precsim H_2$ and $H_2 \precsim H_1$.

AXIOM 2. *Independence*: Convex combinations involving a common act preserve preferences when the common term is removed. (This is where the compounding of chances in lotteries is made value neutral.) Stated formally,

$$\forall[H_1 H_2 H_3, 0 < x \leq 1]\ H_1 \prec H_2 \text{ iff } xH_1 \oplus (1-x)H_3 \prec xH_2 \oplus (1-x)H_3.$$

Next is a technical condition that leads to real-valued (as opposed to lexicographic, or extended real-valued) utilities and probabilities.

AXIOM 3. The *Archimedean condition* for preference:

$$\text{If } H_1 \prec H_2 \prec H_3, \text{ then } \exists 0 < x, y < 1,\ xH_1 \oplus (1-x)H_3 \prec H_2 \prec yH_1 \oplus (1-y)H_3.$$

The fourth axiom is formulated in terms of so-called "constant" acts and "non-null" states, defined as follows. Let H_L denote the horse lottery that awards the same lottery outcome L in each state. Then H_L is the *constant act* with outcome L. (A special case is the constant horse lottery with only one reward, r, for certain. Denote such constant acts by a boldface '$\mathbf{r}$'.) A state s_0 is called *null* provided the agent is indifferent between each pair of acts that agree off s_0. That is, when s_0 is null and $H(s) = H'(s)$, for $s \neq s_0$, then $H \approx H'$. (Under the intended representation, null states carry probability 0. Hence, they contribute 0 expected utility to the valuation of acts.) A state is called *non-null* if it is not null.

AXIOM 4. *State-independent preferences for lotteries*: Given two lotteries L_1 and L_2, let H_1 and H_2 be two horse lotteries that differ on some one non-null state s_j: where $L_{1j} = L_1$ and $L_{2j} = L_2$, while $L_{1k} = L_{2k}$ for $k \neq j$. Then,

$$H_{L1} \prec H_{L2} \quad \text{iff} \quad H_1 \prec H_2.$$

Axiom 4 asserts that the agent's preferences over simple lotteries (expressed in terms of preferences over constant acts) are reproduced within each non-null state, by considering pairs of acts that are the same apart from the one non-null state in question.

Last, to avoid trivialities when all acts are indifferent, some strict preference is assumed.

AXIOM 5. *Non-triviality*: For some pair of acts, $H_1 \prec H_2$.

THEOREM 1.2 (REPRESENTATION THEOREM). The Anscombe–Aumann (subjective expected utility) theorem asserts that these five axioms are necessary and sufficient for the existence of a real-valued (state-independent) cardinal utility U defined over lotteries and one personal probability P defined over states, where:

$$H_1 \precsim H_2 \quad \text{iff} \quad \sum_j P(s_j)U(L_{1j}) \leq \sum_j P(s_j)U(L_{2j}).$$

The utility U satisfies a basic expectation rule: when $L_3 = xL_1 \oplus (1-x)L_2$, then $U(L_3) = x\,U(L_1) + (1-x)\,U(L_2)$. Last, U is unique up to positive linear transformations: $U' = a\,U + b$ (for $a > 0$) is equivalent to U in the representation. Thus, the Representation Theorem identifies a coherent preference over horse-lotteries with a single probability/utility pair (P, U), that indexes acts by their P–expected U–utility.

The Anscombe–Aumann theory, like Savage's theory, incorporates two structural assumptions about rational preferences which are not mandatory (they are not "normative") for expected utility theories generally. The utility U is state-independent because the value assigned to a lottery outcome does not depend upon the (non-null) state under which it is awarded. Regarding the use of state-independent utilities, unfortunately, even when the five axioms are satisfied, hence, even when the representation theorem obtains, that is insufficient reason to attribute the pair (P, U) to the agent as his/her degrees of belief and values for rewards. The representation of $\precsim$ is unique where U is *state-independent*. However, given any probability P' that agrees with P on null states (so that P and P' are mutually absolutely continuous), there is an expected utility representation for $\precsim$ in terms of the probability/(state-dependent) utility, $(P', U'_j\ \{j = 1, \ldots, n\})$ – where $U'_j(L)$ is the (cardinal) utility of lottery L in state s_j. We investigate this problem in Schervish, Seidenfeld, and Kadane (1990).

Also, the representation for $\precsim$ is in terms of states that are probabilistically independent of acts: $P(s_j|H_i) = P(s_j)$. I know of no decision theory for acts that admits (general) act-state dependence and yields an expected utility representation of preference with (even) the uniqueness of the Anscombe–Aumann result.

REMARK 1.2. R.C. Jeffrey's (1965) decision theory allows for act-state independence at the price of confounding acts and states with respect to personal probabilities. That is, in Jeffrey's theory one is obliged to assign probabilities to all propositions, regardless their status as *options* (that which we have the power to choose true) or *states* (that which we cannot so determine). For criticisms of the idea that we may assign personal probabilities to our own current options see Levi (1992) and Kadane (1985). □

Axioms 1, 2, 3, and 5 duplicate the von Neumann–Morgenstern (1947) theory of cardinal utility. Hence, a corollary to the Anscombe–Aumann Representation Theorem is existence of a real-valued value function V, defined on acts, with the cardinal utility properties for representing preference: that is, V is additive over the operation $\oplus$, and unique up to positive linear transformations. Expressed formally:

(i) $H_1 \precsim H_2$ iff $V(H_1) \leq V(H_2)$,

(ii) When $H_3 = xH_1 \oplus (1-x)H_2$, $V(H_3) = xV(H_1) + (1-x)V(H_2)$,

(iii) $V' = aV + b$ $(a > 0)$ also satisfies (i) and (ii).

This observation affords a better appreciation of the content added by Axiom 4. That condition is the requirement that the cardinal utility V of a constant act (the von Neumann–Morgenstern lottery H_L) is duplicated for each family of horse lotteries whose members differ exactly in one (non-null) state. In symbols, $\forall L$, $V(L) = U_j(L)$.

In responding to the challenge of §1.1, what happens when each of these axioms is relaxed? Can we develop a normative theory of preference that permits Pareto efficient compromises for a cooperative group? Obviously, the fifth axiom assures only that rational preference is not fully degenerate, that not all options are indifferent. If this axiom is abandoned, then two problems result: (1) Preference reveals nothing about an agent's degrees of belief; and (2) The (weak) Pareto condition is voided of all content. Thus, we may move directly to examining Axiom 4. Regarding Axiom 4, state-independent utilities, the impossibility of a Pareto efficient compromise does not depend upon this assumption when the domain of preferences is enhanced to allow preferences to reveal state-dependent utilities. See Schervish, M.J., Seidenfeld, T., and Kadane, J.B. (1991) for details.

Regarding Axiom 3, the Archimedean condition, relaxing that (alone) leads to a theory of subjective expected lexicographic utility, e.g., expected utility is vector-valued rather than real-valued. See Fishburn and La Valle (1996), or Fishburn (1974), for details. However, the impossibility result of §1.1 applies, as before. Under a Pareto rule, all but the degenerate "compromises" (that is, where, e.g., one agent controls the leading coordinate in the expected utility vector) are precluded.

There is considerable economic literature about utility theory without the Independence, or the so-called "Sure-Thing," postulate – Axiom 2. (See Machina, 1982, and McClennen, 1990, for extended discussion.) However, I believe that theories which relax only this one axiom mandate unsatisfactory sequential decisions (Seidenfeld, 1988) which other approaches (like that in §1.3) avoid and, therefore, I do not accept this as a way out of the challenge posed in §1.1. That brings us, Sherlock Holmes style, to focus on Axiom 1.

1.3. Outline of a theory of partially ordered preferences. Begin with four definitions that fix the structure of the space of acts: simple horse lotteries on a finite partition. (Our results apply, also, to discrete lotteries on denumerable sets of rewards, but these extensions involve additional mathematical details which I skip for this presentation.)

Let $\mathcal{R} = \{r_i : i = 1, \ldots, k\}$ be a finite set of rewards.
Let $\mathcal{L} = \{L : L \text{ is a lottery}\}$ be the set of probabilities on $\mathcal{R}$.

(Since $\mathcal{R}$ here is finite, each L is a simple lottery.)

Let Π be a finite partition of "states" $\Pi = \{s_1, \ldots, s_n\}$.
Let $\mathcal{H} = \{\text{horse lotteries } H, H : \Pi \to \mathcal{L}\}$.

Our axioms mimic those of the Anscombe–Aumann theory.

AXIOM HL–1. Preference $\prec$ is a strict partial order, being transitive and irreflexive.

When neither $H_1 \prec H_2$ nor $H_2 \prec H_1$, say that acts are not comparable by strict preference, and denote this by $H_1 \sim H_2$. In our theory, typically, non-comparability may not be transitive. It may be that $H_1 \sim H_2$ and $H_2 \sim H_3$, yet $H_1 \prec H_3$.

AXIOM HL–2. *Independence*: $\forall\, [H_1 H_2 H_3, 0 < x \leq 1]$,

$$H_1 \prec H_2 \quad \text{iff} \quad xH_1 \oplus (1-x)H_3 \;\prec\; xH_2 \oplus (1-x)H_3.$$

AXIOM HL–3. Modified Archimedes – also a continuity condition: If $\{H_n\} \Rightarrow H$ and $\{M_n\} \Rightarrow M$ are sequences of acts converging (pointwise), respectively, to acts H and M, then

$$(A3a) \text{ If } \forall n\ (H_n \prec M_n) \text{ and } M \prec N, \quad \text{then } H \prec N.$$

$$(A3b) \text{ If } \forall\, n(H_n \prec M_n) \text{ and } K \prec H, \quad \text{then } K \prec M.$$

REMARK 1.3. The Archimedean axiom of the Anscombe–Aumann theory is inadequate for our needs. Even with a domain of constant acts it is too restrictive, as it is shown in §2.5 of Seidenfeld, Schervish, and Kadane (1995). Also it is insufficient for handling non-simple, discrete acts. See chapter 10 of Fishburn (1970) or Seidenfeld and Schervish (1983) for examples of preferences that satisfy Axioms 1, 2, and 3, but which cannot be represented by a cardinal utility V over *non-simple* lotteries. □

In parallel with the Anscombe–Aumann Axiom 4, call a state s_0 *potentially null* if non-comparability obtains between each pair of acts that agree off s_0. That is, when s_0 is potentially null and $H(s) = H'(s)$, for

$s \neq s_0$, then $H \sim H'$. (Under the intended representation, potentially null states carry probability 0 for some probability in $\mathcal{S}$.)

AXIOM HL–A4. If s_k is not $\prec$–potentially null, then for each quadruple of acts H_{Li}, H_i, $(i = 1, 2)$ as defined in Axiom 4: $H_{L1} \prec H_{L2}$ iff $H_1 \prec H_2$.

We allow for potentially null states through an additional axiom, not reported here, that augments HL–A4. See Seidenfeld, Schervish, and Kadane (1995) p. 2188.

Last, we take $\prec$ to be a non-trivial relation:

POSTULATE HL–A5. For some pair of acts, $H_1 \prec H_2$.

In §2.7 of Seidenfeld, Schervish, and Kadane (1995) we show how to extend a partial order satisfying axioms HL–A1 — HL–A4 to one that is bounded below and above, respectively, by two constant acts $\mathbf{W}$ and $\mathbf{B}$ where $\mathbf{W} \prec \mathbf{B}$. Hence, we satisfy HL–A5 by construction rather than by assumption, and signal this with its label "postulate."

We now summarize our main results. Let $\prec$ be a partial order over simple horse lotteries that satisfies these axioms (and assume that states are not potentially null – a condition that we relax in (1995), as noted above).

THEOREM 1.3. The strict preference relation $\prec$ carries a representation in terms of a set of probability/utility pairs, $\mathcal{S} = \{(P, U)\}$, according to the Pareto rule applied to expected utility:

$$H_1 \prec H_2 \text{ only if } \forall (P, U) \in \mathcal{S}\, E_{P,U}[H_1] < E_{P,U}[H_2].$$

REMARK 1.4. The utilities U are bounded though (typically) they are state-dependent. However, "almost" state-independent agreeing utilities exist, in the following sense. Given $\epsilon > 0$, there exists $(P, U) \in \mathcal{S}$ and a set of states E with $P(E) \geq 1 - \epsilon$, where for each reward r, the U–utility of r varies by no more than ϵ across states in E. Moreover, in §4.3 of Seidenfeld, Schervish, and Kadane (1995) we provide a sufficient condition for existence of agreeing state-independent U's, depending upon closure of the set of agreeing (cardinal utility) value functions V over acts. □

THEOREM 1.4. Parallel to the results for Anscombe–Aumann theory, $\prec$ is represented by a (convex) set of $\mathcal{V}$ cardinal value functions V defined over acts, where V is additive over the operation $\oplus$, and unique up to positive linear transformations. Expressed formally:

(i) $H_1 \prec H_2$ only if $V(H_1) < V(H_2)$, for each $V \in \mathcal{V}$.

(ii) When $H_3 = xH_1 \oplus (1 - x)H_2$, $V(H_3) = xV(H_1) + (1 - x)V(H_2)$.

(iii) $V' = aV + b$ $(a > 0)$ also satisfies (i) and (ii).

REMARK 1.5. The missing "if" in Theorems 1.3 and 1.4 (i) can be supplied by adding preferences to constrain the faces of V, as we explain in Seidenfeld, Schervish, and Kadane (1995), pp. 2204-2205. □

THEOREM 1.5. The HL–axioms are necessary for the representation, above.

THEOREM 1.6. The representation equates the set of conditional probability/utility pairs in $\mathcal{S}$ and the partially ordered preferences on "called-off" families of acts.

That is, consider a maximal subset, a family of horse lotteries $\mathcal{H}_E$ which, pairwise, agree on the states belonging to event E^c (the complement of event E). In other words, acts in $\mathcal{H}_E$ are "called-off" in case E *fails* to obtain. Restricting attention to $\prec$–preferences over $\mathcal{H}_E$ yields us a set $\mathcal{S}_E = \{(P_E, U)\}$ of representing probability/utility pairs, which treat states in E^c as null. Each such probability P_E is the conditional probability $P(\cdot|E)$ for some pair (P, U) belonging to the set $\mathcal{S}$ that represents the partial order $\prec$. Provided $P(E) > 0$ for $(P, U) \in \mathcal{S}$, the converse holds as well.

REMARK 1.6. Because the set $\mathcal{S}$ is (sometimes) disconnected, the proof of these theorems cannot rely on a "separating hyperplanes" technique, familiar in many expected-utility representation of preferences that are weak orderings. Instead, we use induction (in the fashion of Szpilrajn (1930) to identify the non-empty set $\mathcal{S}$ of (all) coherent extensions for the partial order $\prec$. □

2. Dilation of sets of probabilities and some asymptotics of robust Bayesian inference. In this section, I explore two general issues concerning diverging sets of Bayesian (conditional) probabilities – divergence of "posteriors" – that can result with increasing evidence. Consider a set $\mathcal{P}$ of probabilities typically, but not always, based on a set of Bayesian "priors." Incorporating sets of probabilities, rather than relying on a single probability, is a useful way to provide a rigorous mathematical framework for studying sensitivity and robustness in classical and Bayesian inference. (See: Berger, 1984, 1985, 1990; Lavine, 1991; Huber and Strassen, 1973; Walley, 1991 and Wasserman and Kadane, 1990.) Also, sets of probabilities arise in group decision problems, as discussed in §1, above. Third, sets of probabilities are one consequence of weakening traditional axioms for uncertainty. (See: Good, 1952; Smith, 1961; Kyburg, 1961; Levi, 1974; Fishburn, 1986; Seidenfeld, Schervish, and Kadane, 1990; and Walley, 1991.)

Fix E, an event of interest, and X, a random variable to be observed. With respect to a set $\mathcal{P}$, when the set of conditional probabilities for E, given X, is strictly larger than the set of unconditional probabilities for E, for each possible outcome $X = x$, call this phenomenon *dilation* of the set of probabilities [Seidenfeld and Wasserman, 1993].

As a backdrop to the discussion of dilation, I begin by pointing to two well known results about the asymptotic merging of Bayesian posterior probabilities.

2.1. Asymptotic merging. Savage (1954, §3.6) provides an (almost everywhere) approach simultaneously to consensus and to certainty among a few Bayesian investigators, provided:

(1) They investigate finitely many statistical hypotheses $\Theta = \{\theta_1, \ldots, \theta_k\}$.

(2) They use Bayes' rule to update probabilities about Θ given a growing sequence of shared data $\{x_1, \ldots\}$. These data are *identically, independently distributed* (i.i.d.) given θ (where the Bayesians agree on the statistical model parameterized by Θ).

(3) They have prior agreement about *null* events. Specifically (given Condition 2), there is agreement about which parameter values have positive prior probability.

By a simple application of the strong law of large numbers, Savage concludes that:

THEOREM 2.1. Almost surely, the agents' posterior probabilities will converge to 1 for the true value of Θ. Asymptotically, with probability 1, they achieve consensus and certainty about Θ.

Blackwell and Dubins (1962) give an impressive generalization about consensus without using either "i" of Savage's i.i.d. Condition (2). Theirs is a standard martingale convergence result which I summarize next:

Consider a denumerable sequence of sets X_i $(i = 1, \ldots)$ with associated σ–fields $\mathcal{B}_i$. Form the infinite Cartesian product $X = X_1 \otimes \cdots$ of sequences $(x_1, x_2, \ldots) = x \in X$, where $x_i \in X_i$. That is, each x_i is an atom of its algebra $\mathcal{B}_i$. Let the measurable sets in X (the events) be the elements of the σ–algebra $\mathcal{B}$ generated by the set of measurable rectangles. Define the spaces of histories $(H_n, \mathcal{H}_n)$ and futures $(F_n, \mathcal{F}_n)$ where $H_n = X_1 \otimes \cdots \otimes X_n$, $\mathcal{H}_n = \mathcal{B}_1 \otimes \cdots \otimes \mathcal{B}_n$, and where $F_n = X_{n+1} \otimes \cdots$ and $\mathcal{F}_n = \mathcal{B}_{n+1} \otimes \cdots$.

Blackwell and Dubins' argument requires that P is a *predictive*, σ–additive probability on the measure space $(X, \mathcal{B})$. (That P is *predictive* means that there exist conditional probability distributions of events given past events, $P^n(\cdot|\mathcal{H}_n)$.) Consider a probability Q which is in agreement with P about events of measure 0 in $\mathcal{B} : \forall E \in \mathcal{B}$, $P(E) = 0$ iff $Q(E) = 0$. That is, P and Q are *mutually absolutely continuous* (m.a.c.). Then Q, too, is σ–additive and predictive if P is, with conditional distributions $Q^n(\mathcal{F}_n|h_n)$.

Blackwell and Dubins (1962) prove that there is almost certain asymptotic consensus between the conditional probabilities P^n and Q^n.

THEOREM 2.2. For each P^n there is (a version of) Q^n so that, almost surely, the distance between them vanishes with increasing histories: $\lim_{n\to\infty} \rho(P^n, Q^n) = 0$ (a.e. P or Q), where ρ is the uniform distance (total variation) metric between distributions. (That is, with μ and ν defined on the same measure space $(M, \mathcal{M})$, $\rho(\mu, \nu)$ is the least upper bound for events $E \in \mathcal{M}$ of $|\mu(E) - \nu(E)|$.) Thus, the powerful assumption that P and Q are mutually absolutely continuous (Savage's Condition 3) is what drives the merging of the two families of conditional probabilities P^n and Q^n.

2.2. Dilation and short run divergence of posterior probabilities. Throughout, let $\mathcal{P}$ be a (convex) set of probabilities on a (finite) algebra $\mathcal{A}$. For a useful contrast with Savage–styled, or Blackwell–Dubins' styled asymptotic consensus, the following discussion focuses on the short run dynamics of upper and lower conditional probabilities in robust Bayesian models.

For an event E, denote by $P_*(E)$ the "lower" probability of E: $\inf_{\mathcal{P}} \{P(E)\}$ and denote by $P^*(E)$ the "upper" probability of E: $\sup_{\mathcal{P}}\{P(E)\}$. Let $(b_1, \ldots, b_n)$ be a (finite) partition generated by an observable B.

DEFINITION 2.1. The set of conditional probabilities $\{P(E|b_i)\}$ *dilate* if

$$P_*(E|b_i) < P_*(E) \leq P^*(E) < P^*(E|b_i) \quad (i = 1, \ldots, n).$$

That is, dilation occurs provided that, for each event b_i in a partition B, the conditional probabilities for an event E, given b_i, properly include the unconditional probabilities for E.

Here is an elementary illustration of dilation.

HEURISTIC EXAMPLE. Suppose A is a highly "uncertain" event with respect to the set $\mathcal{P}$. That is, $P^*(A) - P_*(A) \approx 1$. Let $\{H, T\}$ indicate the flip of a fair coin whose outcomes are independent of A. That is, $P(A, H) = P(A)/2$ for each $P \in \mathcal{P}$. Define event E by, $E = \{(A, H), (A^c, T)\}$.

It follows, simply, that $P(E) = .5$ for each $P \in \mathcal{P}$. (E is *pivotal* for A.) But then,

$$0 \approx P_*(E|H) < P_*(E) = P^*(E) < P^*(E|H) \approx 1,$$

and

$$0 \approx P_*(E|T) < P_*(E) = P^*(E) < P^*(E|T) \approx 1.$$

Thus, regardless how the coin lands, conditional probability for event E dilates to a large interval, from a determinate unconditional probability of .5. Also, this example mimics Ellsberg's (1961) "paradox," where the mixture of two uncertain events has a determinate probability. □

The next two theorems on existence of dilation serve to motivate using indices of departures from independence to gauge the extent of dilation. (They appear in [Seidenfeld and Wasserman, 1993].)

Independence is sufficient for dilation: Let $\mathcal{Q}$ be a convex set of probabilities on algebra $\mathcal{A}$ and suppose we have access to a "fair" coin which may be flipped repeatedly: algebra $\mathcal{C}$. Assume the coin flips are independent and, with respect to $\mathcal{Q}$, also independent of events in $\mathcal{A}$. Let $\mathcal{P}$ be the resulting convex set of probabilities on $\mathcal{A} \times \mathcal{C}$. (This condition is similar to, e.g., DeGroot's assumption of an extraneous continuous random variable, and is similar to the "fineness" assumptions in the theories of Savage, Ramsey, Jeffrey, etc.)

THEOREM 2.3. If $\mathcal{Q}$ is not a singleton, there is a 2×2 table of the form $(E, E^c) \times (H, T)$ where both:

$$P_*(E|H) < P_*(E) = .5 = P^*(E) < P^*(E|H) ,$$

and

$$P_*(E|T) < P_*(E) = .5 = P^*(E) < P^*(E|T).$$

That is, then dilation occurs.

Independence is necessary for dilation: Let $\mathcal{P}$ be a convex set of probabilities on algebra $\mathcal{A}$. The next result is formulated for subalgebras of 4 atoms: (p_1, p_2, p_3, p_4).

TABLE 2
The case of 2×2 *tables.*

	b_1	b_2
A_1	p_1	p_2
A_2	p_3	p_4

Define the quantity $S_P(A_1, b_1) = p_1/(p_1 + p_2)(p_1 + p_3) = P(A_1, b_1)/P(A_1)P(b_1)$. Thus, $S_P(A_1, b_1) = 1$ iff A and B are independent under P and "S_P" is an index of dependence between events.

LEMMA 2.1. If $\mathcal{P}$ displays dilation in this sub-algebra, then

$$\inf_{\mathcal{P}}\{S_P(A_1, b_1)\} < 1 < \sup_{\mathcal{P}}\{S_P(A_1, b_1)\} .$$

THEOREM 2.4. If $\mathcal{P}$ displays dilation in this sub-algebra, then there exists $P^{\#} \in \mathcal{P}$ such that

$$S_{P^{\#}}(A_1, b_1) = 1\,.$$

Thus, independence is also necessary for dilation.

2.3. The extent of dilation. I begin by reviewing some results obtained for the ϵ–contaminated model. Details are found in Herron, Seidenfeld, and Wasserman (1995). Given probability P and $0 < \epsilon < 1$, define the convex set $\mathcal{P}_\epsilon(P) = \{(1-\epsilon)P + \epsilon Q : Q$ an arbitrary probability$\}$. This model is popular in studies of Bayesian robustness. (See Huber, 1973, 1981; Berger, 1984.)

LEMMA 2.2. In the ϵ–contaminated model, dilation occurs in algebra $\mathcal{A}$ if and only if it occurs in some 2×2 subalgebra of $\mathcal{A}$.

So without loss of generality, the next result is formulated for 2×2 tables.

THEOREM 2.5. $\mathcal{P}_\epsilon(P)$ experiences dilation if and only if:

Case 1: $S_P(A_1, b_1) > 1$
$\epsilon > [S_P(A_1, b_1) - 1] \max\{P(A_1)/P(A_2); P(b_1)/P(b_2)\}$.

Case 2: $S_P(A_1, b_1) < 1$
$\epsilon > [1 - S_P(A_1, b_1)] \max\{1; P(A_1)P(b_1)/P(A_2)P(b_2)\}$.

Case 3: $S_P(A_1, b_1) = 1$
P is internal to the simplex of all distributions.

Thus, dilation occurs in the ϵ–contaminated model if and only if the focal distribution P is close enough (in the tetrahedron of distributions on four atoms) to the saddle-shaped surface of distributions which make A and B independent. Here, S_P provides one relevant index of the proximity of the focal distribution P to the surface of independence.

Generally, for $B \subset \mathcal{B}$ (B not necessarily a binary outcome) define the *extent of dilation* by

$$\Delta(A, B) = \min_{b \in B} [P^*(A|b) - P^*(A) + (P_*(A) - P_*(A|b))]\,.$$

For the ϵ–contamination model we then have

THEOREM 2.6.

$$\Delta(A, B) = \min_{b \in B} \frac{\epsilon(1-\epsilon)P(b^c)}{\epsilon + (1-\epsilon)P(b)}.$$

In this model, $\Delta(A, B)$ does not depend upon the even A. Moreover, the extend of dilation is maximized when $\epsilon = \sqrt{P(b_\Delta)}/(\sqrt{P(b_\Delta)} + 1)$, where $b_\Delta \in B$ solves $\Delta(A, B)$.

Similar findings are obtained for *total variation neighborhoods.*

Given a probability P and $0 < \epsilon < 1$, define the convex set $\mathcal{U}_\epsilon(P) = \{Q : \rho(P, Q) \leq \epsilon\}$. Thus $\mathcal{U}_\epsilon(P)$ is the uniform distance (total variation) neighborhood of P, corresponding to the metric of Blackwell–Dubins' consensus. As before, consider dilation in 2×2 tables. Define a second index of association: $d_P(A, B) = P(AB) - P(A)P(B)$.

THEOREM 2.7 (INFORMAL VERSION). $\mathcal{U}_\epsilon(P)$ experiences dilation if and only if P is sufficiently close to the surface of independence, as indexed by d_P.

And the extent of dilation for the total variation model also may be expressed in terms of the d_P–index, though there are annoying cases depending upon whether the set $\mathcal{U}_\epsilon(P)$ is truncated by the simplex of all distributions.

Whereas, in the previous two models, each of the sets $\mathcal{P}_\epsilon(P)$ and $\mathcal{U}_\epsilon(P)$ has a single distribution that serves as its natural focal point, some sets of probabilities are created through constraints on extreme points directly. For example, consider a model where $\mathcal{P}$ is defined by lower and upper probabilities on the *atoms* of the algebra $\mathcal{A}$. In Section 2 of Herron, Seidenfeld, and Wassermann (1995) we call these *ALUP* models. For convenience, take the algebra to be finite with atoms a_{ij} ($i = 1, 2$; $j = 1, \ldots, n$) and where $A = \cup_j a_{1j}$ and $b_j = \{a_{1j}, a_{2j}\}$. I discuss dilation of the event A given the outcome of the random quantity $B = \{b_1, \ldots, b_n\}$.

Given an event E and a (closed) set $\mathcal{P}$ of probabilities, use the notation $\{P_*(E)\}$ and $\{P^*(E)\}$ for denoting, respectively, the set of probabilities within $\mathcal{P}$ that achieve the lower and upper probability bounds for event E. Next, given events E and F and a probability P, define the (covariance) index $\delta_P(E, F) = P(EF)P(E^cF^c) - P(E^cF)P(EF^c)$.

Within ALUP models, the extent of dilation for A, given $B = b_j$, is provided by the $\delta_P(A, b_j)$ (covariance-) index. Specifically, let P_{1j} be a probability such that $P_{1j} \in \{P^*(A)\} \cap \{P^*(a_{1j})\} \cap \{P_*(a_{2j})\}$, and let P_{2j} be a probability such that, $P_{2j} \in \{P_*(A)\} \cap \{P_*(a_{1j})\} \cap \{P^*(a_{2j})\}$. (Existence of P_{1j} and P_{2j} are demonstrated in §2 of Seidendfeld and Wasserman (1993).) Then a simple calculation shows:

THEOREM 2.8.

$$\Delta(A, B) = \min_j \left\{ \frac{\delta_{P_{1j}}(A, b_j)}{P_{1j}(b_j)} - \frac{\delta_{P_{2j}}(A, b_j)}{P_{2j}(b_j)} \right\}.$$

Thus, as with the ϵ–contamination and total-variation models, the extent of dilation in ALUP models also is a function of an index of probabilistic independence between the events in question.

Observe that the ϵ–contamination models are a special case of the ALUP models: they correspond to ALUP models obtained by specifying the lower probabilities for the atoms and letting the upper probabilities be as large as possible consistent with these constraints on lower probabilities. Then the extent of dilation for a set $\mathcal{P}_\epsilon(P)$ of probabilities may be reported either by attending to the S_P–index for the focal distribution of the set (as in Theorem 2.6), or by attending to the δ_P–index for the extreme points of the set (as in Theorem 2.8).

2.4. Asymptotic dilation for classical and Bayesian interval estimates. In an interesting essay, L. Pericchi and P. Walley (1991, pp. 14–16), calculate the upper and lower probabilities of familiar normal confidence interval estimates under an ϵ–contaminated model for the "prior" of the unknown normal mean. Specifically, they consider data $x = (x_1, \ldots, x_n)$ which are i.i.d. $N(\theta, \sigma^2)$ for an unknown mean θ and known variance σ^2. The "prior" class $\mathcal{P}_\epsilon(P_0)$ is an ϵ–contaminated set $\{(1-\epsilon)P_0 + \epsilon Q\}$, where P_0 is a conjugate normal distribution $N(\mu, \nu^2)$ and Q is arbitrary. Note that pairs of elements of $\mathcal{P}_\epsilon(P_0)$ are *not* all mutually absolutely continuous since Q ranges over one-point distributions that concentrate mass at different values of θ. Hence, Blackwell and Dubins' Theorem 2.2 does not apply.

For $\epsilon = 0$, $\mathcal{P}_\epsilon(P_0)$ is the singleton Bayes' (conjugate) prior, P_0. Then the Bayes' posterior for θ, $P_0(\theta|x)$, is a normal $N(\mu', \tau^2)$; where $\tau^2 = (\nu^{-2} + n\sigma^{-2})^{-1}$, $\mu' = \tau^2[(\mu/\nu^2) + (n\hat{\times}/\sigma^2)]$, and where $\hat{\times}$ is the sample average (of x). The standard 95% confidence interval for θ is $A_n = [\hat{\times} \pm 1.96\sigma/n^{.5}]$. Under the Bayes' prior P_0 (for $\epsilon = 0$), the Bayes' posterior of A_n, $P_0(A_n|x)$, depends upon the data, x. When n is large enough that τ^2 is approximately equal to σ^2/n, i.e., when $\sigma/\nu n^{.5}$ is sufficiently small, then $P_0(A_n|x)$ is close to .95. Otherwise, $P_0(A_n|x)$ may fall to very low values. Thus, asymptotically, the Bayes' posterior for A_n approximates the usual confidence level. However, under the ϵ–contaminated model $\mathcal{P}_\epsilon(P_0)$ (for $\epsilon > 0$), Pericchi and Walley show that, with increasing sample size n, $P^n_*(A_n) \to 0$ while $P^{n*}(A_n) \to 1$. That is, in terms of dilation, the sequence of standard confidence intervals estimates (each at the same fixed confidence level) dilate their unconditional probability or coverage level.

What sequence of confidence levels avoids dilation? That is, if it is required that $P^n_*(A'_n) \geq .95$, how should the intervals A'_n grow as a function of n? Pericchi and Walley (1991, p. 16) report that the sequence of intervals $A'_n = [\hat{\times} \pm \zeta_n\sigma/n^{.5}]$ has a posterior probability which is bounded below, e.g., $P^n_*(A'_n) \geq .95$, provided that ζ_n increases at the rate $(\log n)^{.5}$. They call intervals whose lower probability is bounded above some constant, "credible" intervals.

A connection exists between this rate of growth for ζ_n that makes A'_n credible, due to Walley and Pericchi, and an old but important result due to Sir Harold Jeffreys (1967, p. 248). The connection to Jeffreys' theory offers another interpretation for the lower posterior probabilities $P_*^n(A'_n)$ arising from the ϵ–contaminated class.

Adapt Jeffreys' Bayesian hypothesis testing, as follows. Consider a (simple) "null" hypothesis, H_0: $\theta = \theta_0$, against the (composite) alternative H_0^c: $\theta \neq \theta_0$. Let the prior ratio $P(H_0)/P(H_0^c)$ be specified as γ: $(1-\gamma)$. (Jeffreys uses $\gamma = .5$.) Given H_0, the x_i are i.i.d. $N(\theta_0, \sigma^2)$. Given H_0^c, let the parameter θ be distributed as $N(\mu, \nu^2)$. Then, when the data make $|\hat{\mathsf{x}} - \theta_0|$ large relative to $\sigma/n^{.5}$ the posterior ratio $P(H_0|x)/P(H_0^c|x)$ is smaller than the prior ratio, and when $|\hat{\mathsf{x}} - \theta_0|$ is small relative to $\sigma/n^{.5}$ the posterior odds favor the null hypothesis. But to maintain a constant posterior odds ratio with increasing sample size *rather than being constant – as a fixed significance level would entail* – the quantity $|\hat{\mathsf{x}} - \theta_0|/(\sigma/n^{.5})$ has *to grow* at the rate $(\log n)^{.5}$ though, of course, the difference $|\hat{\mathsf{x}} - \theta_0|$ approaches 0.

In other words, Jeffreys' analysis reveals that, from a Bayesian point of view, the posterior odds for the usual two-sided hypothesis test of H_0 versus the alternative H_0^c depends upon *both* the observed type$_1$ error (or significance level), α and the sample size, n. At a fixed significance level, e.g., at observed significance $\alpha = .05$, larger samples yield ever higher (in fact, unbounded) posterior odds in favor of H_0. To keep posterior odds constant as sample size grows, the observed significance level must decrease towards 0.

It is well known that confidence intervals are the result of inverting on a family of hypothesis tests, generated by varying the "null" hypothesis. That is, the interval $A_n = [\hat{\mathsf{x}} \pm 1.96\sigma/n^{.5}]$, with confidence 95%, corresponds to the family of unrejected null hypotheses: Each value θ belonging to the interval is a null hypothesis that is not rejected on a standard two-sided test at significance level $\alpha = .05$. Consider a family of Jeffreys' hypothesis tests obtained by varying the "null" through the parameter space and, corresponding to each null hypothesis, varying the prior probability which puts mass γ on H_0. Say that a value of θ, $\theta = \theta_0$, is rejected when its posterior probability falls below a threshold, e.g., when $P(H_0|x) < .05$ for the Jeffreys' prior $P(\theta = \theta_0) = \gamma$. The class of probabilities obtained by varying the null hypothesis forms an ϵ–contaminated model: $\{(1-\gamma)P(\theta|H^c) + \gamma Q\}$, with extreme points (for Q) corresponding to all the one-point "null" hypotheses.

Define the interval B_n of null hypotheses, with sample size n, where each survives rejection under Jeffreys' tests. The B_n are the intervals $A'_n = [\hat{\mathsf{x}} \pm \zeta_n \sigma/n^{.5}]$ of Pericchi and Walley's analysis, reported above. What Pericchi and Walley observe, expressed in terms of the required rate of growth of ζ_n for credible intervals (intervals that have a fixed lower posterior probability with respect to the class $\mathcal{P}_\epsilon(P)$) is exactly the result Jeffreys reports about the shrinking α–levels in hypothesis tests in order that posterior probabilities for the "null" be constant, regardless of sample size. In short, credible intervals from the ϵ–contaminated model $\mathcal{P}_\epsilon(P_0)$

are the result of inverting on a family of Jeffreys' hypothesis tests that use a fixed lower bound on posterior odds to form the rejection region of the test.

Continuing with interval estimation of a normal mean, for $0 < \alpha < 1$, let us consider a prior (symmetric) family $\mathcal{S}_\alpha$ of rearrangements of the density $p_\alpha(\theta) = (1-\alpha)\theta^{-\alpha}$. Let the interval estimate of θ be $A_n = [\hat{\theta} - a_n,\ \hat{\theta} + a_n]$, with $\hat{\theta}$ the maximum likelihood estimate. For constants $C > 0$ and d, write $a_n = \{n^{-1}(C + d \log n)\}^{1/2}$.

THEOREM 2.9. For the $\mathcal{S}_\alpha$ model, there is asymptotic dilation of A_n if and only if $d < \alpha$.

2.5. Dilation and Dempster–Shafer belief functions. Within the Artificial Intelligence community, the use of belief functions to represent expert opinion has been received as a useful alternative to adopting a strict (Bayesian) probabilistic model: see, e.g., Hestir, Nguyen, and Rogers (1991). As has long been understood, belief functions are (formally) equivalent to the lower probability envelope from select convex sets of probabilities: Dempster (1966, 1967), Kyburg (1987), Black (1996a,b).

In sharp contrast with the robust Bayesian theory described in this paper, however, belief functions are aligned with a different dynamics, the so-called "Dempster's rule" for updating. Expressed in terms of the convex-set formalism, Dempster's rule amounts to a restrictive version of Bayesian conditioning: Use Bayes' rule within the (convex) proper subset of probabilities that maximize the conditioning event. Hence, for those convex sets of probabilities that are (formally) equivalent to belief functions, Dempster's rule produces a range of probabilities that is never wider than Bayes' updating. Does Dempster's rule create dilations, then? The following provides a simple case where it does.

THEOREM 2.10. (i) In finite sets, the ϵ–contamination models are belief functions and (ii), for them, Dempster–updating yields the same intervals of probabilities as Bayes updating. Hence, belief functions updated by Dempster's rule admit dilation.

PROOF. (i) For a finite set $X = \{x_1, \ldots, x_n\}$, the ϵ–contamination model is equivalent to the (largest) convex set of probabilities achieved by fixing the lower probabilities for the x_i, the elements of X, i.e., the atoms of the algebra. Using the m–function representation for belief functions, we see that each ϵ–contamination model is equivalent to a belief function with $m(x_i) = \underline{P}(x_i)$ and where the rest of the (free) m–mass, $1 - \sum_i m(x_i) = m^*$ is assigned to X, i.e., with $m(y) = 0$ if y is a subset different from an atom or X itself.

(ii) In the ϵ–contamination model, the upper and lower Bayesian conditional probabilities for $P(A|B)$ are given respectively by $\overline{P}(A|B) = \overline{P}(AB)/[\overline{P}(AB) + \underline{P}(A^cB)]$ and $\underline{P}(A|B) = \underline{P}(AB)/[\underline{P}(AB) + \overline{P}(A^cB)]$. These are determined by shifting all the free mass m^* to AB to get $\overline{P}(A|B)$ and to $B \setminus A$ to get the $\underline{P}(A|B)$. However, each of these results in a probability where $P(B) = \overline{P}(B)$; hence, Bayesian and Dempster updating yield the same intervals of probability for the ϵ–contamination model. □

Thus, our previous discussion of dilation for this model carries over, without change, to the belief function theory.

3. Summary and some open issues. In this paper, I reviewed issues relating to static and to dynamic aspects of robust Bayesian theory.

Concerning the representation of preferences for a Pareto efficient, cooperative group, there is reason to doubt the adequacy of traditional (strict) Bayesian theory. For example, in the case of two cooperating Bayesian agents, provided they have some differences in their subjective probabilities for events and cardinal utilities for outcomes, the resulting cooperative (Pareto efficient) partial order admits no Bayesian compromises! Instead, in §1.3 the theory offered relaxes expected utility theory by weakening the "ordering" postulate to require of (strict) preference that it be merely a strict partial order – where not all acts need be comparable by preference. The upshot is a theory of robust preference where representation is in terms of a set $\mathcal{S}$ of probability/utility pairs. To wit: Act A_2 is robustly preferred to act A_1 provided, for each probability/utility pair in $\mathcal{S}$, the subjective expected utility of A_2 is greater than that for A_1.

Among the many open issues for this theory include some basic matters for implementation. One area of application, for example, deals with incomplete elicitation. Suppose we elicit strict preferences from an agent, but in an order that we may control through the order in which we ask questions about preferences between acts. Recognizing that individuals have limited capabilities at introspection, how shall we structure our interview to extract useful preferences? That is, at each stage i of our elicitation of the agent's strict preferences (and assuming answers are reliable) we form a representing set $\mathcal{S}_i$ of the agent's preferences, where $\mathcal{S}_j \subseteq \mathcal{S}_i$, for $i < j$. What is the right order to elicit preferences in order that the representation will be useful to the agent? What is the right order to elicit preferences in order that the representation will be useful for predicting the agent's other choices?

The second issue discussed, relating to the dynamics of robust Bayesian theory, focuses on a counter-intuitive phenomenon that we call "dilation," where new evidence *increases* uncertainty in an event. Specifically, new information about B dilates a set $\mathcal{P}$ of probabilities for A when the updated probabilities for A, given a random variable, have a strictly wider range than do the unconditional probabilities for A. I reported, primarily, on our investigation for the set $\mathcal{P}$ formed with the ϵ–contamination model: We relate dilation to independence between A and the random variable; we examine the extent of dilation; we consider dilation in 2×2 tables; and we study the asymptotics of dilation for credible interval estimates with (conditionally) i.i.d. data. Also, I reported on dilation for belief functions with updating by Dempster's rule in place of Bayes' rule. These results all run counter to the familiar lore of strict Bayesian theory, where asymptotic merging of different opinions is the norm.

Dilation of probabilities changes the basic theory for sequential investigations, since new evidence may fail to be of positive value. Among the open questions, then, are those addressing the theory of experimental design for robust Bayesian inference: some (finite) sequences of experiments may be sure to increase disagreements, rather than settle them. I leave these to future inquiry.

REFERENCES

[1] F.J. ANSCOMBE AND R.J. AUMANN, *A definition of subjective probability*, Ann. Math. Stat., 34 (1963), pp. 199–205.

[2] J.O. BERGER, *The robust Bayesian viewpoint* (with Discussion), Robustness of Bayesian Analysis (J.B. Kadane, ed.), Amsterdam: North–Holland, pp. 63–114, 1984.

[3] J.O. BERGER, *Statistical Decision Theory*, (2nd edition), Springer–Verlag, New York, 1985.

[4] J.O. BERGER, *Robust Bayesian analysis: Sensitivity to the prior*, J. Stat. Planning and Inference, 25 (1990), pp. 303–328.

[5] P. BLACK, *An examination of belief functions and other monotone capacities*, Ph.D. Thesis, Department of Statistics, Carnegie Mellon University, 1996a.

[6] P. BLACK, *Geometric structure of lower probability*, This Volume, pp. 361–383, 1996b.

[7] D. BLACKWELL AND L. DUBINS, *Merging of opinions with increasing information*, Ann. Math. Stat., 33 (1962), pp. 882–887.

[8] B. DEFINETTI, *La prevision: Ses Lois logiques, ses sources subjectives*, Annals de L'Institut Henri Poincaré, 7 (1937), pp. 1–68.

[9] A.P. DEMPSTER, *New methods for reasoning towards posterior distributions based on sample data*, Ann. Math. Stat., 37 (1966), pp. 355–374.

[10] A.P. DEMPSTER, *Upper and lower probabilities induced by a multi–valued mapping*, Ann. Math. Stat., 38 (1967), pp. 325–339.

[11] D. ELLSBERG, *Risk, ambiguity, and the savage axioms*, Quart. J. Econ., 75 (1961), pp. 643–669.

[12] P.C. FISHBURN, *Utility Theory for Decision Making*, Krieger Pub., New York, (1979 ed.), 1970.

[13] P.C. FISHBURN, *Lexicographic orders, utilities, and decision rules: A survey*, Management Science 20 (1974), pp. 1442–1471.

[14] P.C. FISHBURN, *The Axioms of subjective probability*, Statistical Science, 1 (1986), pp. 335–358.

[15] P.C. FISHBURN AND I. LAVALLE, *Subjective expected lexicographic utility: Axioms and assessment*, Preprint, AT&T Labs, Murray Hill, 1996.

[16] I.J. GOOD, *Rational decisions*, J. Royal Stat. Soc., B14 (1952), pp. 107–114.

[17] T. HERRON, T. SEIDENFELD, AND L. WASSERMANN, *Divisive condition*, Technical Report #585, Dept. of Statistics, Carnegie Mellon University, Pittsburgh, Pennsylvania, 1995.

[18] K. HESTIR, H. NGUYEN, AND G. ROGERS, *A random set formalism for evidential reasoning*, Conditional Logic in Expert Systems, (I.R. Goodman, M.M. Gupta, H.T. Nguyen, and G.S. Rogers, eds.), Amsterdam: Elsevier Science, pp. 309–344, 1991.

[19] P.J. HUBER, *The use of Choquet capacities in statistics*, Bull. Int. Stat., 45 (1973), pp. 181–191.
[20] P.J. HUBER, *Robust Statistics*, Wiley and Sons, New York, 1981.
[21] P.J. HUBER AND V. STRASSEN, *Minimax tests and the Neyman–Pearson lemma for capacities*, Annals of Statistics, 1 (1973), pp. 241–263.
[22] R.C. JEFFREY, *The Logic of Decision*, McGraw Hill, New York, 1965.
[23] H. JEFFREYS, *Theory of Probability*, 3rd Edition, Oxford University Press, Oxford, 1967.
[24] J.B. KADANE, *Opposition of interest in subjective Bayesian theory*, Management Science, 31 (1985), pp. 1586–1588.
[25] H.E. KYBURG, *Probability and Logic of Rational Belief*, Wesleyan University Press, Middleton, 1961.
[26] H.E. KYBURG, *Bayesian and non–Bayesian evidential updating*, Artificial Intelligence, 31 (1987), pp. 279–294.
[27] M. LAVINE, *Sensitivity in Bayesian statistics: The prior and the likelihood*, J. Amer. Stat. Assoc., 86 (1991), pp. 396–399.
[28] I. LEVI, *On indeterminate probabilities*, J. Phil., 71 (1974), pp. 391–418.
[29] I. LEVI, *Conflict and social agency*, J. Phil., 79 (1982), pp. 231–247.
[30] I. LEVI, *Imprecision and indeterminacy in probability judgment*, Phil. of Science, 52 (1985), pp. 390–409.
[31] I. LEVI, *Feasibility*, Knowledge, Belief, and Strategic Interaction, (C. Bicchieri and M.L. Dalla Chiara, eds.), Cambridge University Press, Cambridge, 1992.
[32] M. MACHINA, *Expected utility analysis without the independence axiom*, Econometrica, 50 (1982), pp. 277–323.
[33] E.F. MCCLENNEN, *Rationality and Dynamic Choice*, Cambridge University Press, Cambridge, 1990.
[34] L.R. PERICCHI AND P. WALLEY, *Robust Bayesian credible intervals and prior ignorance*, Int. Stat. Review, 58 (1991), pp. 1–23.
[35] F.P. RAMSEY, *Truth and probability*, The Foundations of Mathematics and Other Essays, Kegan, Paul, Trench Trubner and Co. Ltd., London, pp. 156–198, 1931.
[36] L.J. SAVAGE, *The Foundations of Statistics*, Wiley and Sons, New York, 1954.
[37] M.J. SCHERVISH, T. SEIDENFELD, AND J.B. KADANE, *State–Dependent Utilities*, J. Am. Stat. Assoc., 85 (1990), pp. 840–847.
[38] M.J. SCHERVISH, T. SEIDENFELD, AND J.B. KADANE, *Shared preferences and state–dependent utilities*, Management Science, 37 (1991), pp. 1575–1589.
[39] T. Seidenfeld, *Decision theory without "independence" or without "ordering" – What is the difference?*, (with Discussion), Economics and Philosophy, 4 (1988), pp. 267–315.
[40] T. SEIDENFELD AND M.J. SCHERVISH, *Conflict between finite additivity and avoiding Dutch book*, Phil. of Science, 50 (1983), pp. 398–412.
[41] T. SEIDENFELD, M.J. SCHERVISH, AND J.B. KADANE, *Decisions without ordering*, Acting and Reflecting, (W. Sieg, ed.), Kluwer Academic, Dordrecht, pp. 143–170, 1990.
[42] T. SEIDENFELD, M.J. SCHERVISH, AND J.B. KADANE, *A representation of partially ordered preferences*, Annals of Statistics, 23 (1995), pp. 2168–2217.
[43] T. SEIDENFELD AND L. WASSERMAN, *Dilation for sets of probabilities*, Annals of Statistics, 21 (1993), pp. 1139–1154.
[44] C.A.B. SMITH, *Consistency in statistical inference and decisions*, J. Royal Stat. Soc., B23 (1961), pp. 1–25.
[45] E. SZPILRAJN, *Sur l'extension de l'ordre partiel*, Fund. Math., 16 (1930), pp. 386–389.
[46] J. VON NEUMANN AND O. MORGENSTERN, *Theory of Games and Economic Behavior*, (2nd Edition), Princeton University Press, Princeton, New Jersey, 1947.
[47] P. WALLEY, *Statistical Reasoning with Imprecise Probabilities*, Chapman Hall, London, 1991.

[48] L. Wasserman and J.B. Kadane, *Bayes' theorem for Choquet capacities*, Annals of Statistics, 18 (1990), pp. 1328–1339.

LIST OF PARTICIPANTS

Batman, Sinan
Texas A & M University
Wisenbaker Eng. Res. Center
College Station, TX 77843-3407
Email: sinan@ee.tamu.edu

Black, Paul
Neptune and Company, Inc.
1505 15th St., Suite B
Los Alamos, NM 87544
Email: paulb@parsifal.lanl.gov

Chawla, Sanjay
University of Minnesota
Institute for Mathematics
514 Vincent Hall
206 Church St. SE
Minneapolis, MN 55455
Email: chawla@math.utk.edu

Chen, Yidong
NCHGR/NIH
Building 49/Room 4B-24
9000 Rockville Pike
Rockville, MD 20892
Email: yidong@nchgr.nih.gov

Daum, Fred
Raytheon Company
Electronics System Division
1001 Boston Post Road
Mail Stop 1-2-1574
Marlborough, MA 01752
Email: Frederick_E_Daum@ccmail.
ed.ray.com

Dougherty, Edward R.
Texas A & M University
215 Wisenbaker Eng. Res. Center
College Station, TX 77843-3407
Email: edward@ee.tamu.edu

Ferson, Scott
Applied Biomathematics
100 North Country Rd.
Setauket, New York 11733
Email: ramas@gramercy.ios.com

Friedman, Avner
University of Minnesota
Institute for Mathematics
514 Vincent Hall
206 Church St. SE
Minneapolis, MN 55455
Email: friedman@ima.umn.edu

Fristedt, Bert
University of Minnesota
Department of Mathematics
127 Vincent Hall
Minneapolis, MN 55455
Email: fristedt@math.umn.edu

Goodman, I.R.
NCCOSC RDTE DIV Code 4221
Building 600, Room 341A
53118 Gatchell Road
San Diego, CA 92152-7446
Email: goodman@cod.nosc.mil

Goutsias, John
Department of Electrical &
Computer Engineering
Barton Hall, Room 211
The Johns Hopkins University
Baltimore, MD 21218
Email: goutsias@mycenae.ece.jhu.edu

Gulliver, Robert
University of Minnesota
Institute for Mathematics
514 Vincent Hall
206 Church St. SE
Minneapolis, MN 55455
Email: gulliver@ima.umn.edu

Handley, John C.
Xerox Corporation
Mail Stop 128-29E
800 Phillips Road
Webster, NY 14580
Email: jchandle@wrc.xerox.com

Hőhle, Ulrich
Bergische Universität
Fachbereich 7, Mathematik
Gesamthochschule-Wuppertal
D-42097 Wuppertal, Germany
Email: hoehle@wmfa2.math.
uni-wuppertal.de

Jaffray, Jean-Yves
LAFORIA-IBP (Case 169)
University of Paris-VI
4 Place Jussieu
F-75252 Paris Cedex 05, France
Email: jaffray@laforia.ibp.fr

Kastella, Keith
Lockheed Martin Corporation
Tactical Defense Systems
3333 Pilot Knob Road
Eagan, MN 55121
Email: kkastell@eag.unisysgsg.com

Kober, Wolfgang
Data Fusion Corporation
7017 S. Richfield St.
Aurora, CO 80016
Email: wkober@datafusion.com

Kouritzin, Michael
University of Minnesota
Institute for Mathematics
514 Vincent Hall
206 Church St. SE
Minneapolis, MN 55455
Email: kouritzi@ima.umn.edu

Kreinovich, Vladik
Department of Computer Sciences
University of Texas-El Paso
El Paso, TX 79968
Email: vladik@cs.utep.edu

Launer, Robert
U.S. Army Research Office
Mathematics Division
USARO-MCS, POB 12211
Research Triangle Park, NC 27513
Email: launer@aro-emh1.army.mil

Mahler, Ronald
Lockheed Martin Corporation
Tactical Defense Systems
3333 Pilot Knob Road
Eagan, MN 55121
Email: rmahler@tds-eagan.lmco.com

McGirr, S.
NCCOSC RTDE DIV 721
53140 Hull St. Code 721
San Diego, CA 92152-7550
Email: mcgirr@nosc.mil

Molchanov, Ilya
Department of Statistics
University of Glasgow
Glasgow G12 8QW, Scotland
United Kingdom
Email: ilya@stats.gla.ac.uk

Mori, Shozo
Texas Instruments, Inc.
Advanced C3I Systems
1290 Parkmoor Avenue
San Jose, CA 95126
Email: smori@ti.com

Musick, Stan
Wright Lab, USAF
WL/AAAS
2241 Avionics Circle
W-PAFB, OH 45433
Email: musicksh@aa.wpafb.af.mil

Nguyen, Hung T.
Dept. of Mathematical Sciences
New Mexico State University
P.O. Box 30001
Las Cruces, NM 88003-001
Email: hunguyen@nmsu.edu

Sander, William
U.S. Army Research Office
Electronics Division
P.O. Box 12211
Research Triangle Park, NC 27709-2211
Email: sander@aro-emh1.army.mil

Schonfeld, Dan
Department of Electrical Engineering
& Computer Science
University of Illinois
Chicago, IL 60680
Email: ds@eecs.uic.edu

Seidenfeld, Teddy
Department of Statistics
Carnegie-Mellon University
Pittsburgh, PA 15213
Email: teddy@stat.cmu.edu

Sidiropoulos, Nikolaos D.
Department of Electrical Engineering
University of Virginia
Charlottesville, VA 22903
Email: nikos@virginia.edu

Sivakumar, Krishnamoorthy
Texas A & M University
Wisenbaker Eng. Res. Center
College Station, TX 77843-3407
Email: siva@ee.tamu.edu

Snyder, Wesley
U.S. Army Research Office
P.O. Box 12211
Research Triangle Park, NC 27709-2211
Email: wes@eos.ncsu.edu

Stein, Michael
Oasis Research Center Inc.
39 County Rd. 113B
Santa Fe, NM 87501
Email: stein@oasisrc.com

Taylor, Robert
Department of Statistics
University of Georgia
Athens, GA 30602-1952
Email: bob@stat.uga.edu

Walker, Carol
Dept. of Math Sciences
New Mexico State University
NMSU
Las Cruces, NM 88003
Email: cwalker@nmsu.edu

Walker, Elbert
Dept. of Mathematical Sciences
New Mexico State University
Las Cruces, NM 88003-8801
Email: elbert@nmsu.edu

Wang, Tonghui
Dept. of Mathematical Sciences
New Mexico State University
Las Cruces, NM 88003-0001
Email: twang@nmsu.edu

IMA SUMMER PROGRAMS

1987 Robotics
1988 Signal Processing
1989 Robustness, Diagnostics, Computing and Graphics in Statistics
1990 Radar and Sonar (June 18 - June 29)
New Directions in Time Series Analysis (July 2 - July 27)
1991 Semiconductors
1992 Environmental Studies: Mathematical, Computational, and Statistical Analysis
1993 Modeling, Mesh Generation, and Adaptive Numerical Methods for Partial Differential Equations
1994 Molecular Biology
1995 Large Scale Optimizations with Applications to Inverse Problems, Optimal Control and Design, and Molecular and Structural Optimization
1996 Emerging Applications of Number Theory
1997 Statistics in Health Sciences
1998 Coding and Cryptography

SPRINGER LECTURE NOTES FROM THE IMA:

The Mathematics and Physics of Disordered Media
Editors: Barry Hughes and Barry Ninham
(Lecture Notes in Math., Volume 1035, 1983)

Orienting Polymers
Editor: J.L. Ericksen
(Lecture Notes in Math., Volume 1063, 1984)

New Perspectives in Thermodynamics
Editor: James Serrin
(Springer-Verlag, 1986)

Models of Economic Dynamics
Editor: Hugo Sonnenschein
(Lecture Notes in Econ., Volume 264, 1986)

The IMA Volumes in Mathematics and its Applications

Current Volumes:

1 **Homogenization and Effective Moduli of Materials and Media** J. Ericksen, D. Kinderlehrer, R. Kohn, and J.-L. Lions (eds.)

2 **Oscillation Theory, Computation, and Methods of Compensated Compactness** C. Dafermos, J. Ericksen, D. Kinderlehrer, and M. Slemrod (eds.)

3 **Metastability and Incompletely Posed Problems** S. Antman, J. Ericksen, D. Kinderlehrer, and I. Muller (eds.)

4 **Dynamical Problems in Continuum Physics** J. Bona, C. Dafermos, J. Ericksen, and D. Kinderlehrer (eds.)

5 **Theory and Applications of Liquid Crystals** J. Ericksen and D. Kinderlehrer (eds.)

6 **Amorphous Polymers and Non-Newtonian Fluids** C. Dafermos, J. Ericksen, and D. Kinderlehrer (eds.)

7 **Random Media** G. Papanicolaou (ed.)

8 **Percolation Theory and Ergodic Theory of Infinite Particle Systems** H. Kesten (ed.)

9 **Hydrodynamic Behavior and Interacting Particle Systems** G. Papanicolaou (ed.)

10 **Stochastic Differential Systems, Stochastic Control Theory, and Applications** W. Fleming and P.-L. Lions (eds.)

11 **Numerical Simulation in Oil Recovery** M.F. Wheeler (ed.)

12 **Computational Fluid Dynamics and Reacting Gas Flows** B. Engquist, M. Luskin, and A. Majda (eds.)

13 **Numerical Algorithms for Parallel Computer Architectures** M.H. Schultz (ed.)

14 **Mathematical Aspects of Scientific Software** J.R. Rice (ed.)

15 **Mathematical Frontiers in Computational Chemical Physics** D. Truhlar (ed.)

16 **Mathematics in Industrial Problems** A. Friedman

17 **Applications of Combinatorics and Graph Theory to the Biological and Social Sciences** F. Roberts (ed.)

18 ***q*-Series and Partitions** D. Stanton (ed.)

19 **Invariant Theory and Tableaux** D. Stanton (ed.)

20 **Coding Theory and Design Theory Part I: Coding Theory** D. Ray-Chaudhuri (ed.)

21 **Coding Theory and Design Theory Part II: Design Theory** D. Ray-Chaudhuri (ed.)

22 **Signal Processing Part I: Signal Processing Theory** L. Auslander, F.A. Grünbaum, J.W. Helton, T. Kailath, P. Khargonekar, and S. Mitter (eds.)

23	**Signal Processing Part II: Control Theory and Applications of Signal Processing** L. Auslander, F.A. Grünbaum, J.W. Helton, T. Kailath, P. Khargonekar, and S. Mitter (eds.)
24	**Mathematics in Industrial Problems, Part 2** A. Friedman
25	**Solitons in Physics, Mathematics, and Nonlinear Optics** P.J. Olver and D.H. Sattinger (eds.)
26	**Two Phase Flows and Waves** D.D. Joseph and D.G. Schaeffer (eds.)
27	**Nonlinear Evolution Equations that Change Type** B.L. Keyfitz and M. Shearer (eds.)
28	**Computer Aided Proofs in Analysis** K. Meyer and D. Schmidt (eds.)
29	**Multidimensional Hyperbolic Problems and Computations** A. Majda and J. Glimm (eds.)
30	**Microlocal Analysis and Nonlinear Waves** M. Beals, R. Melrose, and J. Rauch (eds.)
31	**Mathematics in Industrial Problems, Part 3** A. Friedman
32	**Radar and Sonar, Part I** R. Blahut, W. Miller, Jr., and C. Wilcox
33	**Directions in Robust Statistics and Diagnostics: Part I** W.A. Stahel and S. Weisberg (eds.)
34	**Directions in Robust Statistics and Diagnostics: Part II** W.A. Stahel and S. Weisberg (eds.)
35	**Dynamical Issues in Combustion Theory** P. Fife, A. Liñán, and F.A. Williams (eds.)
36	**Computing and Graphics in Statistics** A. Buja and P. Tukey (eds.)
37	**Patterns and Dynamics in Reactive Media** H. Swinney, G. Aris, and D. Aronson (eds.)
38	**Mathematics in Industrial Problems, Part 4** A. Friedman
39	**Radar and Sonar, Part II** F.A. Grünbaum, M. Bernfeld, and R.E. Blahut (eds.)
40	**Nonlinear Phenomena in Atmospheric and Oceanic Sciences** G.F. Carnevale and R.T. Pierrehumbert (eds.)
41	**Chaotic Processes in the Geological Sciences** D.A. Yuen (ed.)
42	**Partial Differential Equations with Minimal Smoothness and Applications** B. Dahlberg, E. Fabes, R. Fefferman, D. Jerison, C. Kenig, and J. Pipher (eds.)
43	**On the Evolution of Phase Boundaries** M.E. Gurtin and G.B. McFadden
44	**Twist Mappings and Their Applications** R. McGehee and K.R. Meyer (eds.)
45	**New Directions in Time Series Analysis, Part I** D. Brillinger, P. Caines, J. Geweke, E. Parzen, M. Rosenblatt, and M.S. Taqqu (eds.)

46 **New Directions in Time Series Analysis, Part II**
D. Brillinger, P. Caines, J. Geweke, E. Parzen, M. Rosenblatt, and M.S. Taqqu (eds.)

47 **Degenerate Diffusions**
W.-M. Ni, L.A. Peletier, and J.-L. Vazquez (eds.)

48 **Linear Algebra, Markov Chains, and Queueing Models**
C.D. Meyer and R.J. Plemmons (eds.)

49 **Mathematics in Industrial Problems, Part 5** A. Friedman

50 **Combinatorial and Graph-Theoretic Problems in Linear Algebra**
R.A. Brualdi, S. Friedland, and V. Klee (eds.)

51 **Statistical Thermodynamics and Differential Geometry of Microstructured Materials**
H.T. Davis and J.C.C. Nitsche (eds.)

52 **Shock Induced Transitions and Phase Structures in General Media** J.E. Dunn, R. Fosdick, and M. Slemrod (eds.)

53 **Variational and Free Boundary Problems**
A. Friedman and J. Spruck (eds.)

54 **Microstructure and Phase Transitions**
D. Kinderlehrer, R. James, M. Luskin, and J.L. Ericksen (eds.)

55 **Turbulence in Fluid Flows: A Dynamical Systems Approach**
G.R. Sell, C. Foias, and R. Temam (eds.)

56 **Graph Theory and Sparse Matrix Computation**
A. George, J.R. Gilbert, and J.W.H. Liu (eds.)

57 **Mathematics in Industrial Problems, Part 6** A. Friedman

58 **Semiconductors, Part I**
W.M. Coughran, Jr., J. Cole, P. Lloyd, and J. White (eds.)

59 **Semiconductors, Part II**
W.M. Coughran, Jr., J. Cole, P. Lloyd, and J. White (eds.)

60 **Recent Advances in Iterative Methods**
G. Golub, A. Greenbaum, and M. Luskin (eds.)

61 **Free Boundaries in Viscous Flows**
R.A. Brown and S.H. Davis (eds.)

62 **Linear Algebra for Control Theory**
P. Van Dooren and B. Wyman (eds.)

63 **Hamiltonian Dynamical Systems: History, Theory, and Applications**
H.S. Dumas, K.R. Meyer, and D.S. Schmidt (eds.)

64 **Systems and Control Theory for Power Systems**
J.H. Chow, P.V. Kokotovic, R.J. Thomas (eds.)

65 **Mathematical Finance**
M.H.A. Davis, D. Duffie, W.H. Fleming, and S.E. Shreve (eds.)

66 **Robust Control Theory** B.A. Francis and P.P. Khargonekar (eds.)

67 **Mathematics in Industrial Problems, Part 7** A. Friedman

68 **Flow Control** M.D. Gunzburger (ed.)

69 **Linear Algebra for Signal Processing**
A. Bojanczyk and G. Cybenko (eds.)
70 **Control and Optimal Design of Distributed Parameter Systems**
J.E. Lagnese, D.L. Russell, and L.W. White (eds.)
71 **Stochastic Networks** F.P. Kelly and R.J. Williams (eds.)
72 **Discrete Probability and Algorithms**
D. Aldous, P. Diaconis, J. Spencer, and J.M. Steele (eds.)
73 **Discrete Event Systems, Manufacturing Systems, and Communication Networks**
P.R. Kumar and P.P. Varaiya (eds.)
74 **Adaptive Control, Filtering, and Signal Processing**
K.J. Åström, G.C. Goodwin, and P.R. Kumar (eds.)
75 **Modeling, Mesh Generation, and Adaptive Numerical Methods for Partial Differential Equations** I. Babuska, J.E. Flaherty, W.D. Henshaw, J.E. Hopcroft, J.E. Oliger, and T. Tezduyar (eds.)
76 **Random Discrete Structures** D. Aldous and R. Pemantle (eds.)
77 **Nonlinear Stochastic PDEs: Hydrodynamic Limit and Burgers' Turbulence** T. Funaki and W.A. Woyczynski (eds.)
78 **Nonsmooth Analysis and Geometric Methods in Deterministic Optimal Control** B.S. Mordukhovich and H.J. Sussmann (eds.)
79 **Environmental Studies: Mathematical, Computational, and Statistical Analysis** M.F. Wheeler (ed.)
80 **Image Models (and their Speech Model Cousins)**
S.E. Levinson and L. Shepp (eds.)
81 **Genetic Mapping and DNA Sequencing**
T. Speed and M.S. Waterman (eds.)
82 **Mathematical Approaches to Biomolecular Structure and Dynamics**
J.P. Mesirov, K. Schulten, and D. Sumners (eds.)
83 **Mathematics in Industrial Problems, Part 8** A. Friedman
84 **Classical and Modern Branching Processes**
K.B. Athreya and P. Jagers (eds.)
85 **Stochastic Models in Geosystems**
S.A. Molchanov and W.A. Woyczynski (eds.)
86 **Computational Wave Propagation**
B. Engquist and G.A. Kriegsmann (eds.)
87 **Progress in Population Genetics and Human Evolution**
P. Donnelly and S. Tavaré (eds.)
88 **Mathematics in Industrial Problems, Part 9** A. Friedman
89 **Multiparticle Quantum Scattering With Applications to Nuclear, Atomic and Molecular Physics** D.G. Truhlar and B. Simon (eds.)
90 **Inverse Problems in Wave Propagation** G. Chavent, G. Papanicolau, P. Sacks, and W.W. Symes (eds.)
91 **Singularities and Oscillations** J. Rauch and M. Taylor (eds.)

92 **Large-Scale Optimization with Applications, Part I: Optimization in Inverse Problems and Design**
L.T. Biegler, T.F. Coleman, A.R. Conn, and F. Santosa (eds.)

93 **Large-Scale Optimization with Applications, Part II: Optimal Design and Control**
L.T. Biegler, T.F. Coleman, A.R. Conn, and F. Santosa (eds.)

94 **Large-Scale Optimization with Applications, Part III: Molecular Structure and Optimization**
L.T. Biegler, T.F. Coleman, A.R. Conn, and F. Santosa (eds.)

95 **Quasiclassical Methods**
J. Rauch and B. Simon (eds.)

96 **Wave Propagation in Complex Media**
G. Papanicolau (ed.)

97 **Random Sets: Theory and Applications**
J. Goutsias, R.P.S. Mahler, and H.T. Nguyen (eds.)